GROUPERS OF THE WORLD

A Field and Market Guide

Cephalopholis fulva

IUCN

SSC Species Survival Commission

Groupers & Wrasses Specialist Group

Matthew T Craig
Yvonne J Sadovy de Mitcheson
Phillip C Heemstra

NISC
DIGITAL & PRINT PUBLISHING

Mycteroperca rosacea

Contents

Cephalopholis miniata

Epinephelus ongus

Foreword

Twenty years ago Phil Heemstra and I published the first comprehensive review of groupers. Since that time, new species have been described, distributional records have been added and biological data have accumulated. This interest in grouper biology and ecology is largely a reflection of the increasing trends in harvest of groupers, noted declines in the abundance of many species (particularly larger, more economically valuable species) and the identification of numerous threats to grouper populations.

In response to some of the more alarming trends, the IUCN Groupers & Wrasses Specialist Group set out in 1998 to apply IUCN criteria to all grouper species, to identify major threats, and to accumulate the majority of information on the biology, ecology, distribution and harvest trends for these important fishes. This volume compiles this information into an easy-to-use and informative guide to grouper species. The numerous photographs showing variation in grouper colouration in life and in the market provide an excellent resource for the identification of grouper species that may look quite different depending on their state of preservation.

This volume highlights many important knowledge gaps that, if approached in a rigorous scientific manner, may help efforts to stem the tide of grouper declines worldwide. In addition, this guide will facilitate accurate species identification for market species, a common problem in fisheries biology.

John E Randall
Curator Emeritus
Bishop Museum, Honolulu

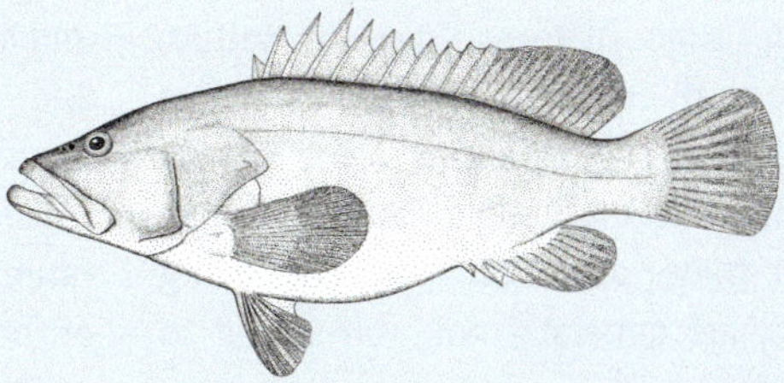

Epinephelus suborbitalis

INTRODUCTION TO GROUPERS

Groupers are not only biologically diverse and fascinating in their own right, but are also among the most commercially important and highly regarded of all fish species associated with reef fisheries globally. While representing but a small percentage of world fisheries landings, groupers nonetheless are a significant component of coastal rocky and coral reef fisheries in much of the tropics and subtropics. According to FAO data, in 2008 more than 300 000 t were landed, an increase of about one-third from the previous decade (almost 200 000 t in 1996) and more than 10 times the production reported in 1950 (http://faostat.fao.org). Globally, these catches comprise about 35 different individual species, mostly medium- to large-sized groupers, as well as unidentified species lumped as 'groupers' or seabasses. While the increasing catch over the last few decades reflects the overall growing demand for seafood, landings recorded by FAO inevitably underestimate total catch. This can be attributed to the nature of the many small-scale, multi-species fisheries that land groupers at multiple sites, unreported home consumption and trade. Many countries where groupers are caught do not report their landings to FAO.

Promoting and facilitating ever higher landings are increasingly efficient methods for finding and catching fish, the development of cash economies in many countries, growth in demand for commercial use, including for international trade and tourism, and a desire for fish as a healthy source of protein. Reflecting these trends, the markets for grouper are also diversifying, from historical use of salted, fresh whole or filleted fish, to frozen and live groupers for high-end restaurants, and in both local and international trade. There is also an ornamental trade, although groupers, being carnivorous, are generally not well-suited to home aquaria.

Given the high value of groupers in restaurants and supermarkets around the world, species once retained for local consumption are often considered to be more valuable as commodities, prompting the development of grouper mariculture (i.e. 'farming') to increase market supply. However, while full-cycle (i.e. hatchery) culture is now possible at commercial levels for only a few species, most notably *Epinephelus coioides, E. lanceolatus, E. fuscoguttatus* and *Cromileptes altivelis*, for most groupers much of the 'culture' still depends on the grow-out of wild-caught fish juveniles and post-settlement fish, most of which goes entirely unrecorded in capture landings records. This activity can increase conflicts for resources because a single grouper population might be exploited at several different life-history phases (e.g. juveniles for culturing as live fish, and adults for both live and chilled fish markets) by very different fishery sectors.

In short, many of the described species of groupers are important socially and economically, yet few are managed, and there is growing concern over the economic and biological viability of many populations. Some species are now even considered threatened with extinction unless action is taken to conserve them (see below). A combination of several of the life-history features that characterise groupers, including longevity, reproductive characteristics and complex social behaviours, combined with the fact that their fisheries are rarely adequately managed, compromises many grouper populations and the benefits they bring to communities and economies. As larger species become overfished, smaller groupers, previously considered of little fishery significance, are targeted. Yet we understand surprisingly little of the life history of most of these fishes, and, even today, few fisheries are monitored to the level of individual species. Thus, there is a growing need for more information.

This field and market guide to groupers of the world strives to provide information that we hope will contribute to meeting today's needs for better understanding and managing this group of fishes in a way that is accessible to a wide range of readers. We hope that the information is useful for fishery managers, students, teachers, conservationists, biologists and divers as a tool to better study, appreciate, record and display groupers. We also hope to encourage and inspire others to enjoy and explore this fascinating group of fishes.

Biology, ecology and distribution

Given the importance of grouper fisheries in many parts of the world, it is surprising how little is known about the biology and ecology of all but a few species. The geographic and depth distributions of most groupers are considerably better known now than they were even a few decades ago as more biologists, photographers and divers document and study marine diversity. As fisheries move into deeper waters in search of untapped fish populations, they encounter deeper-living species, while the loss of larger groupers from shallower fisheries focuses more attention on smaller species.

The earliest studies on groupers were taxonomic, initially simple species descriptions from collections and research cruises. As SCUBA became more widely available, many more species and colour variations were described, and information on food habits collected, while biological and ecological work started to attract the attention of field biologists and fishery scientists. The first regional compilation was produced in 1971 by C Lavett Smith in his revision of American groupers, which helped to clarify the alpha taxonomy of groupers. Smith also worked extensively on sexual patterns, particularly complex in this group of fishes and including many kinds of hermaphroditism. Despite the earlier perception that all groupers are hermaphrodites, it is now understood that several species do not change sex and are, from a functional, if not developmental, standpoint, gonochoristic. He also explored the relationship between phylogeny and sexual pattern, a theme that continues to attract interest today. Smith's was also the first scientific report of a grouper spawning aggregation. Diving on a gathering of Nassau grouper, he estimated the presence of at least 30 000 individuals at a single site in the Bahamas (Smith 1972). In the Indo-Pacific, Jack Randall's collaboration with Phil Heemstra on the extensive revision of groupers (Randall and Heemstra 1991), and a life-time of field exploration and photographic recording, has greatly advanced our understanding and appreciation of the distribution of these fishes.

Overall, the species assessments in this book, along with the citations provided, reveal that our understanding of groupers and their lives is greatest for reproduction, habitat, maximum size, social structure and, for a handful of species, includes age and growth. The expansion of fishing into deeper waters, as shallower areas become more heavily fished, and the growth in popularity of underwater visual census surveys, have helped develop our knowledge of relative abundance and distribution patterns for an ever wider range of species. Histological and age and growth studies reveal that some of the larger species may not become sexually mature until at least 10 years, with longevities exceeding 40 years; even some of the smaller species have surprisingly high longevities. Reproductive systems are highly diverse; smaller groupers, living in distinct social groupings often comprising a male and several females (harems), can spawn for many months of the year while most medium- to larger-size groupers reproduce in gatherings of tens to many thousands of fish. In these aggregations they mate either in pairs or in what appears to be single female-multiple male groups. Such aggregations may last for just a few days each year and are often associated with a distinct lunar phase.

While groupers include some of the largest of all teleosts on coral reefs anywhere, only a few exceed about 1 m in total length. Exceptionally, the giant grouper *Epinephelus lanceolatus* of the Indo-West Pacific, the Pacific goliath grouper *E. quinquefasciatus* of the eastern Pacific, and the Atlantic goliath grouper *E. itajara* of the tropical western Atlantic, can exceed 2 m.

Our knowledge of distribution, population structure and movement patterns is also improving due to the increasing sophistication of tagging technology and advancements in molecular techniques, as well as from a generally growing interest in grouper biology and fisheries. Small- to medium-sized groupers that are not highly mobile as adults typically respond well to the protection conferred by establishing effectively managed marine protected areas (MPAs) because much of their adult life is spent within a few hundred square metres of reef. However, for some of the more mobile medium- to larger-sized groupers that may travel tens to hundreds of kilometres as adults to and from annual spawning sites, the typically small MPAs that prevail today are much less effective for conserving them, and other management measures, such as effort or gear controls, are necessary.

Longevity, size and age of sexual maturation, mortality and reproductive parameters, among other life-history characteristics, are of great value for fishery modelling and management, as well as for assessment of conservation status. As such information accumulates, it is becoming clear why some of the larger groupers were among the first to disappear from many coral reef fisheries; their biological features make them more vulnerable than most other species in the multi-species, multi-gear fisheries they typically form part of. The sexual pattern of female-to-male hermaphroditism found in many groupers is also problematic when fishing selects for the larger, mostly male, fish and when the rate of female-to-male sex change cannot keep pace with the prevailing fishing rate. It is hardly surprising, then, that there is some concern over the current fishery and conservation status of a number of species.

Status of grouper fisheries

Despite the importance of grouper fisheries, both recreational and commercial, few today are monitored adequately or consistently enough to track general population trends over the long term, or to assess the outcomes of conservation action or fisheries management, if any. Fishery-independent data on the abundance of grouper species are needed to evaluate population trends over time, while size information from catches can also provide an indication of population status. Identifying species of conservation concern and the numbers of species that are 'Data Deficient' (see below), according to IUCN criteria, further highlight where research efforts on this group are needed. One challenge to species-specific monitoring which this guide specifically aims to address is species identification; many groupers look similar, some species change colour as they grow or their colour may vary in different areas, while the colours of dead and live animals can vary considerably, posing significant challenges to those working to reliably distinguish grouper species. Note that the species assessments in this guide are not necessarily comprehensive literature reviews, but strive to use key research and fishery data, both published and unpublished, to best evaluate the conservation status of each species, according to the best available information.

Many grouper fisheries are on the decline, both in terms of reducing catch per unit effort (a measure of fish abundance) and also in mean size at capture. Indeed, in the most heavily fished areas, such as parts of South-East Asia and the Caribbean, larger species are becoming uncommon in local catches and a large proportion of the mid- to large-size groupers are taken below their size of sexual maturation — both signs that such fisheries are becoming unsustainable. Juveniles are also intensively sought after to supply mariculture 'grow-out' operations, whereby fish ranging from the stage shortly post-settlement to early adulthood are caught and placed in cages for grow-out, or 'farming', to market size. This activity is now referred to as 'capture-based aquaculture' (CBA) and represents an additional fishing pressure, especially in South-East Asia. Associated with CBA is a large trade in high-priced reef fishes as part of what is known as the live reef food fish trade (LRFFT). Within this trade, groupers are by far the most important type of fish. In the LRFFT, consumers in Chinese restaurants and markets are able to select fish displayed alive in tanks prior to cooking (Figure 1), for which they pay much higher prices than they would for chilled fish. This luxury seafood trade has become a driving force in the intensive fishing and mariculture of several medium- to larger-sized groupers, particularly in parts of South-East Asia and the western Pacific.

© Yvonne J Sadovy de Mitcheson

***Figure 1** Groupers are the most important fishes sold by seafood restaurants and markets in a large live food fish trade centred in Hong Kong and mainland China. They are sourced widely from reefs throughout much of the Indo-Pacific*

This demand for live groupers far outweighs the projected supply from available reef areas, and is expected to grow as China's economy develops. It has also been a factor driving businesses to source fish from countries ever more distant from Hong Kong, the global trading centre for the LRFFT (Figure 2). This and other trends in the expanding use of fish globally are expected to continue to increase pressures on grouper populations, and further emphasise the need for management, monitoring and precautionary use.

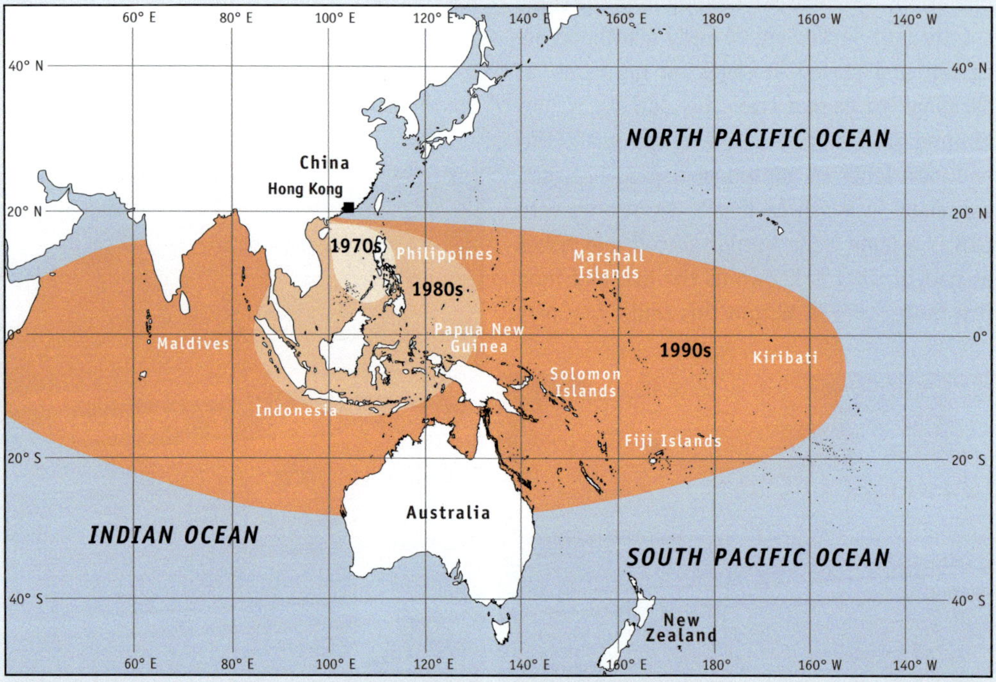

Figure 2 *Map showing expansion of the countries supplying groupers for the live reef food fish trade, with the demand centred in Hong Kong and mainland China. Fish were mainly taken from the northern sector of the South China Sea in the 1970s, extending out to much of South-East Asia through the 1980s and out to the western and central Pacific and into the Indian Ocean in the 1990s and 2000s. Source: Sadovy et al. (2003)*

Conservation status of groupers

The conservation status of groupers has been variously assessed at both the national and international levels over the last decade. The IUCN (International Union for Conservation of Nature) Red List criteria have been applied to all grouper species in relation to their global conservation status by the IUCN Groupers & Wrasses Specialist Group (GWSG) since 1998. These evaluations determined that at least 20 species could be threatened with extinction in the foreseeable future if current trends continue. This does not include the many species that could not be assessed at all due to an absence of data (i.e. IUCN 'Data Deficient' category — see page xv). These assessments are also important because they so clearly highlight the general scarcity of data, both biological and fisheries related, for most species of groupers, and because they identify key threatening factors (GWSG 2007, Sadovy de Mitcheson et al. in press).

Of the 20 threatened species assessed, the principal threat factor was population reduction attributable to levels of harvest that appear to be unsustainable, especially since this situation is typically associated with a lack of fishery management (IUCN 'Criterion A' — see page xv). Body size was clearly also a factor, with larger-sized groupers (>1 m total length) more commonly assessed as 'Threatened' or 'Near Threatened' (Table 2, page xv). Presumably this is partly attributable to the general finding in fishes that larger body size tends to be associated with longevity, delayed sexual maturation and hence lower population replacement rates generally, factors typically associated with heightened vulnerability to fishing. Other notable risk factors identified for certain endemic species were habitat loss due to coastal development, restricted geographic range, and small population size; several deeper-living species are also at risk (see below). Interestingly, many species of grouper considered to be Threatened or Near Threatened have wide geographic ranges, so broad distribution does not necessarily mean higher resilience to fishing pressure for such species.

Whereas the major overall threat factor was undoubtedly overfishing, perhaps the most important single biological characteristic contributing to the decline of several grouper species is their reproductive biology, in particular the habit of aggregation spawning. Such aggregations make very attractive fishing targets, and are rarely effectively managed. In extreme cases, populations have been reduced to such low numbers that aggregations no longer form. The most extreme and best-known case is the now endangered Nassau grouper *Epinephelus striatus*, the massive declines in its populations are largely attributable to aggregation fisheries, but there are also concerns for several other *Mycteroperca* and *Epinephelus* species (Sadovy de Mitcheson et al. 2008, Sadovy de Mitcheson and Colin in press; www.scrfa.org).

While MPAs promise conservation benefits to many reef fish species, for most exploited groupers fishing effort also needs to be controlled. The larger species, in particular, require special attention since many migrate long distances to spawn each year, usually well beyond the boundaries of most MPAs. Hence MPAs should encompass spawning aggregation sites as well as other key habitats for such species. It is also the larger species that are increasingly being landed in their juvenile stage as adults disappear from fisheries. Moreover, deeper-living species need special attention because of the problem of barotrauma (expansion of gases in the gas bladder resulting in haemorrhage and death) associated with hauling fish up from deep waters. Barotrauma is a challenge because it means that when protected species or undersized fish are hauled to the surface, even if released they are unlikely to survive. Given the relatively restricted distribution of such species to specific areas of outer reef slopes, consideration could be given to including such areas within MPA boundaries; currently few MPAs protect such habitats.

The need for a better understanding of the biology, ecology, distribution and patterns of exploitation on groupers is clear from the assessments in this guide. The first step is, of course, to correctly identify individual species. While seemingly a simple task, grouper identification can be challenging, even for those who regularly work with reef-associated fishes and fisheries; many species look similar, there can be geographic and ontogenetic changes in colouration patterns, and dead and live animals might look very different. The definitive guide to the world's groupers, Heemstra and Randall's (1993) FAO guide to groupers of the world, was the first comprehensive compilation of all grouper species and reflected all that was known about this group of fishes at that time. This updated guide builds upon and supplements Heemstra and Randall's earlier work in the hope that it will assist workers to help us better understand and conserve this interesting, important and vulnerable group of fishes.

Systematics of groupers

Grouper taxonomy is in a state of flux. With the now widespread use of molecular data in systematic research, many novel relationships among grouper species and genera have come to light (e.g. Craig and Hastings 2007, Smith and Craig 2007). This has led to new taxonomic arrangements, both within the true 'groupers', and among various other allied genera in the 'seabasses'. Some of these new taxonomic changes are adopted in this book, whereas others are not (see discussion on taxonomy below). The absence of a complete adoption of all newly proposed taxonomic changes is a reflection of a largely theoretical argument among members of the ichthyological community, and is not unique to groupers. Briefly, one school of thought places more emphasis on taxonomic groups being a reflection of monophyly (shared common ancestry), while the other places more emphasis on morphologically similar (or different) features. Both arguments have merit, and further information and study will shed light on decisions pertaining to grouper systematics.

Groupers have traditionally been treated as a subfamily of the family Serranidae. As early as the late 19th century, hypotheses of close relationships between 'groupers' and 'seabasses' were present in the literature (e.g. Jordan and Eigenmann 1890), and through the mid to late 20th century, the placement of genera within the Serranidae and their relationships were regularly changed (Katayama 1959, Gosline 1966, Kendall 1976, 1979, Johnson 1983, 1988, Baldwin and Johnson 1993).

At the turn of the 21st century, molecular techniques became well-established methods for discerning relationships among animal groups. Early studies using these techniques highlighted novel relationships among 'groupers' and 'seabasses' and suggested that some genera (e.g. *Acanthistius, Niphon*) were in fact distantly related percomorph lineages and, moreover, that the 'Serranidae' was not a monophyletic lineage (e.g. Craig et al. 2001, Imamura and Yabe 2002, Dettaï and Lecointre 2004, 2005, Smith and Wheeler 2004). This prompted a large-scale investigation into the limits and relationships of the traditional 'Serranidae' (Smith and Craig 2007), which corroborated many of the findings of earlier studies.

It is now thought that the 'groupers' and remaining members of the traditional 'Serranidae' are not allied groups. Those genera traditionally treated as the subfamily 'Epinephelinae' (groupers, fairy basslets and soapfishes), with the exclusion of the monotypic genus *Niphon*, are now recognised as the family Epinephelidae. The remaining seabasses and anthiines are now regarded as a monophyletic family (Serranidae), with the exclusion of the genus *Acanthistius*. The monophyletic lineages formerly referred to as tribes within the subfamily Epinephelinae (*sensu* Johnson 1983) should now be regarded as subfamilies within the Epinephelidae (Diploprioninae, Epinephelinae, Liopropominae and Grammistinae). The definition of a 'grouper' (from the Portuguese 'garropa'; subfamily Epinephelinae) is any member of the genera *Aethaloperca, Alphestes, Anyperodon, Cephalopholis, Cromileptes, Dermatolepis, Epinephelus, Gonioplectrus, Gracila, Hyporthodus* (newly resurrected genus), *Mycteroperca, Paranthias, Plectropomus, Saloptia, Triso* or *Variola*. Currently, the subfamily Epinephelinae contains 163 species.

Recent taxonomic changes

Craig and Hastings (2007) presented a hypothesis of relationships among grouper species and proposed an aggressive re-organisation of genera within the groupers. Major changes proposed included synonymising the monotypic genera *Anyperodon* and *Cromileptes* with *Epinephelus*, synonymising the genera *Aethaloperca, Gracila* and *Paranthias* with *Cephalopholis*, restricting the genus *Epinephelus*, expanding the genus *Mycteroperca*, and resurrecting the genus *Hyporthodus*.

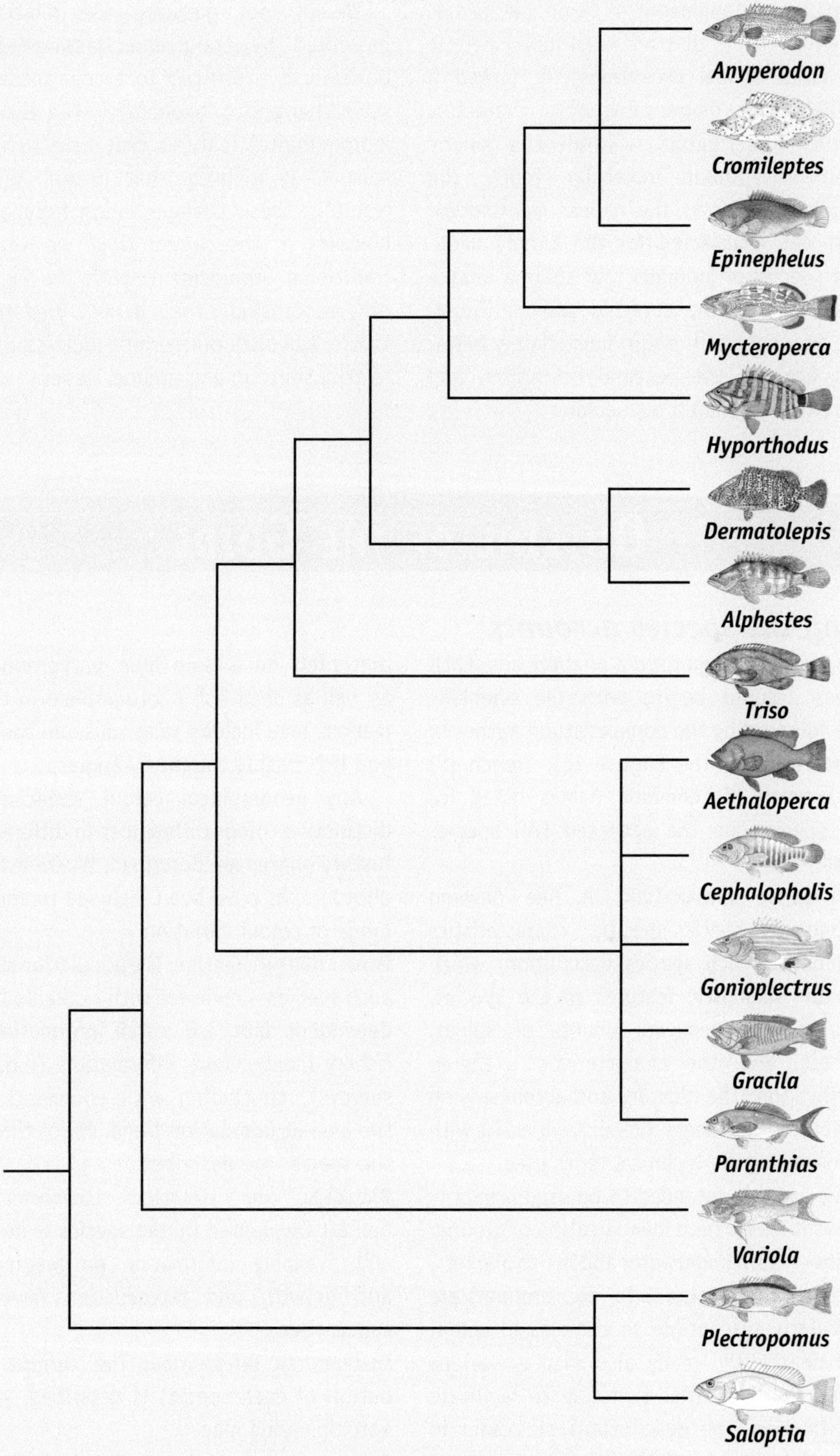

Figure 3 *Currently hypothesised relationships among grouper genera*

While this rearrangement of taxonomic names yielded a scheme that reflected monophyletic taxa (with a few exceptions), it lacked a morphologically diagnosable set of characters by which investigators could identify generic placement without molecular tools. The exception to this is the genus *Hyporthodus*, which was resurrected for the largely deep-water species of groupers that share a unique arrangement of the coracoid and cleithrum, have their pelvic fin origin immediately below or in front of the pectoral fin origin, and generally are a drab brown colour.

We do not question the relationships presented by Craig and Hastings (2007); however, it is difficult to accept these large-scale changes to taxonomy in the absence of morphological features that help to identify species. It is likely that future study will result in these changes being fully applied; however, at the current time, we retain the traditional arrangement, with the exception of recognising the genus *Hyporthodus*. Figure 3 depicts our current understanding of relationships among grouper genera.

OUTLINE OF THE GUIDE

Using the species accounts

The species are organised alphabetically. Each species account begins with the scientific name followed by the nomenclatural authority of that species. The English (E), French (F) and Spanish (S) common names listed for each species are the accepted FAO species names.

IDENTIFICATION INFORMATION: A line drawing highlighting species-specific characteristics accompanies each species description, which lists the diagnostic features of the species, including measurements, counts of spines, rays, etc., and other characteristics useful in identification. The glossary and accompanying illustrations (see pages xvi–xix) will assist with understanding the technical terms used.

This guide is intended to be used primarily as a resource for field identification of grouper species — both underwater and in fish markets, fishery outlets, restaurants, etc. Groupers are particularly susceptible to changes in colour upon death, and many also display various colour morphs in life. Therefore to facilitate identification, the descriptions of 'colour in life' and 'colour in market' are accompanied by photographs of both live and market specimens. Market specimens are invariably presented on a pale blue background and, as well as dead fish photographed in the fish market, may include some museum specimens and fish freshly caught by anglers.

Any geographical colour variations and distinctive colour differences in different life-history phases are described. Where available, photographs have been included to show this range of colour variation.

POPULATION INFORMATION: The population status of each species is reviewed with available fishery-dependent data (e.g. catch information) and fishery-independent information (e.g. diver surveys), concluding with comments about the overall population trend. Major threats to the species are described.

BIOLOGICAL AND ECOLOGICAL INFORMATION: The habitat frequented by the species is described and available information on feeding, age and growth, and reproduction have been summarised.

DISTRIBUTION INFORMATION: The current distribution of each species is described, with an accompanying map.

CONSERVATION INFORMATION: The IUCN Red List status is given for each species (this listing is described in more detail on the next page).

Biological data and IUCN criteria

The biological, ecological and distributional data contained herein were largely compiled by members of the IUCN Species Survival Commission Groupers & Wrasses Specialist Group in the context of applying IUCN criteria for extinction risk. The individuals directly responsible for the compilation of these data are listed in alphabetical order as the 'authors' at the end of each species account.

The five IUCN criteria provide a standardised method for assessing threat levels to animal species of any group (Table 1). Using the threat criteria, species can be placed into eight categories ranging from 'Extinct' to 'Data Deficient' (Table 2).

There are three categories in which a species is considered threatened with extinction if current trends continue: Critically Endangered, Endangered and Vulnerable. 'Near Threatened' indicates that a species is considered to be under threat, but has not met the quantitative threshold for a threat category. A species in the 'Least Concern' category indicates that there is no current evidence to suggest the species is under major risk of extinction. The 'Data Deficient' category indicates that a major threat has been identified, but the information available is insufficient to evaluate the degree to which the threat is affecting populations. It is important to note that the 'Least Concern' and 'Data Deficient' categories do not necessarily imply that a particular species is not experiencing some decline or threat, but rather that the data at present do not meet the IUCN thresholds. A more thorough discussion of IUCN criteria and categories may be found in Mace et al. (2008) or online at www.iucnredlist.org/documents/redlist_cats_crit_en.pdf. Red List Assessments are valid for 10 years.

Table 2 *Summary of IUCN status categories and severity of impact by threat*

Category	Abbreviation	Severity
Extinct	EX	–
Extinct in the Wild	EW	High
Critically Endangered	CR	High
Endangered	EN	High
Vulnerable	VU	High
Near Threatened	NT	Moderate
Least Concern	LC	Low
Data Deficient	DD	Unknown

Table 1 *Summary of IUCN criteria* for listing in a threat category (Vulnerable, Endangered or Critically Endangered)*

Criterion A	**Population Reduction:** Size of population has been observed, estimated or inferred to have declined by a considerable proportion over the past three generations.
Criterion B	**Geographic Range:** Species has a small range and is either (a) severely fragmented, (b) experiencing decline in range area or number of mature individuals, or (c) is experiencing extreme fluctuations in range area or number of mature individuals.
Criterion C	**Small Population Size and Decline:** Number of mature individuals is small and there is continuing decline (minimum 10%) expected over the next three generations or (a) a continuing decline in the number or percent of mature individuals in each subpopulation and/or (b) extreme fluctuations in the number of mature individuals.
Criterion D	**Very Small or Restricted Population:** Number of mature individuals is less than 1000 and/or the area of occupancy is less than 20 km^2 or ≤5 locations.
Criterion E	**Quantitative Analysis:** Quantitative population analysis indicates the probability of extinction in the wild to be ≥10% in the next 100 years.

* Each criterion may have a number of subcriteria (1, 2, etc.; a, b, etc.; i, ii; etc.) that qualify the status of a species falling into any of these three threat categories, e.g. Vulnerable (A4cd)

Glossary of terms

The following is a list of technical terms and measurements used in the descriptions of species in this guide. Pictorial representations of some key characters may be found in the accompanying illustrations.

Anterior – Towards the head or front end of the fish.

Antrorse – Curved or pointing anteriorly.

Axil – The inner side of the base of the pectoral or pelvic fins.

Band – An elongate colour marking, approximately 1 or 2 eye diameters in width (Figure 4).

Bar – An elongate colour mark with vertical or near vertical orientation (Figure 4).

Belly – The lower part of the body from the pelvic fins to the anal fin.

Blotch – A colour mark, larger than the eye and usually with irregular or indistinct borders (Figure 4).

Body width – The greatest width (thickness) of the body, usually occurring at the base of the pectoral fins.

Branchiostegals – Ray-like bones supporting the gill membrane that seals the underside of the gill cavity.

Canine – A prominent, conical tooth (sometimes curved) that is distinctly larger than other teeth in the jaws.

Catch per unit effort – The total catch divided by the total amount of effort used to harvest the catch. For example, 100 fish taken in two hours of fishing would have a catch per unit effort of 50 fish per hour. (abbreviation CPUE)

Caudal fin – The tail fin (Figure 5). The shape of the caudal fin is distinctive for many species of groupers (Figure 6). The caudal-fin length is the length of the longest caudal-fin rays.

Caudal peduncle – The part of the tail that joins the caudal fin to the body. The caudal-peduncle length is the horizontal distance from the rear end of the anal fin base to a vertical at the caudal-fin base (Figure 5). The caudal-peduncle depth is the least depth of the peduncle.

Compressed – Laterally flattened. A compressed fish (e.g. *Anyperodon*) has the body width less than half of the body depth.

Ctenoid scale – A scale with minute spines (ctenii) along the posterior margin. Ctenoid scales feel rough when the fish is stroked towards the head.

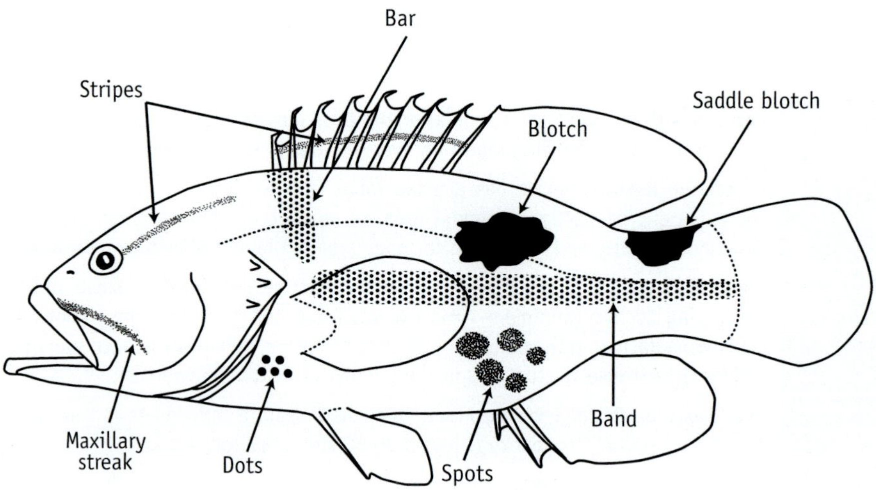

Figure 4 *Terms for colour markings*

DISTAL – Away from the centre of the body or base of a fin.

DORSAL – On or towards the upper part of the fish.

DORSAL FIN – The median fin on the upper part of the body (Figure 5).

DOT – A colour mark about the size of a nostril (Figure 4).

EMARGINATE – Slightly concave margin (Figure 6).

EYE DIAMETER – The horizontal diameter of the fleshy orbit (eye opening) (Figure 5).

FORK LENGTH – A measure of length from the tip of the snout to the "fork" in the caudal fin. (abbreviation FL)

GILL ARCH – One of the 4 arches of bones that bear the gills. If the gill cover or operculum (Figure 7) is lifted, the outermost gill arch is visible in the gill cavity (Figure 8).

GILL FILAMENTS – The soft, bright red filaments attached along the posterior (outer) edge of the gill arches (Figure 8).

GILL RAKERS – The long bony protuberances along the anterior (inner) edge of the gill arches (Figure 8). The gill-raker count is

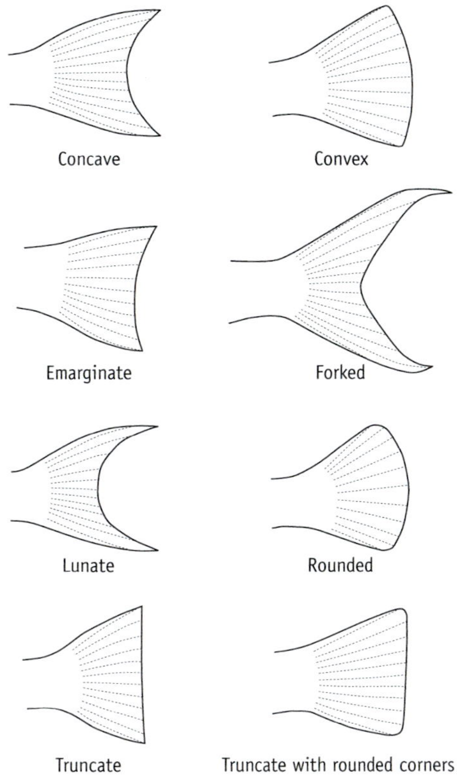

Figure 6 *Types of caudal fins*

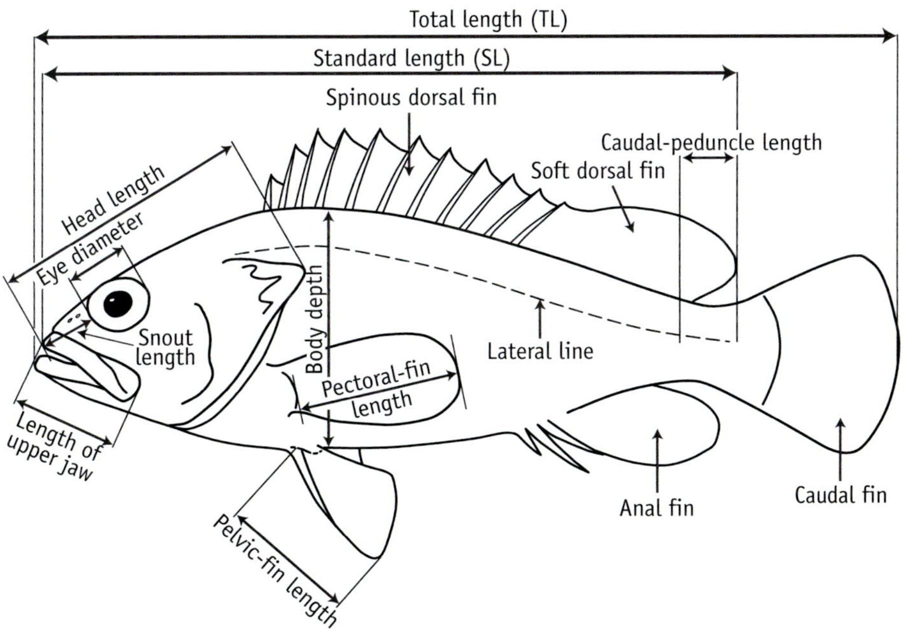

Figure 5 *External morphology and measurements*

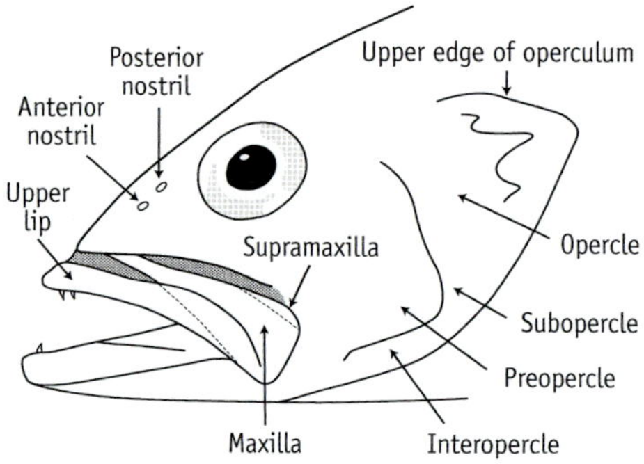

Figure 7 *Parts of the head*

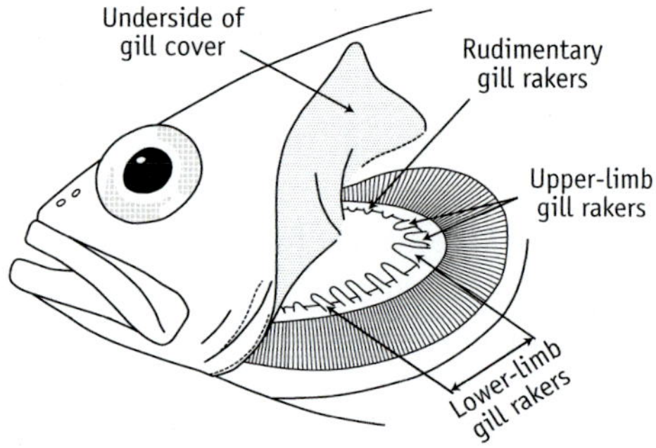

Figure 8 *First gill arch*

often given as two numbers (X + Y), where X is the number of rakers on the upper (dorsal) limb of the first arch, and Y is the number of rakers on the lower (ventral) limb, with the raker at the angle (junction of upper and lower limbs) included in the lower-limb count. The range in number of gill rakers for a species is often given as follows: 7–9 + 16–19 (which means there are 7 to 9 rakers on the upper limb and 16 to 19 rakers on the lower limb). Rudimentary rakers or rudiments (Figure 8) are wider than long (high) and (unless indicated otherwise) are usually included in the gill-raker counts. If the rudiments are difficult to distinguish from intercalated bony tooth plates, the gill-raker count is of 'developed' rakers only.

HEAD LENGTH – The straight-line measurement from the front of the upper lip at the midline (with upper jaw retracted) to the posteriormost point of the operculum (Figure 5).

INTERORBITAL WIDTH – The least distance between the fleshy edges of the orbits.

LATERAL – Of or towards the side.

LATERAL LINE – The series of pored or tubed scales that run from the upper end of the gill cavity to the caudal fin.

LATERAL-SCALE SERIES – The oblique series of scales.

LUNATE – Shaped like a crescent moon; deeply concave margin (Figure 6).

MARINE PROTECTED AREA – An area afforded special protection by a local or national government agency. In some cases, this includes restrictions in the harvest of marine organisms. (abbreviation MPA)

MAXILLA – The posterior, toothless bone of the upper jaw; partly covered by the upper lip when the mouth is closed (Figure 7).

MAXILLARY STREAK – A dark streak along the edge of the maxillary groove (the groove below the eye which is hidden by the maxilla when the mouth is closed) (Figure 4). Also called 'moustache streak'.

MEDIAN FINS – The dorsal, anal and caudal fins.

NAPE – The posterior part of the top of the head.

OPERCLE – The largest bone of the operculum (gill cover); mostly covered by skin and scales (Figure 7).

ORBIT – The fleshy eye opening. In live fish, the transparent cornea of the eye protrudes slightly through the orbit.

PALATINES – The paired bones on each side of the roof of the mouth immediately posterior to the vomer. In most groupers, the palatines bear minute teeth that are visible on the roof of the mouth.

POSTERIOR – Towards the tail or rear end of the fish.

PREMAXILLA – The paired, tooth-bearing bones that run above the lateral line from the upper end of the upper jaw, mostly hidden by the upper lip.

Preopercle – The rounded or angular bone on the front part of the operculum; the posterior edge is exposed and usually serrate (Figure 7).

Preorbital – The anteriormost and largest bone of the circumorbital bones that form the lower edge of the orbit.

Procurrent ray – Small slender fin rays along the upper and lower edges of the tail-fin base; some are segmented, some spinous.

Pterygiophore – The series of 2 or 3 bones that support each of the spines and rays of the dorsal and anal fins. Note that the distal element of each pterygiophore is between the left and right halves of its associated (following) fin ray.

Rays – The rigid or flexible struts that support the fins. Soft rays (usually just called 'rays') are segmented, flexible and usually branched; the rays are also bilaterally paired (this bilateral structure is evident if the fish is cooked). Spinous rays (or fin 'spines') are unsegmented, simple (unbranched) and pungent (stiff and sharp pointed). In the ichthyological literature, the fin spines are usually designated by Roman numerals and the soft rays by Arabic numerals (D IX,14: meaning that the dorsal fin has IX spines and 14 rays). The last ray of the dorsal and anal fins of groupers is usually split to its base and looks like two rays, but it is counted as a single ray because it is supported by a single pterygiophore.

Saddle blotch – A dark saddle-shaped blotch on the dorsal part of the body or caudal peduncle (Figure 4).

Serrate – With a series of small points (serrae) like the teeth of a saw.

Snout – The part of the head in front of the eyes (Figure 5).

Spine – A sharp and stiff projection (see also rays).

Spot – A colour mark that is eye-sized or smaller, but larger than the size of the nostrils and usually with a regular or distinct margin (Figure 4).

Standard length – The straight-line distance from the tip of the snout to a vertical line passing through the base of the caudal fin (taken to be the point of flexure of the caudal fin) (Figure 5). (abbreviation SL)

Stripe – A narrow, oblique or horizontal colour marking, less than an eye diameter in width (Figure 4).

Subadult – A fish that displays a colour pattern that is intermediate between a juvenile and adult. This may include showing characters of both phases, or something different altogether.

Supramaxilla – A small elongate bone lying along the dorsal edge of the maxilla (Figure 7).

Synonym – An alternate name given to a particular species, genus, family, etc. The oldest (senior) synonym that is given to a species is usually considered the valid (correct) scientific name for that species.

Thoracic – Pertaining to the thorax or chest region below the pectoral-fin base and in front of the pelvic fins.

Total length – The straight-line distance from the anteriormost tip of the jaw to a vertical line passing through the posterior tip of the longest caudal-fin ray (Figure 5). (abbreviation TL)

Truncate – Cut square; a caudal fin with a straight rear margin (Figure 6).

Ventral – On or towards the lower part (underside) of fish.

Vertebrae – Bones of axial skeleton; divided into two sections, precaudal (abdominal) and caudal vertebrae.

Vomer – A median bone that lies on the roof of the mouth; often bears teeth.

Species Accounts

Aethaloperca rogaa – Variola louti

Paranthias colonus

Aethaloperca rogaa Forsskål 1775

redmouth grouper (E), mérou noir (F), cherna roga (S)

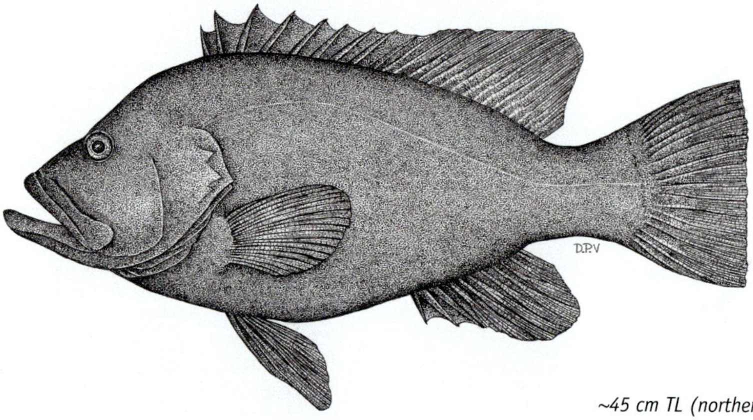

~45 cm TL (northern Mozambique)

Body deep and compressed, the depth greater than the head length and contained 2.1–2.4 times in standard length, the width 2.3–2.8 times in the depth. Head length 2.5–2.7 times in standard length; interorbital area convex; dorsal head profile steep, straight or slightly concave along the snout and distinctly convex from eye to dorsal fin; preorbital depth 6.5–9.2 times in head length; preopercle finely serrate, the lower edge fleshy; subopercle and interopercle smooth or with minute serrae; opercle with 3 flat spines, the middle one closer to the lower spine; upper edge of operculum convex; posterior nostril round or oval, not much larger than anterior one; maxilla reaches past eye; ventro-posterior corner of maxilla with a distinct bony protuberance; supramaxilla slender; small canines at front of jaws, none elsewhere; 2–4 rows of small slender teeth at side of lower jaw; palatine teeth present. Gill rakers 8–10 on upper limb, 15–17 on lower limb, the longest gill raker slightly longer than longest gill filaments. Dorsal fin with 9 spines and 17 or 18 rays, the fin origin over the opercle; dorsal-fin membranes slightly incised between the spines, the 3rd or 4th spine longest; anal fin with 3 spines and 8 or 9 rays; middle dorsal- and anal-fin rays elongate in adults, giving these fins an angular profile, with the rear margin almost vertical; pectoral fins asymmetric, with 17–19 rays, the 5th or 6th ray longest; a well-developed scaly flap of skin joining upper pectoral-fin rays to body; pelvic fins subequal to pectoral fins, reaching to or beyond anus; caudal fin truncate, with 8 branched rays and 9 procurrent rays in upper part and 7 branched rays and 8 procurrent rays in lower part. Scales on body ctenoid, with auxiliary scales; lateral-line scales 48–54; lateral-scale series 94–104.

MAXIMUM RECORDED SIZE 60 cm TL

Aethaloperca rogaa is an Indo-West Pacific species found from the Red Sea to South Africa, extending eastward to Samoa and the Phoenix Islands, and northward to southern Honshu (Japan). It is probably found around all tropical islands of the Indian Ocean (but is not yet recorded from Mauritius).

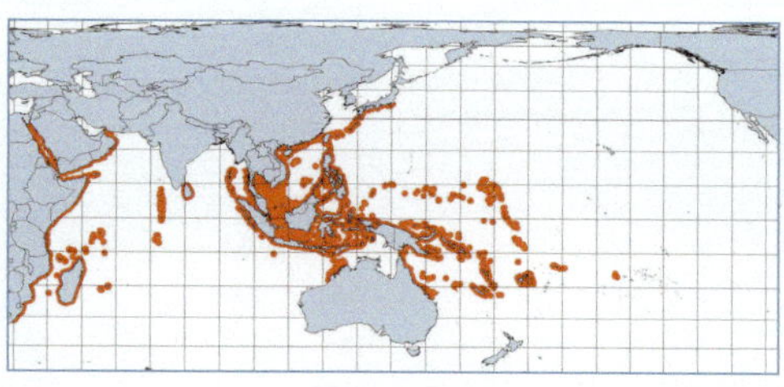

POPULATION **General** *A. rogaa* is widespread, but typically uncommon with variable regional abundance. It is occasionally found in markets, but no detailed population information is available.

POPULATION TREND Unknown

MAJOR THREATS The primary threat to *A. rogaa* is overfishing, and it is captured incidentally by artisanal and small-scale commercial fishers.

HABITAT AND ECOLOGY **General** *A. rogaa* is a reef-associated species found in coastal reefs and lagoons, in or near caves and holes in the reef, in depths of 1–60 m. Small juveniles mimic *Centropyge vrolikii* and *C. nox* until they outgrow these species in size. The species is taken as part of the live reef food fish trade centred in Hong Kong and China. **Feeding** *A. rogaa* feeds mainly on small fishes (including *Pempheris* spp.), stomatopods (*Pseudosquilla* spp.) and crustaceans (Morgans 1982). **Reproduction** Preliminary data indicate that the species spawns at any time of the year and matures at about 35 cm SL (Morgans 1982).

COLOUR IN LIFE Dark brown to black, occasionally with a greenish to reddish cast, usually with a pale vertical bar on the side of the abdomen; distal part of the spinous dorsal fin dark orange to brownish red; inside of the mouth, gill cavity and upper jaw membranes reddish orange. Juveniles with a broad white posterior margin on the caudal fin and a narrow white margin on the soft dorsal fin; Allen and Steene (1987) show a 30 cm TL juvenile from the Maldives with a yellow-green spot on each body scale and the median fins with a pale bluish grey margin posteriorly.

COLOUR IN MARKET Generally dark brown with no distinguishing marks. Its characteristically steep dorsal profile aids in the identification of this species, along with its asymmetrical pectoral fins.

© Matthew T Craig

Adult *A. rogaa* (Okinawa Fish Market, Japan)

Adult *A. rogaa* (Bali, Indonesia)

© Robert F Myers

COMMENTS Craig and Hastings (2007) demonstrated a close, nested relationship between this genus and the genus *Cephalopholis*, and suggested that *A. rogaa* is best treated as a member of the genus *Cephalopholis*. It is distinct from other members of *Cephalopholis* only in its asymmetrical pectoral fins. We retain it here in *Aethaloperca* pending further study.
2008 IUCN RED LIST STATUS Data Deficient

Authors: *AS Cabanban, AS Cornish, PC Heemstra and MA Samoilys*

Alphestes afer (Bloch 1793)

mutton hamlet (E), varech (F), guaseta (S)

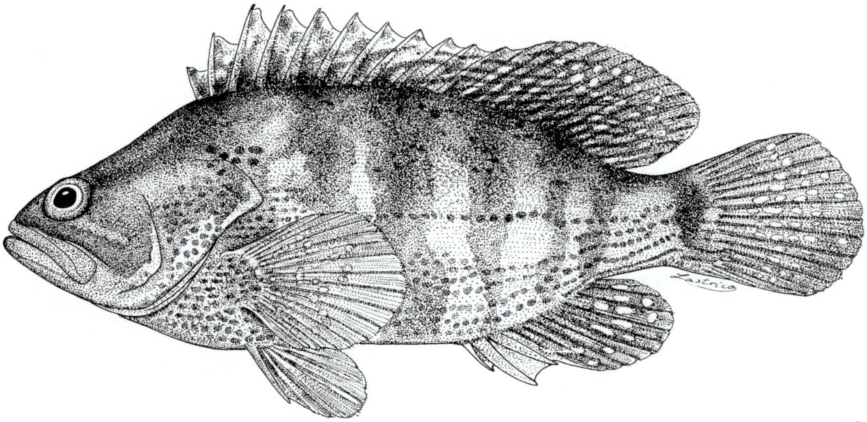

19 cm SL

Body depth slightly less than head length, depth contained 2.4–3.1 times in standard length; caudal-peduncle depth 12–14% of standard length. Eye diameter greater than snout length, eye diameter contained 4.1–5.3 times in head length; preopercle rounded, the lower edge distinctly serrate with a large spine (usually hidden by skin) directed downward and forward at the angle. Gill rakers 6–8 on upper limb, 16 or 17 on lower limb, 22–25 total. Dorsal fin with 11 spines and 17–19 rays; anal fin with 3 spines and 9 rays; pectoral fins with 16 or 17 rays; caudal fin rounded. Scales smooth; lateral-line scales 55–61; lateral-scale series 68–77.
MAXIMUM RECORDED SIZE 33 cm TL

Alphestes afer is distributed in the western Atlantic from Bermuda to south Florida (USA), the Bahamas, Cuba, West Indies, Panama and Venezuela, and southward to the state of São Paulo (Brazil), and in the eastern Atlantic in the Gulf of Guinea, West Africa.

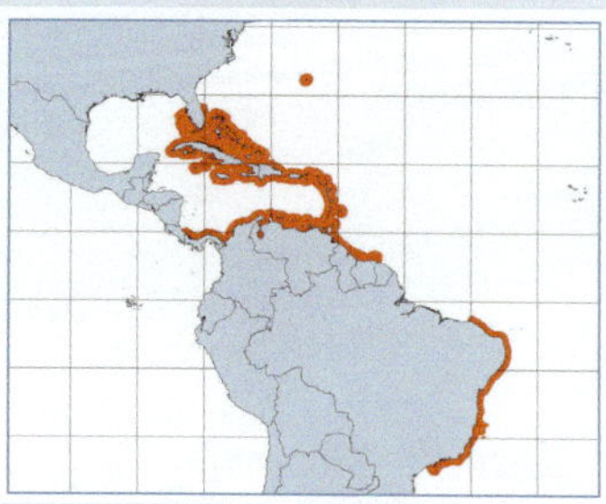

POPULATION **General** *A. afer* is generally uncommon but may be present in high numbers locally. **Fishery dependent** *A. afer* is generally not landed (or at least not reported) in fisheries due to its small size, and therefore accurate data on landings are unavailable at this time. However, the species is taken opportunistically in small-scale, multi-species fisheries throughout its range. It is caught by spear, line and traps. According to data from the REVIZEE/SCORE-Central Program, *A. afer* is uncommon in fisheries between Cabo São Tomé (Rio de Janeiro State) and Salvador (Bahia State) (Brazil; Rocha and Costa 1999). **Fishery independent** Visual censuses indicate that *A. afer* is generally uncommon. **Genetics** Recently, genetic studies have confirmed the presence of *A. afer* in the eastern Atlantic (Gulf of Guinea, West Africa; see comments below and Craig et al. 2006a).

POPULATION TREND Unknown

MAJOR THREATS Fishing pressure and habitat destruction are the primary threats to *A. afer*. The species is caught as part of a mixed-catch fishery by artisanal fisheries (especially trap fisheries) in several locations within its distributional range, and its predilection for seagrass habitats makes it vulnerable to coastal development.

HABITAT AND ECOLOGY **General** *A. afer* is a reef-associated species with a depth range of 1–30 m. It is a solitary species that typically occurs in seagrass habitats. Individuals are sedentary during the day, hiding in crevices or lying among seaweed, sometimes partly covering themselves with sand while lying on their side. **Feeding** *A. afer* is a nocturnal feeder of benthic crustaceans, fishes and invertebrates. **Reproduction** *A. afer* is a sequential hermaphrodite (Randall and

COLOUR IN LIFE Head, body and median fins olivaceous or light brown, with irregular dark brown blotches and bars, and densely covered with orange spots; head, body and all fins with scattered small white spots; pectoral fins orange or yellow with faint brownish reticulations.

COLOUR IN MARKET Similar to above, but general body colour may fade to a pale drab cream, and spots and reticulations may fade.

© Matthew T Craig

Adult *A. afer* (Florida, USA)

© Áthila Bertoncini Andrade

Adult *A. afer*
(Barra Grande, Brazil)

Bishop 2004). Thompson and Munro (1983) estimated the number of eggs in four females to range from 157 512 to 223 706 per fish.

COMMENTS Originally described from the coast of Guinea by Bloch (1793) as *Epinephelus afer*, this species had not been reported from the eastern Atlantic since the original description, and the type locality was long regarded as erroneous (Smith 1971, Heemstra and Randall 1993). In 2004, P Wirtz observed an unfamiliar grouper in a fish market on São Tomé Island (Gulf of Guinea, West Africa). He photographed the fish and obtained a tissue sample for genetic analysis. The fish was identified from the photograph as *A. afer* by PC Heemstra. Subsequent examination of DNA sequences by MT Craig indicated that the specimen was conspecific with western Atlantic specimens of *A. afer* Bloch 1793. The species is now confirmed to have an amphi-Atlantic distribution (Craig et al. 2006a).

2008 IUCN RED LIST STATUS Least Concern

Authors: *AA Bertoncini, JH Choat, MT Craig, BP Ferreira, S Marques and LA Rocha*

Alphestes immaculatus Breder 1936

Pacific mutton hamlet (E), varech pacifique (F), guaseta pacifico (S)

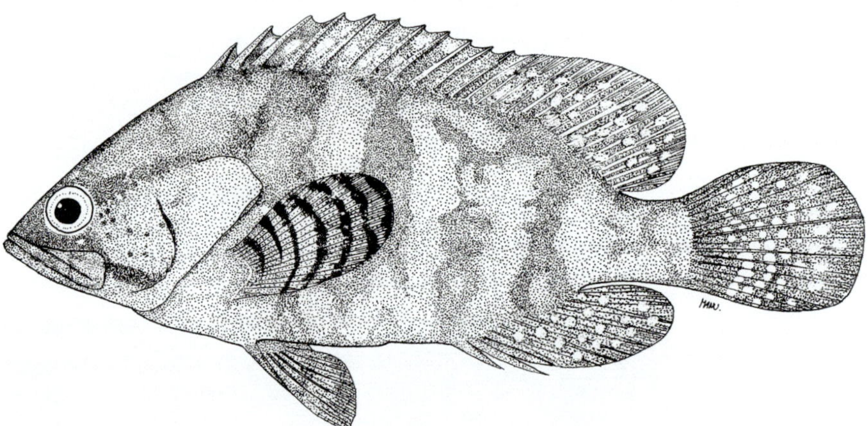

~20 cm TL
© FAO

Body depth less than or equal to head length, depth contained 2.3–2.8 times in standard length; caudal-peduncle depth 11–12% of standard length. Eye diameter subequal to snout length, contained 4.1–5.5 times in head length; preopercle rounded, the lower edge distinctly serrate, with a large spine (usually hidden by skin) directed downward and forward at the angle. Gill rakers 5–7 on upper limb, 14–16 on lower limb, 20–22 total. Dorsal fin with 11 spines and 17–19 rays; anal fin with 3 spines and 9 rays; pectoral fins with 16–18 rays; caudal fin rounded. Scales smooth; lateral-line scales 55–63; lateral-scale series 71–75.

MAXIMUM RECORDED SIZE 30 cm TL

Adult *A. immaculatus* (Gulf of California, Mexico)

COLOUR IN LIFE Head and body marbled reddish to light brown, with obscure bars and small dark spots; 7 or 8 irregular transverse series of small dark spots on the pectoral fins; pale spots and blotches over all of the head, body and median fins.

COLOUR IN MARKET Similar to colour in life but the body colour may fade to a buff. The reticulations on the pectoral fins usually remain.

Adult *A. immaculatus* (Gulf of California, Mexico)

POPULATION GENERAL *A. immaculatus* is more common in the north of its range and becomes less common in the southern extent of its range where *A. multiguttatus* becomes the more common of the two species.

POPULATION TREND Unknown

MAJOR THREATS Fishing pressure is the major threat to *A. immaculatus*. It is not currently a target of any commercial fishery, but it is caught opportunistically in small-scale, multi-species fisheries.

HABITAT AND ECOLOGY GENERAL *A. immaculatus* is found on rocky reefs in depths of 1–50 m. It is a cryptically coloured, secretive species that inhabits algae-dominated and rocky reef areas. Individuals are sedentary during the day, hiding in crevices or lying among algae, and rely on their effective camouflage to escape detection; they sometimes partly cover themselves with sand.

FEEDING *A. immaculatus* is a nocturnal predator that feeds mainly on benthic crustaceans.

Alphestes immaculatus is found in the eastern Pacific from the northern Gulf of California (Mexico) to southern Peru, including the Galápagos Islands.

COMMENTS Two species of *Alphestes* are known from the eastern Pacific: *A. immaculatus* and *A. multiguttatus*. They can be easily distinguished by the number of dark bars on the pectoral fins. *A. immaculatus* has 7 or 8 irregular bars, while *A. multiguttatus* has 5 or 6 more distinct bars. In the northern portion of its range, *A. immaculatus* is more commonly found in shallow depths, with *A. multiguttatus* generally found deeper. In the southern portion of its range, the situation is reversed.

While *A. immaculatus* is presently not a major target of any known fisheries, increased fishing pressure on smaller groupers has been noted as larger, more valuable species are harvested to low levels.

2008 IUCN RED LIST STATUS Least Concern

Authors: *AA Bertoncini, JH Choat, MT Craig, BP Ferreira, PC Heemstra and LA Rocha*

Alphestes multiguttatus (Günther 1867)

rivulated mutton hamlet (E), varech veine (F), guaseta rayada (S)

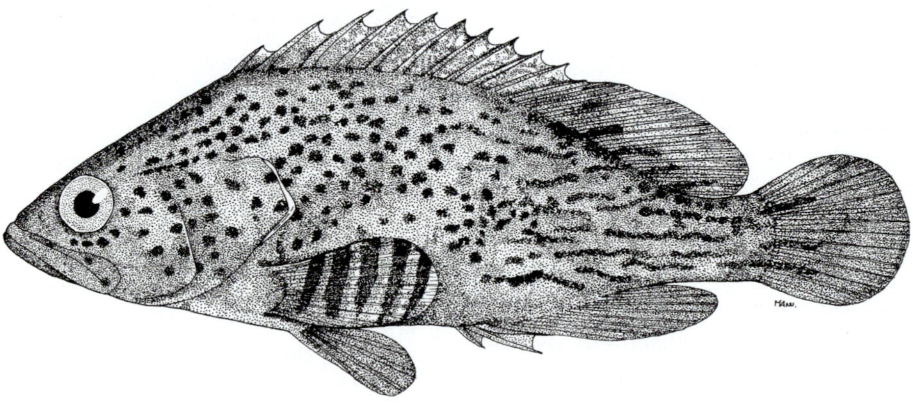

17 cm TL
© FAO

Body depth distinctly less than head length, depth contained 2.7–3.1 times in standard length; caudal-peduncle depth 11–12% of standard length. Eye diameter slightly greater than snout length, contained 4.3–4.7 times in head length; preopercle rounded, the rear edge distinctly serrate with a large spine (usually hidden by skin) directed downward and forward at the angle. Gill rakers 6–8 on upper limb, 14–16 on lower limb, 20–23 total. Dorsal fin with 11 spines and 18–20 rays; anal fin with 3 spines and 9 rays; pectoral fins with 17–19 rays; caudal fin rounded. Scales smooth; lateral-line scales 54–62; lateral-scale series 67–75.

MAXIMUM RECORDED SIZE 25 cm TL

COLOUR IN LIFE Head and body pale to dark brownish, with small dark spots forming horizontal streaks posteriorly and extending onto the soft dorsal and caudal fins; pectoral fins pale, with 5 or 6 broad dark crossbars, which are darker and more distinct distally.
COLOUR IN MARKET Generally the same as in life although the body colour may fade to a drab brown hue. The spots and crossbars on the pectoral fins usually remain.

Adult *A. multiguttatus* (Mazatlán, Mexico)

POPULATION GENERAL *A. multiguttatus* is more common in the southern portion of its range and becomes less common in the northern portion of its range where *A. immaculatus* becomes the more common of the two species.
POPULATION TREND Unknown
MAJOR THREATS Fishing pressure is the major threat to *A. multiguttatus*. While it is not currently targeted by any known commercial fishery, it is caught opportunistically in small-scale, multi-species fisheries.
HABITAT AND ECOLOGY GENERAL *A. multiguttatus* is a demersal, rocky reef and shallow-water species that is cryptically coloured and secretive, and is known to depths of at least 30 m. The species is sedentary during the day, hiding in crevices or lying among seaweed, and relies on its camouflage to escape detection (sometimes lying on its side and partly covering itself with sand). FEEDING *A. multiguttatus* is a nocturnal predator feeding mainly on benthic crustaceans.

Alphestes multiguttatus is an eastern Pacific species found from the Gulf of California (Mexico) to extreme northern Peru, including the offshore islands of Cocos (Costa Rica) and Malpelo (Colombia).

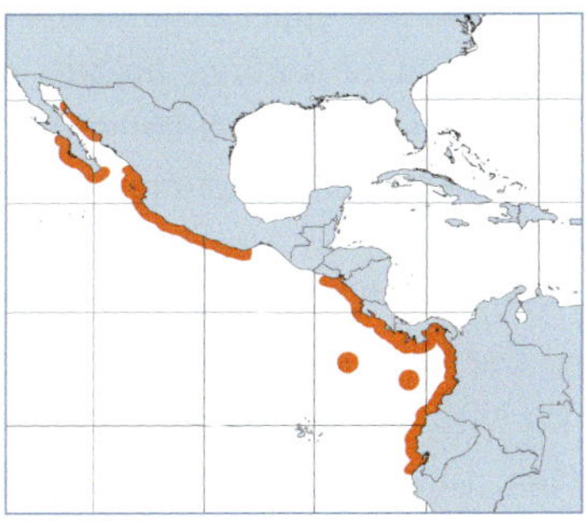

COMMENTS Two species of *Alphestes* are known from the eastern Pacific: *A. immaculatus* and *A. multiguttatus*. They can be distinguished by the number of dark bars on the pectoral fins. *A. immaculatus* has 7 or 8 irregular bars, while *A. multiguttatus* has 5 or 6 more distinct bars. In the northern portion of their range, *A. immaculatus* is more commonly found in shallow depths, with *A. multiguttatus* generally found deeper. In the southern portion of their range, the situation is reversed.

While *A. multiguttatus* is presently not a major target of any known fisheries, fishing pressure on this smaller grouper will increase as larger, more valuable species are harvested to low levels.
2008 IUCN RED LIST STATUS Least Concern

Authors: *AA Bertoncini, JH Choat, MT Craig, BP Ferreira, PC Heemstra and LA Rocha*

Anyperodon leucogrammicus

Valenciennes 1828

slender grouper (E), cheval de brois (F), mero elegante (S)

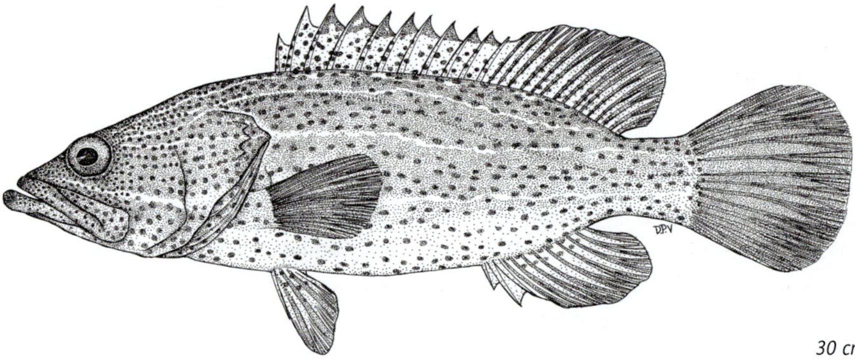

30 cm SL (Sudan)
© SAIAB

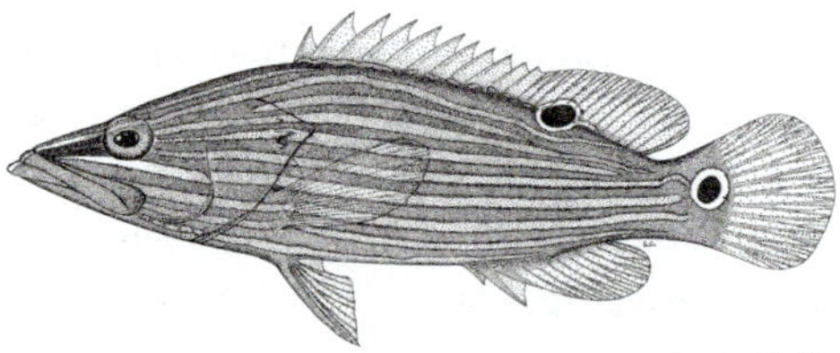

~8 cm SL (Indonesia)
© SAIAB

Body and head elongate and markedly compressed, the depth distinctly less than head length and contained 3.1–3.7 times in standard length (for fish 10–40 cm SL), the body width 2.3–2.8 times in the depth. Head pointed, its length contained 2.3–2.5 times in standard length; interorbital region narrow, dorsal head profile almost straight; preorbital narrow, its depth 14–17 times in head length; snout distinctly longer than eye; preopercle rounded, finely serrate, the lower serrae only slightly enlarged, the lower edge fleshy; ventral edge of interopercle with a shallow indentation; opercular spines small, the upper edge of operculum distinctly convex; diameter of posterior nostril about twice that of anterior nostril; maxilla reaches well past eye, the exposed part covered with tiny cycloid scales; no bony process on rear end of maxilla; supramaxilla well developed; no teeth on palatines; canines at front of jaws rudimentary or absent; teeth at sides of lower jaw subequal, in 2 or 3 irregular rows. Gill rakers 7–9 on upper limb, 14–17 on lower limb, the longest gill raker subequal to longest gill filaments. Dorsal fin with 11 spines and 14–16 rays, the fin origin over the opercle, the membranes distinctly incised between the spines; anal fin with 3 spines and 8 or 9 rays; soft dorsal and anal fins rounded; pectoral and pelvic fins small, subequal, their length contained 1.9–2.4 times in head length; pectoral fins thin, transparent and symmetrical; pelvic fins not reaching anus; caudal fin rounded, with 8 branched rays and in upper part and 7 branched rays in lower part. Midlateral-body scales ctenoid, with auxiliary scales; lateral-line scales 61–72; lateral-scale series 106–125.

MAXIMUM RECORDED SIZE 65 cm TL

COLOUR IN LIFE Adults greenish to brownish grey with numerous orange-red spots on the head, body, dorsal fin, and basally on the caudal fin; 4 longitudinal whitish bands or series of streaks often visible on the postorbital head and body; membranes of the soft dorsal, anal and caudal fins clear. Juveniles (less than 8 cm TL) with longitudinal dark-edged pale bluish grey stripes on a gold background; a blue-edged black spot (or double spot) at the base of the caudal fin.

COLOUR IN MARKET Body colour fades to a pale drab grey and coloured spots become less visible. Longitudinal lines may remain if the specimen had prominent lines in life.

© Matthew T Craig

Adult *A. leucogrammicus* (Okinawa Fish Market, Japan)

© Luiz A Rocha

Juvenile *A. leucogrammicus* (Seychelles)

Adult *A. leucogrammicus* (Bali, Indonesia)

© Robert F Myers

Adult *A. leucogrammicus* (Big Drop Off, Palau)

© Robert F Myers

Anyperodon leucogrammicus is an Indo-Pacific species which is widely distributed from the Red Sea to Mozambique, the Comoros Islands, Madagascar, Mauritius, Réunion, Seychelles, Oman and Chagos, extending east to the Phoenix Islands, north to Japan, and south to Australia. Its distribution probably includes all the islands of the tropical Indian Ocean.

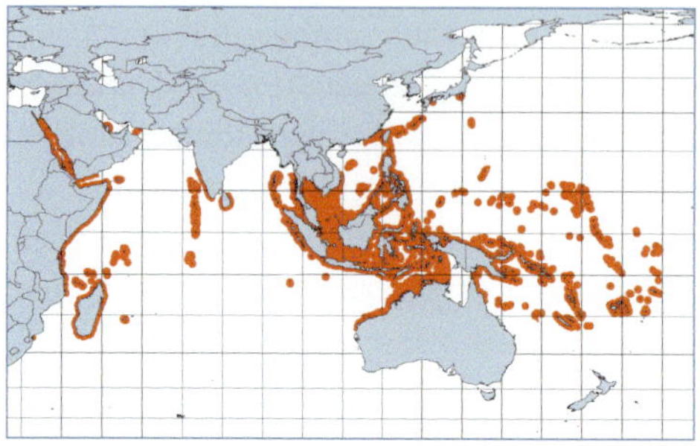

POPULATION GENERAL *A. leucogrammicus* is widespread, but not known to be abundant throughout its range. **FISHERY DEPENDENT** The slender grouper was noted as uncommon in catches in Pohnpei, where it was shown to contribute only to the spearfishery and occurred in only 3.2% of catches overall (Rhodes and Tupper 2007).

POPULATION TREND Unknown

MAJOR THREATS No major threats have been identified for this species. However, it is traded internationally as part of the valuable live reef food fish trade which is unmanaged.

HABITAT AND ECOLOGY GENERAL *A. leucogrammicus* is a reef-associated species that inhabits coral-rich areas at depths of 5–80 m and clear waters of lagoons and seaward reefs. The distinctive blue-and-gold-striped juveniles are mimics of the wrasse *Halichoeres purpurescens*. **FEEDING** Feeds mainly on fishes and probably on crustaceans.

COMMENTS Craig and Hastings (2007) demonstrated a nested relationship between this species and species of the genus *Epinephelus*, and suggested it is best regarded as a member of *Epinephelus*. It differs from other members of *Epinephelus* only in its pronounced slender body and absence of teeth on the palatines. We retain it here in *Anyperodon* pending further study.

2008 IUCN RED LIST STATUS Least Concern

Authors: *JH Choat, PC Heemstra, DA Pollard, MA Samoilys and BM Yeeting*

Cephalopholis aitha Heemstra and Randall 1991

rusty hind (E), vieille rouillée (F), cherna (S)

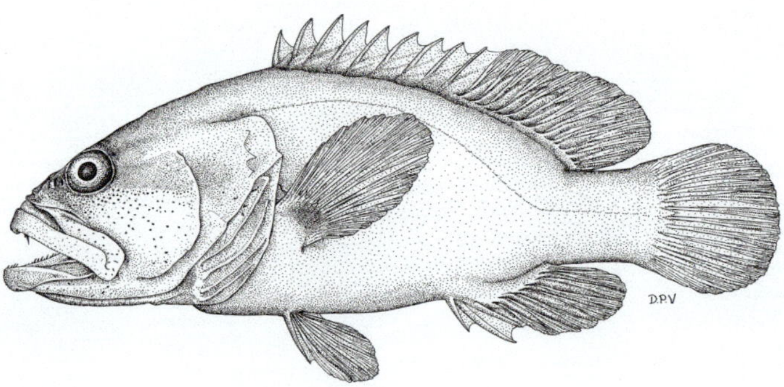

13 cm SL (Philippines)

Body depth contained 2.5–2.9 times in standard length. Head length 2.1 to 2.3 times in standard length; preopercle rounded, the posterior edge finely serrate, the lower edge fleshy; upper edge of operculum very convex; maxilla extends past eye; teeth noticeably large. Gill rakers 8–9 on upper limb, 16–19 on lower limb. Dorsal fin with 9 spines and 14 rays, the rays distinctly longer than the spines; anal fin with 3 spines and 8 rays; pectoral-fin rays 15 or 16; pelvic-fin length contained 2.3–2.45 times in head length. Midlateral-body scales ctenoid; lateral-line scales 46–49; lateral-scale series 82–99.

MAXIMUM RECORDED SIZE 14 cm SL

COLOUR IN LIFE Reddish brown, the edges of the dorsal body scales dark brown; snout dark brown; postorbital head with pale blue network; pale blue line along the rear edge of the orbit; upper jaw and cheeks with scattered pale blue dots and short lines; fins reddish; margin of the soft dorsal, anal and caudal fins pale bluish.

COLOUR IN MARKET Generally retains similar colouration in death.

Adult *C. aitha* (Papua New Guinea)

POPULATION GENERAL *C. aitha* is a small, poorly known species with a restricted known distribution.

POPULATION TREND Unknown

MAJOR THREATS No major threats have been identified for this species.

HABITAT AND ECOLOGY GENERAL *C. aitha* is a reef-associated species that inhabits protected or silty reef areas, in depths to at least 30 m. **REPRODUCTION** A mature female (containing hydrated eggs) of 14 cm was recorded by Randall and Heemstra (1991).

COMMENTS *C. aitha* is a poorly known species. It likely has a larger distribution than presently known, and, while it is currently not a major target of any known fishery, it likely occurs in small-scale, multi-species fisheries.

2008 IUCN RED LIST STATUS Data Deficient

Cephalopholis aitha is a western Pacific species found in the Philippines, Indonesia (Sulawesi, Flores and East Java, but not reported in the Raja Ampat Islands although likely present) and northern Papua New Guinea.

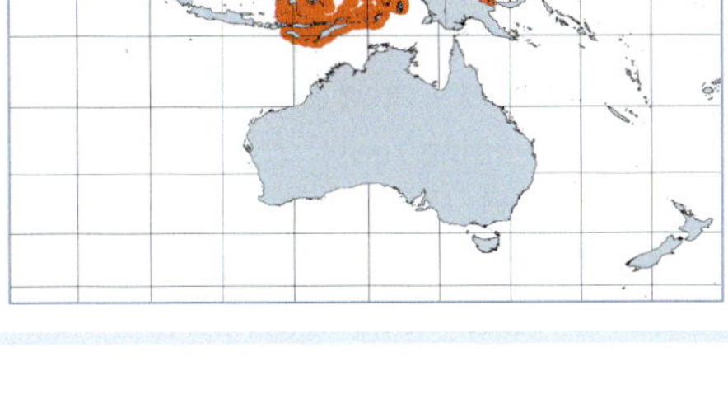

Authors: *PC Heemstra and M Liu*

Cephalopholis argus Schneider 1801

peacock grouper (E), mérou celeste (F), cherna pavo real (S)

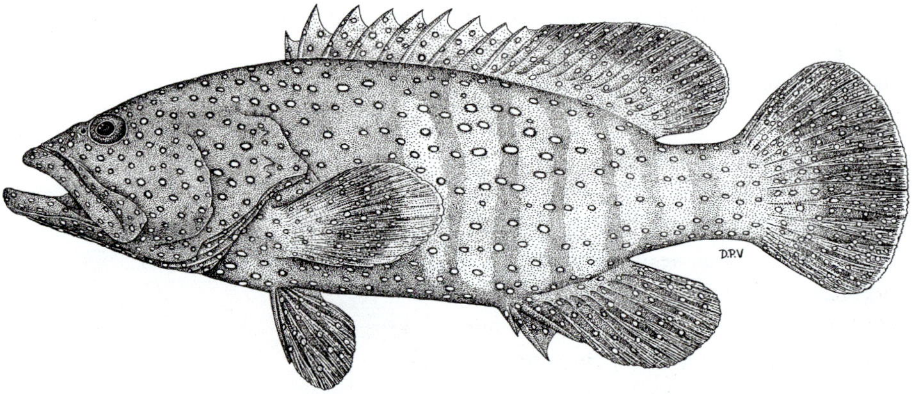

30 cm SL (composite)
© SAIAB

Body depth distinctly less than head length, depth contained 2.7–3.2 times in standard length. Head length 2.4–2.7 times in standard length; eye small, its diameter distinctly less than snout length; interorbital area flat to slightly convex; preopercle rounded, finely serrate in young, virtually smooth in large adults, the lower edge fleshy; subopercle and interopercle smooth; maxilla scaly, reaching well past eye. Gill rakers 9–11 on upper limb, 17–19 on lower limb, the rudimentary rakers on lower limb difficult to distinguish from intercalated bony plates. Dorsal fin with 9 spines and 15–17 rays, the membranes indented between the spines; anal fin with 3 spines and 9 rays; pectoral-fin rays 16–18; pectoral fins clearly longer than pelvic fins, pectoral-fin length contained 1.4–2.0 times in head length; pelvic fins not reaching anus, their length 1.9–2.4 times in head length; caudal fin well rounded. Lateral-body scales ctenoid, with a few auxiliary scales in adults; lateral-line scales 46–51; lateral-scale series 95–110.

MAXIMUM RECORDED SIZE 50 cm TL

Cephalopholis argus is the most widely distributed of the groupers, occurring from the Red Sea to South Africa, extending east to French Polynesia and the Pitcairn group (including northern Australia), south to Lord Howe Island, and north to southern Japan. It was introduced to the Hawaiian Islands in the 1950s as a sport fish.

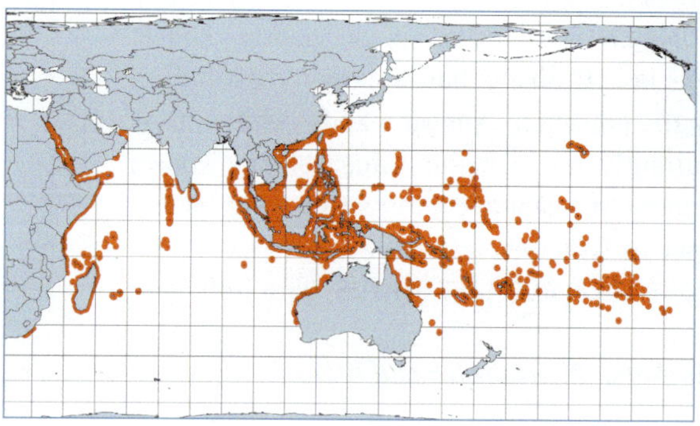

COLOUR IN LIFE Dark brown, covered with small black-edged blue ocelli; often with 5 or 6 pale bars on the rear part of the body and a large pale area over the chest; posterior margin of the median fins usually with a narrow white edge; distal part of the pectoral fins sometimes maroon-brown; triangular membranes at the tips of the dorsal-fin spines orange-gold.

COLOUR IN MARKET Dark brown, the blue edges of the ocelli often prominent. The white bars are generally absent, but may remain as faint pale sections on the posterior body.

© Amanda Ackiss

Adult *C. argus* (Ternate, Indonesia)

© Robert F Myers

Adult *C. argus* (Bali, Indonesia)

Adult *C. argus* (Bali, Indonesia)

© Robert F Myers

POPULATION GENERAL *C. argus* is the most widely distributed grouper and is abundant throughout its range despite noted declines in fisheries locally. FISHERY DEPENDENT In Saudi Arabia, landings ranged from 11 to 72 t during 2000–2004 (www.fao.org). In the Maldives, there has been a slight reduction in average size since exploitation began (Sattar and Adam 2005). This species is a strong coloniser and has established large populations in Hawaii after deliberate introduction from Moorea and the Marquesas Islands (Randall 1987). The species is noted as one of the most frequently speared groupers in reef fisheries in Pohnpei (Rhodes and Tupper 2007). FISHERY INDEPENDENT On the Great Barrier Reef (Australia) the highest abundances are in the northern regions (3.6 fish per 1000 m^2; Pears 2005). Replicated estimates of abundance derived from belt transects are also available for the Seychelles (4.5–7 fish per 1000 m^2), and American Samoa (1.7–14.8 fish per 1000 m^2) (A Green pers. comm.).

POPULATION TREND Stable

MAJOR THREATS Fishing pressure is the primary known threat to *C. argus*. The species is taken both for chilled (dead) and live fish trade, but marketing is often difficult due to its association with ciguatera poisoning in many places.

HABITAT AND ECOLOGY GENERAL *C. argus* occurs in a variety of habitats, but is most commonly associated with exposed reef-front habitats in depths of up to 40 m. *C. argus* causes ciguatera at some islands in the Pacific Ocean, such as French Polynesia, the Marshall Islands and Hawaii (Randall 1980, 1987, Hokama et al. 1994, Buslé et al. 1998). FEEDING *C. argus* is a crepuscular feeder (evening and early morning) and feeds on fishes and crustaceans. AGE AND GROWTH The demography of this species is highly plastic and shows great variation over its distributional range especially in terms of estimated maximum longevity (14–40 years), estimated and observed maximum sizes (L_∞ = 272–440 mm SL; L_{max} = 305–471 mm SL) and estimated growth parameters (k = 0.20–0.36) (Pears 2005, JH Choat unpublished data). REPRODUCTION This species occurs in social groups comprising a single male and several females. Male territory size is positively associated with fish size. In the Gulf of Aqaba, *C. argus* is found in social units comprising up to 12 adults, including one dominant male. Each group occupies a specific area (up to 2000 m^2) that is defended by the territorial male and subdivided into secondary territories, each inhabited by a single female (Shpigel and Fishelson 1991). A similar social organisation has been observed in populations in Hawaii where it was introduced in the 1950s (Meyer 2004).

COMMENTS *C. argus* is often confused with *C. cyanostigma*, which has a generally similar colour pattern of small blue ocelli on a dark brown or reddish background. *C. cyanostigma* differs in that the pectoral fins are yellow-orange at least distally, with a black margin and blue spots only proximally; also, the body often has pale spots containing 1–3 small blue spots and 5 or 6 chain-like dark bars. In addition to differences in colour pattern, *C. cyanostigma* has 8 anal-fin rays and fewer gill rakers (7–9 on upper limb and 14–18 on lower limb).

2008 IUCN RED LIST STATUS Least Concern

Authors: *JH Choat, M Liu and DA Pollard*

Cephalopholis aurantia (Valenciennes 1828)

golden hind (E), cuisinier (F), cherna dorada (S)

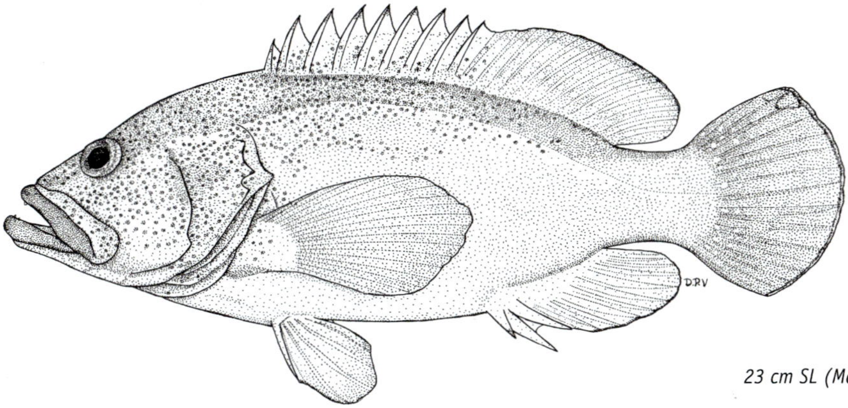

23 cm SL (Mauritius)
© SAIAB

Body depth less than head length, depth contained 2.6–2.9 times in standard length. Head length 2.4–2.6 times in standard length; interorbital area flat; preopercle rounded, finely serrate, the lower edge fleshy; subopercle and interopercle serrate; maxilla scaly, reaching to or beyond vertical at rear edge of eye. Gill rakers 7–9 on upper limb, 14–17 on lower limb. Dorsal fin with 9 spines and 14–16 rays, the membranes distinctly incised between the spines; anal fin with 3 spines and 9 (rarely 8) soft rays; pectoral-fin rays 17–19; pectoral fins distinctly longer than pelvic fins, pectoral-fin length contained 1.3–1.5 times in head length; pelvic fins reach to or slightly short of anus, their length 1.6–2.0 times in head length; caudal fin well rounded. Lateral-body scales ctenoid; lateral-line scales 47–53; lateral-scale series 94–121.

MAXIMUM RECORDED SIZE 30 cm TL

COLOUR IN LIFE Pale orange-red to orange-yellow or golden with small red or yellow dots on the head, antero-dorsally on the body, and on the base of the dorsal fin. Posterior margin of the caudal fin with a pale blue or white edge and with or without a black submarginal line; posterior parts of the dorsal and anal fins often with a narrow pale blue or white edge. May display 4–5 pale white bars on the posterior body.

COLOUR IN MARKET Generally pale orange to yellow, dots on the head may fade, posterior margin of the caudal fin with a white or blue edge. Pale bars on the posterior body are often present as faded irregular patches on the body.

Adult *C. aurantia*
(Papua New Guinea)

Cephalopholis aurantia is an Indo-Pacific species found around the islands of the western Indian Ocean to Japan and the central Pacific, except for a single specimen caught off the coast of KwaZulu-Natal (South Africa; Heemstra and Randall 1993). The species likely occurs in deep water all along the east coast of Africa. The record of *C. aurantia* reported by Morgans (1982) is a misidentification of *C. nigripinnis*.

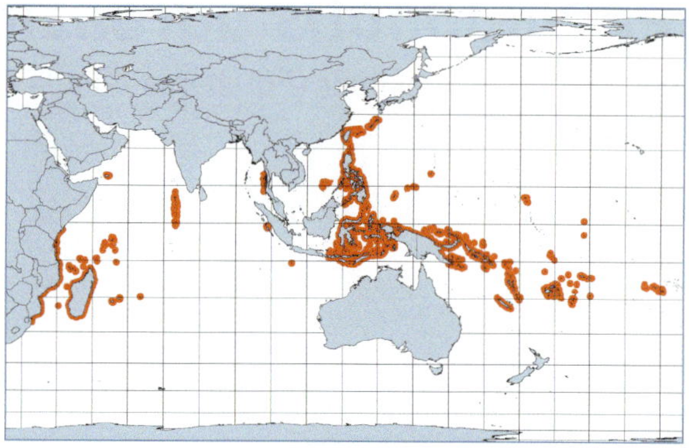

POPULATION GENERAL *C. aurantia* is a poorly known deep-water grouper.

POPULATION TREND Unknown

MAJOR THREATS No major threats have been identified for this species. *C. aurantia* is unknown from major fisheries, but likely occurs in small-scale, multi-species fisheries.

HABITAT AND ECOLOGY GENERAL *C. aurantia* is a reef-associated species found on steep seaward reefs, usually at depths of 100–250 m, but can occur in shallower water.

COMMENTS *C. aurantia* is easily misidentified as *C. spiloparaea* and, to a lesser degree, as *C. sonnerati*. In *C. spiloparaea,* the colour differs (generally pale reddish, mottled and blotched with dark red or brownish red) and the pelvic fins are usually shorter (not reaching the anus). *C. sonnerati* differs in colour pattern (generally darker red body colour), has a greater body depth (contained 2.3–2.7 times in standard length) and more numerous scales (lateral-line scales 66–80). The steep profile of the head is to some degree more pronounced in *C. sonnerati*. In the Indian Ocean, *C. aurantia* was formerly known as *C. analis*.

2008 IUCN RED LIST STATUS Data Deficient

Authors: *PC Heemstra and M Liu*

Cephalopholis boenak (Bloch 1790)

chocolate hind (E), vielle chocolat (F), cherna chocolate (S)

20 cm SL (southern Mozambique)

Body depth less than head length, depth contained 2.6–3.0 times in standard length. Head length 2.3–2.7 times in standard length; preorbital depth usually less than half maxilla width, 2.5–3.6% of standard length; interorbital area flat; preopercle rounded, very finely serrate, no enlarged spines at the angle, the lower edge fleshy; subopercle and interopercle weakly serrate; maxilla naked, reaching approximately to vertical at rear edge of eye. Gill rakers 7–9 on upper limb, 14–17 on lower limb. Dorsal fin with 9 spines and 15–17 rays; the membranes indented between the spines; anal fin with 3 spines and 8 rays, the 2nd spine 15–18% of standard length, reaching past tip of 3rd spine; pectoral-fin rays 16 or 17; pectoral fins longer than pelvic fins, pectoral-fin length contained 1.3–1.6 times in head length; pelvic fins usually reaching to or beyond anus, pelvic-fin length 1.7–2.2 times in head length; caudal fin well rounded. Body scales ctenoid, without auxiliary scales; lateral-line scales 46–51; lateral-scale series 86–100.
MAXIMUM RECORDED SIZE 26 cm TL

COLOUR IN LIFE Dark brown (occasionally dark reddish brown), usually with 7 or 8 dark bars on the body; some fish with dark brown bands radiating from the eye; black spot between the upper and middle opercular spines; soft dorsal, anal and caudal fins darker distally, with a pale bluish line on the edge (except on the central part of the caudal fin). **COLOUR IN MARKET** Generally drab brown, body may become blotchy. The bars fade, especially on the anterior portion of the body, but the black spot between opercular spines remains prominent. Blue edges on the fins are generally not present.

© Gerald R Allen

Juvenile *C. boenak* (Raja Ampat Islands, Indonesia)

© Matthew T Craig

Adult *C. boenak* (Sulawesi, Indonesia)

© Robert F Myers

Adult *C. boenak* (Helen Atoll, Palau)

Cephalopholis boenak is an Indo-West Pacific species with a broad distribution from East Africa (Kenya to southern Mozambique) to southern India and the western Pacific, including the Ryukyu Islands (Japan), Taiwan and China, and extending south to New Caledonia and northern Australia. It is unknown from the Red Sea, Persian Gulf and the islands of Micronesia, except for Palau.

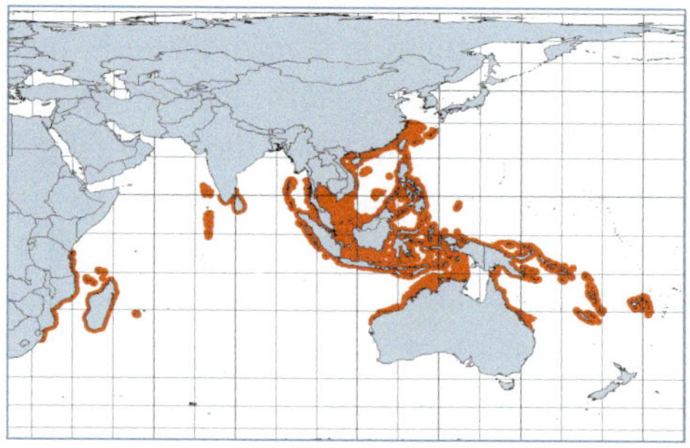

POPULATION **GENERAL** Although widespread, the geographic range of *C. boenak* is centred in South-East Asia where it is relatively common. **FISHERY DEPENDENT** Chan and Sadovy (2002) described *C. boenak* as the last remaining commercially important grouper in Hong Kong, where it is being heavily fished in its smaller size and age classes; half of all specimens sampled from local municipal markets within the last few years were small (between 112 mm and 143 mm SL) and young (ages 2–4 years). Despite its importance, there are no official records of landings in Hong Kong (Chan and Sadovy 2002).

POPULATION TREND Unknown

MAJOR THREATS Fishing pressure is the major threat to *C. boenak*.

HABITAT AND ECOLOGY **GENERAL** *C. boenak* is a reef-associated species mainly found in coastal areas to depths of 30 m. The species inhabits silty dead reefs in protected waters and is secretive, but may also be seen on live coral (Sadovy and Cornish 2000, Liu and Sadovy 2005). **FEEDING** *C. boenak* feeds mainly on fishes and crustaceans. **AGE AND GROWTH** *C. boenak* reaches at least 11 years. In Hong Kong, the species grows according the relationship $L = 147.66(1 - e^{-0.49(\text{Age} + 1.52)})$ where L is the standard length in mm (Chan and Sadovy 2002). **REPRODUCTION** *C. boenak* is a diandric protogynous hermaphrodite, with the smallest female and male maturing at 80 mm SL and 116 mm SL, respectively (Chan and Sadovy 2002). The natural spawning season in Hong Kong was reported by Chan and Sadovy (2002) to be from April to October; however, spawning in captivity has been recorded in August and September irrespective of lunar phase (Liu and Sadovy 2001). Among other similar-sized *Cephalopholis* species, *C. boenak* has the smallest size of female maturation and sex change, and a faster growth rate possibly related to high fishing pressure (Chan and Sadovy 2002); the species also exhibits plasticity of sexual development in both juveniles and adults which can be mediated by social conditions (Liu and Sadovy 2004a, 2004b).

COMMENTS *C. boenak* was once considered too small to be of commercial importance. However, following the declines of larger groupers in the Hong Kong area, this species has become a significant component of the grouper catch. It is also now listed independently in FAO fisheries statistics, and over the last decade represented the largest catch by weight for any grouper in these records (www.fao.org).

The reproductive characteristics of *C. boenak* may make the species more resilient to fishing pressure than other grouper species that take longer to mature, or that exhibit monandry. Given the recent increase in take of this species in some locations, population declines may be likely in the near future.

2008 IUCN RED LIST STATUS Least Concern

Authors: *KE Carpenter, T-C Chan, M Kulbicki, DA Pollard, YJ Sadovy de Mitcheson and MA Samoilys*

Cephalopholis cruentata (Lacepède 1802)

graysby (E), côné essaim (F), cherna enjambre (S)

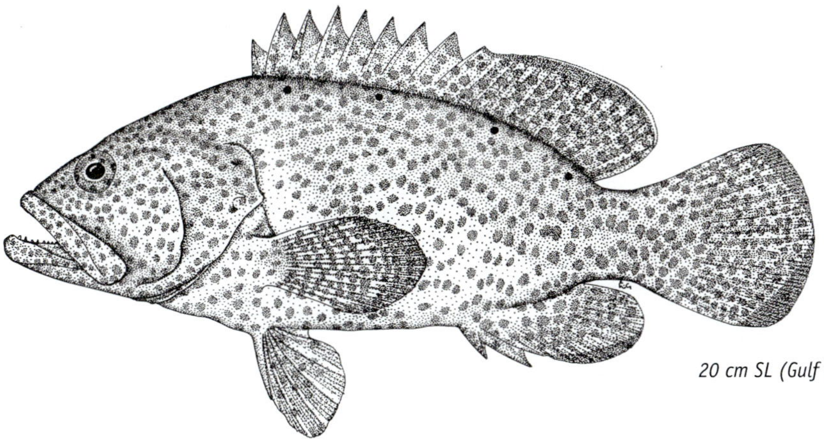

20 cm SL (Gulf of Mexico)
© SAIAB

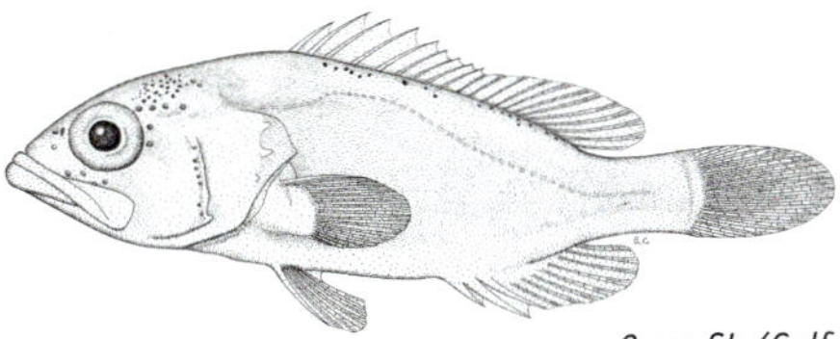

2 cm SL (Gulf of Mexico)
© SAIAB

Body depth distinctly less than head length, depth contained 2.5–2.9 times in standard length. Head length 2.4–2.6 times in standard length; interorbital area flat to slightly convex; preopercle rounded, finely serrate, with shallow notch above the angle; posterior and anterior nostrils small and subequal; maxilla scaly, reaching past vertical at rear edge of eye. Gill rakers 18–25 (total). Dorsal fin with 9 spines and 13–15 rays, the 4th or 5th spines longest and the membrane distinctly indented between all the spines; anal fin with 3 spines and 8 rays; pectoral-fin rays 16; caudal fin rounded. Lateral-body scales distinctly ctenoid; lateral-line scales 47–51; lateral-scale series 69–81.

MAXIMUM RECORDED SIZE 32 cm TL

POPULATION **GENERAL** *C. cruentata* is a wide-ranging species that is locally abundant in portions of its range. **FISHERY DEPENDENT** Few fishery-dependent data are available for this species. In localised fisheries it may represent a significant portion of total landings (e.g. in Curaçao where it has represented up to 11% of the total grouper catch; Nagelkerken et al. 2005). **FISHERY INDEPENDENT** Visual surveys in Curaçao have shown this species to be relatively uncommon (10–22 fish per 1000 m^2), with a stable population from 1973 to 2003 (Nagelkerken et al. 2005). Other work has shown a reduction in the maximum size and age of populations under higher levels of fishing pressure. For example, under heavy fishing pressure in Belize, the maximum

Cephalopholis cruentata is found in the western North Atlantic and within the Caribbean (Greater Antilles and Lesser Antilles), Trinidad, Gulf of Mexico, USA (North Carolina to southern Florida), Bahamas and Bermuda. There are unsubstantiated records from Brazil.

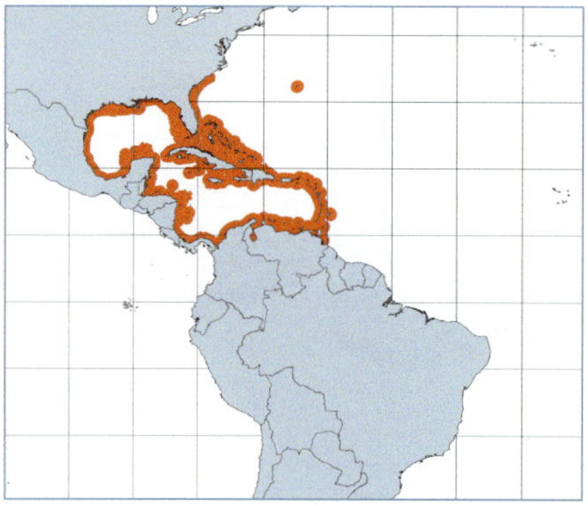

age of fish was 8 years, with a maximum length of 266 mm SL, whereas in protected Los Roques the maximum observed age was 14 years, and maximum length 281 mm SL (K Ranatung unpublished data).

POPULATION TREND Stable

MAJOR THREATS Fishing pressure and habitat loss are the major threats to this species.

HABITAT AND ECOLOGY **GENERAL** *C. cruentata* is found on coral reefs from the shore to 170 m. In the eastern Gulf of Mexico, it occurs on rocky reef ledges in depths greater than 27 m. It is a small, secretive species that usually stays near hiding places in the reef during the day. **FEEDING** *C. cruentata* is a crepuscular predator, and adults feed mainly on fishes. **AGE AND GROWTH** *C. cruentata* may reach up to 8 cm in the first year of growth (Nagelkerken 1979). **REPRODUCTION** *C. cruentata*

COLOUR IN LIFE Head, body and fins pale grey, brown or olive green, covered with orange-brown or reddish spots; 4 distinct spots, which can change rapidly from black to white or back again, are present at the base of the dorsal fin; a mid-dorsal white stripe is sometimes present from the tip of the lower jaw to the nape, and is prominent in juveniles.

COLOUR IN MARKET Pale (nearly white) orange, grey or brown, with reddish spots prominent. The black spots on the upper body below the dorsal fin remain prominent. The mid-dorsal stripe is generally not visible, especially in adults.

© Matthew T Craig

Adult *C. cruentata* (Dry Tortugas, Florida, USA)

© Robert F Myers

Adult *C. cruentata* (Utila, Honduras)

is a protogynous hermaphrodite (Nagelkerken 1979, Thompson and Munro 1983, Potts and Manooch 1999). Thompson and Munro (1978) estimate the number of eggs per spawning at 262 000–604 000 for a fish of 29 cm TL. Hawkins et al. (2007) estimate size at first female maturity for the St Lucia population at 16 cm FL with 100% maturity at 26 cm FL.

2008 IUCN RED LIST STATUS Least Concern

Authors: *AA Bertoncini, JH Choat, MT Craig, BP Ferreira and LA Rocha*

Cephalopholis cyanostigma (Valenciennes 1828)

bluespotted hind (E), vielle étoiles bleues (F), cherna de pintas azules (S)

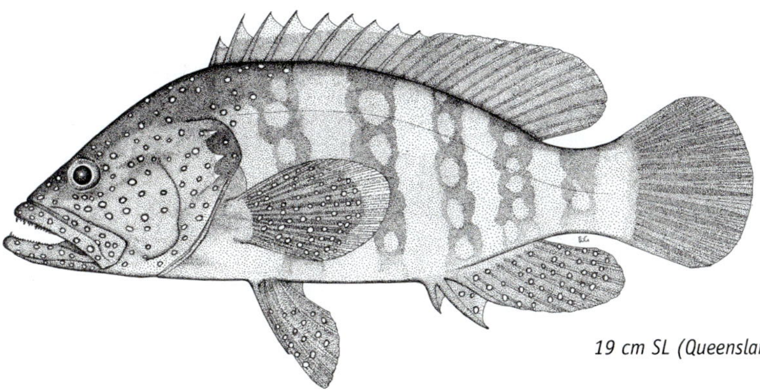

19 cm SL (Queensland, Australia)
© SAIAB

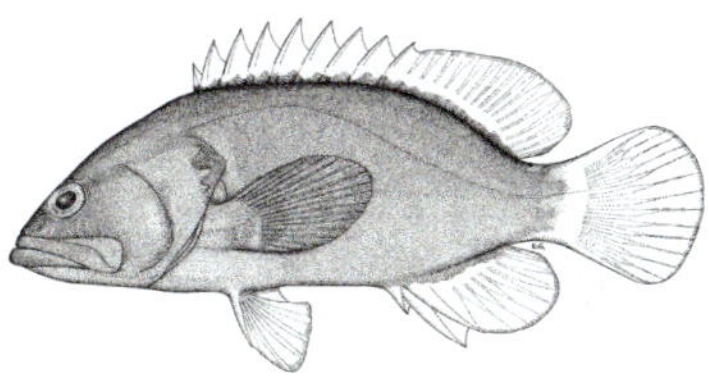

9 cm SL (Solomon Islands)
© SAIAB

Body somewhat compressed, the width contained 2.0–2.3 times in the depth; depth less than head length, 2.6–3.0 times in standard length. Head length 2.3–2.6 times in standard length; interorbital area flat to slightly convex, its width subequal to eye diameter and maxilla width; preopercle rounded, finely serrate, the lower edge fleshy; subopercle and interopercle smooth or with a few serrae; upper edge of operculum very convex, the rear edge almost vertical; maxilla scaly, reaching well past eye. Gill rakers 7–9 on upper limb, 14–18 on lower limb. Dorsal fin with 9 spines and 15–17 rays, the fin membrane indented between the spines; anal fin with 3 spines and 8 rays; pectoral-fin rays 15–18; pectoral fins distinctly longer than pelvic fins, pectoral-fin length contained 1.5–1.8 times in head length; pelvic fins usually not reaching anus, their length 1.8–2.1 times in head length; caudal fin well rounded. Body scales ctenoid, with a few auxiliary scales in adults; lateral-line scales 46–50; lateral-scale series 92–106.

MAXIMUM RECORDED SIZE 30 cm TL

COLOUR IN LIFE Adults brown to brownish red, with numerous small black-edged blue ocelli on the head, body and fins, those on the head and chest larger and more distinctly black-edged than those on the upper part of the body, those on the fins smaller (except basally on the pectoral fins) and also distinctly black-edged (spots entirely dark in preservative, those on the body faint); body often with dark chain-like bars; distal part of the pectoral fins orange-yellow with a blackish edge or white edge and black submarginal line; proximal two-thirds of the pectoral fins with blue ocelli becoming smaller distally; median fins darker than the body, the posterior margin bluish white with a submarginal blackish band. Juveniles with head, body and pectoral fins bluish brown to purple, the other fins bright yellow; blue ocelli on the head and front of the body indistinct or absent.

COLOUR IN MARKET A lighter shade of brown (may become blotchy). Spots on the entire body generally remain clear. The distal portion of the pectoral fins is markedly paler than the proximal portion.

© Matthew T Craig

Adult *C. cyanostigma* (Los Angeles fish store, USA)

© Robert F Myers

Adult *C. cyanostigma* (Bali, Indonesia)

© Robert F Myers

Adult *C. cyanostigma* (Bali, Indonesia)

© Robert F Myers

Juvenile *C. cyanostigma* (New Britain, Papua New Guinea)

POPULATION **GENERAL** *C. cyanostigma* is a broadly distributed species that may be common regionally. It is the most abundant small serranid on the midshelf of the Great Barrier Reef (Australia) (Pears 2005). **FISHERY INDEPENDENT** Visual census surveys indicate that this species may no longer be found in parts of its historical range (e.g. in Sabah, Malaysia) (AS Cabanban pers. obs.).

Cephalopholis cyanostigma is an Indo-West Pacific species found from eastern Sumatra, extending west to the Solomon Islands, and from the Philippines extending south to northern Australia.

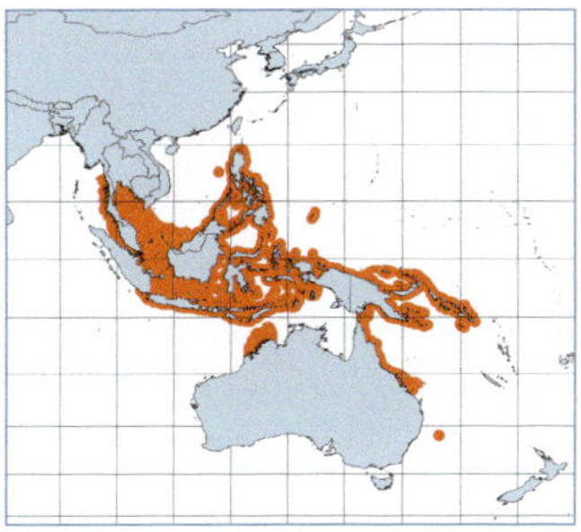

POPULATION TREND Unknown

MAJOR THREATS The main threat to *C. cyanostigma* is fishing pressure.

HABITAT AND ECOLOGY **GENERAL** *C. cyanostigma* is found on protected coastal reefs, seagrass beds and coral-rich areas to depths of up to 50 m. **FEEDING** *C. cyanostigma* feeds on small fishes and crustaceans. **AGE AND GROWTH** *C. cyanostigma* reaches at least 40 years (Mosse et al. 2002). **REPRODUCTION** *C. cyanostigma* is a protogynous hermaphrodite (Mosse et al. 2002).

COMMENTS *C. cyanostigma* may be confused with other species of the genus in having numerous blue spots. It differs from *C. miniata* in having a brown rather than red body colour, and from *C. argus* in that it lacks the characteristic white bars and chest patch of *C. argus*.

2008 IUCN RED LIST STATUS Least Concern

Authors: *AS Cabanban, JH Choat, ST Fennessy and RF Myers*

Cephalopholis formosa (Shaw 1812)

bluelined hind (E), vielle lignes bleues (F), cherna rayada (S)

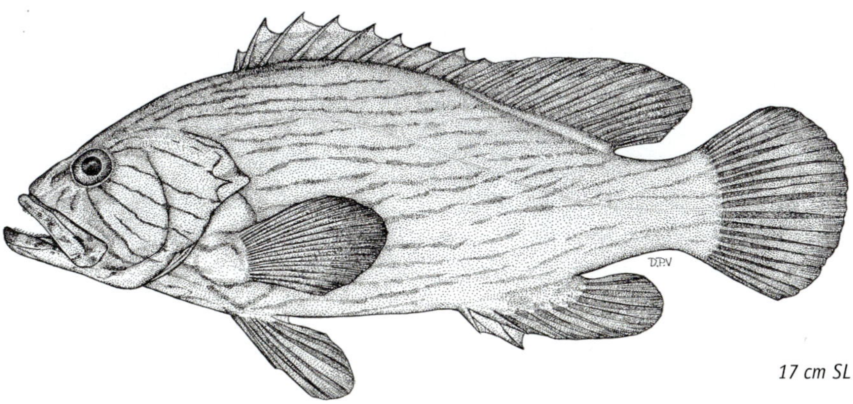

17 cm SL (India)

Body depth contained 2.5–2.9 times in standard length. Head length 2.4–2.6 times in standard length; preorbital depth 3.6–4.3% of standard length, more than half maxilla width; interorbital area convex; preopercle rounded, finely serrate, the lower edge fleshy; subopercle and interopercle smooth; maxilla naked, reaching to or just past vertical at rear edge of eye. Gill rakers 8–10 on upper limb, 14–18 on lower limb. Dorsal fin with 9 spines and 15–17 rays, the membranes indented between the spines; anal fin with 3 spines and 8 (rarely 7) rays, the 2nd spine 12–15% of standard length, not reaching past tip of 3rd spine; pectoral-fin rays 16–18; pectoral fins usually longer than pelvic fins, pectoral-fin length contained 1.5–1.8 times in head length; pelvic fins reaching about to anus, their length 1.6–2.0 times in head length; caudal fin well rounded. Body scales ctenoid, without auxiliary scales; lateral-line scales 47–51; lateral-scale series 91–109.

MAXIMUM RECORDED SIZE 34 cm TL

COLOUR IN LIFE Dark brown to yellowish brown, with slightly irregular dark blue lines on the head, body and fins (those on the body and fins are sometimes absent); snout, lips and ventral part of the head and chest with small dark blue spots.

COLOUR IN MARKET Generally dark brown and blotchy. Dark blue lines fade anteriorly, but usually remain visible posteriorly.

Subadult *C. formosa* (Los Angeles fish store, USA)

Adult *C. formosa* (Phuket Aquarium, Thailand)

POPULATION GENERAL The population status of *C. formosa* is unknown, but the species may be common regionally, for example in western India (Sluka and Lazarus 2010).

POPULATION TREND Unknown

MAJOR THREATS No major threats are presently known.

HABITAT AND ECOLOGY GENERAL *C. formosa* is a reef-associated species that prefers dead or silty reefs in 10–30 m depth. It is found

Cephalopholis formosa is an Indo-West Pacific species distributed from western India to the Philippines, extending north to southern Japan (Honshu), and south to northern Western Australia.

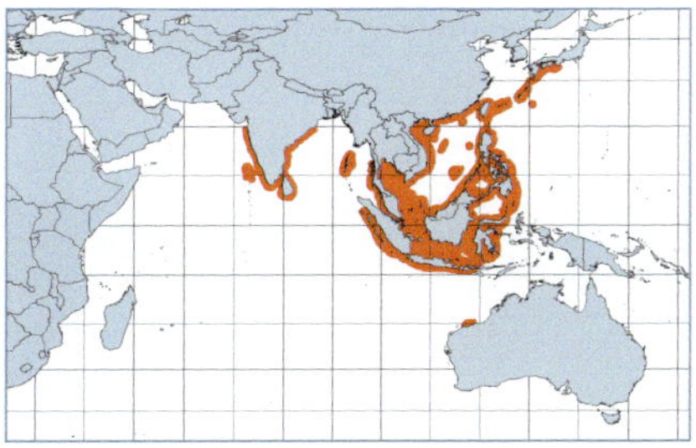

primarily off continental coasts rather than around offshore islands.

COMMENTS Records of *C. formosa* from Madagascar, Réunion and Mauritius are probably based on misidentifications of *C. polleni*. The species is sometimes confused with *C. boenak*, but it lacks the prominent spot on its operculum.

2008 IUCN RED LIST STATUS Least Concern

Authors: *ST Fennessy, PC Heemstra, RF Myers and DA Pollard*

Cephalopholis fulva (Linnaeus 1758)

coney (E), côné ouatalibi (F), cherna cabrilla (S)

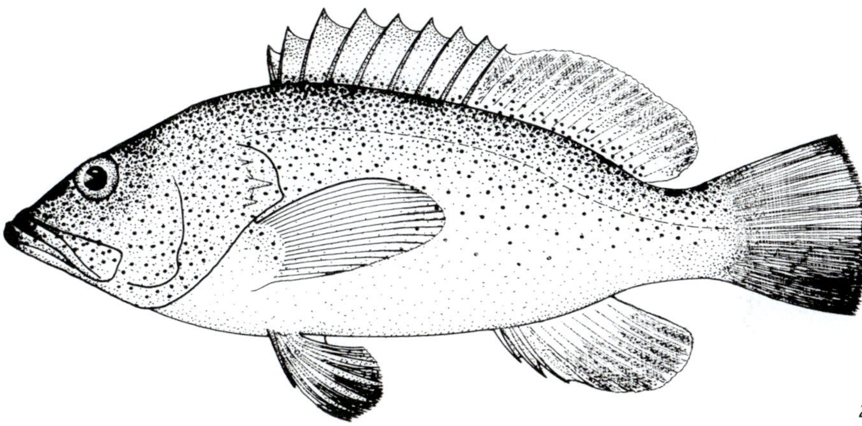

23 cm TL

Body depth distinctly less than head length, depth contained 2.6–2.9 times in standard length. Head length 2.3–2.5 times in standard length; interorbital area flat; preopercle rounded, with a shallow notch above the angle with upper edge finely serrate and moderately enlarged, the lower edge fleshy; subopercle and interopercle smooth; posterior and anterior nostrils small and subequal; maxilla scaly, reaching to or beyond vertical at rear edge of eye. Gill rakers 7–9 on upper limb, 16–18 on lower limb, 23–27 total. Dorsal fin with 9 spines and 14–16 rays, the membranes distinctly notched between the spines; anal fin with 3 spines and 9 rays; pectoral-fin rays 17–19; caudal fin convex posteriorly, but the corners are angular. Lateral-body scales ctenoid; lateral-line scales 46–54; lateral-scale series 90–97.

MAXIMUM RECORDED SIZE 41 cm TL

COLOUR IN LIFE There are three colour phases in this species: (1) red (deep water); (2) orange-brown or bicoloured with the upper part of the fish of normal colour and the lower part abruptly pale (shallow water); and (3) xanthic, a yellow morph that may be found in shallow or deep water. In the first two colour morphs, the head and body are covered with small dark-edged pale blue spots. In the xanthic form, these spots are fewer and confined to the head and anterior body. In all three morphs, there are 2 small black spots on the top of the caudal peduncle and another 2 at the tip of the lower jaw. The bicoloured pattern of the shallow-water morph is transient. The night pattern is generally pale, with irregular vertical bars and a dark 'tuning-fork' mark on the interorbital region. The nuptial male has a horizontal dark brown band from the lower end of the pectoral-fin base to the end of the caudal fin, a black margin on the soft dorsal fin, a dark stripe through the eye, and a white spot on the body near the middle of the dorsal-fin base.

COLOUR IN MARKET Generally resembles the 'normal' colour pattern described above (i.e. rarely a bicoloured phase in death). Spots are dark and usually prominent.

Juvenile *C. fulva* (Barra Grande, Brazil)

Adult *C. fulva* (bicoloured phase; Bonaire, Netherlands Antilles)

Adult *C. fulva* (yellow phase; Utila, Honduras)

Adult *C. fulva* (red phase; Hogsty Reef, Bahamas)

POPULATION **General** *C. fulva* is a common species throughout its range. **Fishery dependent** The species is relatively common in fisheries throughout its range. For example, it may represent up to 2.5% of the reported total catch in Brazil (Frédou et al. 2006). Unreported catches may, however, be much larger. Klippel *et al.* (2005) estimated landings of *C. fulva* in the Abrolhos-Vitória area as 1116 t per year, almost six times higher than that reported in official statistics. **Fishery independent** *C. fulva* was noted as the most abundant grouper on the Brazilian coast using underwater visual census (Ferreira et al. 2004). **Genetics** Genetic data on population connectivity are available from Brazil and indicate a high level of connectivity between Atol das Rocas and the north-eastern coast of Brazil (Freitas et al. 2003).

POPULATION TREND Decreasing

MAJOR THREATS The major threat to *C. fulva* is fishing pressure.

Cephalopholis fulva is a western Atlantic species ranging from South Carolina (USA) and Bermuda to southern Brazil, including the offshore islands.

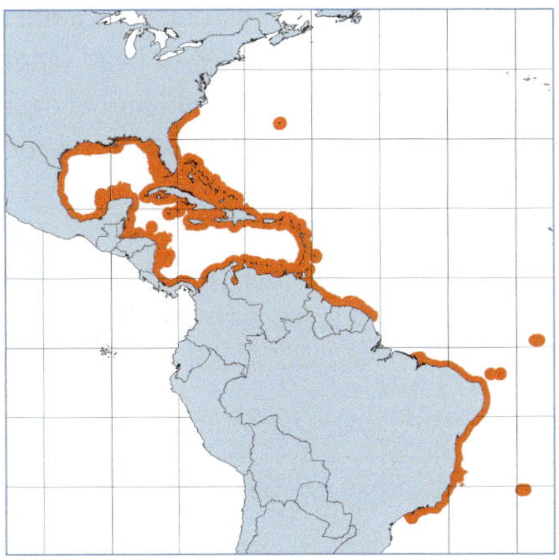

HABITAT AND ECOLOGY **General** *C. fulva* is a reef-associated species that prefers coral reefs and clear water in depths of 1–40 m. It is generally found above rocks and coral heads, and when threatened hides in caves or under ledges. **Feeding** *C. fulva* feeds mainly on small fishes and crustaceans, and has been observed following moray and snake eels to feed on flushed prey. Juveniles are known to mimic damselfish and feed opportunistically on fish within damselfish schools (Sazima et al. 2005). **Age and growth** In the eastern United States, the maximum reported age is 11 years (Potts and Manooch 1999). This species is a relatively fast-growing grouper, achieving up to 60% of its maximum size in the first year (Araújo and Martins 2006). **Reproduction** *C. fulva* is a protogynous hermaphrodite, with females maturing at 16 cm TL and transforming to males at about 20 cm (Coehlo 2001). Males are territorial and form harems. Pelagic spawning occurs just before sunset over several days. A male will spawn daily with each of the several females in his harem (YJ Sadovy de Mitcheson pers. comm.). Fecundity estimates range from about 150 000 to 282 000 eggs per female. In Puerto Rico, spawning activity appears to be limited to several days around the last quarter and new moon phases from January to March (Figuerola et al. 1997, YJ Sadovy de Mitcheson unpublished data). In some areas, the spawning season may be protracted (up to 10 months in Brazil; Coelho 2001).

COMMENTS *C. fulva* is known to hybridise with the creolefish *Paranthias furcifer* (Smith 1965, Thompson and Munro 1978, Bostrom et al. 2002). Bostrom et al. (2002) report that the hybrid is viable and capable of reproduction.

2008 IUCN RED LIST STATUS Least Concern

Authors: *AA Bertoncini, JH Choat, MT Craig, BP Ferreira, ALB Gaspar, S Marques and LA Rocha*

Cephalopholis hemistiktos (Rüppell 1860)

yellowfin hind (E), vielle d'Arabie (F), cherna arábiga (S)

20 cm SL (Bahrain)

Body depth less than head length, the depth contained 2.7–3.0 times in standard length (for fish 10–20 cm SL). Head length 2.4–2.6 times in standard length; interorbital area flat; preopercle rounded, finely serrate, the lower edge fleshy; upper edge of operculum distinctly convex; maxilla scaly, reaching to or somewhat beyond vertical at rear edge of eye. Gill rakers 6–8 on upper limb, 13–15 on lower limb. Dorsal fin with 9 spines and 14 or rarely 15 rays, the membranes distinctly indented between the spines; anal fin with 3 spines and 8–10 rays, the posterior margin of the fin pointed in adults; pectoral-fin rays 16–18; pectoral fins pointed, distinctly longer than pelvic fins, pectoral-fin length contained 1.4–1.6 times in head length; pelvic fins nearly or just reaching anus, pelvic-fin length 1.7–2.0 times in head length; caudal fin well rounded. Body scales ctenoid, no auxiliary scales; lateral-line scales 47–52; lateral-scale series 95–121.

MAXIMUM RECORDED SIZE 35 cm TL

Cephalopholis hemistiktos has a disjunct distribution. It is known with certainty from the Red Sea, and from the northern coast of Oman to the Arabian Gulf and coast of Pakistan. Recent anecdotal records indicate the presence of this species in Socotra (Yemen) and mainland Somalia.

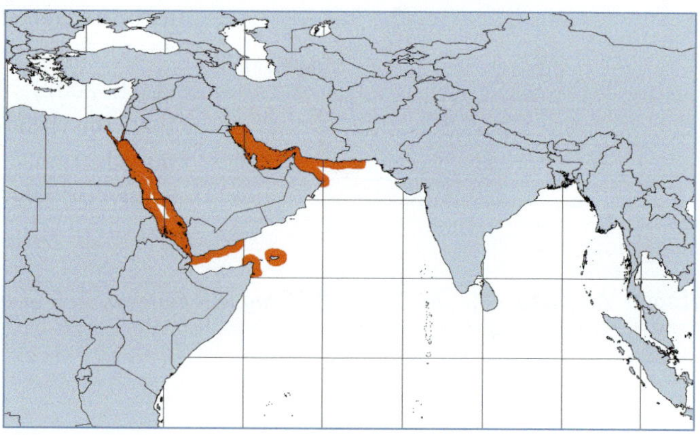

Adult *C. hemistiktos* (Red Sea)

COLOUR IN LIFE The colour of shallow-water individuals is brownish, those from depths of 15–20 m are brownish red, and fish from deep water are primarily reddish; dark-edged blue ocelli on the head and lower body are more numerous ventrally; the caudal fin and posterior part of dorsal and anal fins are generally darker than the body and are covered with small blue ocelli, their margins having a pale blue line; the triangular membrane at the tip of each dorsal-fin spine is orange; the pectoral fins are mostly brown to reddish brown, with a few small blue ocelli at the base, the outer margin broadly and abruptly yellow; pelvic fins dark red to reddish brown, the rays darker than the membranes. Some fish have a large yellow or buff area below the soft dorsal fin.

COLOUR IN MARKET Body generally reddish and may become blotchy; the dorsal and ventral portions of the body are the same shade. The caudal fin is darker than the general body colour, and the pectoral fins remain pale. The spots on the body are less visible.

Adult *C. hemistiktos* (Red Sea)

Adult *C. hemistiktos* (Red Sea)

POPULATION **General** While variable by location, *C. hemistiktos* is one of the most abundant species of *Cephalopholis* recorded in the Red Sea. This species is absent from reefs of the Arabian Sea south of Ras al Hadd. Elsewhere, little population information is available. **Fishery dependent** Despite its small size, *C. hemistiktos* is a heavily fished and important component of the fisheries in some portions of its range. For example, in central Oman, this species makes up more than one-third of the total grouper catch and is taken by line and trapping (JL McIlwain unpublished data). The species is not as heavily fished as larger co-occurring groupers at present, but as these larger species decline and become no longer economically viable to target, it is expected that fisheries will focus more on this species. Logbooks filled out by Omani traditional fishermen from Barka (35 km north of Muscat) suggest that *C. hemistiktos* accounts for 7.5% of the total catch (all fish species combined) and 42% of grouper species.

POPULATION TREND Decreasing

MAJOR THREATS The major threats to *C. hemistiktos* are overfishing, lack of management and habitat loss.

HABITAT AND ECOLOGY **GENERAL** *C. hemistiktos* is found on patchy reef areas in depths of up to 55 m. **FEEDING** *C. hemistiktos* is a diurnal, ambush predator that feeds throughout the day on fishes and crustaceans. **REPRODUCTION** *C. hemistiktos* is a monogamous species. Pairs will jointly defend a common territory of up to 62 m² (Shpigel and Fishelson 1991).

COMMENTS *C. hemistiktos* is easily confused with *C. miniata* and *C. sexmaculata*. These species differ in colour pattern. In *C. hemistiktos*, there are no small blue spots on the dorsal part of the body, while the other two species have the entire body covered with blue spots. The caudal fin and posterior parts of the dorsal and anal fins are distinctly darker than the body in *C. hemistiktos*, but these fins are not darker in the other species. The outer margin of the pectoral fins is abruptly yellow in *C. hemistiktos*, but gradually orange-yellow in *C. miniata* and uniformly coloured in *C. sexmaculata*. The number of dorsal-fin rays is usually 14 in *C. hemistiktos*, while in the other species the usual count is 15.

2008 IUCN RED LIST STATUS Near Threatened

Authors: *AA Bertoncini, JH Choat, MT Craig, BP Ferreira and LA Rocha*

Cephalopholis igarashiensis Katayama 1957

garish hind (E), vielle voyant (F), cherna chillona (S)

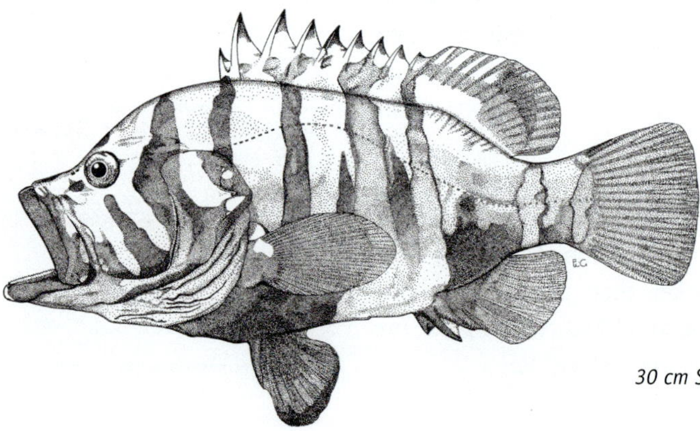

30 cm SL (Okinawa, Japan)
© SAIAB

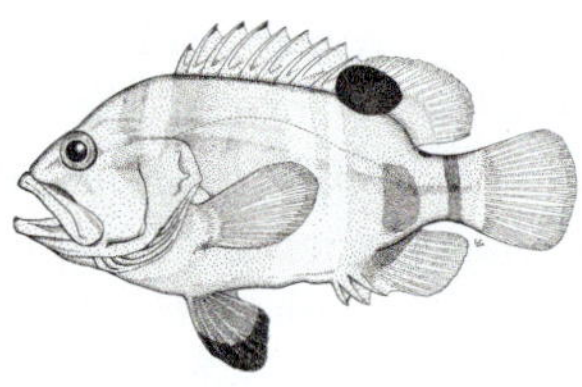

14 cm SL (Taiwan)
© SAIAB

Body depth greater than head length, the depth contained 2.0–2.4 times in standard length; body width 2.3–2.6 times in the depth; caudal-peduncle depth equal to or greater than its length. Head length 2.3–2.5 times in standard length; dorsal head profile straight or slightly concave to above eye, the nape markedly convex; interorbital area flat; preopercle rounded, finely serrate, the ventral serrae slightly enlarged, the lower edge irregular but hidden by skin; subopercle and interopercle finely serrate; upper edge of operculum distinctly convex; maxilla naked, reaching vertical at rear edge of eye, a prominent knob on the ventral rear corner. Gill rakers 8 or 9 on upper limb, 16 or 17 on lower limb. Dorsal fin with 9 spines and 14 rays, the membranes distinctly incised; pectoral-fin rays 18 or 19; pelvic fins reach to or beyond anus, their length contained 1.7–2.0 times in head length; anal fin with 3 spines and 9 rays; pectoral-fin length 1.6–1.8 times in head length; caudal fin rounded. Lateral line noticeably arched over pectoral fin. Body scales ctenoid, without auxiliaries; lateral-line scales 60–65; lateral-scale series 101–117.

MAXIMUM RECORDED SIZE 43 cm TL

COLOUR IN LIFE Head, body and fins reddish orange with 7 yellow bars on the dorsal part of the body and extending onto the dorsal fin; 3 broad yellow bands radiating from the eye; fins red, except for extension of yellow bars into the dorsal fin and orange tip of the dorsal-fin membranes. Juveniles yellowish with a large black spot in the dorsal fin; pelvic fins and tips of the interspinous dorsal-fin membranes dark.

COLOUR IN MARKET *C. igarashiensis* generally retains its life colour pattern in death. The bars may fade, but are often visible.

Juvenile *C. igarashiensis* (Guam)

Adult *C. igarashiensis* (Okinawa Fish Market, Japan)

Cephalopholis igarashiensis is distributed from southern Japan to Fiji and French Polynesia.

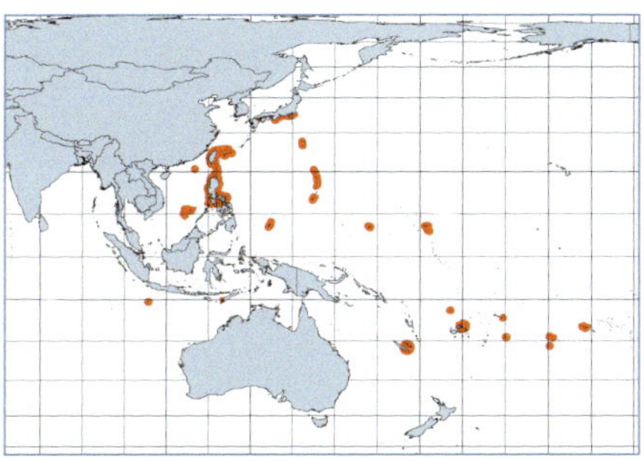

POPULATION General *C. igarashiensis* is considered rare throughout its range. **Fishery dependent** Little fishery-dependent information is available for this species. It has been recorded in the catch statistics in Guam.

POPULATION TREND Unknown

MAJOR THREATS Fishing pressure is the major threat to *C. igarashiensis*.

HABITAT AND ECOLOGY General *C. igarashiensis* is a deep-water species associated with steep drop-offs, seamounts and offshore banks in depths of 80–250 m.

FEEDING *C. igarashiensis* is known to feed on fishes and crustaceans.

2008 IUCN RED LIST STATUS Data Deficient

Authors: *AS Cabanban, ST Fennessy, RF Myers and BC Russell*

Cephalopholis leopardus (Lacepède 1801)

leopard hind (E), vielle léopard (F), cherna leopardo (S)

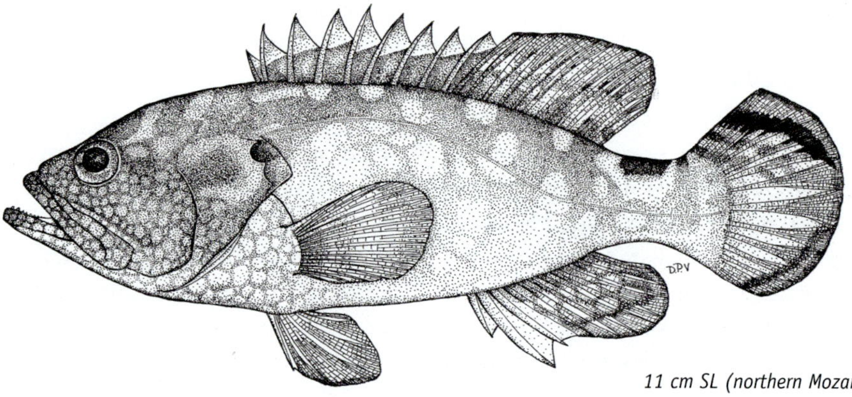

11 cm SL (northern Mozambique)
© SAIAB

Body depth less than head length, contained 2.5–2.8 times in standard length. Head length 2.2–2.4 times in standard length; interorbital area flat; preopercle rounded, finely serrate, the lower edge fleshy; subopercle and interopercle smooth; upper edge of operculum distinctly convex; maxilla extends past eye. Gill rakers 7–9 on upper limb, 15–17 on lower limb. Dorsal fin with 9 spines and 13–15 rays; anal fin with 3 spines and 9 or 10 rays; pectoral-fin rays 16–18; pectoral-fin length contained 1.4–1.6 times in head length; pelvic fins not reaching anus, their length 2.0–2.3 times in head length; caudal fin well rounded. Body scales ctenoid, no auxiliary scales; lateral-line scales 47–50; lateral-scale series 79–88.

MAXIMUM RECORDED SIZE 20 cm TL

Cephalopholis leopardus is distributed in the Indo-Pacific from East Africa to French Polynesia, extending north to the Ryukyu Islands (Japan), and south to northern Australia. It is absent from the Red Sea, the Persian Gulf and South Africa.

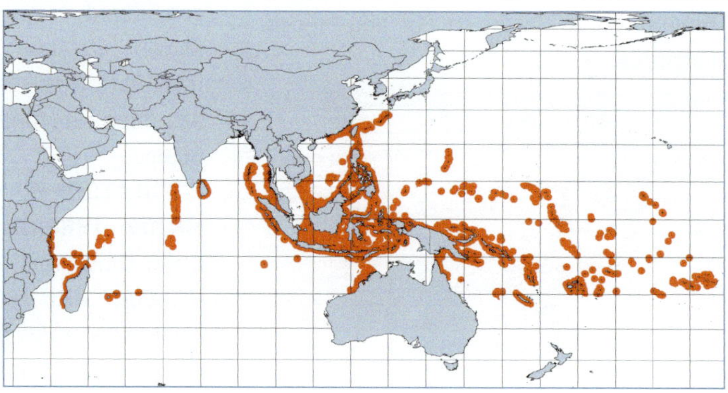

COLOUR IN LIFE Head and body mottled red-brown with a paler ventral surface; numerous red-orange or pinkish red spots present with a dark brown saddle on the caudal peduncle and a similar, but much smaller, spot just behind it. A dark brown or reddish submarginal streak on the upper part of the caudal fin and a fainter reddish streak on the lower part of the fin are usually present. The operculum has a dark brown spot at its posterior end. Pectoral fins may have a yellowish tint.

COLOUR IN MARKET Generally reddish brown with faded spots. The yellow tint on the pectoral fins often fades, but the fins appear lighter in colour. The saddle on the caudal peduncle is usually present along with the dark spot on the operculum.

© Richard Winterbottom

Juvenile *C. leopardus* (Comoros Islands)

© Patrice Marker

Adult *C. leopardus* (Fiji)

© Robert F Myers

Adult *C. leopardus* (Guam)

POPULATION **General** *C. leopardus* has variable abundance across its range. It is common in the western Indian Ocean, but uncommon elsewhere (e.g. on the Great Barrier Reef [Australia] and at Réunion).

POPULATION TREND Unknown

MAJOR THREATS The major threat to *C. leopardus* is habitat degradation from dynamite fishing, sedimentation and eutrophication.

HABITAT AND ECOLOGY **General** *C. leopardus* is a reef-associated species found at depths of 1–40 m. The species is solitary and secretive, usually hiding in caves and crevices in the reefs. **Feeding** *C. leopardus* is omnivorous with a diet composed of fishes and crustaceans. **Age and growth** This species may reach an age of at least 12 years (Mellin et al. 2006). **Reproduction** Little information on reproduction is available for *C. leopardus*, but anecdotal reports indicate a relatively low egg production per individual.

COMMENTS *C. leopardus* may be confused with *C. urodeta* which is also generally small with a red background colour. The latter, however, does not possess the dark saddle on the caudal peduncle. While *C. leopardus* is not a major target of commercial fisheries, it is fished for subsistence purposes throughout its range and has been noted in the live food fish trade from the Maldives.

2008 IUCN RED LIST STATUS Least Concern

Authors: *AS Cabanban, PC Heemstra, M Kulbicki and MA Samoilys*

Cephalopholis microprion (Bleeker 1852)

freckled hind (E), vielle tache de rousseur (F), cherna pecosa (S)

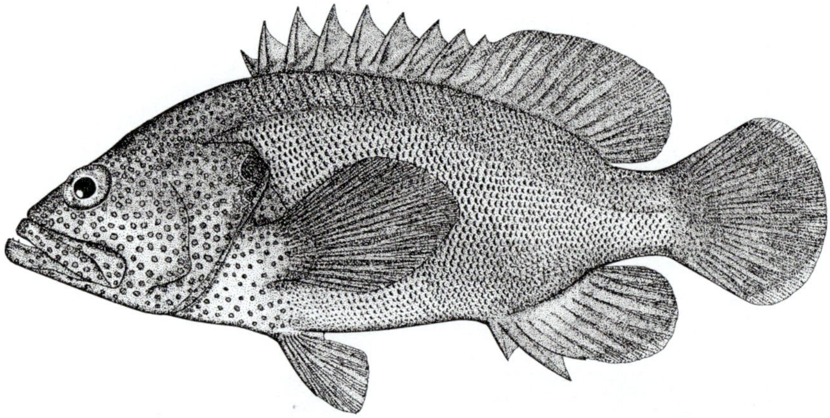

12 cm SL (Guadalcanal, Solomon Islands)
© SAIAB

Body depth contained 2.5–2.8 times in standard length. Head length 2.3–2.5 times in standard length; interorbital area flat or slightly concave; preopercle rounded, finely serrate, the lower edge fleshy; subopercle and interopercle smooth or finely serrate; upper edge of operculum very convex; maxilla naked, extending past eye. Gill rakers 7–9 on upper limb, 14–17 on lower limb. Dorsal fin with 9 spines and 14–16 rays; anal fin with 3 spines and 8 rays; pectoral-fin rays 14–16; pectoral fins usually reaching past vertical at anus, their length contained 1.4–1.6 times in head length; pelvic-fin length 1.9–2.1 times in head length; caudal fin well rounded. Body scales ctenoid, without auxiliary scales; lateral-line scales 45–51; lateral-scale series 84–98.
MAXIMUM RECORDED SIZE 24 cm TL

COLOUR IN LIFE Dark brown, the proximal part of each scale darker than the posterior part producing a very fine reticulated pattern; some fish with indistinct dark bars on the body; head and anterior part of the body with numerous small dark-edged blue ocelli, rarely occurring posterior to the pectoral-fin base; some fish also have blue ocelli on the pectoral-fin base; soft dorsal, anal and caudal fins usually with a greyish blue margin (narrow or absent at the tips of the central caudal rays).
COLOUR IN MARKET Generally dark brown with fine reticulations due to scale pigmentation. Ocelli remain prominent on the head. Median fins may retain blue hue.

© Robert F Myers

Adult *C. microprion* (Batangas, Philippines)

© Matthew T Craig

Adult *C. microprion* (Dongala, Sulawesi, Indonesia)

POPULATION **GENERAL** Little population information is available for *C. microprion*. It is known to be common on the Great Barrier Reef (Australia; Pears 2005) and is taken as part of multi-species reef fisheries.

Cephalopholis microprion is a western Pacific species distributed from the Gulf of Thailand to New Caledonia, north to Luzon (Philippines), and south to the Great Barrier Reef (Australia).

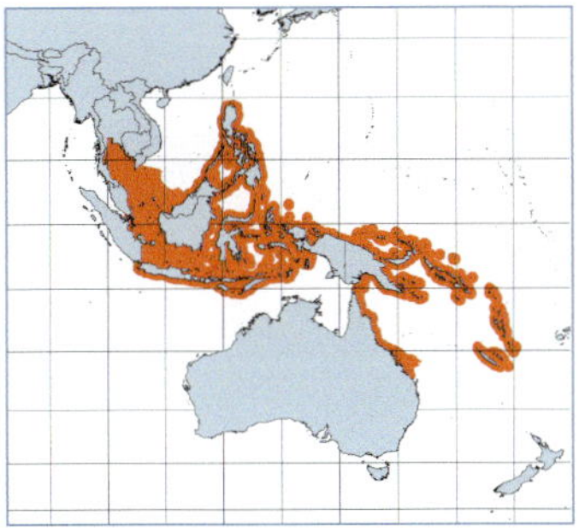

POPULATION TREND Unknown

MAJOR THREATS Although not specifically targeted, fishing pressure is potentially a threat to this species.

HABITAT AND ECOLOGY **GENERAL** *C. microprion* is a shallow, reef-associated species, usually found on silty reefs, midshelf lagoons and back-reef areas in depths of up to 30 m. Juveniles are often found under corals surrounded by rubble substrate. **FEEDING** *C. microprion* likely feeds on fishes and crustaceans.

COMMENTS *C. microprion* may be confused with *C. boenak*. Randall (1987) recognised *C. microprion* as a valid species, and distinguished it from *C. boenak* by the presence of its characteristic spots.

2008 IUCN RED LIST STATUS Least Concern

Authors: *ST Fennessy, PC Heemstra, M Kulbicki, DA Pollard, BC Russell and BM Yeeting*

Cephalopholis miniata (Forsskål 1775)

coral hind (E), vielle de corail (F), cherna estrellada (S)

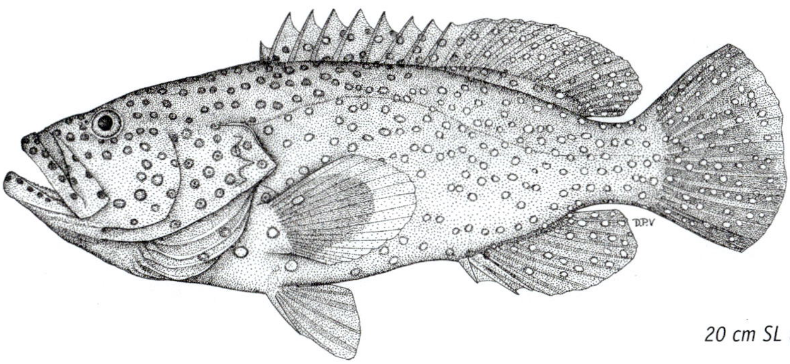

20 cm SL (South Africa)
© SAIAB

Body depth contained 2.6–3.0 times in standard length. Head length 2.4–2.6 times in standard length; interorbital area flat to slightly convex; preopercle rounded, finely serrate, the lower edge fleshy; upper edge of operculum very convex; maxilla scaly, reaching to or beyond vertical at rear edge of eye. Gill rakers 7–9 on upper limb, 14–16 on lower limb. Dorsal fin with 9 spines and 14–16 rays, the membranes distinctly indented between the spines; anal fin with 3 spines and 9 (rarely 8) rays, the fin margin rounded in adults; pectoral-fin rays 17 or 18, pectoral-fin length contained 1.4–1.75 times in head length; pelvic fins usually not reaching anus, their length 1.9–2.3 times in head length; caudal fin well rounded. Midlateral-body scales ctenoid, no auxiliary scales; lateral-line scales 47–56; lateral-scale series 94–114.

MAXIMUM RECORDED SIZE 45 cm TL

COLOUR IN LIFE Light to dark red, sometimes orange, and covered with small (usually dark-edged) pale blue-grey spots. The pectoral fins are lighter distally. Soft dorsal, caudal and anal fins often have a narrow blue margin and blackish submarginal line. The distal edge of the pelvic fins is dark bluish grey. Juveniles are more yellowish, with faint pale blue spots, fewer than in adults. During certain stages of excitement this species may display 5 light bars on its body.

COLOUR IN MARKET Retains reddish colour and spotting pattern but may become blotchy. Pectoral fins retain lighter hue in death. No bars generally visible in death.

© Robert F Myers

Juvenile *C. miniata* (Bali, Indonesia)

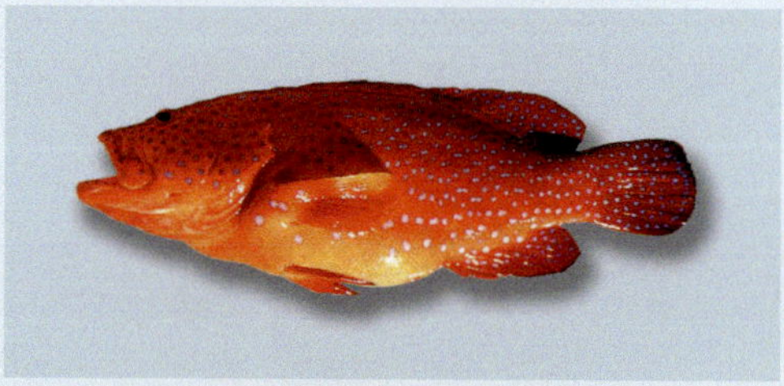

© Matthew T Craig

Adult *C. miniata* (Sodwana Bay, South Africa)

© Patrick Louisy

Subadult *C. miniata* (Marsá al Alam, Egypt, Red Sea)

© Matthew T Craig

Adult *C. miniata* (Okinawa Fish Market, Japan)

© Robert F Myers

Adult *C. miniata* (Bali, Indonesia)

Cephalopholis miniata is an Indo-Pacific species recorded from the Red Sea south to Durban (South Africa) and eastward to the Line Islands. The species is absent from the Persian Gulf and the Gulf of Oman. It is also recorded from the East Andaman Sea, Thailand (Allen and Stone 2005), and south-western India (RF Myers distributional database).

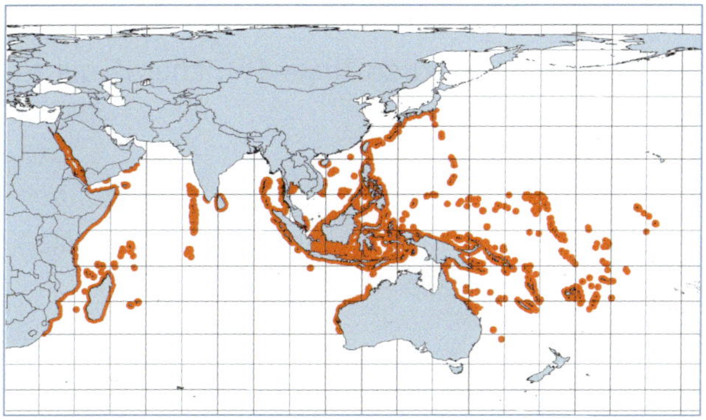

POPULATION **General** *C. miniata* is common in most coral-rich areas within its range. **Fishery dependent** *C. miniata* is a locally important member of small-scale, multi-species fisheries, taken by handlines and spears among other gears. For example, in New Caledonia, *C. miniata* constitutes 3% of the grouper catch (M Kulbicki pers. obs.), while in Pohnpei it constitutes less than 1% of the total catch (Rhodes and Tupper 2007). It occasionally is included in the Hong Kong-based live reef food fish trade. **Fishery independent** While *C. miniata* is a ubiquitous member of coral reef communities, fishery-independent surveys indicate that it is not particularly dense. For example, in New Caledonia, it reaches a maximum density of 0.8 individuals per 1000 m^2 (M Kulbicki pers. obs.).

POPULATION TREND Decreasing

MAJOR THREATS The major threat to this species is overfishing.

HABITAT AND ECOLOGY **General** *C. miniata* inhabits clear waters of coral reefs and is more often found in exposed rather than protected reef areas (Fischer et al. 1990). On the Great Barrier Reef (Australia), it is only found in outer-shelf, back-reef areas (Newman et al. 1997). **Feeding** *C. miniata* feeds on fishes and crustaceans, and in some areas seems to show a preference for *Pseudanthias* spp. **Reproduction** *C. miniata* forms social and mating groups comprising a dominant male and 2–12 females. These groups occupy territories of up to 475 m^2 subdivided into secondary territories defended by a single female (Shpigel and Fishelson 1991). Females mature around 25 cm TL in New Caledonia (M Kulbicki pers. obs.).

COMMENTS *C. miniata* is easily confused with *C. hemistiktos* and *C. sexmaculata*. These species differ in colour pattern. In *C. hemistiktos*, there are no small blue spots on the dorsal part of the body, while the other two species have the entire body covered with blue spots. The caudal fin and posterior parts of the dorsal and anal fins are distinctly darker than the body in *C. hemistiktos,* but these fins are not darker in the other species. The outer margin of the pectoral fins is abruptly yellow in *C. hemistiktos*, but gradually orange-yellow in *C. miniata* and uniformly coloured in *C. sexmaculata*. This species is common in the aquarium trade.

2008 IUCN RED LIST STATUS Least Concern

Authors: *AS Cabanban, ST Fennessy, PC Heemstra, M Kulbicki, DA Pollard and BM Yeeting*

Cephalopholis nigri (Günther 1859)

Niger hind (E), mérou du Niger (F), cherna del Níger (S)

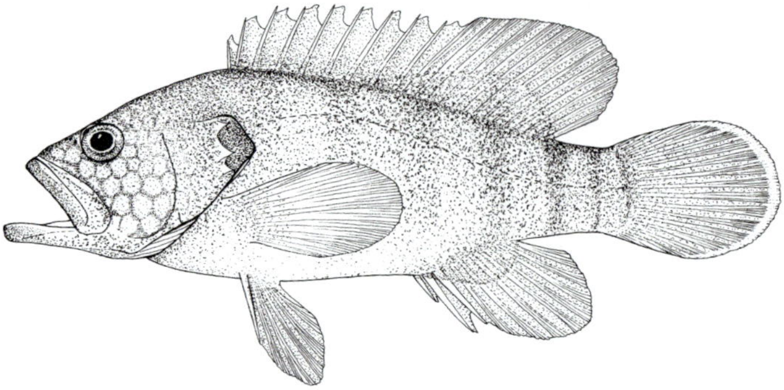

~25 cm TL
© FAO

Body depth distinctly less than head length, the depth contained 2.8–3.0 times in standard length. Head length 2.5–2.6 times in standard length; eye diameter greater than interorbital width, eye diameter 4.6–5.5 times in head length; interorbital area flat; preopercle rounded, finely serrate, the lower edge fleshy; subopercle and interopercle smooth; upper edge of operculum convex; maxilla scaly, reaching well past eye. Gill rakers 8–10 on upper limb, 14–17 on lower limb; 5 or 6 rudiments on each limb. Dorsal fin with 9 spines and 14 or 15 rays, the membranes distinctly indented between the spines; anal fin with 3 spines and 8 rays; pectoral-fin rays 17 or 18; pectoral-fin length contained 1.4–1.7 times in head length; pelvic fins reach anus, their length 1.6–1.9 times in head length; caudal fin rounded. Lateral-body scales ctenoid, without auxiliary scales; lateral-line scales 45–50; lateral-scale series 73–80.

MAXIMUM RECORDED SIZE 30 cm TL

COLOUR IN LIFE Dark brownish, with 3 or 4 indistinct dark bars on the body (extending onto the dorsal fin) and 2 more on the caudal peduncle. Belly reddish, head with close-set reddish spots forming a reticulated pattern of dark lines that diminish posteriorly. Reddish spots on the body which become smaller and less distinct towards the rear of the body. Pelvic and anal fins dusky, with small pale spots; tips of the interspinous dorsal-fin membranes black. Juveniles entirely covered (head and body) with red spots.

COLOUR IN MARKET Body colour may become pale, although it generally retains its brown hue. The bars on the midbody are usually absent, the posterior 2 or 3 may remain. The reticulated pattern on the face is generally prominent.

Adult *C. nigri* (São Tomé)

© Luiz A Rocha

Cephalopholis nigri is an eastern Atlantic species ranging from Senegal to Angola, including the Canary Islands (Spain) and islands of the Gulf of Guinea.

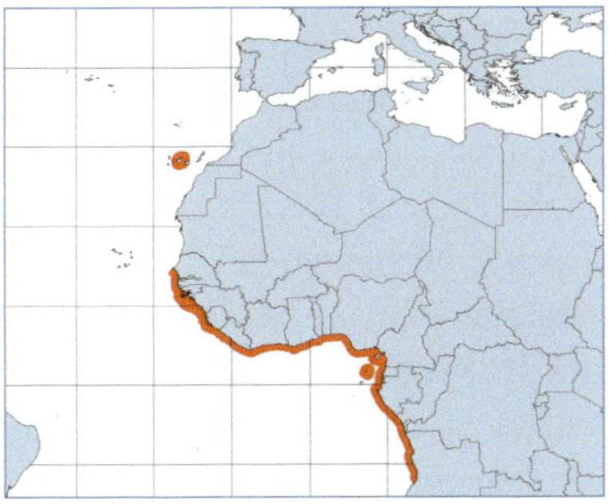

POPULATION **GENERAL** *C. nigri* is a common species throughout its range although it rarely reaches high densities.

POPULATION TREND Unknown

MAJOR THREATS No major threats have been identified for this species.

HABITAT AND ECOLOGY **GENERAL** *C. nigri* occurs in a variety of habitats including mud, sand and reef to depths of 100 m.

2008 IUCN RED LIST STATUS Least Concern

Authors: *AA Bertoncini, JH Choat, MT Craig, BP Ferreira and LA Rocha*

Cephalopholis nigripinnis

(Valenciennes in Cuv. and Val. 1828)

darkfin hind (E), mérou tacheté (F), cherna alinegra (S)

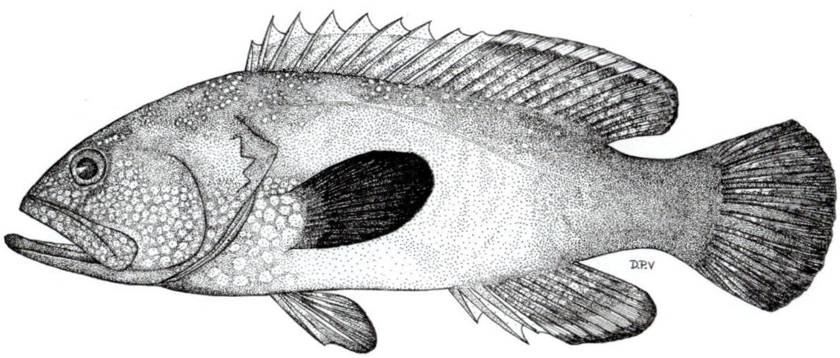

13 cm SL (northern Mozambique)
© SAIAB

Body depth less than head length, depth contained 2.7–3.3 times in standard length. Head length 2.4–2.7 times in standard length; interorbital area convex; preopercle rounded, finely serrate, the lower edge fleshy; subopercle and interopercle usually smooth; maxilla reaching well past eye. Gill rakers 7–9 on upper limb, 15–17 on lower limb. Dorsal fin with 9 spines and 14–16 rays; anal fin with 3 spines and 9 (rarely 8) rays; pectoral-fin rays 17–19; pectoral fins distinctly longer than pelvic fins, pectoral-fin length contained 1.3–1.7 times in head length; pelvic fins usually not reaching anus, pelvic-fin length 1.8–2.3 times in head length; caudal fin rounded. Snout and maxilla with minute cycloid scales; lateral-body scales ctenoid, without auxiliary scales; lateral-line scales 54–68; lateral-scale series 88–108.

MAXIMUM RECORDED SIZE 28 cm TL

COLOUR IN LIFE Body generally reddish brown to brownish red, darker posteriorly, sometimes with small pale spots and/or 4 faint irregular dark bars that bifurcate ventrally and another 2 dark bars on the caudal peduncle. The head has numerous small close-set orange-red spots, sometimes with irregular reddish brown blotches. A dark spot between the upper and middle opercular spines may be present. A pair of small dark spots on the edge of the lower lip in line with the pair of anterior canine teeth may be pronounced in some individuals. The soft dorsal and anal fins have small orange-red spots, the outer triangular part of the interspinous dorsal-fin membranes orange. The pelvic fins are orange-red, usually with a dark blue edge. The caudal fin is dark reddish brown to almost black and covered with small pale spots. Pectoral fins are coloured like the body basally, becoming dark brown to almost black distally.

COLOUR IN MARKET Retains nearly identical colouration in death, although colours are duller in appearance.

© Luiz A Rocha

Adult *C. nigripinnis* (Seychelles)

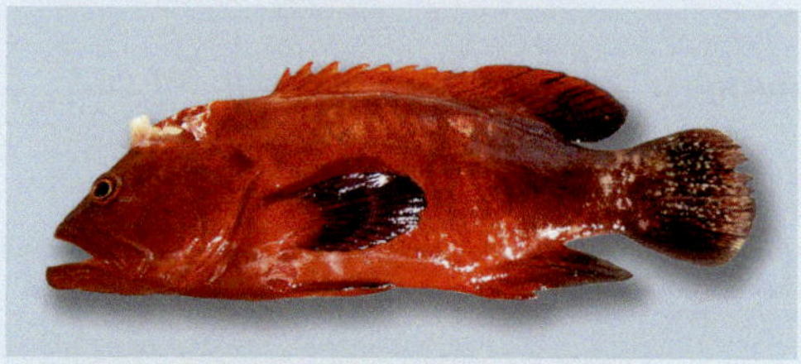
© Matthew T Craig

Adult *C. nigripinnis* (Sodwana Bay, South Africa)

© Robert F Myers

Adult *C. nigripinnis* (Maldives)

Cephalopholis nigripinnis is distributed in the western Indian Ocean from the east coast of Africa (South Africa to Kenya) to Jakarta, including the Comoros Islands, Maldives, Sri Lanka, Seychelles, Mascarene Islands (Mauritius, Réunion and Rodrígues) and Chagos.

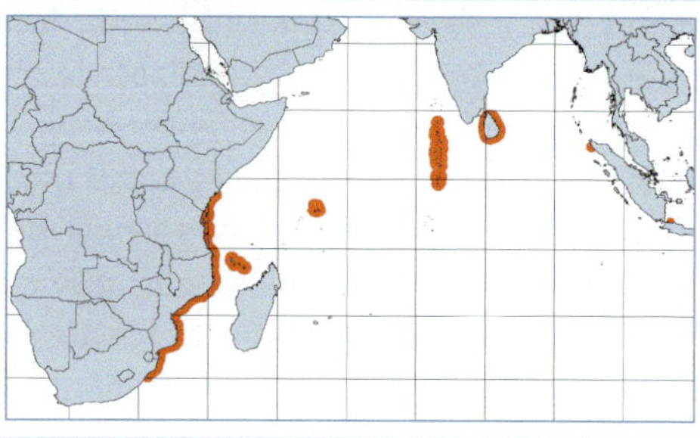

POPULATION GENERAL *C. nigripinnis* is a common species throughout its range.
POPULATION TREND Unknown
MAJOR THREATS No major threats have been identified for this species.
HABITAT AND ECOLOGY GENERAL *C. nigripinnis* is generally found in coral-rich areas to depths of 60 m (although usually not deeper than 15 m). FEEDING Feeds mainly on small fishes and crustaceans.
COMMENTS *C. nigripinnis* is easily confused with its Pacific Ocean sister species *C. urodeta*. The former has a caudal fin that lacks the two converging white lines seen in the latter. *C. nigripinnis* was previously considered conspecific with *C. urodeta*. However, based on the distributions of the two colour morphs and differences in genetic data, they are currently recognised as distinct species (Craig and Hastings 2007).
2008 IUCN RED LIST STATUS Data Deficient

Author: *MT Craig*

Cephalopholis oligosticta

Randall and Ben-Tuvia 1983

vermilion hind (E), vielle de la Mer Rouge (F), cherna del mar rojo (S)

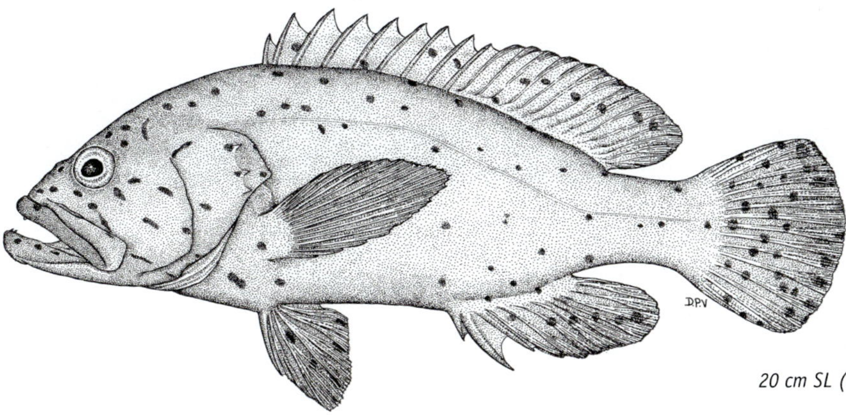

20 cm SL (Red Sea)
© SAIAB

Body depth less than head length, depth contained 2.6–3.0 times in standard length. Head length 2.4–2.6 times in standard length; interorbital area slightly convex, the width equal to orbit diameter on a fish of 19 cm SL; preopercle rounded, finely serrate posteriorly, the lower edge more coarsely serrate; subopercle and interopercle also completely serrate; maxilla reaches well past eye. Gill rakers 7 or 8 on upper limb, 14 or 15 on lower limb. Dorsal fin with 9 spines and 14 or 15 rays; anal fin with 3 spines and 9 rays; pectoral-fin rays 16–18; pectoral-fin length contained 1.4–1.6 times in head length; pelvic fins reach about to anus, their length 1.6–2.0 times in head length; caudal fin rounded. Lateral-body scales ctenoid, without auxiliary scales; lateral-line scales 58–71; lateral-scale series 103–123.
MAXIMUM RECORDED SIZE 30 cm TL

COLOUR IN LIFE The body is orange-red, with widely scattered, small pale blue spots on the head, body and fins. The soft dorsal, anal and caudal fins usually have a narrow pale blue margin.

COLOUR IN MARKET Pale red to orange, spots usually visible although in a much duller hue.

© John E Randall

Adult *C. oligosticta* (Red Sea)

© Dieter Eichler

Adult *C. oligosticta* (Red Sea)

Cephalopholis oligosticta is endemic to the Red Sea and ranges from Eilat in the northern Gulf of Aqaba to the Farasan Islands off the southern end of Saudi Arabia.

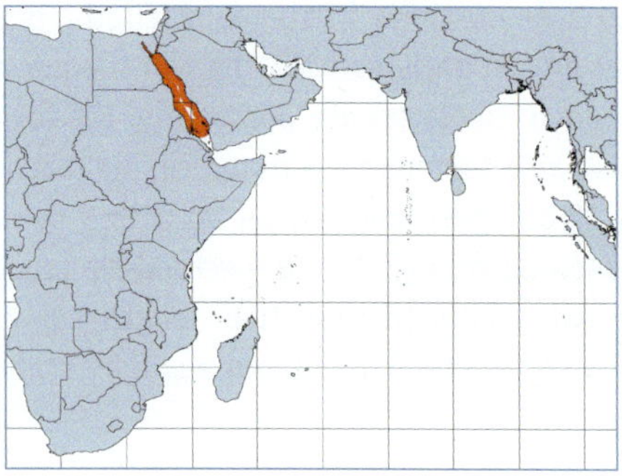

POPULATION GENERAL *C. oligosticta* is generally uncommon throughout its range. **FISHERY INDEPENDENT** Little fishery-independent data are available. Underwater visual surveys in Hurghada (Egypt) have, however, recorded up to 10 individuals per 250 m^2 (Lieske and Myers 2004).

POPULATION TREND Unknown

MAJOR THREATS No major threats have been identified for this species.

HABITAT AND ECOLOGY GENERAL *C. oligosticta* is a reef-associated species that inhabits dead reefs in silty areas, as well as silty lagoons from depths of 15 m to at least 50 m. It is often found in or near caves as adults, while juveniles are more common in coral rubble areas. **REPRODUCTION** In their original description of the species, Randall and Ben-Tuvia (1983) reported mature females (i.e. had well-developed oocytes) in fish 17–19 cm SL, and males with well-developed testes in individuals 22 cm SL.

COMMENTS *C. oligosticta* may easily be confused with *C. miniata*. The former species has fewer blue spots, more lateral-line scales (58–71), and usually longer pelvic fins (contained 1.6–2.0 times in head length). Both species may be found at the same locality in the Red Sea (e.g. Eilat), but they occur in different habitats.

2008 IUCN RED LIST STATUS Least Concern

Authors: *KE Carpenter, ST Fennessy, PC Heemstra, RF Myers and MA Samoilys*

Cephalopholis panamensis

(Steindachner 1877)

Pacific graysby (E), cône bandé (F), enjambre (S)

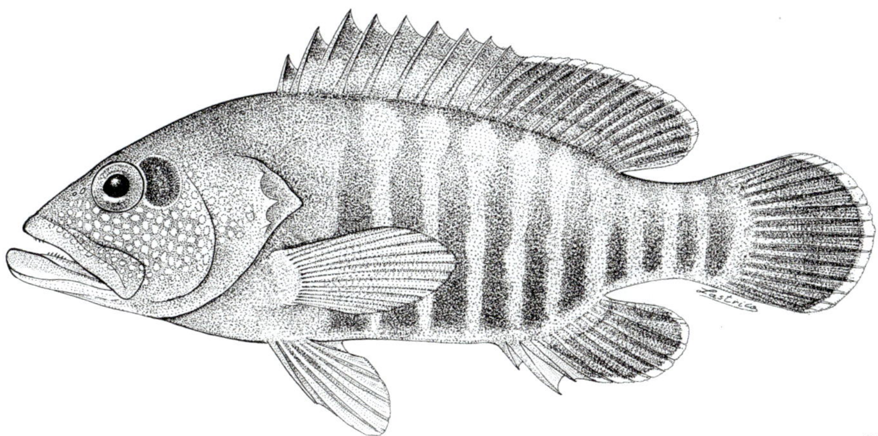

~25 cm TL
© FAO

Body depth distinctly less than head length, depth contained 2.7–2.9 times in standard length. Head length 2.3–2.5 times in standard length; preopercle rounded, finely serrate; posterior and anterior nostrils subequal. Gill rakers 16–19 (total). Dorsal fin with 9 spines, 14 or 15 soft rays, 3rd spine longest, the membranes incised between the anterior spines; anal fin with 3 spines, 8 rays; pectoral-fin rays 17 or 18; pectoral fins clearly longer than pelvic fins, pectoral-fin length contained 1.5–1.7 times in head length; caudal fin rounded. Lateral-body scales strongly ctenoid; lateral-line scales 48 or 49; lateral-scale series 80–83.

MAXIMUM RECORDED SIZE 39 cm TL

COLOUR IN LIFE Head and body dark brown/grey, with 8–10 dark bars, wider and more distinct ventrally. Head, maxilla and upper lip with close-set reddish orange spots that form a bluish green reticulum. Adults have a large dark brown spot immediately behind the eye, and the median fins are often edged with blue.

COLOUR IN MARKET Generally the same as in life, but the dark bars on the body may fade. The reticulated pattern on the head is usually prominent.

© Matthew T Craig

Adult *C. panamensis* (Los Islotes, La Paz, Baja California, Mexico)

© Matthew T Craig

Adult *C. panamensis* (Loreto, Mexico)

POPULATION **GENERAL** *C. panamensis* is a common species throughout its range. **FISHERY INDEPENDENT** Fishery-independent data are limited, but indicate a stable population in the Gulf of California (B Erisman unpublished data).

Cephalopholis panamensis is distributed from the Gulf of California to Ecuador, including the Galápagos Islands.

POPULATION TREND Stable

MAJOR THREATS None

HABITAT AND ECOLOGY **GENERAL** *C. panamensis* is a common, shy and cryptic reef-associated species that is found in rocky shore habitats in depths to at least 30 m. **FEEDING** Feeds on fishes and crustaceans (B Erisman and MT Craig unpublished data). **AGE AND GROWTH** *C. panamensis* reaches at least 25 years (Craig et al. 1999). **REPRODUCTION** Recently confirmed as a protogynous hermaphrodite, with spawning peaks in July and August in the Gulf of California (Erisman et al. 2010).

COMMENTS *C. panamensis* is common throughout its range. Although not directly targeted, it is often caught opportunistically in small-scale, multi-species fisheries.

2008 IUCN RED LIST STATUS Least Concern

Author: *MT Craig*

Cephalopholis polleni (Bleeker 1868)

harlequin hind (E), vielle arlequín (F), cherna arlequin (S)

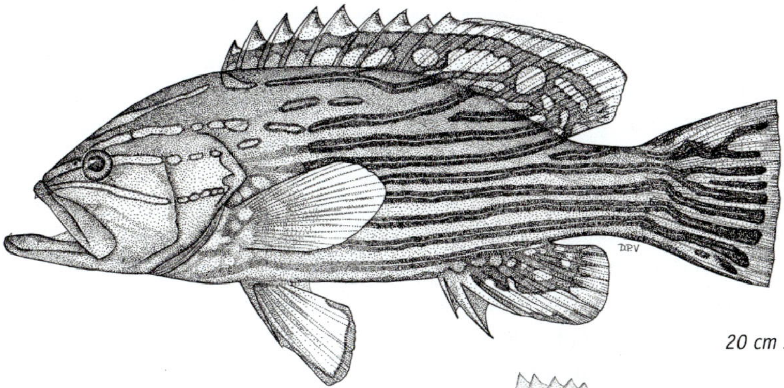

20 cm SL (Mauritius)
© SAIAB

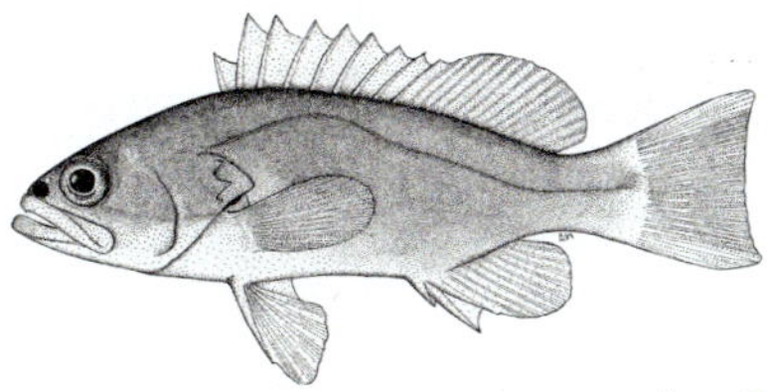

4 cm SL (Indonesia)
© SAIAB

Body depth subequal to head length, depth contained 2.7–3.1 times in standard length. Eye diameter subequal to interorbital width and about twice the depth of the preorbital; preopercle rounded, finely serrate, the lower edge with 1–3 broad serrae; subopercle and interopercle finely serrate; maxilla reaches approximately to vertical at rear edge of eye. Gill rakers 7 or 8 on upper limb, 14–17 on lower limb. Dorsal fin with 9 spines (with the fin membrane distinctly indented between them) and 14–16 rays; anal fin with 3 spines and 9 (rarely 8) rays; pectoral-fin rays 17–19; pectoral fins subequal to pelvic fins, pectoral-fin length contained 1.5–1.9 times in head length; caudal fin truncate to slightly emarginate. Lateral-body scales ctenoid, without auxiliary scales; lateral-line scales 66–72; lateral-scale series 112–135.

MAXIMUM RECORDED SIZE 43 cm TL

Cephalopholis polleni is an Indo-Pacific species that is scattered among insular localities from the Comoros Islands to the Line Islands and French Polynesia. It is unknown from the Red Sea, the coast of East Africa, Sri Lanka, Australia, New Guinea, and the larger islands of Indonesia.

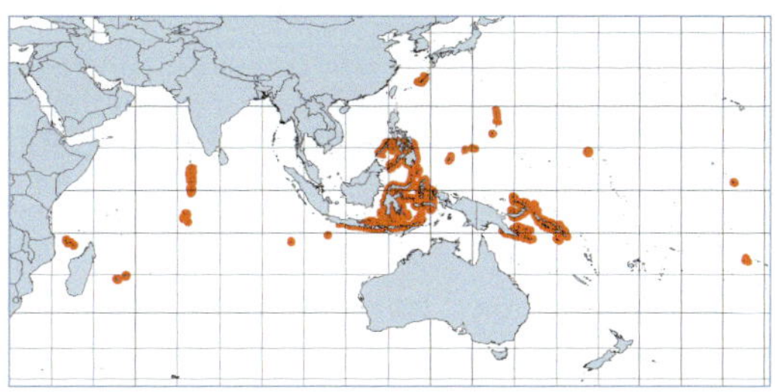

COLOUR IN LIFE Adults mostly yellow, with 10–12 bright blue or violet horizontal stripes on the body, the upper 2 or 3 broken into segments and extending posteriorly onto the dorsal fin; the stripes running onto the peduncle bifurcate and continue onto the caudal fin almost to the posterior margin; head with 3–6 horizontal blue stripes; dark brown maxillary streak present; pelvic fins yellow with blue stripes. Small juveniles (4 cm SL) yellowish brown, shading to lavender ventrally and to yellow at the base of the caudal fin, the yellow continuing as a broad band in each caudal lobe; snout yellow, with a large black spot on each side in front of the nostrils. Larger juveniles (10 cm SL) brownish orange with longitudinal purple stripes on the head and body.

COLOUR IN MARKET Colours remain as in life but are very dull in appearance.

© Phillip C Heemstra

Juvenile *C. polleni* (Mauritius)

© Robert F Myers

Adult *C. polleni* (Palau)

POPULATION GENERAL *C. polleni* is an uncommon species throughout its range.

POPULATION TREND Unknown

MAJOR THREATS No major threats have been identified for this species.

HABITAT AND ECOLOGY GENERAL *C. polleni* is a rare deep-water species found at depths of up to 120 m.

2008 IUCN RED LIST STATUS Least Concern

Authors: *ST Fennessy, PC Heemstra, M Kulbicki and RF Myers*

Cephalopholis polyspila

Randall and Satapoomin 2000

Andaman hind (E), vielle Andaman (F), cherna Andamán (S)

Body depth contained 2.9 times in standard length; body compressed, the width 2.05 times in the depth. Head length 2.3–2.4 times in standard length; dorsal profile of head convex with a slight indentation above posterior edge of orbit; snout length 4.0–4.3 times in head length; orbit diameter 4.7–6.2 times in head length; interorbital width 8.2–10.0 times in head length; caudal peduncle depth 3.1–3.5 times in head length. Gill rakers 8–9 on upper limb, 14–16 on lower limb. Dorsal fin with 11 spines and 15 (15–16) rays; anal fin with 3 spines and 8 rays; all dorsal and anal soft rays branched, the last to base; pectoral rays 16 (16–17), the uppermost not branched; caudal branched rays 15, with 8 procurrent rays in the upper part and 7 procurrent rays in the lower part. Lateral-line scales 47–50 plus 3–4 pored scales on caudal-fin base.*

MAXIMUM RECORDED SIZE 17 cm SL

COLOUR IN LIFE Brown, becoming pale reddish brown on the abdomen, thorax and lower part of the head; numerous small black-edged blue spots on the head, body and fins; margin of the soft portions of the dorsal and anal fins and posterior margin of the caudal fin pale bluish grey with a broad dark brown submarginal band free of blue spots; pectoral fins reddish brown, the black-edged blue spots progressively smaller posteriorly; pelvic fins with brownish red rays and clear membranes with a few dark-edged pale blue spots, the leading edge and tip dark brown. When at rest, 4 dark bands may appear on the posterior half of the body.

COLOUR IN MARKET Maintains similar colouration in death; spots remain prominent, although the body colour may take on a lighter hue.

© John E Randall

Adult *C. polyspila* (Thailand)

© Robert F Myers

Adult *C. polyspila* (Similan Island, Andaman Sea)

* *No line drawing available*

POPULATION General Little is known about the population of this species although it was reportedly common at the type locality (Andaman Sea).

POPULATION TREND Unknown

MAJOR THREATS Threats to this species include incidental catch and habitat destruction.

HABITAT AND ECOLOGY General *C. polyspila* is a reef-associated species found in mixed coral and rubble in depths of 3–20 m.

Cephalopholis polyspila is an eastern Indian Ocean species found in Thailand (near Phuket in the East Andaman Sea, as well as the Similan Islands) and Indonesia (off Sumatra).

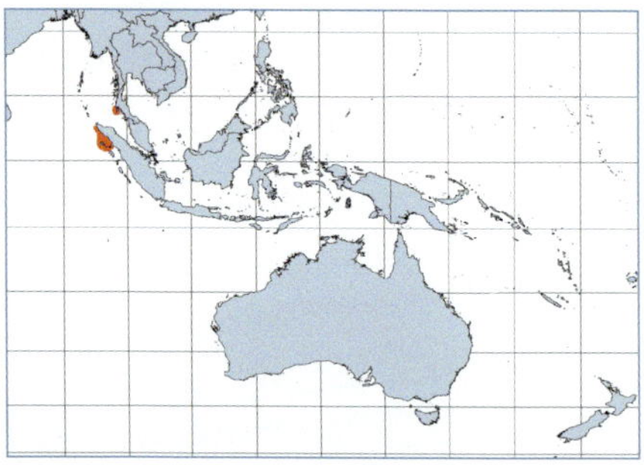

Reproduction In their original description of the species, Randall and Satapoomin (2000) report a mature female (i.e. with visibly hydrated ovaries) of 13 cm SL.

COMMENTS *C. polyspila* is easily confused with *C. cyanostigma* and *C. microprion*. It differs from *C. cyanostigma* primarily in life colour: *C. cyanostigma* has pale spots superimposed on the brown-barred, blue-spotted pattern of its body; it usually has a more reddish to orangish brown ground colour; the pectoral fins are orange-yellow on the outer portion; it has a prominent black spot on the opercular membrane; and the median and pelvic fins are bright yellow in subadults. *C. cyanostigma* also has 16 dorsal soft rays modally whereas *C. polyspila* has 15. *C. polyspila* differs from *C. microprion* in having small blue spots on the body and fins, 16 or 17 (instead of 15–16) pectoral rays, a higher count of scales in the longitudinal series (91–103 in *C. polyspila*, 84–98 in *C. microprion*), small scales dorsally on the snout extending nearly to the upper lip, a more elongate body, and a shorter pectoral fin (contained 1.5–1.8 times in standard length in *C. polyspila*, 1.4–1.5 times in *C. microprion*).

2008 IUCN RED LIST STATUS Least Concern

Authors: *PC Heemstra and RF Myers*

Cephalopholis sexmaculata (Rüppell 1830)

sixblotch hind (E), vielle six tache (F), cherna de seis manchas (S)

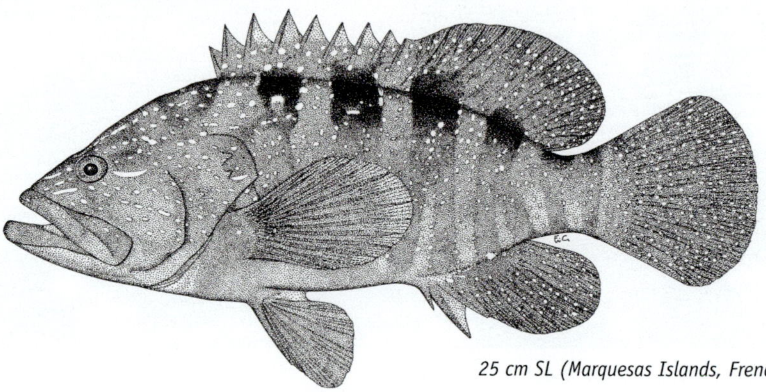

25 cm SL (Marquesas Islands, French Polynesia)

Body depth contained 2.5–3.0 times in standard length. Head length 2.3–2.5 times in standard length; interorbital area flat to slightly convex; dorsal head profile of large specimens (more than 30 cm SL) distinctly concave above the eyes; preopercle rounded, finely serrate in young, virtually smooth in large adults, the lower edge fleshy; subopercle and interopercle with a few small serrae mostly hidden by skin; maxilla scaly, reaching past eye. Gill rakers 7–9 on upper limb, 14–16 on lower limb. Dorsal fin with 9 spines (the fin membranes distinctly indented between them) and 14–16 rays; anal fin rounded, with 3 spines and 9 rays; pectoral-fin rays 16–18; pectoral fins clearly longer than pelvic fins, pectoral-fin length contained 1.4–1.6 times in head length; pelvic-fin length 1.9–2.2 times in head length; caudal fin well rounded. Lateral-body scales ctenoid, without auxiliary scales (just a few in very large specimens); lateral-line scales 49–54; lateral-scale series 95–108.

MAXIMUM RECORDED SIZE 50 cm TL

COLOUR IN LIFE Body orange-red, with small blue spots that are denser on the head and median fins. Head with elongated blue spots and lines; body with 4 dark bars (sometimes very faint) merging dorsally with blackish blotches at the base of the dorsal fin and extending onto the fin. Caudal peduncle with 2 similar but smaller dark bars/blotches, and the spaces between the dark body bars sometimes very pale.

COLOUR IN MARKET Body colour fades to pale orange and spots on body become less prominent. The bars on the body are nearly always absent, but the blackish blotches at the base of the dorsal fin remain for a short time after death.

© John E Randall

Adult *C. sexmaculata* (Marquesas Islands, French Polynesia)

© Robert F Myers

Adult *C. sexmaculata* (Fiji)

Cephalopholis sexmaculata is an Indo-Pacific species, occurring from the Red Sea to South Africa, and eastward to Tahiti (French Polynesia). It has been reported from the Arafura Sea (Russell and Houston 1989) and Yemen (Rouphael et al. 1998). Reports by Heemstra and Randall (1984) from the Gulf of Oman, Pakistan, India and Sri Lanka are unsubstantiated. It is absent from the Persian Gulf and is not yet known from Lakshadweep Islands (India).

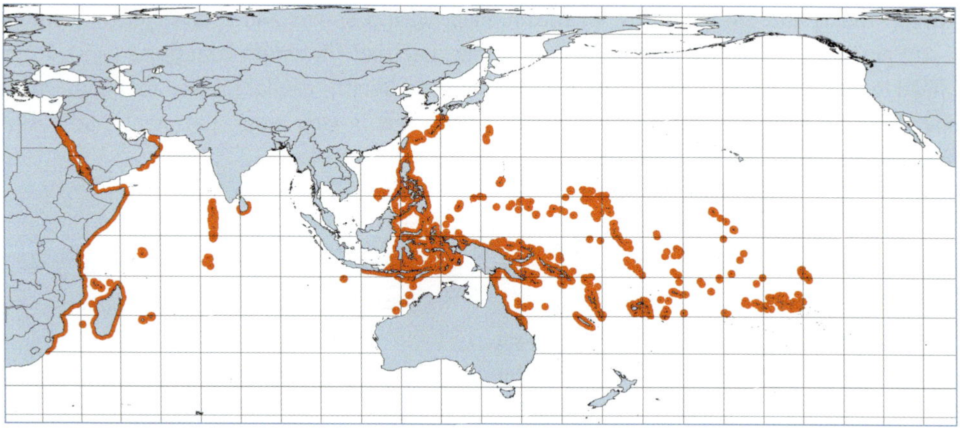

POPULATION **GENERAL** *C. sexmaculata* is a rare species across its range that is locally abundant in specific habitats. **FISHERY DEPENDENT** Fishery-dependent data indicate declines in catches where the species is abundant. In the northern Mariana Islands, landings decreased from 120 kg per year in the 1950s to 25 kg per year in the 1980s, where they remained stable through 2002 (Zeller 2006). *C. sexmaculata* is taken in small-scale fisheries and occasionally found in the live reef fish trade. It is captured for food in artisanal and subsistence fisheries as juveniles (14.4 cm) in Kenya (Mangi and Roberts 2006). **FISHERY INDEPENDENT** Underwater visual censuses in several South Pacific islands indicate that this species rarely reaches a density of more than 0.2 fish per 1000 m^2 (M Kulbicki pers. obs.).

POPULATION TREND Decreasing

MAJOR THREATS The major threat to this species is overfishing.

HABITAT AND ECOLOGY **GENERAL** *C. sexmaculata* is a reef-associated species inhabiting depths of 6–150 m. It is found in coastal and outer reefs, usually along deep walls with caves, with heavy invertebrate growth such as large sponges or soft corals (Kuiter and Tonozuka 2001). The species may also be found inside lagoons (Sluka 2002). *C. sexmaculata* is a secretive species that is active nocturnally in shallow water and diurnally in deeper water (Shpigel and Fishelson 1991b). **FEEDING** Feeds mainly on fishes (Randall and Brock 1960). **AGE AND GROWTH** *C. sexmaculata* reaches at least 8 years (for fishes up to 24 cm SL; Fry et al. 2006). **REPRODUCTION** *C. sexmaculata* reaches sexual maturity at 25 cm SL (Mangi and Roberts 2006).

COMMENTS This species may be confused with *C. miniata* and *C. oligosticta*, especially in markets when dead. *C. miniata* lacks the radiating bars of colour from the eye, and *C. oligosticta* and *C. miniata* both lack the bifurcating body bars present in *C. sexmaculata*.

2008 IUCN RED LIST STATUS Least Concern

Authors: *AS Cabanban, ST Fennessy, M Kulbicki, RF Myers, DA Pollard and BM Yeeting*

Cephalopholis sonnerati (Valenciennes 1828)

tomato hind (E), vielle ananas (F), cherna piña (S)

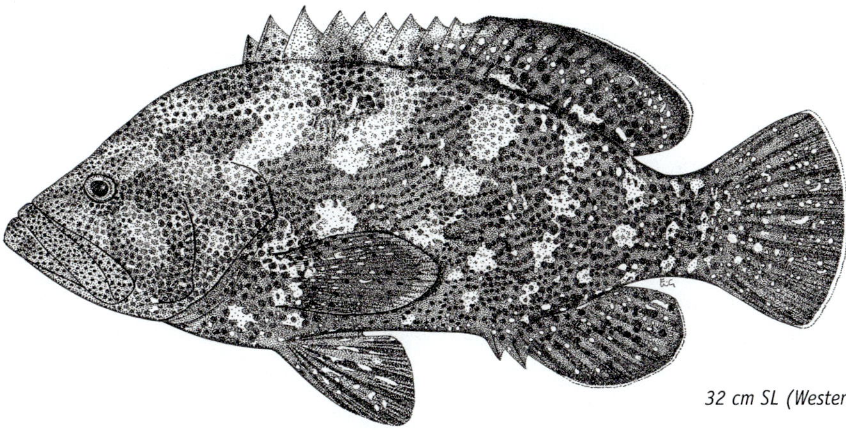

32 cm SL (Western Australia)

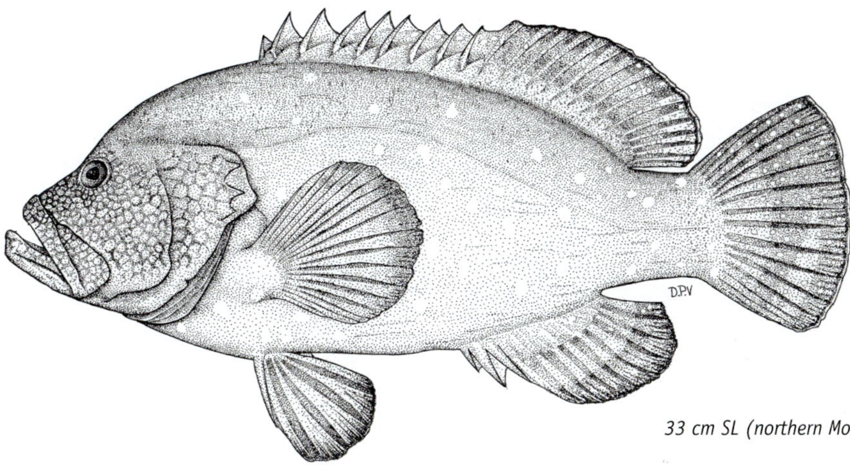

33 cm SL (northern Mozambique)

Body depth greater than or subequal to head length, depth contained 2.3–2.8 times in standard length. Head length 2.5–2.7 times in standard length; anterior head profile of adults concave, the nape distinctly convex; interorbital area flat to slightly convex; preopercle rounded, finely serrate, with a shallow notch just above the angle, the lower edge serrate or irregular with broad spinules, usually exposed; subopercle and interopercle finely serrate; maxilla reaches past eye. Gill rakers 7–9 on upper limb, 14–16 on lower limb. Dorsal fin with 9 spines and 14–16 rays, the membranes distinctly indented between the spines; anal fin with 3 spines and 9 rays; pectoral fins with 18–20 rays; pectoral fins subequal to pelvic fins, pectoral-fin length contained 1.5–1.7 times in head length for fish 9–30 cm SL (in fish larger than 30 cm SL, the pelvic fins are longer than the pectoral fins); pelvic fins reaching or extending beyond anus; caudal fin rounded. Lateral-body scales ctenoid; lateral-line scales 66–80; lateral-scale series 115–134.

MAXIMUM RECORDED SIZE 57 cm TL

COLOUR IN LIFE There are two colour morphs in this species: one that is generally found in the Indian Ocean, and one in the Pacific Ocean. Adults from the Indian Ocean are orange-red to reddish brown, and often have scattered, small whitish or purple spots. There is a purple network of lines on the head, maxilla and lips. The pectoral fins are orange distally, and the membranes of soft dorsal, caudal, anal and pelvic fins are dusky. The dorsal-fin rays are orange distally, and the pelvic-fin tips are blackish. Adults from the Pacific Ocean are generally pale reddish to yellowish brown and covered with small brownish red or dark brown spots. Juveniles are dark reddish brown to nearly black, with the posterior margin of the caudal fin, and sometimes the pectoral fins, whitish. Larger juveniles are brownish orange with scattered pale greenish spots on the body and a black spot between the upper and middle opercular spines.

COLOUR IN MARKET In death, both colour morphs lose their distinguishing features (e.g. spots) and fade to a drab background colour. They are easily distinguished from other groupers by the concave dorsal profile of the head.

© John E Randall

Adult *C. sonnerati* (Mauritius)

© Luiz A Rocha

Adult *C. sonnerati* (spotted phase; Seychelles)

© Robert F Myers

Adult *C. sonnerati* (West Fayu, Federated States of Micronesia)

© Robert F Myers

Subadult *C. sonnerati* (Bali, Indonesia)

Cephalopholis sonnerati is widespread in the Indo-Pacific, occurring from the east coast of Africa to the Line Islands, extending north to southern Japan, and south to southern Queensland (Australia). The species is not reported from the Red Sea or the Persian Gulf.

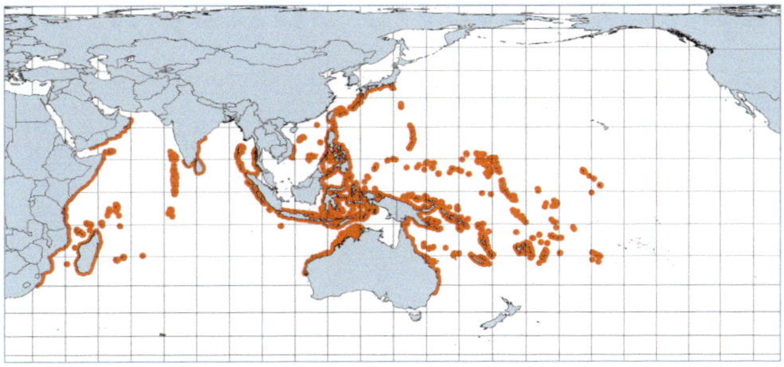

POPULATION **GENERAL** *C. sonnerati* is considered to be a common species throughout its range. **FISHERY DEPENDENT** Fishery-dependent data are numerous for this species in areas that maintain records. It has been noted as a prominent component of fisheries in Mozambique, Somalia, northern Kenya and Madagascar (Morgans 1964, Dengo and David 1993, Darar 1994, van der Elst et al. 1994). While an important component of fisheries, there is no indication of reduced landings over the past few decades. **FISHERY INDEPENDENT** Fishery-independent surveys indicate that this species is rather rare but few surveys have included deep habitat where this species is most commonly found.

POPULATION TREND Stable

MAJOR THREATS The major threat to this species is overfishing.

HABITAT AND ECOLOGY **GENERAL** *C. sonnerati* is a reef-associated species that generally occurs in deep lagoon reefs and steep outer reef slopes in depths of up to 100 m. Juveniles are often found near sponges or coral heads. **FEEDING** *C. sonnerati* feeds on crustaceans and fishes. **AGE AND GROWTH** Kulbicki et al. (2005a) report a length–weight relationship of $W = 0.0145 \times L^{3.058}$ where W is the weight in g and L is the fork length in cm. **REPRODUCTION** Reproductively active fishes have been observed throughout the year in the eastern Indian Ocean, indicating that this species may have a particularly protracted spawning season (Morgans 1964, Nzioka 1979). Shakeel and Ahmed (1997) report that females mature at 28 cm SL, while males mature at 34 cm SL.

COMMENTS This species is easily distinguished from other grouper species by the concave dorsal profile of the head.

2008 IUCN RED LIST STATUS Least Concern

Authors: *ST Fennessy, M Liu and BC Russell*

Cephalopholis spiloparaea

(Valenciennes 1828)

strawberry hind (E), vielle fraise (F), cherna frutillera (S)

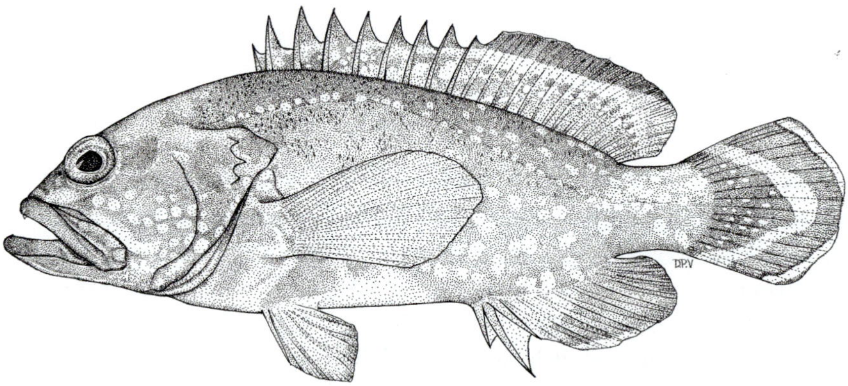

13 cm SL (Mauritius)

Body depth distinctly less than head length, depth contained 2.6–3.2 times in standard length. Head length 2.3–2.5 times in standard length; eye large, its diameter about twice the depth of the preorbital; interorbital area flat; preopercle rounded, very finely serrate, with a shallow notch, the lower edge fleshy; subopercle and interopercle smooth or with minute serrae; maxilla scaly, reaching to or well beyond vertical at rear edge of eye. Gill rakers 7–9 on upper limb, 14–16 on lower limb. Dorsal fin with 9 spines and 14–16 rays; anal fin with 3 spines and 9 rays; pectoral-fin rays 17–19, rarely 19; pectoral fins clearly longer than pelvic fins, pectoral-fin length contained 1.3–1.6 times in head length; pelvic fins not reaching anus, their length 1.9–2.2 times in head length; caudal fin rounded. Lateral-body scales ctenoid, without auxiliary scales; lateral-line scales 47–53; lateral-scale series 84–103.

MAXIMUM RECORDED SIZE 30 cm TL

Cephalopholis spiloparaea is a widespread Indo-Pacific species ranging from East Africa to French Polynesia and the Pitcairn Islands, extending north to the Ryukyu Islands (Japan), and south to Heron Island at the southern end of the Great Barrier Reef (Australia).

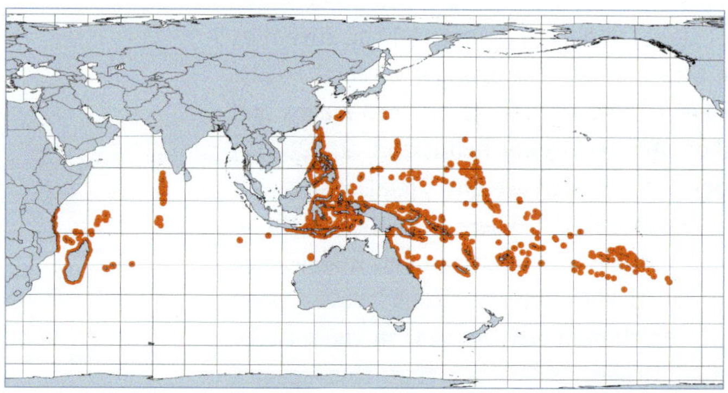

COLOUR IN LIFE Body is pale reddish orange, mottled and blotched. There are faint pale spots usually present on the head, body and median fins. The caudal fin is usually coloured like the body (but distinctly yellowish in some fish from the Comoros Islands), with a bluish white submarginal band at the corners, becoming narrow and marginal (or even disappearing) at the centre of the fin. The ventral margin of the soft portion of the anal fin (and to a lesser extent the dorsal fin) is bluish. Some fish have about 8 faint dark saddle blotches at the base of the dorsal fin and another on the front of the caudal peduncle.

COLOUR IN MARKET The body is a duller shade of red to orange. The spots generally fade, but the posterior margin of the caudal fin retains its dark hue. The saddle blotches at the base of the dorsal fin are sometimes absent.

© Phillip C Heemstra

Adult *C. spiloparaea* (Grand Comoro Island, Comoros)

© Robert F Myers

Juvenile *C. spiloparaea* (Palau)

Adult *C. spiloparaea* (Guam)

© Robert F Myers

POPULATION GENERAL *C. spiloparaea* is a common member of deep reef communities throughout its range. **FISHERY DEPENDENT** Fishery-dependent data are scarce for this species. In Guam, *C. spiloparaea* represented 0.25% of the total offshore catch from 1985 to 2002 (Zeller 2006). **FISHERY INDEPENDENT** Fishery-independent data indicate that this species may be common, for example, in the southern Mariana Islands.

POPULATION TREND Unknown
MAJOR THREATS No major threats have been identified for this species.
HABITAT AND ECOLOGY **GENERAL** *C. spiloparaea* is a reef-associated species found at depths of 16–108 m. The species is known primarily from insular localities. **REPRODUCTION** Donaldson (1995a) described courtship and spawning behaviour of *C. spiloparaea* from Rota, Mariana Islands. The species had male-dominated haremic groups. Daily courtship behaviour began late in the afternoon and proceeded until after sunset. Males repeatedly visited females in single-male, multiple-female mating groups during each period and engaged in courtship bouts. Foraging behaviour by both sexes was virtually absent during daylight and pre-courtship periods.
COMMENTS *C. spiloparaea* is often misidentified as *C. aurantia*. These two species differ in colour (*C. aurantia* is generally orange-yellow or golden, with red to orange dots on the head and dorsally on the body), and the pelvic fins of *C. aurantia* are longer, usually reaching the anus, their length being contained 1.6–2.0 times in head length.
2008 IUCN RED LIST STATUS Least Concern

Authors: *ALB Gaspar, RF Myers and BP Ferreira*

Cephalopholis taeniops (Valenciennes 1828)

African hind (E), mérou africain (F), cherna colorada (S)

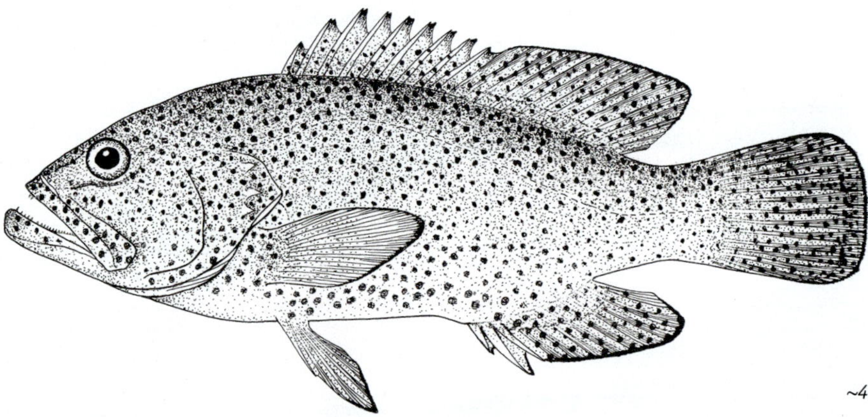

~40 cm TL
© FAO

Body depth less than head length, depth contained 2.8–3.2 times in standard length. Head length 2.7–3.0 times in standard length; interorbital area convex; preopercle rounded, finely serrate, with a shallow notch, the lower edge fleshy; subopercle and interopercle with a few small serrae covered by skin; maxilla reaching vertical at rear edge of eye or beyond. Gill rakers 8 on upper limb, 15 or 16 on lower limb. Dorsal fin with 9 spines and 15 or 16 rays, the fin membranes distinctly indented between the spines; anal fin with 3 spines and 9 or 10 rays; pectoral-fin rays 18 or 19; pectoral fins longer than pelvic fins, pectoral-fin length contained 1.5–1.7 times in head length; pelvic fins reaching or nearly reaching anus; caudal fin rounded. Lateral-body scales ctenoid, without auxiliary scales; lateral-line scales 68–72; lateral-scale series 114–122.

MAXIMUM RECORDED SIZE 40 cm TL

COLOUR IN LIFE Generally reddish orange, the head, body and the median fins covered with small blue spots. The fins are blackish distally, and the soft dorsal, caudal and anal fins have a narrow bluish margin. There is a horizontal blue line just below the eye. According to Séret (1981) there is a black variety that also has blue spots but it is much less common than the orange variety.

COLOUR IN MARKET Generally a pale orange to pink; the spots on the median fins remain and the dark colour on their posterior margin is usually present.

© Rui Freitas

Adult *C. taeniops* (Cape Verde)

Adult *C. taeniops* (São Tomé)

© Luiz A Rocha

Cephalopholis taeniops is an eastern Atlantic species found from the western Sahara to Angola, including Cape Verde, São Tomé and Príncipe islands. It was recently reported in the Mediterranean (Ben Abdallah et al. 2007).

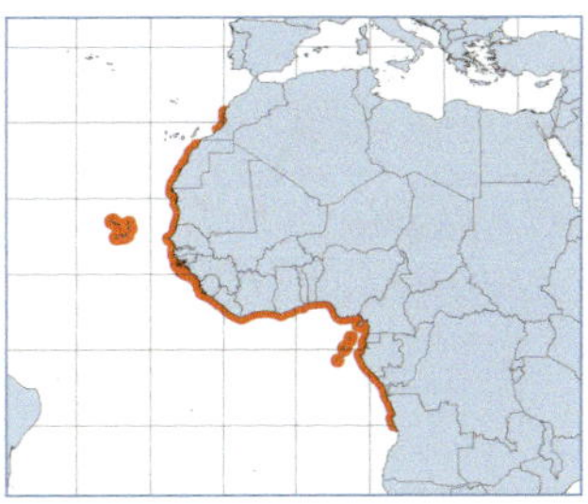

POPULATION GENERAL *C. taeniops* is relatively common and abundant at São Tomé and Cape Verde. Elsewhere in its range there are no data on its population. FISHERY DEPENDENT No explicit fishery-dependent data are currently available, but the species is of some importance to commercial fisheries in Senegal and Dakar.

POPULATION TREND Unknown

MAJOR THREATS There are no major threats identified for this species.

HABITAT AND ECOLOGY GENERAL *C. taeniops* is found on tropical rocky reefs and sandy bottoms in depths of 20–200 m. REPRODUCTION *C. taeniops* is a diandric protogynous hermaphrodite (Siau 1994). Females attain sexual maturity at 18 cm TL. In Senegal, the spawning season is from July to October, peaking in August when the water temperature reaches 26 °C or higher.

2008 IUCN RED LIST STATUS Data Deficient

Authors: *AA Bertoncini, JH Choat, MT Craig, BP Ferreira, M Liu and LA Rocha*

Cephalopholis urodeta (Forster 1801)

tailstriped hind (E), vielle aile noire (F), cherna alinegra (S)

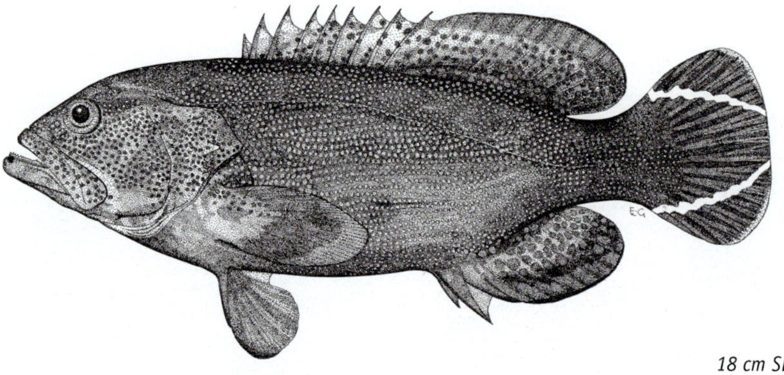

18 cm SL (Taiwan)
© SAIAB

Body depth less than head length, depth contained 2.7–3.3 times in standard length. Head length 2.4–2.7 times in standard length; interorbital area convex; preopercle rounded, finely serrate, the lower edge fleshy; subopercle and interopercle usually smooth; maxilla reaching well past eye. Gill rakers 7–9 on upper limb, 15–17 on lower limb. Dorsal fin with 9 spines and 14–16 rays; anal fin with 3 spines and 9 (rarely 8) rays; pectoral-fin rays 17–19; pectoral fins distinctly longer than pelvic fins, pectoral-fin length contained 1.3–1.7 times in head length; pelvic fins usually not reaching anus, pelvic-fin length 1.8–2.3 times in head length; caudal fin rounded. Snout and maxilla with minute cycloid scales, lateral-body scales ctenoid, without auxiliary scales; lateral-line scales 54–68; lateral-scale series 88–108.

MAXIMUM RECORDED SIZE 28 cm TL

Cephalopholis urodeta is widely distributed within the Pacific Ocean, occurring eastward to French Polynesia and the Pitcairn Islands, and to Christmas Island in the Indian Ocean.

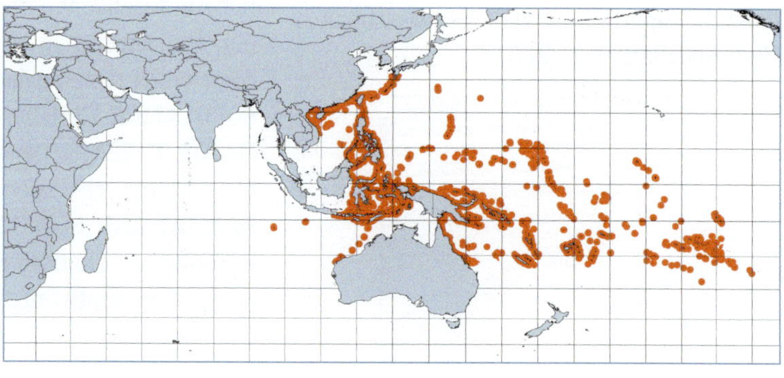

COLOUR IN LIFE Reddish brown to brownish red, darker posteriorly; body sometimes with small pale spots and/or 4 faint irregular dark bars that split ventrally and another 2 dark bars on the caudal peduncle; head with numerous small close-set orange-red spots (more evident on the paler parts of the head) and sometimes with irregular reddish brown blotches; dark spot between the upper and middle opercular spines (more evident on juveniles); a pair of small dark spots on the edge of the lower lip in line with the pair of anterior canine teeth; soft dorsal and anal fins with small orange-red spots; outer triangular part of the interspinous dorsal-fin membranes orange; pelvic fins orange-red, usually with a dark blue edge. Caudal fin (of fish larger than 4 cm SL) with 2 white to bluish white bands that converge posteriorly, the area between the bands dark, often with pale spots, the outer corners of the fin red, with a whitish margin. Pectoral fins of Pacific fish reddish brown basally, becoming yellow distally.
COLOUR IN MARKET Retains similar colour although the colour may fade to dull red or brown.

Adult *C. urodeta* (Okinawa Fish Market, Japan)

Adult *C. urodeta* (Marquesas Islands, French Polynesia)

Adult *C. urodeta* (Fiji)

POPULATION **General** *C. urodeta* is one of the most abundant species of groupers throughout its range. **Fishery dependent** While small in comparative size, *C. urodeta* is a component of both the live and freshly dead food trade. **Fishery independent** Fishery-independent data are scarce but in some areas local declines are indicated (e.g. Philippines).
POPULATION TREND Unknown
MAJOR THREATS The major threat to this species is habitat destruction. *C. urodeta* shows a clear preference for healthy coral

reefs in shallow areas, and, as such, is greatly affected by reef degradation. Fishing pressure is a second major threat to this species.

HABITAT AND ECOLOGY GENERAL *C. urodeta* is a reef-associated species that is not tolerant of habitat degradation. The species prefers clear, shallow waters of outer reef areas to 60 m. It is occasionally found in lagoons, back-reef areas and on the reef-top. FEEDING Feeds mainly on small fishes and crustaceans. REPRODUCTION Sexual maturity occurs by at least 17 cm TL (Nakai and Sano 2002).

COMMENTS The Indian Ocean population of this species was recently confirmed to be a distinct species (*C. nigripinnis*; Craig and Hastings 2007). That species lacks the white 'V' on the caudal fin and shows a distinct genetic signature. *C. urodeta* was introduced into Hawaii from 1958 to 1961 as a means of increasing sportfishing, but it did not become established and is no longer present there (Randall 1987).

2008 IUCN RED LIST STATUS Least Concern

Authors: *MT Craig, PC Heemstra, M Liu, DA Pollard, BC Russell and BM Yeeting*

Cromileptes altivelis (Valenciennes 1828)

humpback grouper (E), mérou bossu (F), mero jorobado (S)

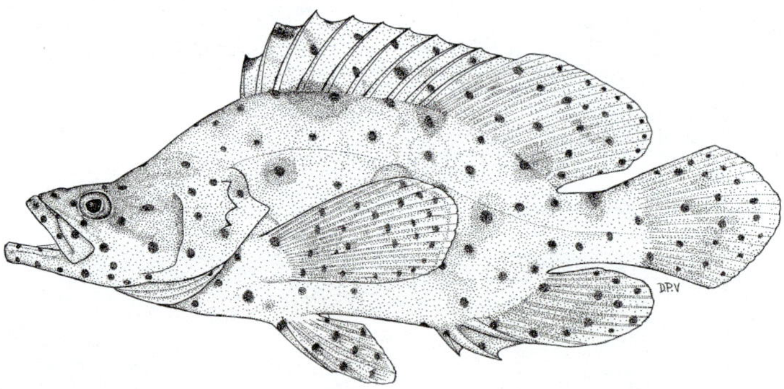

~30 cm SL
© SAIAB

Body compressed, the depth less than head length and contained 2.6–3.0 times in standard length (for fish 12–37 cm SL); body width 2.4–2.6 times in the depth. Head length 2.5–2.8 times in standard length; dorsal head profile distinctly concave, rising steeply at the nape; preorbital very narrow, its depth about 5 times in eye diameter and 30–32 times in head length; preopercle finely serrate, the serrae at the angle slightly enlarged, the lower edge smooth; opercle with middle spine inconspicuous, the upper and lower spines rudimentary; posterior nostril a large, crescentic, vertical slit; maxilla extending to below rear half of eye, no step or knob on ventral edge of maxilla; supramaxilla well developed; jaws with bands of villiform teeth; no canines; vomer and palatines with teeth. Gill rakers short, 8–11 on upper limb, 13–17 on lower limb. Dorsal fin with 10 spines and 17–19 rays, the fin origin over the opercle, the fin membranes not incised between the spines, the posterior spines longest and the soft rays even longer; anal fin with 3 spines and 9 or 10 rays; pectoral fins rounded, with 17 or 18 rays, the middle rays longest; caudal fin rounded, with 8 branched rays and 8 procurrent rays in the upper part and 7 branched rays and 8 procurrent rays in the lower part. Scales on body smooth (the ctenii greatly reduced); lateral-line scales 54–62; lateral-scale series 106–122. Pyloric caeca 13.

MAXIMUM RECORDED SIZE 71 cm TL

COLOUR IN LIFE Pale greenish brown to white, with widely spaced round black spots on the head, body and fins; some spots on the body and base of the median fins overlain by a large dusky blotch. Black spots on juveniles fewer than on adults and may be as large as or larger than the eye.

COLOUR IN MARKET Body colour generally pale white to grey with prominent black spots.

© Patrice Marker

Juvenile *C. altivelis*
(Lambeh Strait, Sulawesi, Indonesia)

© Matthew T Craig

Adult *C. altivelis* (Okinawa Fish Market, Japan)

© Robert F Myers

Adult *C. altivelis* (Dampier Strait, West Papua, Indonesia)

POPULATION **GENERAL** *C. altivelis* is naturally uncommon over most of its range. The species is a favoured target for the live reef food fish trade; it is also commonly found in the aquarium trade. **FISHERY DEPENDENT** Fishery-dependent data are scarce as many countries do not record species-specific data. In 1998, only 0.02% of the live fish export from Papua New Guinea was comprised of *C. altivelis*. In Hong Kong, the volume of imported *C. altivelis* for the live reef food fish trade ranged from over 133 kg to over 8700 kg for the period 1997–2005, with an estimated import value of HK$9 950 000 (~US$1 276 000) at its peak (Hong Kong Census and Statistics Department). **FISHERY INDEPENDENT** Pears (2005) notes the difficulty of obtaining robust population density estimates for this somewhat cryptic species. Density values recorded on the Great Barrier Reef (Australia) range from 0.2 to 0.5 fish per 1000 m^2 (Pears 2005).

POPULATION TREND Decreasing

MAJOR THREATS The major threat to this apparently uncommon species is fishing pressure; it is one of the most highly valued species in the live fish trade (Sadovy et al. 2003) and is heavily sought after because it provides a high profit margin for traders.

HABITAT AND ECOLOGY **GENERAL** *C. altivelis* is found in lagoons and seaward reefs in dead or silty areas adjacent to well-developed coral habitat. It also occurs in tide pools and can be caught at depths of up to 40 m. Juveniles (<15 cm) are found inshore, in lagoons and in fringing reefs and seagrass. **FEEDING** Feeds on fishes and crustaceans. **AGE AND GROWTH** Davies et al. (2006) provide age and growth data for fishes on the Great Barrier Reef (Australia); samples ranged in size from 350 to 710 mm FL and were between 1 and 19 years of age. Von Bertalanffy growth parameters were calculated as follows: L_{max} = 610 mm FL, L_{∞} = 738 mm FL, k = 0.08 y^{-1}, and Z = 0.28. **REPRODUCTION** *C. altivelis* is a monandric protogynous hermaphrodite (Gardner et al. 2005) and females mature at between 30 and 40 cm TL, and between 1 and 2 years (Lau and Li 2000, Davies et al. 2006, Liu and Sadovy de Mitcheson 2009a). *C. altivelis* can spawn many times within a reproductive season (Pomeroy et al. 2006). The species can be successfully hatchery-produced at commercial levels, mainly for the aquarium trade; its growth appears to be rather slow to regularly supply the live food trade. Spawning activities in floating cages among 15 males and 24 females (in two cages) were observed in Komodo, Indonesia, from October 2000 to July 2003. Spawning occurred in pairs between 21:00 and midnight from the third quarter to the first quarter of the moon phase. Cage spawning was observed to last for at least eight consecutive days (Sudaryanto et al. 2004).

Cromileptes altivelis is known in the western Pacific from southern Japan (Ogasawara) southwards to southern Queensland (Australia) and northern Western Australia, and eastwards to Fiji. Records from Hawaii are probably based on released aquarium fish (Heemstra and Randall 1993).

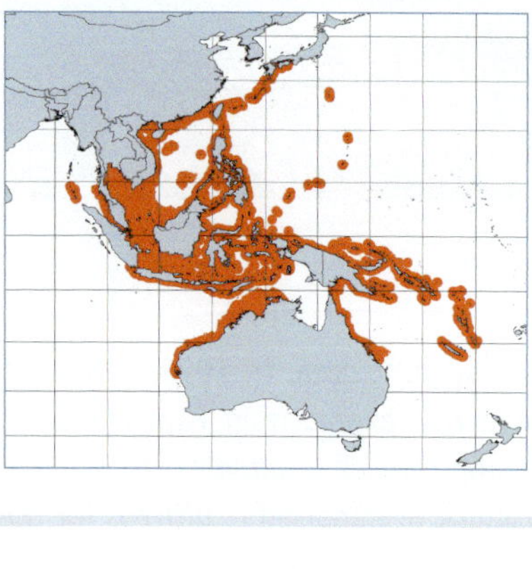

COMMENTS Craig and Hastings (2007) demonstrate a close, nested relationship between this species and members of the genus *Epinephelus*, and suggest that this species is best regarded as a member of that genus. The obvious differences in skull configuration, the presence of 10 dorsal fin spines and the absence of canines indicate that this species is no more allied to *Epinephelus* than it is to *Cephalopholis* or *Dermatolepis*. These apomorphic differences are confounding given the close relationships to *Epinephelus* indicated by molecular analysis and the topic is in need of further study (Craig and Hastings 2007).

2008 IUCN RED LIST STATUS Vulnerable (A4cd)

Authors: *AS Cabanban, T-C Chan, JH Choat and YJ Sadovy de Mitcheson*

Dermatolepis dermatolepis

(Boulenger 1895)

leather bass (E), mérou cuir (F), mero coriáceo (S)

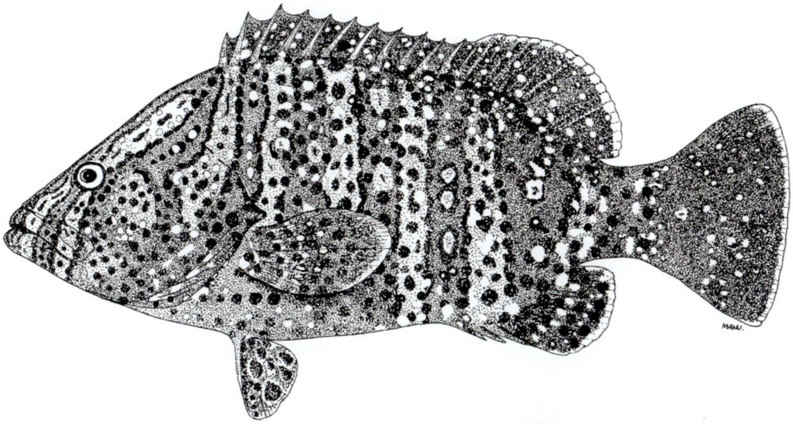

~80 cm TL
© FAO

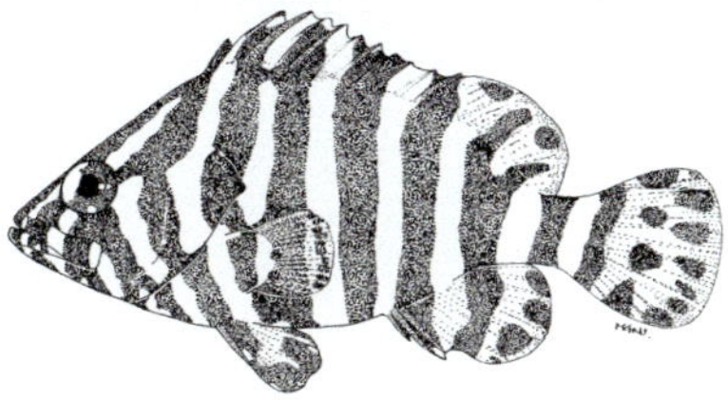

~30 cm TL
© FAO

Body depth usually greater than head length, depth contained 2.1–2.5 times in standard length. Head steep in dorsal profile. Eye diameter much less than snout length, contained 5.2–8.0 times in head length. Gill rakers 21–24 (total). Dorsal fin with 11 spines and 18–20 rays; anal fin with 3 spines and 9 rays; pectoral fins with 19 or 20 rays; pectoral fins short, their length 18–26% of standard length; caudal fin rounded. Scales smooth, mostly covered by skin; lateral-line scales 62–67; lateral-scale series difficult to count.

MAXIMUM RECORDED SIZE 100 cm TL

Dermatolepis dermatolepis is an eastern Pacific species distributed from southern California (USA) to Ecuador, including the Revillagigedo and Galápagos Islands, and Cocos, Malpelo and Clipperton islands.

POPULATION GENERAL *D. dermatolepis* is locally abundant throughout its range, especially at insular locations.

POPULATION TREND Unknown

MAJOR THREATS No major threats have been identified for this species. It is, however, recreationally targeted and incidentally caught along with other grouper species. At present, fishing is apparently not at a level constituting a major threat. If spawning aggregations are targeted in the future, this could represent a major threat, particularly given the small size of the aggregations.

COLOUR IN LIFE Adults with ground colour grey or brownish, with irregular white spots and blotches on dark bars and small dark spots on pale interspaces; large adults with a yellow margin on the posterior parts of the median fins. Juveniles white with several black bars on the body and extending into the dorsal and anal fins; head with similar black bands; paired fins and distal parts of the median fins with black spots.

COLOUR IN MARKET Generally mottled grey to brown with irregular white spots and blotches; the yellow margin on the median fins fades but remains visible.

© Octavio Aburto

Adult *D. dermatolepis* (Gulf of California, Mexico)

© Octavio Aburto

Juvenile *D. dermatolepis* (Gulf of California, Mexico)

© Octavio Aburto

Subadult *D. dermatolepis* (Gulf of California, Mexico)

HABITAT AND ECOLOGY **General** *D. dermatolepis* is a reef-associated species inhabiting rocky reefs in depths of up to 40 m. **Feeding** *D. dermatolepis* is a diurnal predator that feeds on small benthic fishes and occasionally on crustaceans. **Reproduction** *D. dermatolepis* forms small spawning aggregations during winter months (November–February; Aburto-Oropeza and Hull 2008, Erisman et al. 2009). **2008 IUCN RED LIST STATUS** Least Concern

Authors: *JH Choat, MT Craig, BP Ferreira and LA Rocha*

Dermatolepis inermis (Valenciennes 1833)

marbled grouper (E), mérou marbré (F), mero mármol (S)

~60 cm TL
© FAO

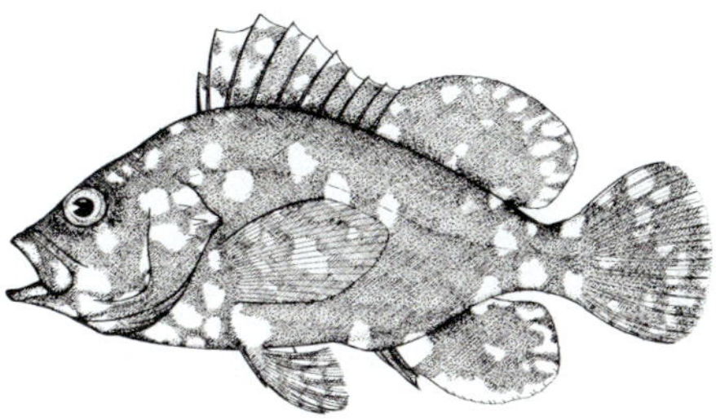

~25 cm TL
© FAO

Body deep and compressed, the greatest depth distinctly more than head length, contained 2.2–2.5 times in standard length; body width 2.2–2.9 times in the depth. Head length 2.5–2.8 times in standard length; preopercle rounded, finely serrate, the serrae at the angle slightly enlarged. Gill rakers 19–22 (total). Dorsal fin with 11 spines and 18–20 rays, the membranes only slightly incised between the spines and the 3rd or 4th spine longest; anal fin with 3 spines and 8–10 rays, the posterior margin falcate in large adults; pectoral-fin length 29–35% of standard length, pectoral-fin rays 18 or 19; caudal fin rounded in juveniles, truncate or concave in adults. Scales smooth and deeply embedded.

MAXIMUM RECORDED SIZE 91 cm TL

POPULATION **GENERAL** *D. inermis* is relatively rare throughout its range and rarely seen in markets. **FISHERY DEPENDENT** *D. inermis* has shown steady declines in landings for some time. According to the US National Marine Fisheries Service records of annual landings of commercial catches in the USA from 1986 to 2004, the quantity caught ranged from 0.4 t in 1987 to 19.7 t in 1993. From 1999 to 2004, the average annual landings were less than 1.4 t. **FISHERY INDEPENDENT** Fishery-independent data confirm the natural rarity of this species.

COLOUR IN LIFE Adults mottled greyish brown with white speckles and small black spots that tend to be arranged in rings; juveniles black or dark brown covered with irregular white spots and blotches.

COLOUR IN MARKET *D. inermis* maintains nearly identical colouration in death; large adults may be uniform brown to grey and lack irregular blotches.

© NOAA

Adult *D. inermis* (Gulf of Mexico)

© Matthew T Craig

Adult *D. inermis* (Florida, USA)

© Áthila Bertoncini Andrade

Adult *D. inermis* (locality unknown)

Dermatolepis inermis is a western Atlantic species ranging from North Carolina (USA) to Venezuela, including the Gulf of Mexico and the Caribbean, and from northern Brazil to Rio de Janeiro, including the offshore islands.

POPULATION TREND Decreasing

MAJOR THREATS Fishing pressure is the major threat to this species. While it supports only a small fishery, existing data indicate that the species is undergoing decline and is not under effective management. Although the species is widespread and has a large depth range, it is rare, and apparently may only be found in part of that range today. It is subjected to fishing pressure including on at least one spawning aggregation off the Louisiana coast (USA).

HABITAT AND ECOLOGY **General** *D. inermis* is a reef-associated species usually found on deep ledges at depths of up to 210 m.

2008 IUCN RED LIST STATUS Near Threatened

Authors: *AA Bertoncini, JH Choat, MT Craig, BP Ferreira, LA Rocha and CLS Sampaio*

Dermatolepis striolata (Playfair 1867)

smooth grouper (E), mérou lisse (F), mero liso (S)

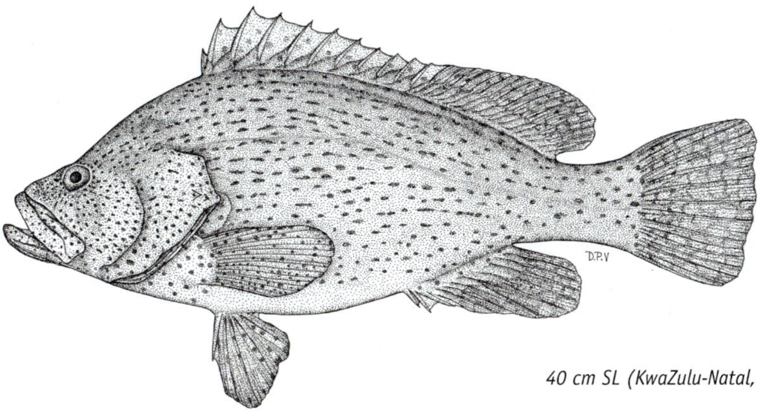

40 cm SL (KwaZulu-Natal, South Africa)

Body depth greater than head length, depth contained 2.4–2.6 times in standard length. Eye diameter much less than snout length, contained 7.2–8.1 times in head length; preopercle serrate, but the serrae are covered by skin; opercular spines not readily visible. Dorsal head profile slightly concave or straight. Gill rakers 5–7 on upper limb, 13–16 on lower limb, 18–22 total. Dorsal fin with 11 spines and 18 or 19 rays, the fin origin over the opercle and the membranes not incised between the spines; anal fin with 3 spines and 9 or 10 rays; pectoral fins with 17–19 rays, pectoral-fin length 23–27% of standard length; caudal fin rounded or truncate with rounded corners. Scales smooth, mostly covered by skin and auxiliary scales; lateral-line scales 69–71; lateral-scale series difficult to count, 118–124.

MAXIMUM RECORDED SIZE 85 cm TL

COLOUR IN LIFE Yellowish to reddish brown, paler ventrally; adults with small round dark spots all over the head, body and fins; juveniles with numerous small dark brown spots on the head and body where they are often horizontally elongate (sometimes forming short lines), and on the fins where they are less distinct; head and body also with fairly distinct irregular pale blotches or rosettes of various sizes.

COLOUR IN MARKET *D. striolata* maintains nearly identical colouration in death, but may lose some of the irregular blotches on the body.

© Margaret M Smith

Adult *D. striolata* (locality unknown)

Adult *D. striolata* (Oman)

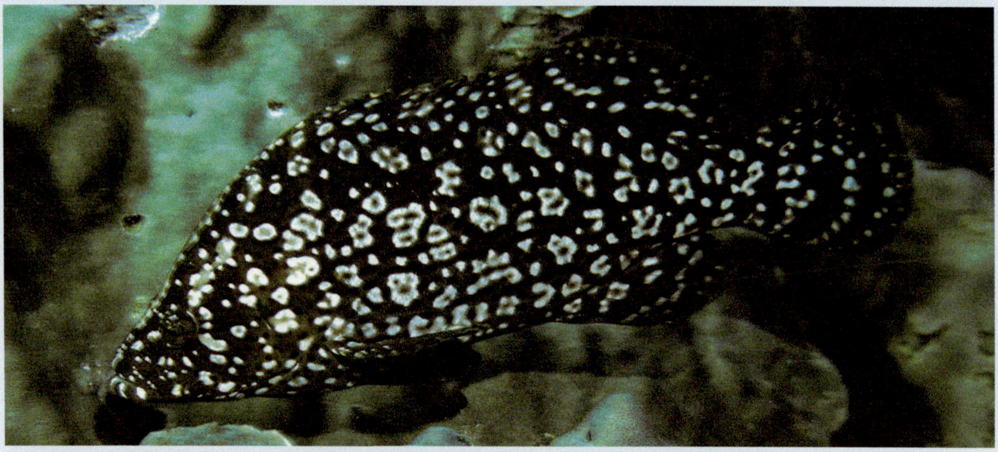

© Klaus Fiedler

Dermatolepis striolata is known in the western Indian Ocean from the Gulf of Oman and southern coast of the Arabian Peninsula, Aldabra (Seychelles), Comoros Islands, Madagascar, and the coast of East Africa from Kenya to South Africa (Durban).

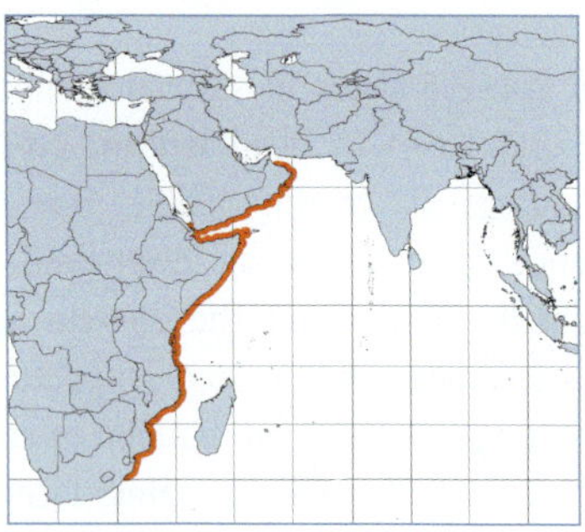

POPULATION GENERAL *D. striolata* is a rare species with locally abundant populations (e.g. Oman).

POPULATION TREND Unknown

MAJOR THREATS No major threats have been identified for this species.

HABITAT AND ECOLOGY GENERAL *D. striolata* inhabits hard substrates in depths to at least 15 m.

2004 IUCN RED LIST STATUS Data Deficient

Author: *AS Cornish*

Epinephelus adscensionis (Osbeck 1765)

rock hind (E), mérou oualioua (F), mero cabrilla (S)

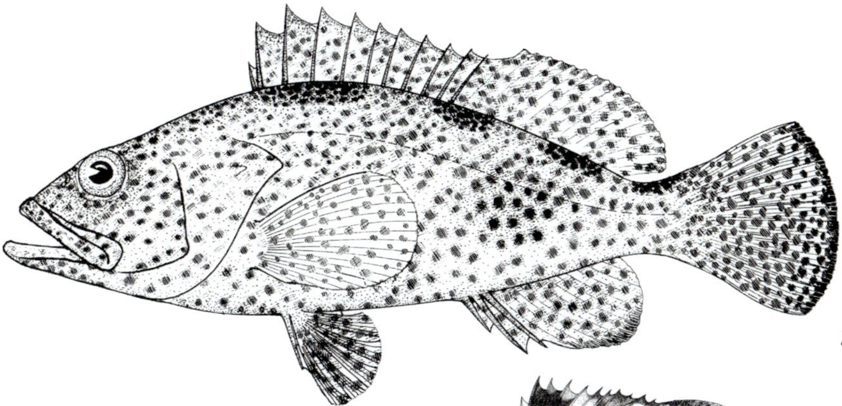

29 cm TL
© FAO

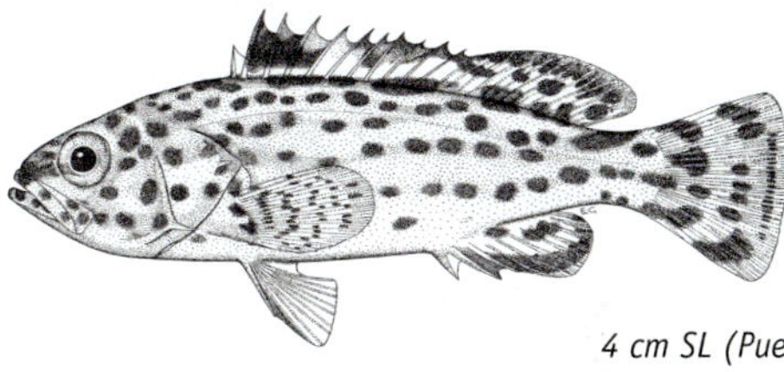

4 cm SL (Puerto Rico)
© SAIAB

Body depth less than head length, depth contained 2.6–3.2 times in standard length. Head length 2.1–2.5 times in standard length; interorbital area flat or slightly concave; preopercle evenly serrate, without obvious angle; subopercle and interopercle smooth; posterior and anterior nostrils subequal. Gill rakers 7–9 on upper limb, 16–19 on lower limb, 23–28 total. Dorsal fin with 11 spines and 16–18 rays, the 4th or 5th spine longest and the interspinous membranes distinctly incised; anal fin with 3 spines and 8 rays; pectoral-fin rays 18–20; pectoral fins longer than pelvic fins, pectoral-fin length contained 1.5–2.1 times in head length; pelvic-fin length 1.8–2.3 times in head length for fish 10–19 cm SL, 2.2–2.7 times in head length for fish 20–38 cm SL; posterior margin of caudal fin convex. Lateral-body scales distinctly ctenoid, with auxiliary scales; lateral-line scales 48–53; lateral-scale series 92–108.

MAXIMUM RECORDED SIZE 60 cm TL

Epinephelus adscensionis is a wide-ranging species in the Atlantic. It is known from Ascension and St Helena islands in the eastern central Atlantic, and from Massachusetts (USA) to southern Brazil in the western Atlantic (including Vila dos Remedios, Abrolhos, Fernando de Noronha and Trinidade islands). It has also been recorded in the eastern Atlantic from São Tomé in the Gulf of Guinea (Heemstra and Randall 1993) and from the Azores (Portugal) (RF Myers unpublished data).

POPULATION **GENERAL** *E. adscensionis* is not a common species in general but may be locally abundant. **FISHERY DEPENDENT** *E. adscensionis* is not commonly taken by major fisheries except in isolated areas where it is abundant (e.g. St Helena and Ascension islands). **FISHERY INDEPENDENT** Fishery-independent data suggest that this species is common but may undergo depletions from fishing pressure. **GENETICS** Genetic data suggest a strong population separation between the USA (Florida) and other Atlantic localities (Carlin et al. 2003).

POPULATION TREND Decreasing

MAJOR THREATS The major threat to this species is fishing which is likely to intensify as larger species of groupers become overfished.

HABITAT AND ECOLOGY **GENERAL** *E. adscensionis* occurs on rocky reefs in depths of 2–100 m. **FEEDING** Feeds mainly on crabs and fishes.

COLOUR IN LIFE Head, body and fins generally buff or pale greenish, covered with reddish brown spots and scattered pale blotches. Three to 5 dark brown blotches (groups of dark spots) at the base of the dorsal fin and a blackish brown blotch on top of the caudal peduncle. Posterior edge of caudal fin with a row of dark brown spots forming a dark margin; small juveniles with fewer but larger dark spots on the head, body and fins.

COLOUR IN MARKET Body colour pale with prominent dark spots. Black blotches at the base of the dorsal fin and on the caudal peduncle are distinct.

© Matthew T Craig

Adult *E. adscensionis* (Dry Tortugas, Florida, USA)

© Yvonne J Sadovy de Mitcheson

Juvenile *E. adscensionis* (Puerto Rico)

© Robert F Myers

Adult *E. adscensionis* (Tamandare, Brazil)

Age and growth *E. adscensionis* may reach ages of 11 years. The von Bertalanffy growth equation was reported as $L_t = 499.4(1 - e^{-0.167(t + 2.495)})$, where *t* is the age in years and *L* is the total length in mm, and the length-weight relationship as $W = 6 \times 10^{-9}L^{3.193}$, where *W* (whole weight) is in kg and *L* (total length) is in mm (Potts and Manooch 1995). **Reproduction** Reproductively active females of 28–36 cm TL have been observed from January through June in Florida (Bullock and Smith 1991).
2008 IUCN RED LIST STATUS Least Concern

Authors: *AA Bertoncini, JH Choat, MT Craig, BP Ferreira and LA Rocha*

Epinephelus aeneus (Geoffroy St Hillaire 1817)

white grouper (E), mérou blanc (F), cherna de ley (S)

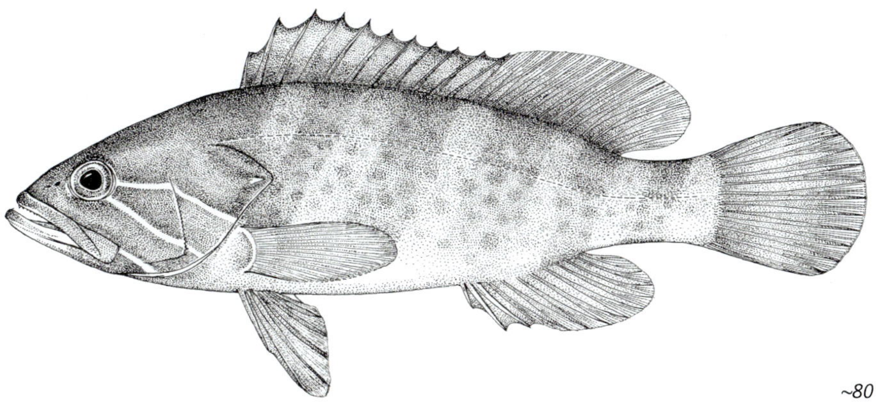

~80 cm TL
© FAO

Body depth distinctly less than head length, depth contained 3.0–3.6 times in standard length. Head length 2.5–2.9 times in standard length; interorbital area convex; preopercle angular, with 3–6 large spines at the angle, the lowermost directed ventrally; posterior nostril slightly bigger than anterior nostril; maxilla reaches nearly to vertical at rear edge of eye. Gill rakers 8–10 on upper limb, 15–17 on lower limb, 23–26 total. Dorsal fin with 11 spines and 14–16 rays, 3rd or 4th spine longest, the interspinous membrane only slightly incised between the spines; anal fin with 3 spines and 7–9 rays; pectoral-fin rays 18 or 19, longest contained 1.5–1.7 times in head length; pelvic-fin origin below base of pectoral fins; caudal fin rounded. Body scales ctenoid; lateral-line scales 67–72; lateral-scale series 98–102. Pyloric caeca 12–14.
MAXIMUM RECORDED SIZE 120 cm TL

POPULATION General *E. aeneus* was once considered a common species throughout its range. However, in the last few decades, marked declines in abundance have been noted, primarily due to fishing effort for this valuable species. **Fishery dependent** *E. aeneus* is a particularly popular species in fisheries throughout its range. Reliable catch data have shown reduction in productivity of more than 60% despite increased effort in Senegal (Gascuel and Ménar 1997), and similar trends are reported elsewhere. *E. aeneus* is considered to be overexploited (Gascuel and Ménar 1997), with the fishery operating above the maximum sustainable yield.
POPULATION TREND Decreasing

Epinephelus aeneus occurs throughout the southern Mediterranean and along the western coast of Africa to southern Angola, including the islands in the Gulf of Guinea.

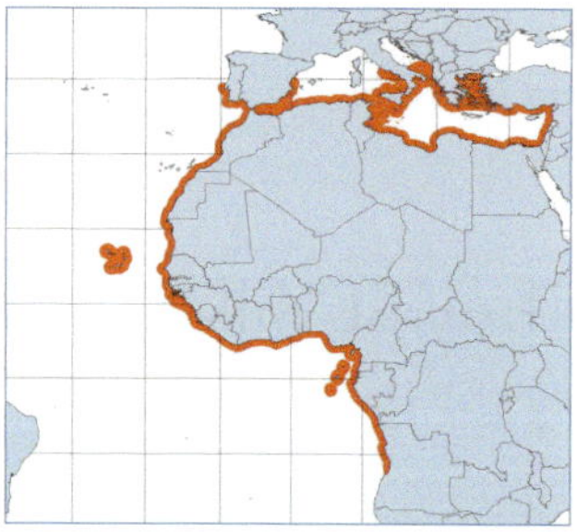

MAJOR THREATS The major threat to *E. aeneus* is overfishing and lack of effective management.

HABITAT AND ECOLOGY **GENERAL** *E. aeneus* occurs on rocky or mud-sand bottoms to depths of 200 m; juveniles have been taken in coastal lagoons and estuaries. **FEEDING** Longhurst (1960) reported the diet of *E. aeneus* in West African waters as fish (58%), stomatopods (21%), crabs (10%), and cephalopods (10%). In Senegal, *E. aeneus* feeds on fishes and invertebrates (Diatta et al. 2003). **AGE AND GROWTH** Age and growth studies in the Gulf of Gabès (Tunisia) estimate the von Bertalanffy parameters as

COLOUR IN LIFE Greenish bronze, the fins darker, brownish violet, bordered with white or pale mauve. Three or 4 pale blue or white lines across the operculum, the lowest from the posterior end of the maxilla to the interopercle, the next from the eye across the preopercle just above the angle and onto the subopercle, the uppermost line from the eye to the upper end of the preopercle where it usually bifurcates and continues to the posterior edge of the operculum. Some adults are uniformly greenish silver, with the white lines on the head very faint. Juveniles with faint dark spots on the body forming 5 indistinct dark bars. Fins also with faint dark spots. In large adults the white lines on the head may be indistinct.

COLOUR IN MARKET Generally dark brown, the median and pelvic fins remain dark. Light blotches may develop. Pale lines on the head are markedly visible and generally do not fade. Abdomen is generally white.

© Peter Wirtz

Adult *E. aeneus* (São Tomé)

© Peter Wirtz

Adult *E. aeneus* (São Tomé)

L_{∞} = 204.34 cm, k = 0.039 y^{-1}, t_0 = −0.767 (Bouain 1986). **REPRODUCTION** *E. aeneus* is a protogynous hermaphrodite. Females mature first at 50–60 cm TL and weigh about 3 kg. Sex change to males occurs at about 9 kg, but smaller males (3–5 kg) are occasionally found (Bruslé 1985). *E. aeneus* spawns in June and July in south-east Tunisia (Bouain and Siau 1983) and is reported by divers and fishers to form a spawning aggregation in Senegal (Luckhurst 2009).

COMMENTS *E. aeneus* is known to undergo a seasonal migration from Mauritania to Senegal (Cury and Worms 1982). Cury and Roy (1988) suggest that this is due to changes in water temperature associated with seasonal upwelling.

2008 IUCN RED LIST STATUS Near Threatened

Authors: *AA Bertoncini, T-C Chan, JH Choat, MT Craig, ST Fennessy, BP Ferreira, DA Pollard and LA Rocha*

Epinephelus akaara

(Temminck and Schlegel 1842)

Hong Kong grouper (E), mérou rouge tacheté (F), mero de pintas rojas (S)

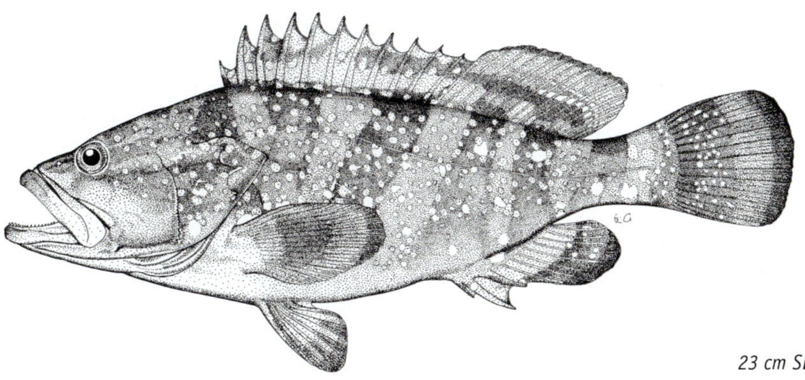

23 cm SL (Taiwan)

Body depth less than head length, depth contained 2.7–3.2 times in standard length. Head length 2.3–2.6 times in standard length; preopercle with enlarged serrae at angle; upper edge of operculum straight; posterior and anterior nostrils subequal; maxilla reaching about to vertical at rear edge of eye; midside of lower jaw with 2 rows of teeth. Gill rakers 8 or 9 on upper limb, 15–17 on lower limb, 23–25 total. Dorsal fin with 11 spines and 15–17 rays, the 3rd to 6th spines longest, their length contained 2.4–3.8 times in head length, the interspinous membranes incised; anal fin with 3 spines and 8 rays; pectoral-fin rays 17–19, pectoral-fin length contained 1.5–2.1 times in head length; pelvic fins not reaching anus, their length 1.9–2.3 times in head length; caudal fin rounded. Lateral-body scales ctenoid, with auxiliary scales in adults; lateral-line scales 50–54; lateral-scale series 92–106.

MAXIMUM RECORDED SIZE 58 cm TL

© Allen W-L To

Adult *E. akaara* (locality unknown)

COLOUR IN LIFE Head and body pale brownish grey, covered with small red, orange or gold spots except on the ventral side; 6 faint oblique dark bars sometimes visible on the body; a dark brown or black blotch on the body at the base of the last 3 dorsal-fin spines; dorsal-fin margin yellow or orange; a row of dusky yellow or orange spots (one per membrane) along the middle of the spinous dorsal fin and another row along the base of the fin; soft dorsal, caudal and anal fins with faint red or orange spots basally, the distal parts of these fins dusky with small faint white spots.

COLOUR IN MARKET Head and body pale grey covered with spots which remain vivid. Dark bars on the body usually fade, particularly on the anterior portion of the body. The dark edges of the fins usually remain distinct, and the dark spot at the base of the posterior dorsal fin remains prominent.

© Andrew S Cornish

Adult *E. akaara* (Hong Kong)

© Liu Min

Adult *E. akaara* (Taiwan)

POPULATION **General** This once-common species has experienced severe population declines due to fishing pressure since the 1970s. **Fishery dependent** Landings of *E. akaara* have declined substantially in recent years, by as much as 90% in China, and by at least 50% in Japan, as well as elsewhere throughout its range (Liu and Sadovy de Mitcheson 2009b).

POPULATION TREND Decreasing

MAJOR THREATS The major threat to *E. akaara* is fishing pressure on both adults, taken for the live food fish trade, and juveniles that are taken for capture-based mariculture. *E. akaara* is fished throughout its range where there is no effective management for the species.

HABITAT AND ECOLOGY **General** *E. akaara* is a rocky-reef species inhabiting depths to at least 55 m. Juveniles are often found shallower than 10 m. **Age and growth** *E. akaara* reaches at least 19 years (Dai et al. 1988). **Reproduction** Tseng and Ho (1988) report that sexual maturation in females occurs at about 23–24 cm SL, or about 3 years old. Ukawa et al. (1966) report that *E. akaara* spawns in pairs in culture ponds; however, anecdotal reports for wild populations indicate that the species may also aggregate to spawn and have a spawning season of several months duration. Sadovy de Mitcheson and Liu (2008) report on bidirectional sex change in this species which was previously described as protogynous.

Epinephelus akaara is a western Pacific species, where it occurs in southern Japan, Republic of Korea, Taiwan and southern China.

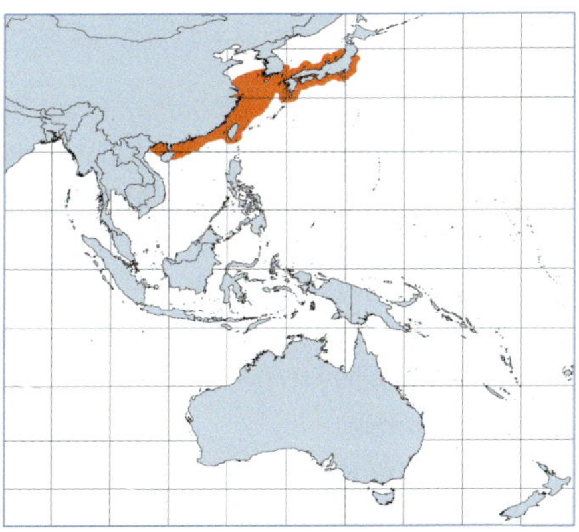

COMMENTS *E. akaara* is very similar to the sympatric *E. fasciatomaculosus* with the main difference observed in its colour pattern. The spots on the body of *E. fasciatomaculosus* are dark brown, yellowish brown or russet (dark in alcohol), and the dark bars on the body are darker, with the first 2 extending to the margin of the spinous dorsal fin (they do not extend onto the dorsal fin in *E. akaara*). There are also no dusky yellow or orange spots on the spinous dorsal fin.

In the early 1980s *E. akaara* was the preferred species for capture-based mariculture in southern China. By the mid-1990s, however, the production of juveniles was low and culture does not appear to be commercially viable at a large scale even though the species can be hatchery-reared (i.e. full-cycle aquaculture) (Sadovy 2000, Liu and Sadovy de Mitcheson 2009b). The species is the most expensive grouper for sale in Hong Kong and mainland China. A recent status review of the species throughout its range confirms the seriously reduced levels (although data are sparse from Korea) (Liu and Sadovy de Mitcheson 2009b).

2003 IUCN RED LIST STATUS Endangered (A2d)

Authors: *AS Cornish and M Liu*

Epinephelus albomarginatus

Boulenger 1903

white-edged grouper (E), mérou bord blanc (F), mero bordiblanco (S)

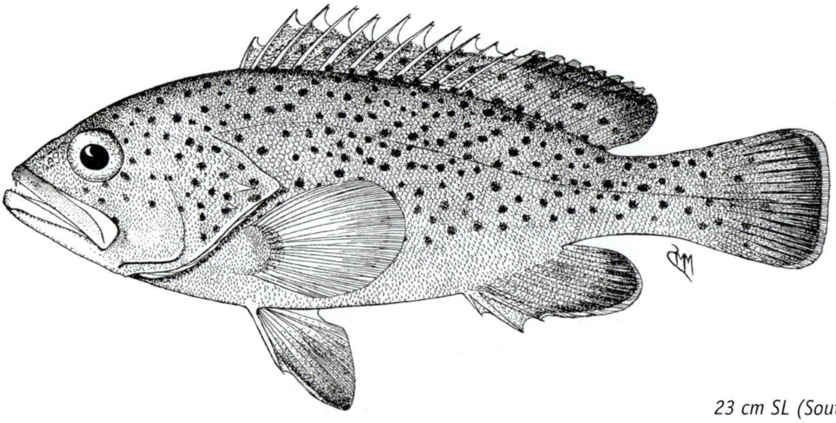

23 cm SL (South Africa)
© SAIAB

Body depth less than head length, depth contained 2.6–3.0 times in standard length. Head length 2.3–2.5 times in standard length; preopercle angular, with 2 or 3 enlarged serrae at the angle; upper edge of operculum straight or convex; diameter of posterior nostril about twice that of anterior nostril; maxilla naked, without a step on ventral edge; maxilla reaches shorter or longer than vertical at rear edge of eye. Gill rakers 8–10 on upper limb, 14–16 on lower limb, 22–26 total. Dorsal fin with 11 spines and 14 rays, the 3rd or 4th spine longest, contained 2.6–3.4 times in head length, the interspinous membranes distinctly incised; anal fin with 3 spines and 8 rays; pectoral fin with 17 or 18 rays, equal to or slightly longer than pelvic fins, and contained 1.9–2.2 times in head length; pelvic fins not reaching anus; caudal fin truncate, with rounded corners in adults, the posterior edge convex in juveniles. Lateral-body scales ctenoid, with a few auxiliary scales in adults; lateral-line scales 56–66; lateral-scale series 92–106.

MAXIMUM RECORDED SIZE 100 cm TL

Epinephelus albomarginatus occurs along the east coast of South Africa (KwaZulu-Natal and Eastern Cape provinces to 33° S), extending into southern Mozambique (24°50′ S).

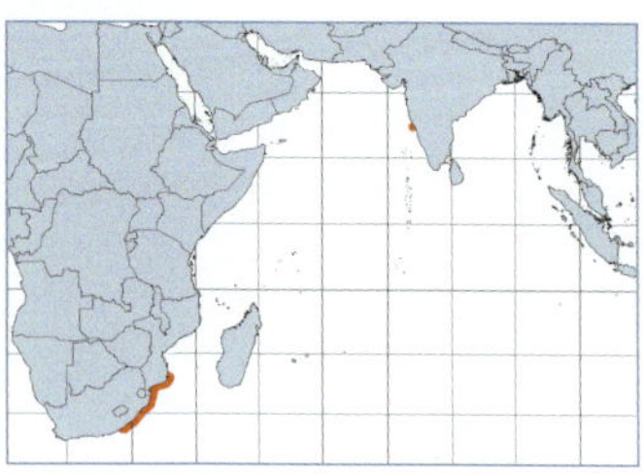

POPULATION **GENERAL** *E. albomarginatus* is a relatively uncommon species across its small range. **FISHERY DEPENDENT** Fennessy (2000) reports marked declines in spawning biomass and mean size from the mid-1980s to the mid-1990s in KwaZulu-Natal (South Africa) where there is an active fishery which catches this species along with other reef fish.

POPULATION TREND Decreasing

MAJOR THREATS Fishing pressure is the major threat to *E. albomarginatus* throughout most of its limited geographic range.

COLOUR IN LIFE Head, body, dorsal and caudal fins pale brown, with numerous small dark brown spots; ventral parts of the head and body without spots; on juveniles, many of the spots merge to form double spots; prominent dark brown streak along the upper edge of the maxillary groove. Margin of the interspinous dorsal-fin membranes yellow or gold; soft dorsal and anal fins dusky distally, with a prominent white edge; upper and lower edges of the posterior part of the caudal fin also white-edged; pectoral fins yellowish or reddish orange.

COLOUR IN MARKET Pale brown head and body; the most distinguishing features fade with the exception of the prominent white edges of the fins which become buff or pale yellow.

© Robert Kyle

Adult *E. albomarginatus* (Cape Vidal, South Africa)

© Dennis King

Juvenile *E. albomarginatus* (Cape Vidal, South Africa)

HABITAT AND ECOLOGY **GENERAL** *E. albomarginatus* is a reef-associated species mainly inhabiting depths of 30–80 m. **AGE AND GROWTH** *E. albomarginatus* has shown relatively slow growth (k = 0.214 y^{-1}) compared with other grouper species, and reaches a maximum age of at least 16 years (Fennessy 2000). The length–weight relationship of $W = 0.0000257 \times L^{2.915}$, where W is the weight in g and L is total length in mm, is provided by Fennessy (2000).

REPRODUCTION There is evidence that spawning in *E. albomarginatus* does not occur in the southern part of its distribution. Histological evidence indicates that *E. albomarginatus* is likely a protogynous hermaphrodite. In Mozambique, 50% of females are mature by 34 cm (3 years), and the smallest mature males are about 59 cm TL (6 years) (Fennessy 2000). **2004 IUCN RED LIST STATUS** Vulnerable (A2d)

Author: *ST Fennessy*

Epinephelus amblycephalus (Bleeker 1857)

banded grouper (E), mérou bande (F), cabrilla venda (S)

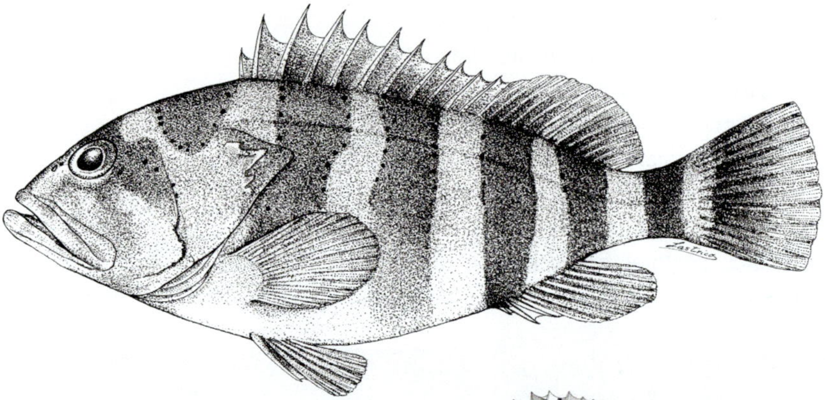

37 cm SL
© FAO

6 cm SL (northern Australia)
© SAIAB

Body depth less than head length, depth contained 2.5–3.0 times in standard length. Head length 2.1–2.4 times in standard length; preopercle rounded, with 3–6 enlarged serrae at the angle; upper edge of operculum slightly convex; nostrils subequal or posterior nostril larger than anterior one; maxilla scaly, reaching about to vertical at rear edge of eye; midside of lower jaw with 2 rows of teeth. Gill rakers 8 on upper limb, 14–16 on lower limb, 22–24 total. Dorsal fin with 11 spines and 15 or 16 rays, the 3rd or 4th spine longest, contained 2.7–3.2 times in head length, the interspinous membranes only slightly incised; anal fin with 3 spines and 8 rays; pectoral fins with 18 or 19 rays, pectoral-fin length equal to pelvic-fin length, contained 1.7–2.3 times in head length; caudal fin rounded. Lateral-body scales ctenoid, with a few auxiliary scales in adults; lateral-line scales 47–52; lateral-scale series 90–121.

MAXIMUM RECORDED SIZE 50 cm TL

Epinephelus amblycephalus is a western Pacific species distributed from the Andaman Sea to southern Japan, Taiwan, China, Philippines, Vietnam, Malaysia, Thailand, Indonesia, New Guinea, the Arafura Sea, north-western Australia and Fiji.

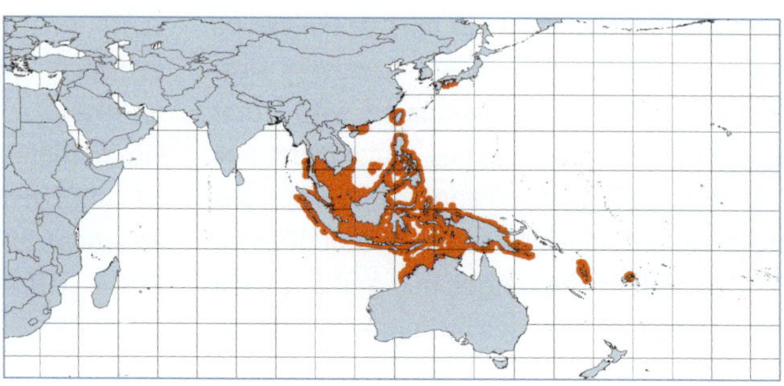

COLOUR IN LIFE Head, body and median fins pale grey; body with 5 broad dark brown bars, the first 2 extending into the spinous part of the dorsal fin, the 3rd and 4th bars extending into the soft dorsal and anal fins, and the 5th dark bar at the base of the caudal fin; small black spots along the edges of the dorsal parts of the body bars; dark brown saddle blotch on the nape, with small black spots mainly along the edge of this blotch; cheeks, snout, interorbital area, jaws and chest mostly dark brown, with 2 or 3 white bands radiating from the eye; irregular black bar on the middle of the caudal fin; maxillary groove yellow; ventral parts of the head and body often with a pinkish tinge. Small juveniles (6 cm SL) white, with black bars on the body as described for adults, the fins pale yellow with small black spots.

COLOUR IN MARKET Body generally pale brown or grey. Body bars may fade, but usually remain faint; small spots on edges of bars generally remain prominent. Yellow of the maxillary groove usually prominent.

© Matthew T Craig

Adult *E. amblycephalus* (Jimbaran Fish Market, Bali, Indonesia)

© Phillip C Heemstra

Adult *E. amblycephalus* (aquarium, Penghu Islands, Taiwan)

© Matthew T Craig

Juvenile *E. amblycephalus* (Jimbaran Fish Market, Bali, Indonesia)

POPULATION GENERAL *E. amblycephalus* is a poorly known species from deep waters of the western Pacific. **FISHERY DEPENDENT** Little is known about the fishery for this species other than that it is a popular species in Hong Kong markets.

POPULATION TREND Unknown

MAJOR THREATS While not quantified, fishing pressure likely poses a threat to this species.

HABITAT AND ECOLOGY GENERAL *E. amblycephalus* is a deep-water species inhabiting reefs from 80 to 130 m.

2008 IUCN RED LIST STATUS Data Deficient

Authors: *RF Myers and BC Russell*

Epinephelus analogus Gill 1863

spotted grouper (E), mérou cabrilla (F), mero moteado (S)

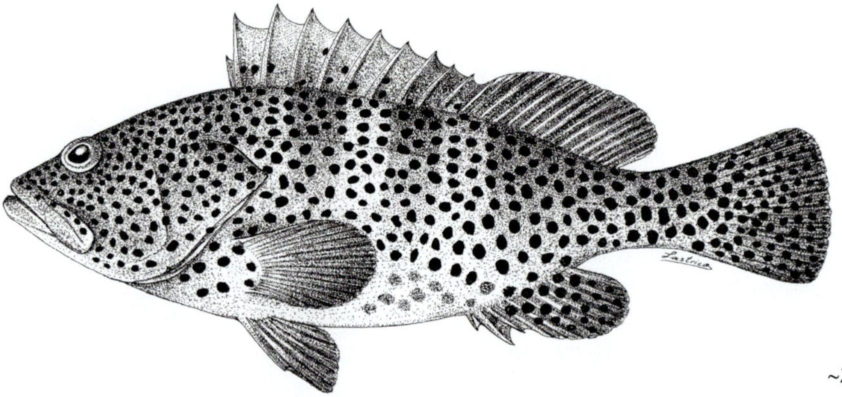

~75 cm TL
© FAO

Body depth distinctly less than head length, depth contained 2.6–3.0 times in standard length. Head length 2.3–2.5 times in standard length; preopercle with a shallow notch above the angle, the serrae at angle slightly enlarged; upper edge of operculum slightly convex; posterior and anterior nostrils subequal. Gill rakers 26–28 (total). Dorsal fin with 10 spines and 16–18 rays, the 3rd spine longest and the membranes moderately incised between the spines; anal fin with 3 spines and 8 rays; pectoral fins with 19 or 20 rays; pelvic fins distinctly shorter than pectoral fins and their origin below or behind lower end of pectoral-fin base; caudal fin rounded. Midlateral-body scales ctenoid, with numerous auxiliary scales; lateral-line scales 53–69; lateral-scale series 96–112.

MAXIMUM RECORDED SIZE 114 cm TL

COLOUR IN LIFE Generally pale reddish brown covered with numerous dark brown spots; body with 5 faint oblique dark bars that are prominent during periods of excitement.

COLOUR IN MARKET Generally pale, the numerous spots remain prominent. Body bars may fade into irregular patterns.

© Matthew T Craig

Adult *E. analogus* (Mazatlán, Mexico)

© Daniel W Gotshall

Adult *E. analogus* (Bahía Magdalena, Baja California, Mexico)

POPULATION GENERAL *E. analogus* is common throughout its range and the most abundant small grouper in the northern Gulf of California, Mexico. FISHERY DEPENDENT *E. analogus* is taken in substantial numbers throughout its range. For example, there is an estimated 18–20 t of annual landings of *E. analogus* in Ecuador (IUCN Red List Assessment 2008) but no indication of declines.

Epinephelus analogus occurs in the eastern Pacific from southern California (USA) to Peru, including the Revillagigedo, Clipperton and Galápagos islands.

POPULATION TREND Stable

MAJOR THREATS There are no major threats known to this species.

HABITAT AND ECOLOGY GENERAL *E. analogus* is a reef-associated species inhabiting depths of at least 50 m. FEEDING Feeds on small fishes and crustaceans; during winter, individuals move closer to the shore and feed heavily on swarms of pelagic red crab *Pleuroncodes planipes*.

2008 IUCN RED LIST STATUS Least Concern

Authors: *JH Choat, MT Craig, BP Ferreira and LA Rocha*

Epinephelus andersoni Boulenger 1903

catface grouper (E), mérou chat (F), mero gato (S)

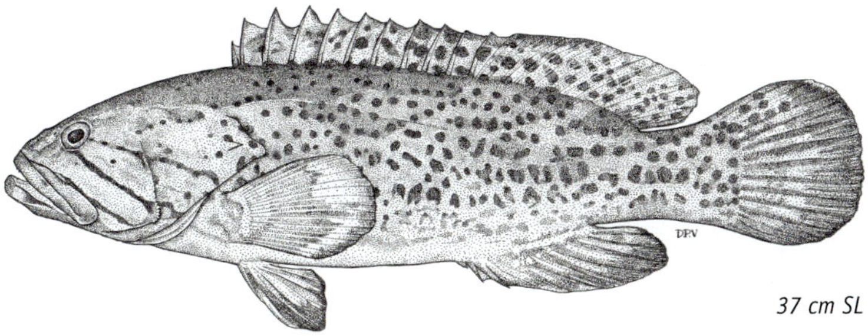

37 cm SL (South Africa)

Body elongate, the depth contained 3.2–3.7 times in standard length; greatest body width 1.4–1.8 times in body depth. Head length 2.4–2.7 times in standard length; preopercle angular, with several enlarged serrae at the angle; upper edge of operculum slightly convex; posterior and anterior nostrils equal; maxilla scaly, reaching to or past vertical at rear edge of eye; midside of lower jaw with 2 or 3 rows of teeth. Gill rakers shorter than gill filaments, 8–11 on upper limb, 14–17 on lower limb, 22–27 total. Dorsal fin with 11 spines and 13–15 rays, the 3rd or 4th spine longest, its length contained 2.9–3.9 times in head length, the interspinous membranes moderately incised; anal fin with 3 spines and 8 rays; pectoral fins with 17–19 rays, their length contained 1.5–2.0 times in head length; pelvic-fin length 2.1–2.6 times in head length; caudal fin well rounded. Midlateral-body scales ctenoid (although mostly embedded), with auxiliary scales; lateral-line scales 66–74; lateral-scale series 97–106.

MAXIMUM RECORDED SIZE 80 cm TL

Epinephelus andersoni occurs along the east coast of South Africa (KwaZulu-Natal and Eastern Cape provinces), extending into southern Mozambique and Madagascar.

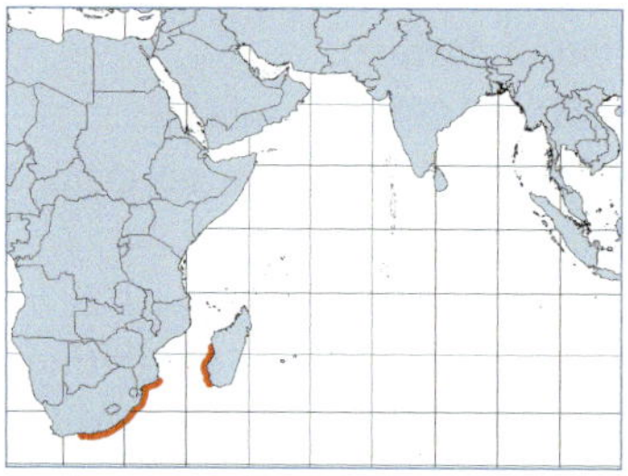

POPULATION **GENERAL** *E. andersoni* is a common species, particularly in the southern portion of its range. **FISHERY DEPENDENT** *E. andersoni* is a common fishery species that has shown signs of a depleted fishery. For example, Fennessy (2000) reports that spawning biomass was reduced to 40% by the mid-1990s.

POPULATION TREND Unknown

MAJOR THREATS Fishing pressure is the major threat to *E. andersoni*.

HABITAT AND ECOLOGY **GENERAL** *E. andersoni* is a rocky-reef species inhabiting depths to at least 101 m. **FEEDING** Feeds on small fishes

COLOUR IN LIFE Head, body and fins brown; body, caudal and dorsal fins with numerous small close-set irregular dark brown spots; 2 dark brown stripes running posteriorly from the eye across the operculum, and another continuing from the maxillary groove to the lower edge of the preopercle. Juveniles with longitudinal dark stripes that break up into spots posteriorly on the body; black blotch at the base of the last few dorsal-fin spines; 2 small black blotches at the base of the dorsal-fin rays, and a 3rd on top of the caudal peduncle; these black blotches are separated by 4 or 5 white spots.

COLOUR IN MARKET Drab brown spots remain prominent. Lines on the face generally remain as the most distinguishing marks after death.

© Matthew T Craig

Adult *E. andersoni* (Umkomaas, South Africa)

© Jade Maggs

Adult *E. andersoni* (uShaka Marine World, Durban, South Africa)

and crustaceans. **Age and growth** *E. andersoni* is a relatively fast-growing grouper (k = 0.24 y^{-1}) and reaches a maximum age of at least 11 years (Fennessy 2000); the length–weight relationship $W = 0.0000226 \times L^{2.905}$, where W is the weight in g and L is the total length in mm, was given by Fennessy (2000). **Reproduction** *E. andersoni* is a diandric protogynous hermaphrodite. Spawning occurs from November to January. Fennessy and Sadovy (2002) report minimum lengths at maturity of 42 cm TL for females and 38 cm TL for males.

COMMENTS *E. andersoni* is harvested in considerable numbers and according to a stock assessment in the mid-1990s it was overexploited. There is no information on the current stock status of this species.

2004 IUCN RED LIST STATUS Near Threatened

Author: *ST Fennessy*

Epinephelus areolatus (Forsskål 1775)

areolate grouper (E), mérou aréolé (F), mero areolado (S)

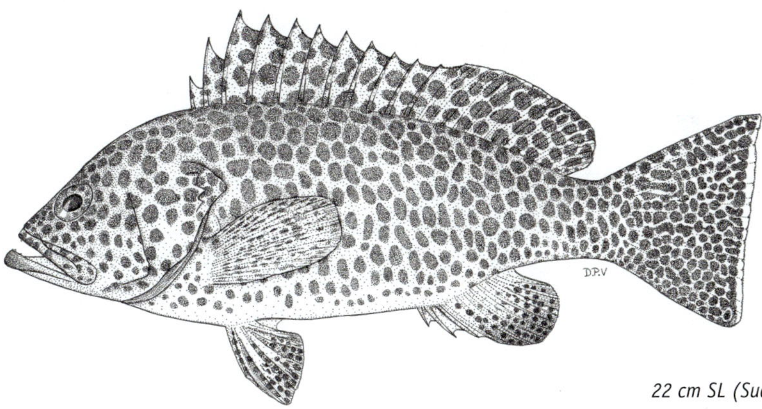

22 cm SL (Sudan, Red Sea)
© SAIAB

Body depth less than head length, depth contained 2.8–3.3 times in standard length. Head length 2.4–2.8 times in standard length; interorbital area convex; preopercle angular, with 2–7 enlarged serrae at the angle; upper edge of operculum straight or slightly convex; posterior and anterior nostrils subequal; maxilla extending to below rear half of eye, the lower edge with a distinct step posteriorly; maxilla, lower jaw and gular area scaly; midlateral part of lower jaw with 2 rows of teeth. Gill rakers 8–10 on upper limb, 14–16 on lower limb, 23–25 total. Dorsal fin with 11 spines and 15–17 rays, the 3rd or 4th spine longest and subequal to longest ray, the interspinous membranes moderately incised; anal fin with 3 spines and 8 rays, the margin rounded or angular; pectoral-fin rays 17–19; pectoral fins longer than pelvic fins, pectoral-fin length contained 1.5–1.8 times in head length; pelvic-fin length 1.7–2.1 times in head length, reaching to or nearly to anus; caudal fin slightly convex in juveniles, truncate or slightly emarginate in adults. Lateral-body scales ctenoid, with auxiliary scales in adults; lateral-line scales 49–53; lateral-scale series 97–116.

MAXIMUM RECORDED SIZE 47 cm TL

COLOUR IN LIFE Head, body and fins pale; covered with numerous close-set brown, brownish yellow or greenish yellow spots, the largest about the size of the pupil, those on the front of the head smaller than those on the operculum. Pectoral fins pale, with small dark spots on the rays. Posterior edge of the caudal fin with a distinct white margin.

COLOUR IN MARKET Head and body pale brown to white on underside. Yellow spots remain particularly prominent in this species following death.

Adult *E. areolatus* (Sulawesi, Indonesia)

Adult *E. areolatus* (Bali, Indonesia)

Epinephelus areolatus is a very widespread Indo-Pacific species that ranges from the Red Sea and the Persian Gulf to KwaZulu-Natal (South Africa), extending east to Fiji, north to Japan, and south to the Arafura Sea and northern Australia. Recently it has been recorded from Tonga (Randall et al. 2003), but appears absent from Micronesia, Polynesia and the islands of the western Indian Ocean.

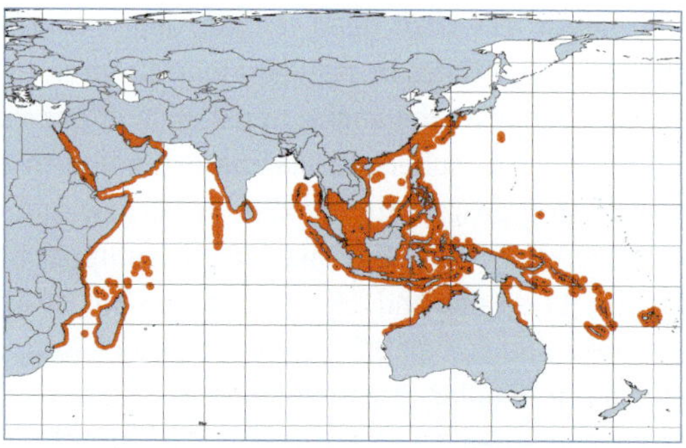

POPULATION **GENERAL** *E. areolatus* is widespread and reasonably common throughout its range. **FISHERY DEPENDENT** *E. areolatus* is one of the most common groupers taken in the fishery of New Caledonia where it represents 8% of the total catch (M Kulbicki unpublished data). It is also taken in considerable numbers in Saudi Arabia and the South China Sea. Once commonly taken in Hong Kong, the species is now uncommon locally and over 50% of fish marketed are immature (Sadovy and Cornish 2000, To and Sadovy de Mitcheson 2009).

POPULATION TREND Unknown

MAJOR THREATS Fishing pressure is the major threat to this species which is taken both as adults and as juveniles for grow-out mariculture.

HABITAT AND ECOLOGY **General** *E. areolatus* is found in seagrass beds or on silty sand bottoms around isolated small rock outcrops, dead coral or soft coral in depths of 2–200 m. Juveniles are common at water depths to 80 m (Kailola et al. 1993). **Feeding** *E. areolatus* feeds on fishes and benthic invertebrates, primarily prawns and crabs (Parrish 1987, Randall and Heemstra 1991, Salini et al. 1994, Kulbicki et al. 2005b). **Age and growth** The maximum reported age for *E. areolatus* is at least 15 years (Loubens 1980b). **Reproduction** In New Caledonia, the sex ratio is reported as 1:6 males to females, with a size at maturity of females at 19.5 cm TL and 29 cm TL for males (M Kulbicki pers. obs.).

2008 IUCN RED LIST STATUS Least Concern

Authors: *AS Cornish, ST Fennessy, M Kulbicki, DA Pollard, BC Russell and BM Yeeting*

Epinephelus awoara

(Temminck and Schlegel 1842)

yellow grouper (E), mérou jaune (F), mero amarillo (S)

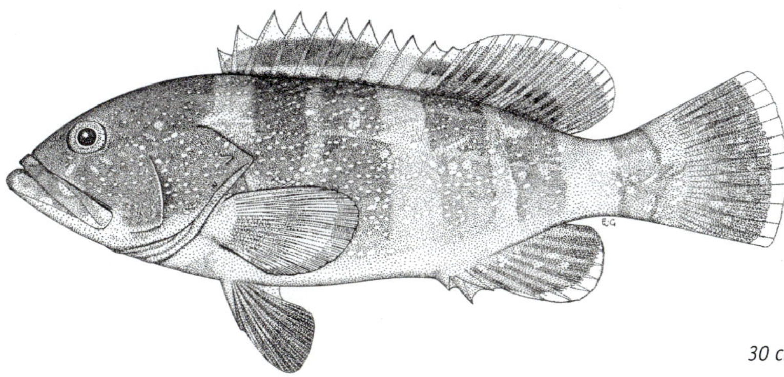

30 cm SL (Taiwan)

Body depth contained 2.7–3.3 times in standard length. Head length 2.3–2.6 times in standard length; interorbital area convex; dorsal head profile strongly convex; preopercle subangular, with 2–5 strong spines at the angle; upper edge of operculum straight, the uppermost spine rudimentary; maxilla reaches about to vertical at rear edge of eye; midlateral part of lower jaw with 2 rows of small, subequal teeth. Gill rakers 8 or 9 on upper limb, 16–18 on lower limb, 22–26 total. Dorsal fin with 11 spines and 15 or 16 rays, the 3rd or 4th spine longest but shorter than the longest ray, the interspinous membranes deeply incised; anal fin with 3 spines and 8 rays; pectoral-fin rays 17–19; pectoral fins longer than pelvic fins, pectoral-fin length contained 1.6–1.9 times in head length; caudal fin convex. Lateral-body scales distinctly ctenoid, with auxiliary scales in fish larger than 30 cm SL; lateral-line scales 49–55; lateral-scale series 92–109. Pyloric caeca 12, in 3 bundles.

MAXIMUM RECORDED SIZE 60 cm TL

COLOUR IN LIFE Head and body pale greyish brown dorsally, usually golden yellow ventrally; 4 broad dark bars on the dorsal part of the body, one on the caudal peduncle and another often visible on the nape; head and body with numerous small yellow spots; body and median fins also with small greyish white spots; soft dorsal and caudal fins, and sometimes the anal fin, with a prominent yellow margin; paired fins dusky yellow. Yellow 'moustache' in the maxillary groove. Dark bars on the body may be faint or absent in large adults.

COLOUR IN MARKET Dark bars on the body may fade or be absent in large adults. Yellow edges of fins and 'moustache' usually remain distinct.

© Matthew T Craig

Adult *E. awoara* (Tsukiji Fish Market, Tokyo, Japan)

© Allen W-L To

Juvenile *E. awoara* (Shelter Island, Hong Kong)

© Allen W-L To

Adult *E. awoara* (locality unknown)

POPULATION **GENERAL** *E. awoara* is a moderately common species throughout its range. It is one of the few grouper species with landings remaining at commercial levels in Hong Kong despite intense fishing pressure; larger species are now uncommon (Sadovy and Cornish 2000). It is also extensively fished in the South and East China seas. **FISHERY DEPENDENT** Despite its use in fisheries, little information is available to quantify the levels of harvest of *E. awoara*. Port surveys in Hong Kong estimated 9–10 t per year were taken in 2001/2002 (Agriculture, Fisheries and Conservation Department, Hong Kong). Juveniles are caught and used extensively as seed for grow-out in capture-based mariculture in southern China and almost all *E. awoara* in retail sales are immature (Sadovy de Mitcheson 2009, To and Sadovy de Mitcheson 2009).

POPULATION TREND Unknown

MAJOR THREATS Fishing pressure, both on adults and juveniles, is the major threat to *E. awoara*.

Epinephelus awoara occurs in the north-western Pacific Ocean, including North and South Korea, Japan, China, Vietnam, and islands in the South China Sea. It is also known from the eastern Indian Ocean.

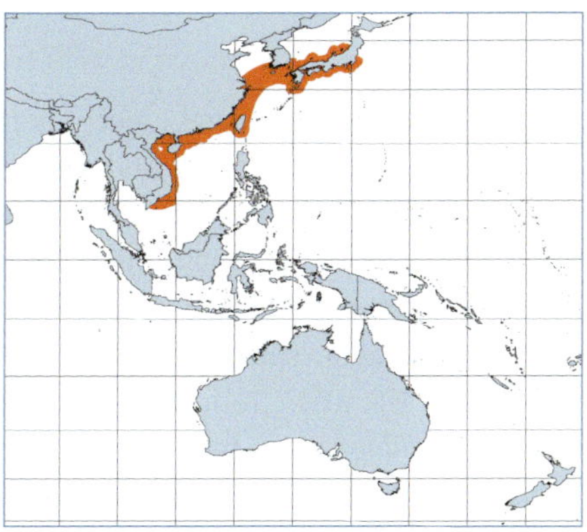

HABITAT AND ECOLOGY **GENERAL** *E. awoara* is found in rocky areas and over adjacent sand-mud bottoms. Juveniles are common in tide pools and adults are caught in depths of 10–50 m. **FEEDING** *E. awoara* feeds on fishes and crustaceans (Zhu 1998). **REPRODUCTION** *E. awoara* is a protogynous hermaphrodite (To 2009). Spawning season lasts from February/March to May in Hong Kong (Liu 1971, To 2009) and from June to July in Taiwan (Shao 2005). Sexual maturation (50%) is attained at about 24 cm SL for females in Hong Kong, and mature males are found from about 26 cm SL (To 2009).

COMMENTS There is little information on the level of exploitation in fisheries for this species. It is known, however, that mean market sizes in Hong Kong are between 15–25 cm TL (in 2004/2005) and about 8–15 cm TL in southern China. Given the size at maturity, the focus of the fishery in this region is largely on juveniles.

2006 IUCN RED LIST STATUS Data Deficient

Authors: *M Liu, W-C Ng and C-M Chan*

Epinephelus bilobatus Randall and Allen 1987

twinspot grouper (E), mérou gemine (F), mero de pintas gemelas (S)

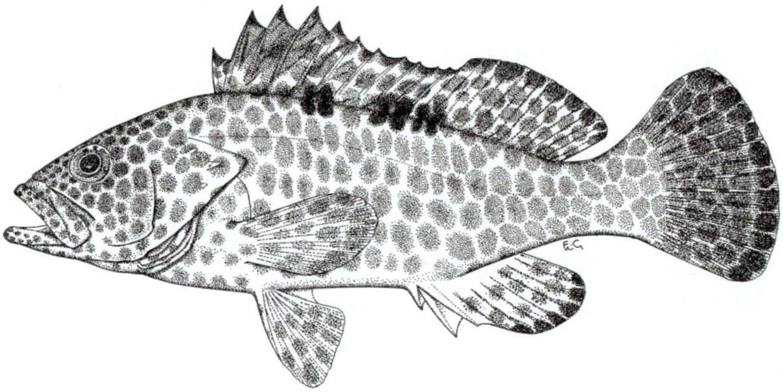

16 cm SL (Western Australia)

Body depth contained 2.8–3.3 times in standard length. Head length 2.4–2.6 times in standard length; preopercle subangular, with the serrae at the angle slightly enlarged; opercular spines inconspicuous; upper edge of operculum straight or slightly concave; nostrils subequal in small fish, but posterior nostril diameter about twice that of anterior nostril in a fish of 27 cm SL; maxilla reaches below rear half of eye; midlateral part of lower jaw with 2 rows of small teeth. Gill rakers 7–9 on upper limb, 14–16 on lower limb, 23–25 total. Dorsal fin with 11 spines and 17 or 18 rays, the 3rd or 4th spine longest (2.1–2.5 times in head length) and slightly longer than the longest dorsal-fin ray, the interspinous membranes slightly to moderately incised; anal fin with 3 spines and 8 rays, the 2nd and 3rd spines subequal, their length contained 2.5–3.2 times in head length; pectoral-fin rays 17–19; pectoral fins longer than pelvic fins, pectoral-fin length 1.5–1.8 times in head length; caudal-peduncle depth 3.2–3.6 times in head length; caudal fin rounded. Lateral-body scales ctenoid, auxiliary scales present or absent; lateral-line scales 48–52; lateral-scale series 94–102.

MAXIMUM RECORDED SIZE 33 cm TL

POPULATION **GENERAL** *E. bilobatus* is a poorly known species. It is unclear whether this species is common or its distribution is wider than currently known.

POPULATION TREND Unknown

MAJOR THREATS No major threats are known to this species.

HABITAT AND ECOLOGY **GENERAL** *Epinephelus bilobatus* is a poorly known, reef-associated species found on coral reefs or rocky bottoms at depths of up to 50 m.

COMMENTS *E. bilobatus* is one of a group of nine shallow-water coral reef species that have a rounded caudal fin and close-set dark brown spots with the pale interspaces forming a network on the body. These species have been much confused in the literature, and many museum specimens have been misidentified. The other 'reticulated groupers' differ from *E. bilobatus* as follows:

- *E. faveatus* has smooth lateral-body scales (except for the area covered by the pectoral fins), shorter dorsal-fin spines (length of 4th spine contained 2.6–3.9 times in head

COLOUR IN LIFE Head and body pale, covered with indistinctly defined, dark orange-brown spots; dark spots extend onto the soft dorsal, caudal and anal fins where the pale interspaces form a network pattern; the spots along the margin of these fins are darker than the other spots; a series of 3 bilobed dark blotches or close-set pairs of dark brown to black spots along the base of the dorsal fin and adjacent body; dorsal part of body of fresh specimens with a pale bluish grey tinge; spinous dorsal fin dusky, with indistinct dark spots and a blackish margin; paired fins also with dark brown spots. Dark brown line in the maxillary groove.

COLOUR IN MARKET Head and body a pale brown. Spots generally remain prominent, especially the darkest set on the dorsal and caudal fin margins.

© Stephen Newman

Adult *E. bilobatus* (Kimberley coast, Western Australia)

Adult *E. bilobatus* (Raja Ampat Islands, Indonesia)

© Gerald R Allen

Juvenile *E. bilobatus* (Raja Ampat Islands, Indonesia)

© Gerald R Allen

length), and the dark spots on the body at the dorsal-fin base are not bilobed or in pairs.

- *E. hexagonatus* has conspicuous white dots on the body between the dark spots, the dark spots on the body at the dorsal-fin base are not bilobed or in pairs; there are 15–17 dorsal-fin rays, and the 2nd anal-fin spine is much longer than the 3rd spine.
- *E. macrospilos* has mostly smooth lateral-body scales, pectoral fins are dusky with a narrow white edge, the dark spots on the body at the dorsal-fin base are not bilobed or in pairs, and the rear edge of the caudal fin is pale yellow or white.
- *E. maculatus* has the dorsal-fin membranes not incised between the spines; juveniles are yellowish brown, with a few small black spots (mainly on head and fins) and a few large irregular white blotches on the head, body and fins; there are 15–17 dorsal-fin rays and 102–120 lateral-scale series.
- *E. melanostigma* has a single black blotch at the base of the last 4 dorsal-fin spines, 14–16 dorsal-fin rays, 56–68 lateral-line scales, and the midlateral part of the lower jaw with 3–5 rows of teeth.
- *E. merra* has dark spots on the body at the dorsal-fin base that are not bilobed or in pairs, and pectoral fins have small black spots that are largely confined to the rays; there are 15–17 dorsal-fin rays and the 4th to 9th dorsal-fin spines are subequal (the longest contained 2.6–2.9 times in head length).
- *E. quoyanus* has dark spots on the body at the dorsal-fin base not bilobed or in pairs, pectoral fins with indistinct dark brown spots, the base with a large semicircular dark brown spot edged posteriorly with white, 2 oblique dark brown bands or blotches linked by bands on the sides of the chest, pectoral-fin length more than 25% of standard length and contained 1.2–1.7 times in head length, 3rd to 5th dorsal-fin spines longest and contained 2.3–3.0 times in head length.
- *E. spilotoceps* has 14–16 dorsal-fin rays, 59–69 lateral-line scales, 4th dorsal-fin spine usually longest (its length contained 2.7–3.5 times in head length), 3 brownish black blotches or spots on the body at the base of the dorsal fin but these are not bilobed or in pairs, and the caudal-peduncle depth is contained 3.7–4.3 times in head length.

2008 IUCN RED LIST STATUS Data Deficient

Epinephelus bilobatus is an eastern Indian Ocean species known only from Western Australia (Port Wolcott [20°39′ S] to Yampi [16°08′ S]) and eastern Indonesia.

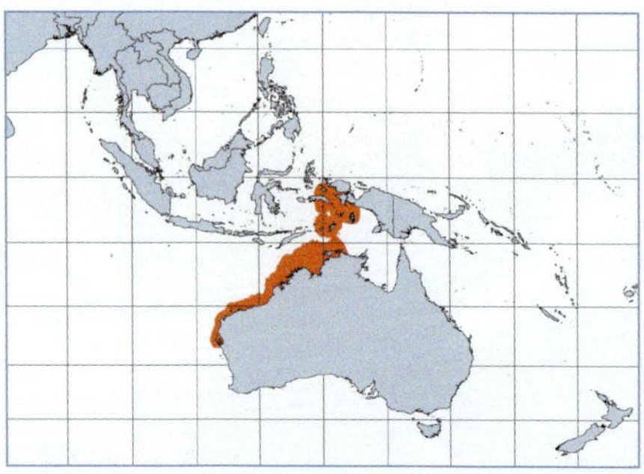

Authors: *DA Pollard, KL Rhodes and BC Russell*

Epinephelus bleekeri (Vaillant 1878)

duskytail grouper (E), mérou demi-deuil (F), mero medioluto (S)

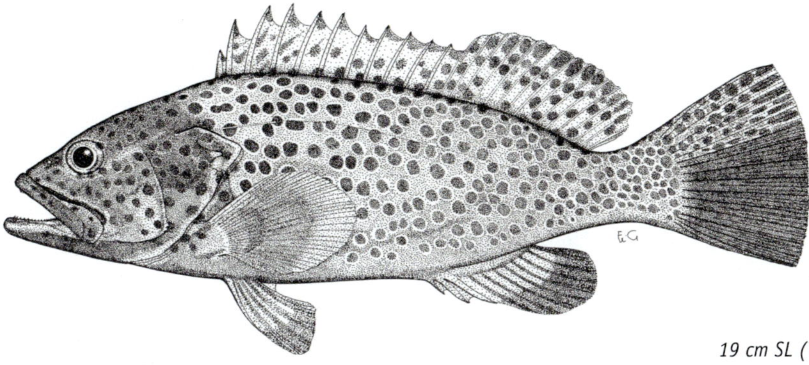

19 cm SL (Taiwan)

Body elongate, the depth contained 3.0–3.5 times in standard length. Head length 2.4–2.7 times in standard length; interorbital area flat to slightly convex; preopercle angle with 2–9 enlarged serrae, adults with a notch above the angle; upper edge of operculum straight; maxilla scaly, reaching to or beyond vertical at rear edge of eye; midlateral part of lower jaw with 2 rows of subequal teeth. Gill rakers 9–11 on upper limb, 16–18 on lower limb, 25–28 total. Dorsal fin with 11 spines and 16–18 rays, 3rd to 5th spines longest, the interspinous membranes incised; anal fin with 3 spines and 8 or 9 (rarely 9) rays; pectoral-fin rays 17–19; pectoral-fin length contained 1.6–2.1 times in head length; pelvic-fin length 1.9–2.5 times in head length; caudal fin truncate or slightly convex. Lateral-body scales ctenoid, with a few small auxiliary scales in adults; lateral-line scales 49–53; lateral-scale series 99–104.

MAXIMUM RECORDED SIZE 76 cm TL

Epinephelus bleekeri is an Indo-West Pacific species ranging from the Persian Gulf to Indonesia, extending north to Taiwan and south to the northern coast of Australia.

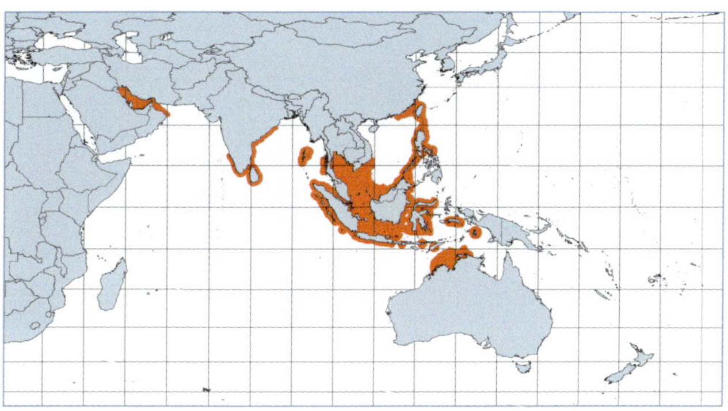

COLOUR IN LIFE Head and body brownish, reddish brown or purplish grey, covered (except ventrally) with numerous reddish orange, gold or yellow spots; dorsal fin and upper third of the caudal fin with spots like those on the body; lower two-thirds of the caudal fin dusky; spots on the body of some fish with a faint dark margin; pectoral and pelvic fins and distal part of the anal fin dusky; dark streak along the maxillary groove. Juveniles and small adults with 7 faint dark bars dorsally on the body, the first 2 on the nape, the last on the caudal peduncle; all bars more or less demarcated by small dark spots; no dark spots on the head or fins.

COLOUR IN MARKET Body generally pale brown or grey. Spots on the body usually remain prominent. Caudal fin retains bicoloured appearance.

© Allen W-L To

Adult *E. bleekeri* (Sai Kung Market, Hong Kong)

© Allen W-L To

Adult *E. bleekeri* (Hong Kong)

POPULATION **General** *E. bleekeri* is a widespread species once considered common. It has become rare in many parts of its range. **Fishery dependent** *E. bleekeri* is an important component of the capture-based mariculture industry in South-East Asia, Hong Kong and mainland China. Advanced settlement-stage larvae and juveniles numbering in the millions are annually exported from Thailand and elsewhere to Hong Kong for grow-out mariculture (Sadovy de Mitcheson and Liu 2008).

POPULATION TREND Decreasing
MAJOR THREATS Fishing pressure, particularly on juveniles, is the major threat to *E. bleekeri*.
HABITAT AND ECOLOGY **GENERAL** *E. bleekeri* is found on rocky reefs and adjacent soft substrate in depths of 30–105 m.
COMMENTS *E. bleekeri* is a widespread species that is locally abundant in some areas. However, it is heavily sought after as adults as food and also as late-stage larvae and juveniles for grow-out mariculture. Unrestricted exploitation appears to be a major factor in widespread local declines in South-East Asia.
2008 IUCN RED LIST STATUS Near Threatened

Authors: *KE Carpenter, AS Cornish, M Liu, BC Russell and MA Samoilys*

Epinephelus bontoides (Bleeker 1855)

palemargin grouper (E), mérou bord clair (F), mero de márgenes pálidos (S)

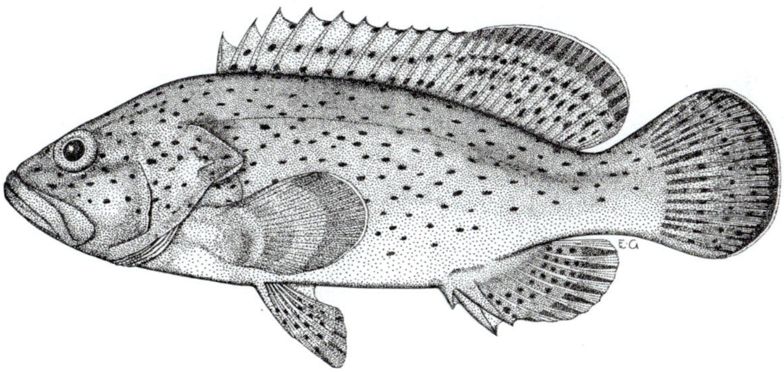

16 cm SL (Indonesia)

Body depth contained 2.8–3.2 times in standard length. Head length 2.3–2.6 times in standard length; interorbital area flat to slightly convex, the dorsal head profile almost straight; preopercle rounded, finely serrate, the ventral serrae slightly enlarged; upper edge of operculum straight; maxilla reaches past rear edge of eye; midlateral part of lower jaw with 2 or 3 rows of teeth. Gill rakers 6–8 on upper limb, 13–15 on lower limb, 21 total. Dorsal fin with 11 spines and 16 or 17 rays, the 3rd to 11th spines subequal and shorter than longest rays, the interspinous membranes incised; anal fin with 3 spines and 8 rays; pectoral-fin rays 18–20; pectoral-fin length contained 1.6–2.0 times in head length; pelvic-fin length 1.9–2.3 times in head length; caudal fin rounded. Midlateral-body scales ctenoid, with auxiliary scales in adults; lateral-line scales 48–51; lateral-scale series 82–86. Pyloric caeca about 18.
MAXIMUM RECORDED SIZE at least 30 cm TL

COLOUR IN LIFE Head and body greyish brown, covered (except ventrally) with well-separated dark reddish brown to black spots, smaller than the pupil and many horizontally elongate; fins darker than the body, the median fins with small dark spots; pectoral, soft dorsal and caudal fins, and sometimes anal fin, dark with a narrow pale yellow or white margin; spinous dorsal fin with 3 rows of oblique black spots. Faint bars are sometimes present on the body in live specimens.

COLOUR IN MARKET Head and body light grey to brown with pale underside. Reddish brown spots remain prominent, and a dark 'moustache' behind the maxilla is often visible.

Adult *E. bontoides* (Dumaguete, Philippines)

Adult *E. bontoides* (Bali, Indonesia)

Epinephelus bontoides is a western Pacific species that occurs in Indonesia (from Java eastwards), Philippines, Taiwan, Solomon Islands and New Britain (Papua New Guinea).

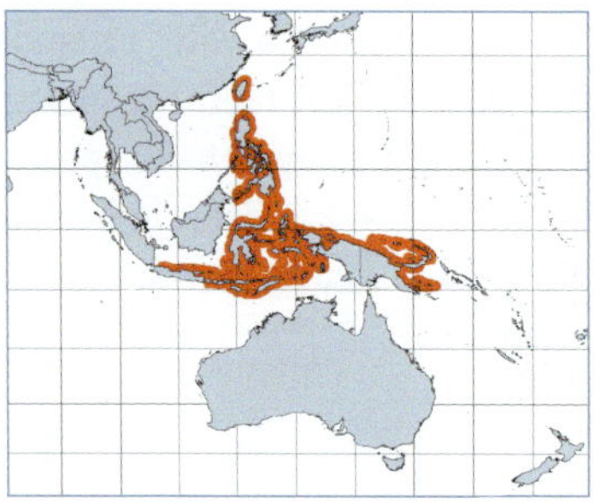

POPULATION **GENERAL** *E. bontoides* is poorly known and patchily distributed throughout its range. It may be locally abundant in suitable habitat (e.g. Bali, Indonesia).

POPULATION TREND Unknown

MAJOR THREATS No major threats are known to *E. bontoides*.

HABITAT AND ECOLOGY **GENERAL** *E. bontoides* is found on mud, rock or cobble bottoms in depths to 30 m.

2008 IUCN RED LIST STATUS Data Deficient

Authors: *PC Heemstra and RF Myers*

Epinephelus bruneus Bloch 1793

longtooth grouper (E), mérou longues dents (F), mero diente (S)

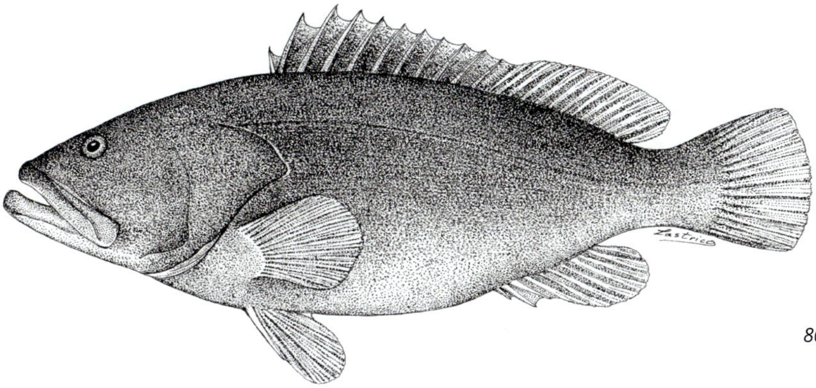

80 cm SL
© FAO

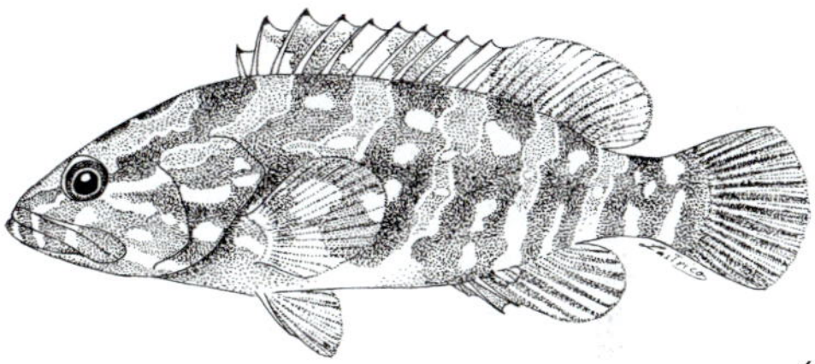

40 cm SL
© FAO

Body elongate, the depth contained 3.0–3.6 times in standard length. Head length 2.3–2.5 times in standard length; interorbital area convex; preopercle angular, with the serrae at angle distinctly enlarged; dorsal spine on opercle inconspicuous; upper edge of operculum convex; posterior and anterior nostrils small and subequal; maxilla reaches past vertical at rear edge of eye, tiny embedded scales on maxilla; no step on ventral edge of maxilla; midlateral part of lower jaw with 2 rows of well-developed canines, those along sides of upper jaw slightly smaller. Gill rakers shorter than gill filaments, 9–11 on upper limb, 16–18 on lower limb, 24–27 total. Dorsal fin with 11 spines and 13–15 rays, the 3rd or 4th spine longest, but shorter than the longest ray, the interspinous membranes deeply incised; anal fin with 3 spines and 8 rays; pectoral-fin rays 17–19; pectoral-fin length contained 1.9–2.5 times in head length; pelvic-fin length 2.2–2.6 times in head length and ending well short of anus; caudal fin rounded. Midlateral-body scales ctenoid, no auxiliary scales on body; lateral-line scales 64–72, some anterior lateral-line scales on adults with branched tubes; lateral-scale series 93–126.

MAXIMUM RECORDED SIZE 136 cm TL

POPULATION GENERAL Little information is currently available for this species across most of its range, although it has been previously abundant in regional fisheries. **FISHERY DEPENDENT** *E. bruneus* was formerly one of the most abundant species caught in Hong Kong waters; it was landed throughout the year (especially abundant from October to May) and was taken by longlines, handlines and trawlers (Chan 1968). In a recent study, almost all *E. bruneus* in retail markets were found to be juveniles (To and Sadovy de

Epinephelus bruneus is narrowly distributed along the coasts of the Republic of Korea, Japan (north to Hegura-jima Island, 37°50′ N), China (north to Shanghai, south to Hong Kong and Hainan Island) and Taiwan.

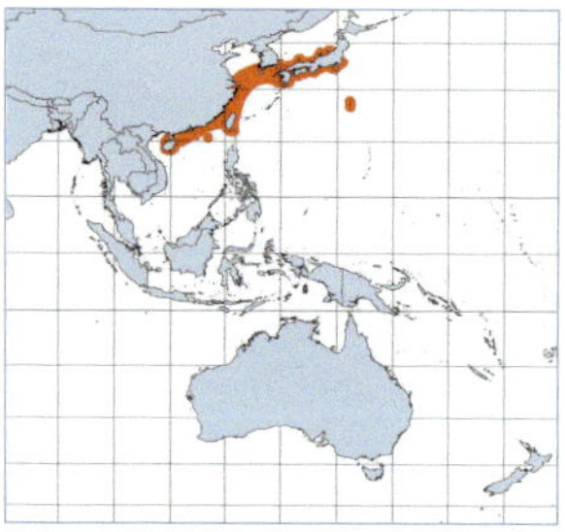

Mitcheson 2009). Landings of *E. bruneus* have shown declines in recent years in South Korea (Song et al. 2005). **FISHERY INDEPENDENT** Wilson (2003) frequently observed *E. bruneus* on Hong Kong artificial reefs. Sadovy and Cornish (2000) described *E. bruneus* as one of the rarest groupers in Hong Kong, especially for fish larger than 50 cm TL.

POPULATION TREND Decreasing

MAJOR THREATS The major threat to *E. bruneus* is fishing pressure, including on juveniles that are captured for mariculture grow-out. It is a relatively large species with a limited distribution, and is a desirable food fish that is unmanaged.

COLOUR IN LIFE Adults (larger than 40 cm SL) dark greyish brown, the bars and bands replaced by faint dorsal blotches or absent altogether, the body covered with small pale grey spots forming short horizontal lines and a mottled pattern; lower margin of the anal fin and lower corner of the caudal fin with a white edge. Juveniles pale yellowish brown, the body with 6 irregular oblique dark bars containing irregular pale spots, first bar extends from the nape to the eye, the last on the caudal peduncle; 3 dark brown bands radiating from the lower part of the eye; some fish with the distal part of interspinous membranes greenish yellow.

COLOUR IN MARKET Generally a lighter shade of brown, bars and bands usually fade and may not be visible, especially in large individuals. Pale grey spots become paler and are irregularly present. Underside of fish becomes markedly pale.

© Matthew T Craig

Adult *E. bruneus* (Okinawa Fish Market, Japan)

© Robert F Myers

Adult *E. bruneus* (Nagoya Aquarium, Japan)

HABITAT AND ECOLOGY **General** *E. bruneus* inhabits rocky reefs and mud bottoms in depths of 20–200 m. Juveniles occur in shallow water. **Reproduction** Chan (1968) reports that gonads of fish taken in summer months were not reproductively active.

COMMENTS *E. bruneus* is one of the most economically important species for aquaculture in Japan, but high mortality during early life stages affects the success of artificial larval rearing (Kato et al. 2004). This partly explains why mariculture usually involves grow-out of wild-caught juveniles (Sadovy de Mitcheson and Liu 2008). *E. bruneus* was often misidentified as *E. moara* in previous literature.

2008 IUCN RED LIST STATUS Vulnerable (A4d)

Authors: *T-C Chan, M Liu, YJ Sadovy de Mitcheson and W-L To*

Epinephelus caninus (Valenciennes 1843)

dogtooth grouper (E), mérou gris (F), mero dentón (S)

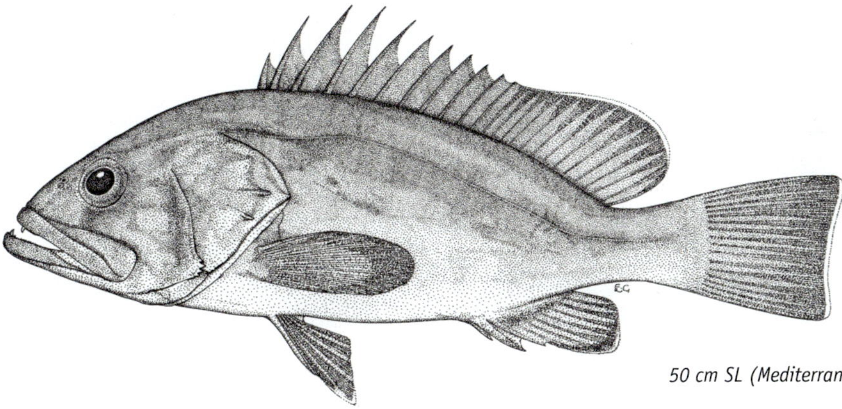

50 cm SL (Mediterranean Sea)

Body depth contained 2.7–3.0 times in standard length. Head length 2.3–2.5 times in standard length; interorbital area convex; preopercle angular, with 3–5 distinctly enlarged serrae at the angle and usually a slender, antrorse or ventrally directed spine on ventral edge near the angle; subopercle and interopercle serrate; upper edge of operculum distinctly convex; eye diameter more than interorbital width in fish less than 30 cm SL, but noticeably less than interorbital in fish more than 45 cm SL; posterior nostril 2 or 3 times size of anterior nostril; maxilla scaly, reaching to or beyond vertical at rear edge of eye; midlateral part of lower jaw with 2 rows of teeth, the inner teeth larger; canines at front of jaws well developed. Gill rakers 8–10 on upper limb, 15–17 on lower limb, 23–27 total. Dorsal fin with 11 spines and 13 or 14 rays, the membranes deeply incised between the spines; anal fin with 3 spines and 8 rays; pectoral fins with 17 or 18 rays, distinctly longer than pelvic fins, pectoral-fin length contained 1.7–2.2 times in head length; pelvic fins clearly not reaching anus; caudal fin truncate, the corners slightly rounded. Body scales distinctly ctenoid, with numerous auxiliary scales; lateral-line scales 70–79; lateral-scale series about 120–135.

MAXIMUM RECORDED SIZE 164 cm TL

COLOUR IN LIFE Uniformly dark reddish brown or greyish violet to yellowish grey; posterior parts of median fins with a distinct white edge. Usually 2 or 3 dark bands radiating posteriorly from the eye, the uppermost extends from the eye to the lower opercular spine, the 2nd runs from the lower edge of the eye across the angle of preopercle to the juncture of the interopercle and subopercle, and the 3rd band (usually the faintest) extends from the dark moustache streak at the upper edge of the maxilla to the lower edge of the preopercle. The dark bands on the head are not discernible in fish larger than 45 cm SL.

COLOUR IN MARKET Generally pale brown or grey with few distinct markings. Dark bands radiating from the eye are faint if present at all. White edges of median fins may or may not be present.

Adult *E. caninus* (Ibiza, Balearic Islands, Mediterranean)

Epinephelus caninus is an eastern Atlantic species that is distributed from Portugal to Angola, including the Mediterranean and Canary Islands (Spain). There are no records from the northern Mediterranean coasts of Spain and France, or the north-western coast of Italy.

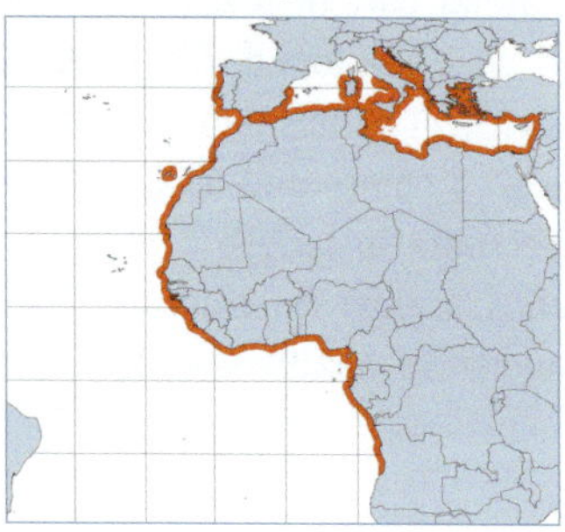

POPULATION GENERAL Although harvested in the Mediterranean and along the West Coast of Africa, *E. caninus* is a poorly known species and little information is available on its population. **FISHERY DEPENDENT** The species was reported as common in the fish markets of Dakar (Heemstra and Randall 1993). **FISHERY INDEPENDENT** Fishery-independent surveys at protected and unprotected areas in the Mediterranean indicated that the species is relatively rare; the highest abundance observed was 0.2 individual per 250 m² (García Charton et al. 2000).

POPULATION TREND Unknown

MAJOR THREATS The major threat to *E. caninus* is overfishing.

HABITAT AND ECOLOGY **GENERAL** *E. caninus* occurs on sand-mud bottoms in depths of 30–400 m. **FEEDING** Feeds on fishes and crustaceans. **AGE AND GROWTH** A single specimen 164 cm TL and 67 kg from Majorca was aged at 75 years (Morales-Nin et al. 2005).

COMMENTS *E. caninus* is similar to *E. goreensis,* which has 16 dorsal-fin rays, a smaller head (its length contained 2.5–2.7 times in standard length), and in which the upper edge of the operculum is almost straight. Another similar species, *Hyporthodus haifensis*, has a rounded caudal fin, 9 anal-fin rays, body depth usually greater (depth contained 2.4–2.8 times in standard length), pelvic fins subequal to pectoral fins and reaching to or beyond the anus, no scales on the maxilla, and fewer lateral-scale series (104–112).

2008 IUCN RED LIST STATUS Data Deficient

Authors: *AA Bertoncini, JH Choat, MT Craig, BP Ferreira, PC Heemstra, DA Pollard and LA Rocha*

Epinephelus chabaudi (Castlenau 1861)

moustache grouper (E), mérou moustache (F), mero bigotudo (S)

24 cm SL (South Africa)

Body depth contained 2.5–2.8 times in standard length. Head length 2.3–2.6 times in standard length; interorbital area convex; preopercle angular, the serrae at angle enlarged and on some specimens they continue onto posterior part of lower limb; upper edge of operculum convex; diameter of posterior nostril 2–4 times larger than anterior nostril; maxilla scaly, reaching to or slightly past vertical at rear edge of eye; midlateral part of lower jaw with 2 rows of teeth in fish 20–56 cm SL and 3 or 4 rows in fish 70–90 cm SL. Gill rakers 8 or 9 on upper limb, 15–17 on lower limb, 22–25 total. Dorsal fin with 11 spines and 13 or 14 rays, the 3rd or 4th spine longest and usually slightly longer than longest ray, the interspinous membranes deeply incised; anal fin with 3 spines and 9 (rarely 8) rays; pectoral-fin rays 17–18; pectoral-fin length contained 1.7–2.0 times in standard length; pelvic fins not reaching anus, their length 1.9–2.4 times in standard length; caudal fin truncate. Lateral-body scales distinctly ctenoid, auxiliary scales sparse or absent on body scales; lateral-line scales 61–69; lateral-scale series 100–114.

MAXIMUM RECORDED SIZE 137 cm TL

COLOUR IN LIFE This species has variable colouration depending on size and depth of capture. For specimens up to 35 cm SL: head and body greyish brown, shading to pale purplish grey ventrally, with 3 broad dark bars between the dorsal fin and lateral line, continuing faintly below the lateral line; broad dark blotch dorsally on the caudal peduncle, reaching about halfway to the lateral line; the nape and dorsal part of the head faintly yellowish; prominent dark brown streak along the edge of the maxillary groove; fins greyish brown, the tips of the spinous dorsal-fin membranes dark reddish brown; rays of median and pelvic fins paler than the membranes; pectoral-fin rays brown, the membranes pale. Colour of specimens 88–115 cm SL from deep water (off Kenya; Morgans 1982): rosy slate or chocolate brown; throat and hidden membranes of head pinkish grey; eye nondescript silvery; no bars, spots, blotches, marginal coloured bands, or conspicuous moustache streak.

COLOUR IN MARKET Generally a drab grey to brown with the upper body a darker shade than the belly. Three or 4 dark bars may be faintly visible. A dark maxillary streak is generally present.

© Matthew T Craig

Adult *E. chabaudi* (damaged caudal fin; Seychelles)

© Phillip C Heemstra

Juvenile *E. chabaudi* (KwaZulu-Natal, South Africa)

Epinephelus chabaudi is known from the east coast of Africa (from Kenya to Knysna, South Africa, 34°3′ S), Seychelles and the Kerala coast of India; however, there are no records between Kenya and Durban (South Africa).

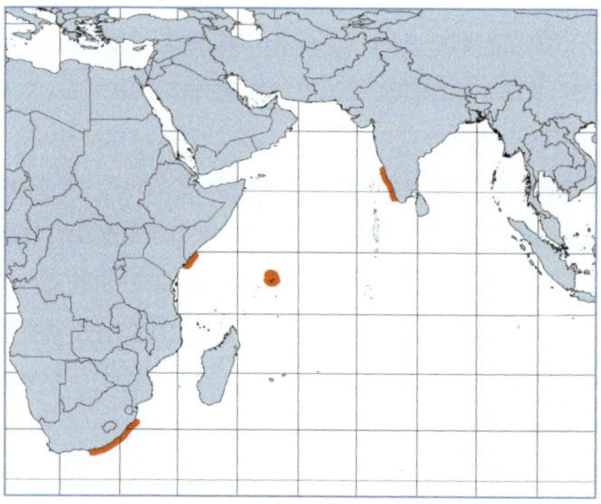

POPULATION GENERAL Little is known about this species; however, it was reported to be common in trawl catches off the Kerala coast (India; Talwar and Kacker 1984).

POPULATION TREND Unknown

MAJOR THREATS No major threats have been identified for this species.

HABITAT AND ECOLOGY GENERAL *E. chabaudi* is a deep-water species that occurs over rocky bottoms to depths of up to 300 m.

2011 IUCN RED LIST STATUS Data Deficient

Authors: *MT Craig and ST Fennessy*

Epinephelus chlorocephalus

(Valenciennes 1830)

Tonga grouper (E), mérou Tonga (F), mero Tonga (S)

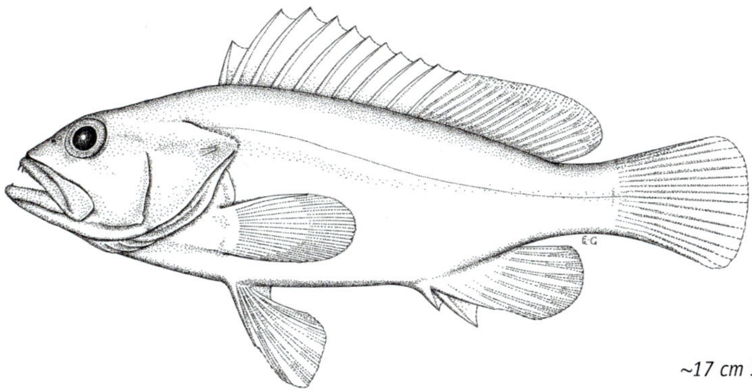

~17 cm SL (Tonga Islands)

Body depth contained 3.4 times in standard length (one specimen known, 171 mm SL). Head length 2.6 times in standard length; head pointed, the interorbital area flat, the dorsal profile slightly convex; preopercle subangular, with 5 distinctly enlarged serrae at the angle; upper edge of operculum straight; maxilla not reaching past vertical at rear edge of eye; midlateral part of lower jaw with 2 rows of teeth. Gill rakers 8 on upper limb, 15 on lower limb. Dorsal fin with 11 spines and 17 rays, the 4th spine longest, its length contained 2.45 times in head length; anal fin with 3 spines and 8 rays; pectoral fins with 18 rays, pectoral-fin length 1.75 times in head length; pelvic fins not reaching anus, their length 1.9 times in head length; caudal fin rounded. Lateral-body scales ctenoid; lateral-line scales 52–54.

MAXIMUM RECORDED SIZE 17 cm TL

COLOUR IN LIFE Dorsal part of the head and snout green, the cheeks pale green; body greenish with 7–8 alternating stripes of greenish brown and yellow-orange; dorsal and anal fins greenish, spotted with reddish brown at their base; paired fins light greenish without spots; caudal fin greenish with faint reddish bars.

COLOUR IN MARKET No market specimens are known.

Epinephelus chlorocephalus is known only from the Tonga Islands.

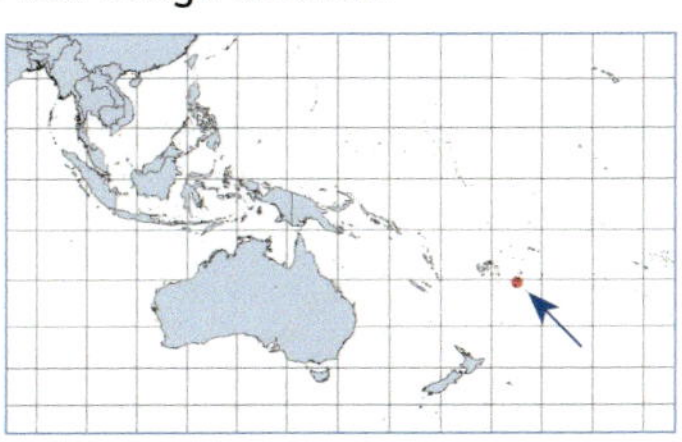

POPULATION GENERAL *E. chlorocephalus* is known from a single specimen. The species is apparently rare.

POPULATION TREND Unknown

MAJOR THREATS Unknown

HABITAT AND ECOLOGY No information on habitat and biology is available for this species.

2008 IUCN RED LIST STATUS Data Deficient

Authors: *PC Heemstra, M Kulbicki and KL Rhodes*

Epinephelus chlorostigma

(Valenciennes 1828)

brownspotted grouper (E), mérou pintade (F), mero pintado (S)

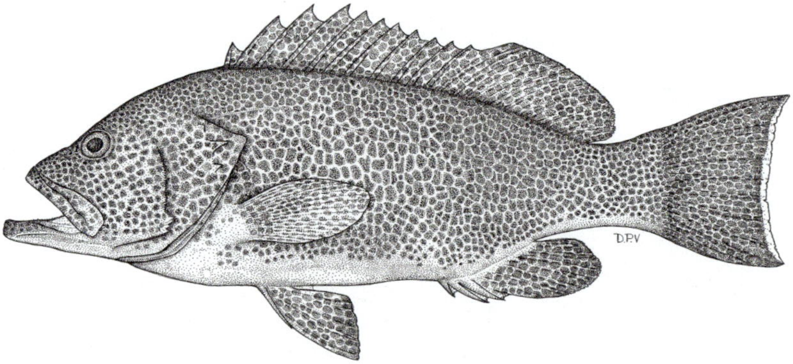

32 cm SL (northern Mozambique)

Body depth contained 2.8–3.3 times in standard length; body width 1.8–2.2 times in the depth. Head length 2.4–2.7 times in standard length; interorbital slightly convex; preopercle angle rounded, with 4–7 enlarged serrae at angle below a slight notch; upper edge of operculum straight; posterior nostril not noticeably larger than anterior nostril; maxilla reaches about to vertical at rear edge of eye; maxilla scaly, with a low step on posterior part of ventral edge; midlateral part of lower jaw with 2–4 rows of teeth, the inner ones about twice the size of outer teeth. Gill rakers 8–11 on upper limb, 15–18 on lower limb, 23–29 total; gill rakers longer than gill filaments. Dorsal fin with 11 spines and 16–18 rays, the 3rd or 4th spine longest, its length contained 2.4–3.2 times in head length, the interspinous membranes slightly to moderately incised; anal fin rounded or angular, with 3 spines and 8 rays, the 3rd spine longer than 2nd; pectoral-fin rays 17–19; pectoral fins usually slightly longer than pelvic fins, pectoral-fin length contained 1.6–2.0 times in head length; pelvic-fin length 1.8–2.3 times in head length; caudal-peduncle depth 3.0–3.6 times in head length; caudal fin truncate or slightly emarginate. Lateral-body scales ctenoid, with auxiliary scales; lateral-line scales 48–53; lateral-scale series 96–122. Pyloric caeca 26–52.

MAXIMUM RECORDED SIZE 80 cm TL

Epinephelus chlorostigma is an Indo-Pacific species that ranges from the Red Sea to KwaZulu-Natal (South Africa) in the west, extending east (including Madagascar) to the western Pacific, north to southern Japan, and south to New Caledonia.

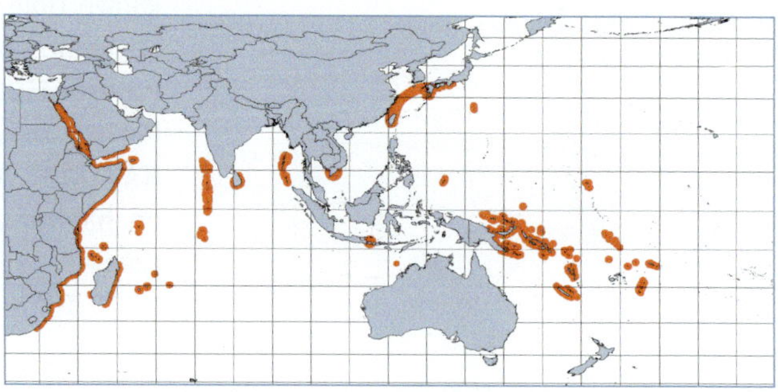

COLOUR IN LIFE Head, body and fins with small irregular close-set dark brown spots, the ground colour forming a pale network; caudal fin usually with a white line along its posterior margin; spots on the pectoral fins mainly confined to the rays.

COLOUR IN MARKET Generally maintains similar colouration as in life although light blotches may form if the fish is left on the deck of a boat or the floor of a fish market.

© Stephen Newman

Adult *E. chlorostigma* (Scott Reef, Western Australia)

© Roger Steene

Adult *E. chlorostigma* (Gulf of Aqaba, Red Sea)

POPULATION **General** *E. chlorostigma* is moderately abundant throughout its range. **Fishery dependent** *E. chlorostigma* varies in its importance to fisheries depending on locality. It is commonly taken in eastern Madagascar (up to 20% of grouper catch, about 30 t per year; Refrigépêche, Madagascar), and was once the most common grouper taken in Somalia (Darar 1994). In Seychelles, the species was heavily fished and catch rates declined until closure of the 'mothership-dory' fishery in 1994 (Mees 1996). **Fishery independent** There are few quantitative fishery-independent data for *E. chlorostigma*. Published surveys indicate, however, that it is not commonly seen by divers (Chater et al. 1995, Robertson et al. 1996, Pereira 2003).

POPULATION TREND Stable

MAJOR THREATS No major threats are known for this species.

HABITAT AND ECOLOGY **General** *E. chlorostigma* is a deep-water, reef-associated species found in a wide range of habitats in depths to 280 m. **Feeding** Feeds on fishes and invertebrates. **Age and growth** *E. chlorostigma* reaches at least 26 years; the von Bertalanffy

growth function from the Seychelles is $L = 386.8(1 - e^{-0.277(Age+0.141)})$, where L is total length in cm (Grandcourt 2002). **REPRODUCTION** *E. chlorostigma* is a protogynous hermaphrodite (Moussac 1996). Females mature as small as 25 cm TL and sex change occurs from 34 cm to 56 cm TL. Spawning period appears protracted and variable throughout the range of this species.
2008 IUCN RED LIST STATUS Least Concern

Authors: *AA Bertoncini, JH Choat, MT Craig, ST Fennessy, BP Ferreira, J Robinson and LA Rocha*

Epinphelus cifuentesi

Lavenberg and Grove in Heemstra and Randall 1993

olive grouper (E), mérou poule (F), gallina (S)

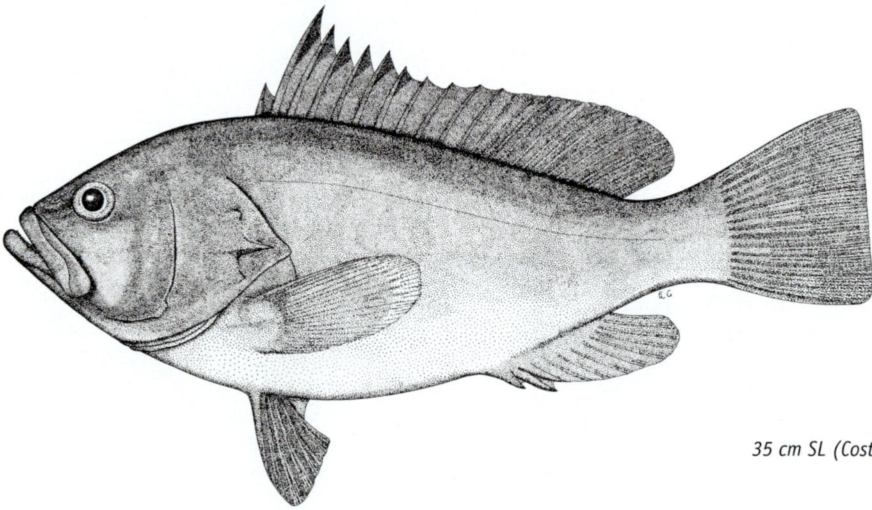

35 cm SL (Costa Rica)

Body depth contained 2.6–3.0 times in standard length (for fish 15–54 cm SL). Head length 2.5–2.8 times in standard length; interorbital area convex, the width greater than eye diameter for fish larger than 23 cm SL; preopercle subangular, the serrae at angle slightly enlarged; upper edge of operculum slightly convex; nostrils subequal or posterior nostril slightly larger; maxilla reaches past vertical through centre of eye; 2 rows of teeth at midside of lower jaw. Gill rakers 9–11 on upper limb, 17–20 on lower limb, 27–31 total. Dorsal fin with 11 spines and 14–16 rays, the 2nd spine longest and the interspinous membranes distinctly incised; anal fin with 3 spines and 9 rays; pectoral-fin rays 17–19, the fin length contained 1.7–1.9 times in head length; pelvic-fin length 2.0–2.3 times in head length, pelvic-fin origin slightly in front of lower end of pectoral-fin base; caudal fin truncate with rounded corners in adults, convex in juveniles. Midlateral-body scales distinctly ctenoid, with auxiliary scales in large adults; lateral-line scales 71–76; lateral-scale series 131–144.
MAXIMUM RECORDED SIZE 100 cm TL

COLOUR IN LIFE Body and head pale brown, with a distinct greenish sheen on the body; lips, tip of lower jaw and dorsal part of the head darker. Fins brown, generally darker than the body, with a blue-green sheen, particularly evident on the pectoral fins, which are cream-coloured distally; pelvic fins pale brown with dusky membranes; dorsal fin dusky distally in fish 57–61 cm TL, the margin between 1st to 5th spines nearly black in fish 33–43 cm TL; anal fin dusky distally with white edge (in fish 33–62 cm TL). Juveniles (15 cm SL) have no distinctive markings except for a dark maxillary groove. Myomeres (muscle bands) are visible throughout the midbody musculature.

COLOUR IN MARKET Pale brown to cream with no distinctive features except for the black maxillary groove and visible myomeres on the body.

Adult *E. cifuentesi* (Galápagos Islands, Ecuador)

POPULATION **General** *E. cifuentesi* is apparently rare outside of the Galápagos Islands. **Fishery dependent** *E. cifuentesi* was once the most common species taken in the deep-water fishery of the Galápagos Islands.

Epinephelus cifuentesi is an eastern Pacific species found from the Gulf of Tehuantepec (Mexico) to Ecuador. It is more common at island localities and is known from the Galápagos Islands, Cocos Island (Costa Rica) and the Paramount Seamount.

Annual landings of *E. cifuentesi* decreased from a peak of 14.0 t in 1998 to an average of 2.7 t from 2000 to 2003 (Murillo et al. 2002, Nicolaides et al. 2002). **Fishery independent** Fishery-independent surveys in the Galápagos Islands have shown evidence of decline in the standing biomass of this species; however, as of 2003, it appears to have remained stable (Murillo et al. 2002, Nicolaides et al. 2002).

POPULATION TREND Unknown

MAJOR THREATS The major threat to *E. cifuentesi* is fishing pressure, particularly in the Galápagos Islands.

HABITAT AND ECOLOGY **General** *E. cifuentesi* is a deep-water species found on rocky reefs in depths of 40–120 m.

COMMENTS At present, the fishery of the Galápagos Islands has shifted effort towards the take of sea cucumbers, thus probably relieving pressure from *E. cifuentesi* and other demersal fish species. However, monitoring of the status of this species is advisable given its size and limited geographic range.

2008 IUCN RED LIST STATUS Near Threatened

Authors: *JH Choat, MT Craig, BP Ferreira and LA Rocha*

Epinephelus clippertonensis

Allen and Robertson 1999

Clipperton grouper (E), mérou Clipperton (F), mero de Clipperton (S)

Body robust and compressed; end of upper jaw exposed; edge of operculum with 3 broad, flat spines; preoperculum without spines. Dorsal fin with 11 spines and 15–16 rays, 3rd and 4th spines longest; anal fin with 3 spines and 8 rays; pectoral-fin rays 18–20; caudal fin rounded. Lateral line complete, ending at base of tail and smoothly arched. Scales rough and relatively small; 48–51 lateral-line scales; 46–50 scales around base of caudal fin, and 16–18 above the lateral line.*
MAXIMUM RECORDED SIZE 61 cm TL

COLOUR IN LIFE Pale grey to grey-brown, with white spots and flecks on the head and body; large fish can switch to a bicoloured pattern by darkening the rear half and lightening the front half of the body. Tips of dorsal-fin spines often edged with red and white. Juveniles have numerous small dark spots.

COLOUR IN MARKET Body shade darkens to brown with irregular light blotches. Underside of the fish lighter with distinct blotches tinted with iridescence.

© Russell D Vetter

Adult *E. clippertonensis* (Clipperton Atoll)

© Gerald R Allen

Adult *E. clippertonensis* (Clipperton Atoll)

* *No line drawing available*

POPULATION **General** *E. clippertonensis* is a common species at insular localities throughout its range. **Fishery independent** *E. clippertonensis* was reported as common in the original species description at Clipperton Atoll (Allen and Robertson 1999). **Genetics** *E. clippertonensis* appears to have genetically isolated populations across its range (Craig et al. 2006b).

Epinephelus clippertonensis is an eastern Pacific species occurring from the Pacific coast of Baja California (Mexico) to Clipperton Atoll and the Revillagigedos Islands.

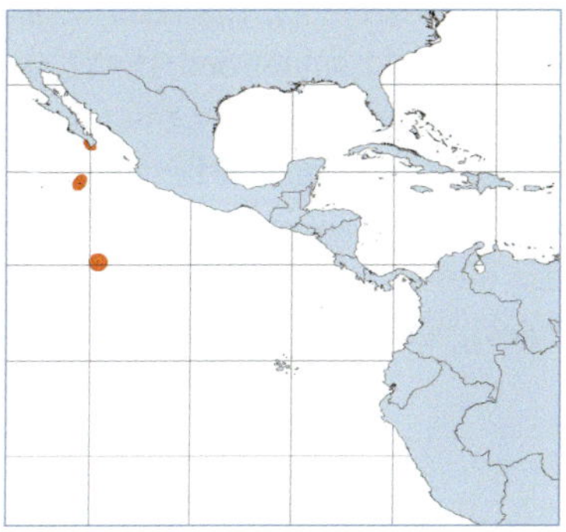

POPULATION TREND Stable

MAJOR THREATS There are no major threats known to this species.

HABITAT AND ECOLOGY **General** *E. clippertonensis* is an insular, reef-associated species inhabiting depths to at least 30 m.

COMMENTS *E. clippertonensis* is very similar to *E. labriformis* (Craig and Hastings 2007) and was often reported as *E. labriformis* or *Epinephelus* sp. prior to its description in 1999. The main difference is colour pattern, with *E. clippertonensis* lacking the olive green body colour and distinct white blotches that are prominent in *E. labriformis*, and in scale counts.

2008 IUCN RED LIST STATUS Least Concern

Authors: *AA Bertoncini, JH Choat, MT Craig, BP Ferreira and LA Rocha*

Epinephelus coeruleopunctatus

(Bloch 1790)

whitespotted grouper (E), mérou taches blanches (F), mero nevero (S)

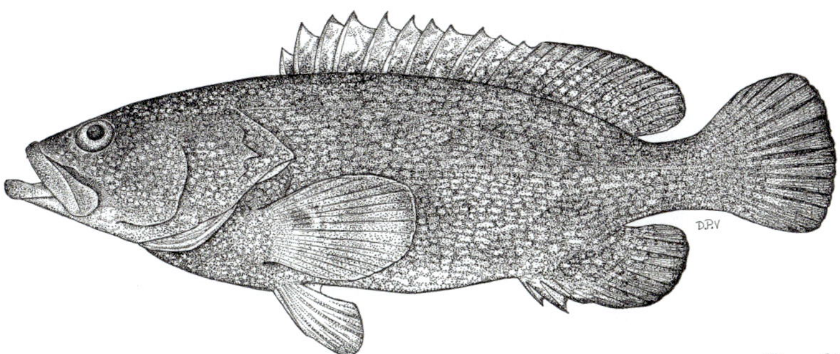

43 cm SL (South Africa)
© SAIAB

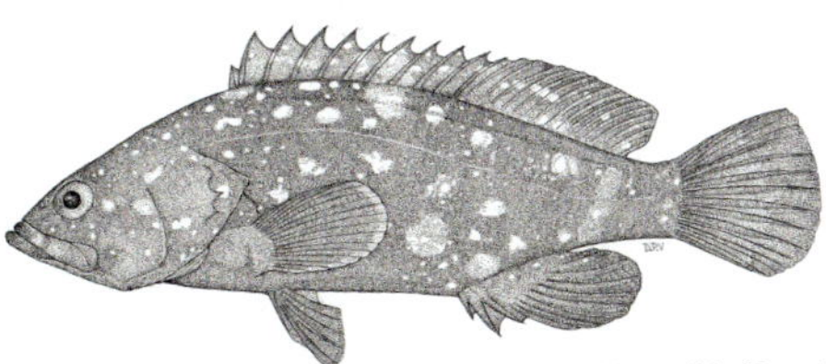

17 cm SL (South Africa)
© SAIAB

3 cm SL (South Africa)
© SAIAB

Body depth distinctly less than head length, depth contained 2.9–3.4 times in standard length. Head length 2.3–2.5 times in standard length; head pointed, the interorbital area usually flat (sometimes concave or slightly convex), and the dorsal profile almost straight; preopercle rounded, finely serrate; opercular spines inconspicuous; upper edge of operculum straight, sinuous or slightly convex; in fish more than 35 cm SL, the posterior nostril becomes vertically elongated, its length 5 or 6 times greater than diameter of anterior nostril; maxilla reaches short of or slightly beyond vertical at rear edge of eye; maxilla naked, mostly covered by upper lip; canines at front of jaws small or absent; midlateral part of lower jaw with 3–5 rows of small teeth. Gill rakers 8–10 on upper limb, 13–17 on lower limb; fish larger than 20 cm SL with only 4–8 developed rakers on lower limb, and the rudiments are difficult to count because of intercalated bony tooth plates; gill rakers shorter than gill filaments, the raker at angle of gill arch about twice as long as adjacent rakers. Dorsal fin with 11 spines and 15–17 rays, the 3rd or 4th spine longest, its length contained 2.7–3.6 times in head length, the interspinous membranes incised; anal fin with 3 spines and 8 rays; pectoral fins large and fleshy, with 17–19 rays; pectoral-fin length contained 1.5–2.1 times in head length; pelvic fins end well short of anus, their length 2.0–2.7 times in head length; caudal fin rounded. Lateral-body scales ctenoid (1 fish, 43 cm SL, with 'mostly smooth' lateral-body scales), with auxiliary scales; lateral-line scales 51–61; lateral-scale series 86–109.

MAXIMUM RECORDED SIZE 76 cm TL

Epinephelus coeruleopunctatus is widespread throughout the Indo-Pacific and ranges from East Africa south to East London (South Africa) and east to Tonga. It is unknown from the Red Sea, but it does occur in the Persian Gulf and Oman.

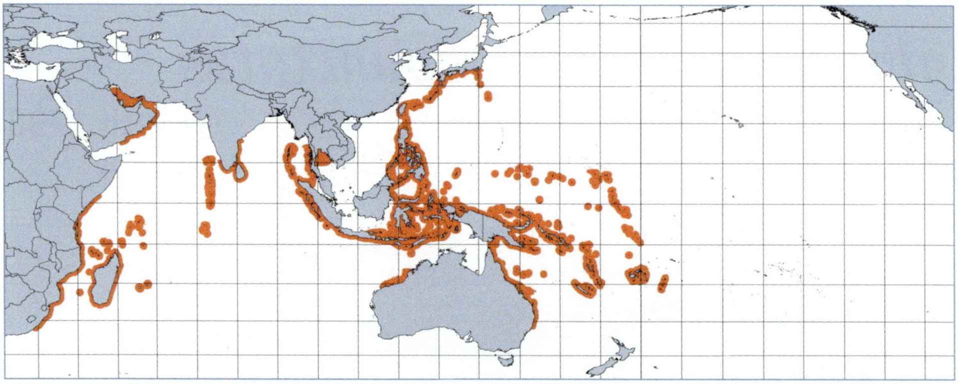

POPULATION **GENERAL** *E. coeruleopunctatus* is a widespread but relatively uncommon species throughout its range. **FISHERY DEPENDENT** *E. coeruleopunctatus* forms part of artisanal and small-scale commercial fisheries in much of its range. It is targeted for both aquarium use and by the live reef food fish trade but no information is available on levels of take.

POPULATION TREND Unknown

MAJOR THREATS The major threat to this species is fishing pressure.

HABITAT AND ECOLOGY **GENERAL** *E. coeruleopunctatus* is a reef-associated species that occurs in rocky or coral-rich areas of deep lagoons, channels and outer reef slopes, usually in or near caves, to depths of at least 30 m. **FEEDING** Feeds mostly on crustaceans and to a lesser degree on fishes.

COLOUR IN LIFE Adults brownish grey, the body covered with small pale spots overlain with large pale blotches; oblique black saddle on the rear half of the peduncle; 4 or 5 indistinct black blotches at the base of the dorsal fin; prominent black streak on the maxillary groove. Large adults (over 40 cm SL) brownish, covered with indistinct contiguous small pale spots. Juveniles (less than 20 cm SL) dark grey to black, covered with prominent pupil-size white spots and smaller white dots. **COLOUR IN MARKET** Body is generally light grey to black with a patchy appearance of lighter blotches. Small spots generally present. Prominent black streak in the maxillary groove.

Adult *E. coeruleopunctatus* (Diego Garcia, Chagos Archipelago)

Adult *E. coeruleopunctatus* (Okinawa Fish Market, Japan)

Adult *E. coeruleopunctatus* (Big Drop Off, Palau)

COMMENTS *E. coeruleopunctatus* is often confused with three other species: *E. ongus, E. summana* and *E. corallicola*. Adults of *E. ongus* also have a pattern of pale spots and blotches and a prominent black maxillary streak; but the white spots tend to form irregular longitudinal bands on the body, the soft dorsal, caudal and anal fins have a blackish brown margin with a narrow white edge, and there is no oblique dark spot on the caudal peduncle. The juveniles (less than 15 cm SL) have close-set uniformly small white spots covering the body and most of the median fins, but *E. coeruleopunctactus* juveniles have scattered prominent irregular white spots of various sizes, with only a few on the median fins. *E. ongus* has 15–17 pectoral-fin rays, usually 15 dorsal-fin rays, and 48–53 lateral-line scales. In *E. ongus* the upper edge of the operculum is very convex, in contrast to the nearly straight or slightly convex margin of *E. coeruleopunctatus*. *E. summana* has the body and median fins dark brown or brownish grey, covered with very small white spots, and superimposed on the body are several pale brown blotches. *E. summana* also has fewer dorsal- and pectoral-fin rays (14–16 and 16–18, respectively) and lateral-line scales (49–54). The shape of the upper edge of the operculum is intermediate between *E. coeruleopunctatus* and *E. ongus*. Adults of *E. corallicola* do not have white spots, but the juveniles have large prominent black-edged white spots. *E. corallicola* also differs in having the posterior nostril noticeably elongated at a smaller size (15 cm SL vs 35 cm SL in *E. coeruleopunctatus*).

2008 IUCN RED LIST STATUS Least Concern

Authors: *ST Fennessy, M Kulbicki, RF Myers, KL Rhodes, BC Russell and BM Yeeting*

Epinephelus coioides (Hamilton 1822)

orangespotted grouper (E), mérou taches oranges (F), mero de pintas naranjas (S)

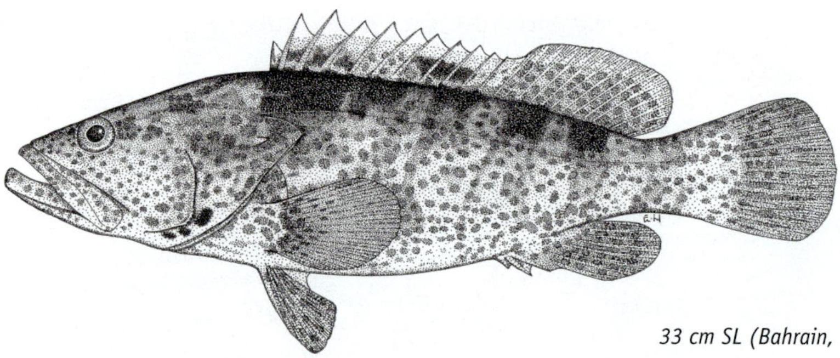

33 cm SL (Bahrain, Persian Gulf)
© SAIAB

Body elongate, the depth contained 2.9–3.7 times in standard length; body width 1.4–2.0 times in the depth. Head length 2.3–2.6 times in standard length; snout length 1.8–1.9 times in upper-jaw length; interorbital flat or slightly convex, interorbital width 5.0–6.2 times in head length and 2.1–3.2 times in upper-jaw length; preopercle subangular, with enlarged serrae at the angle and a broad shallow notch just above the angle; upper edge of operculum straight or somewhat convex; posterior and anterior nostrils subequal; maxilla reaches to or slightly past a vertical at rear edge of eye, maxilla width 4.2–5.5% of standard length; upper-jaw length 17–20% of standard length, midlateral part of lower jaw with 2 or 3 rows of subequal teeth. Gill rakers 8–10 on upper limb, 14–17 on lower limb, 23–26 total; adults with small bony platelets on lateral side of first gill arch. Dorsal fin with 11 spines and 14–16 rays, the 3rd or 4th spine longest, contained 2.9–4.0 times in head length, the interspinous membranes distinctly incised; anal fin with 3 spines and 8 rays, the 3rd spine usually longer than the 2nd, the fin margin rounded; pectoral-fin length 1.6–2.2 times in head length; pectoral-fin rays 18–20; pelvic-fin length 1.9–2.7 times in head length; caudal fin rounded. Lateral-body scales ctenoid, with minute auxiliary scales; lateral-line scales 58–65; lateral-line tubes of anterior scales branched in adults; lateral-scale series 100–118. Pyloric caeca numerous (about 50–60).

MAXIMUM RECORDED SIZE 114 cm TL

COLOUR IN LIFE Head and body tan dorsally, shading to whitish ventrally; numerous small brownish orange or reddish brown spots on head, body and median fins; the spots become darker, smaller and more numerous with growth; body with 5 faint irregular oblique dark bars which bifurcate ventrally; first dark bar below the anterior dorsal-fin spines, last bar on the caudal peduncle; 2 dark spots on the interopercle and another 1 or 2 at the subopercle–interopercle junction.

COLOUR IN MARKET Body colour usually fades to a light tan and orange spots are prominent. Tips of fins may take on a darker hue than the rest of the body.

© Stephen Newman

Adult *E. coioides* (Kimberley coast, Western Australia)

© Matthew Pember

Juvenile *E. coioides* (Port Smith, Western Australia)

Juvenile *E. coioides* (Abu Dhabi, United Arab Emirates)

© Edwin Grandcourt

Adult *E. coioides* (Townsville, Australia)

© Luiz A Rocha

Epinephelus coioides occurs in the Red Sea and south to at least Durban (South Africa), eastwards to Palau and Fiji, north to the Ryukyu Islands (Japan), and south to the Arafura Sea and northern Australia (Heemstra and Randall 1993). It has also migrated through the Suez Canal to the eastern Mediterranean (Randall 1995a).

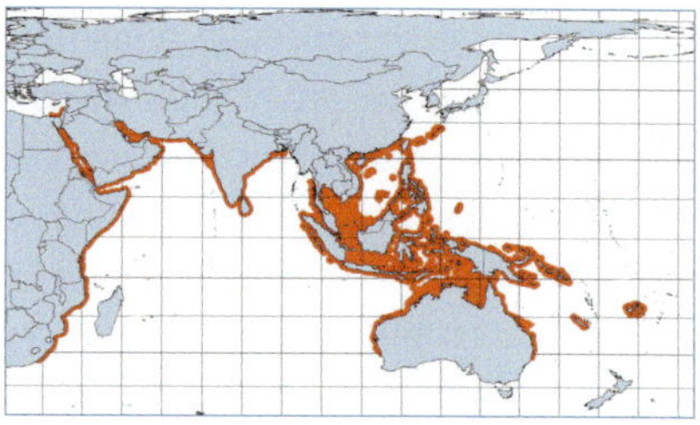

POPULATION **GENERAL** *E. coidoides* is a widespread and common species. **FISHERY DEPENDENT** Although explicit fishery-dependent data are unavailable, the species is commonly taken in many fisheries and is considered an important component of fisheries in the Persian Gulf, India, Singapore, Hong Kong and Taiwan. Imports of 'plate-sized', wild-caught fish into the Hong Kong marketplace have declined substantially over the last decade. In southern China, and from several countries in South-East Asia including the Philippines and Vietnam, large numbers of wild-caught *E. coioides* of 2–10 cm TL are used for capture-based mariculture (Liu and Sadovy de Mitcheson 2009a). The species is also one of the few groupers to be produced by hatcheries at commercial levels. Despite the availability of full-cycle (larvae to adult) hatchery production, juveniles are still taken from the wild for 'fattening' when it is more cost effective to do so (YJ Sadovy de Mitcheson pers. comm.).

POPULATION TREND Decreasing

MAJOR THREATS Overfishing is the major threat to this species as it is taken not only as adults but also in large numbers as juveniles for the 'grow-out' mariculture industry. However, given its occurrence in brackish water and the noted increase in coastal development, habitat destruction may also threaten *E. coioides.*

HABITAT AND ECOLOGY **GENERAL** *E. coioides* inhabits a wide range of habitats including shallow reefs and lagoons with brackish and/or turbid waters in depths to at least 30 m. Juveniles are commonly found in shallow waters of estuaries over sand, mud and gravel and among mangroves. **FEEDING** Feeds on fishes and invertebrates. **AGE AND GROWTH** Reaches a maximum age of at least 22 years (Matthews and Samuel 1991). **REPRODUCTION** *E. coioides* is a diandric protogynous hermaphrodite, in which initial sexual differentiation can be influenced by social factors (Liu and Sadovy de Mitcheson 2009a), and is an aggregate spawner (Grandcourt et al. 2009, Liu and Sadovy de Mitcheson 2009a). In the Persian Gulf, females are mature at 25–30 cm TL (2–3 years old), and sexual transition occurs at a length of 55–75 cm (Mathews et al. 1986), while the major spawning period is from March to June. In the southern Persian Gulf, spawning has been documented from March to May (Grandcourt et al. 2009). In New Caledonia, spawning aggregations form in late October to early December (M Kulbicki, pers. comm., 2004) and are known to occur inshore.

COMMENTS *E. coioides* is an important component of many fisheries and as such has been used considerably in the aquaculture industry, particularly in South-East Asia.

E. coioides is often mistaken for *E. malabaricus* and *E. tauvina*. The colour patterns of all 3 species are similar, but the dark spots of *E. malabaricus* are smaller and blackish brown (not reddish brown or brownish orange, as on *E. coioides*). *E. malabaricus* also has irregular white spots on the head and body (no white spots on *E. coioides*). *E. tauvina* often has a black blotch (larger than eye) on the body at the base of the last 4 dorsal-fin spines and extending onto the lower part of the fin, and juveniles have the dark spots on the median fins so close-set that the intervening pale spaces appear as a pale reticulum. *E. tauvina* also has a longer jaw (upper jaw length 21–24% of standard length vs 17–20% in *E. coioides*), usually more gill rakers (17–20 on lower limb vs 14–17 in *E. coioides*), and no small bony platelets on the lateral side of the 1st gill arch.

2004 IUCN RED LIST STATUS Near Threatened

Authors: *AS Cornish, ML Harmelin-Vivien, M Liu and DA Pollard*

Epinephelus corallicola (Valenciennes 1828)

coral grouper (E), mérou corail (F), mero de coral (S)

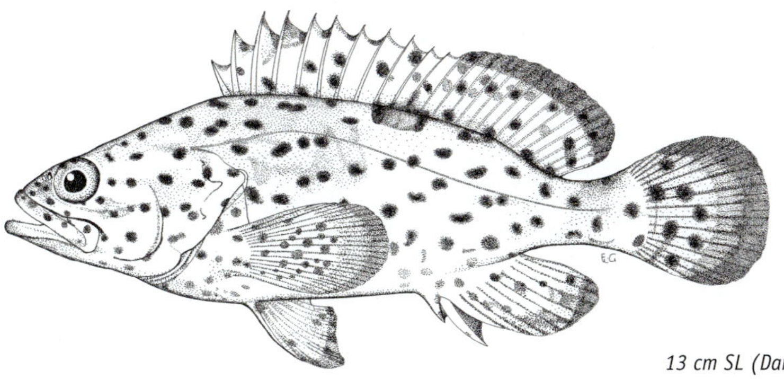

13 cm SL (Darwin, Australia)

2 cm SL (Western Australia)

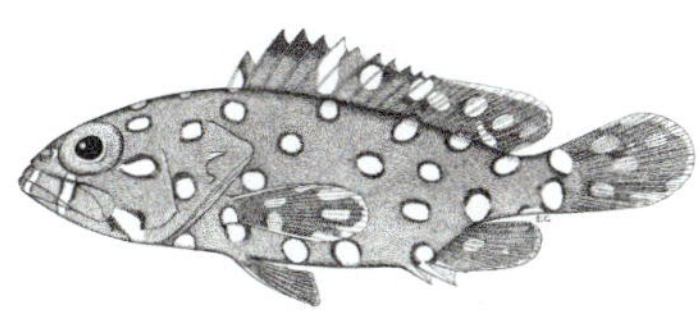

3 cm SL (Western Australia)

Body depth less than head length, depth contained 2.7–3.2 times in standard length. Head length 2.3–2.5 times in standard length; head pointed, the interorbital area flat, and the dorsal head profile almost straight; preopercle rounded, finely serrate, with slight notch; upper edge of operculum convex; posterior nostril vertically elongated in fish larger than 15 cm SL, its length 5 or 6 times larger than diameter of anterior nostril; maxilla reaches about to vertical at rear edge of eye, maxilla naked, mostly covered by upper lip; canines at front of jaws small or absent; midlateral part of lower jaw with 4–6 rows of teeth (2 or 3 rows in juveniles). Gill rakers 8–10 on upper limb, 14–17 on lower limb (with 7–10 developed rakers on lower limb); gill rakers shorter than gill filaments; gill raker at angle of first arch subequal to adjacent rakers. Dorsal fin with 11 spines and 15–17 rays, the 3rd or 4th spine longest, its length contained 2.8–3.8 times in head length, the interspinous membranes moderately incised; anal fin with 3 spines and 8 rays; pectoral-fin rays 17–20; pectoral-fin length contained 1.5–1.8 times in head length; pelvic fins not reaching past anus, their length 1.9–2.4 times in head length; caudal fin rounded. Lateral-body scales ctenoid (smooth in large adults), with auxiliary scales; lateral-line scales 53–63; lateral-scale series 88–109. Pyloric caeca 9–13.

MAXIMUM RECORDED SIZE 49 cm TL

COLOUR IN LIFE Head and body brownish to greenish grey; widely spaced small black spots over all of the head, body and fins; 3 dark blotches on the body at the base of the dorsal fin, the first (at the base of the last 2 or 3 dorsal-fin spines) largest and most distinct; dark saddle blotch on the caudal peduncle; black streak hidden by the upper edge of the maxilla when the mouth is closed; white line usually present along the rear edge of the median and pectoral fins. Juveniles 3–6 cm SL with prominent white spots on the head, body and fins, those on the head and body with black edges (especially along the dorsal and ventral edges of the white spots) or 2 contiguous black spots at the upper and lower edges of the white spots; juveniles 7–10 cm SL with black edges dividing to form 2–4 black spots as the white spots become fainter; juveniles of 13 or 14 cm SL with white spots almost completely faded.

COLOUR IN MARKET Body colour a pale grey with prominent pale black spots; blotches at the base of dorsal fin remain visible but may fade.

Adult *E. corallicola* (Okinawa Fish Market, Japan)

Juvenile *E. corallicola* (Darwin, Australia)

Adult *E. corallicola* (Raja Ampat Islands, Indonesia)

POPULATION **General** Little is known regarding population densities of *E. corallicola*. It is generally considered uncommon but may be locally abundant. **Fishery dependent** *E. corallicola* represented less than 1% of the total fishery landings in Pohnpei (Rhodes and Tupper 2007).

POPULATION TREND Unknown

MAJOR THREATS Fishing pressure is the major threat to this species.

HABITAT AND ECOLOGY **General** *E. corallicola* is a reef-associated species common in shallow silty reefs, and sometimes in estuarine areas. **Reproduction** Fisher interviews in Indonesia and Papua New Guinea provide evidence of spawning aggregations from reports of high catches of gravid females (www.scrfa.org database).

COMMENTS Adults of *E. corallicola* are often misidentified as *E. macrospilos*, which differs in having the dark spots on the body larger and more closely set, the lower jaw projecting, and the posterior nostrils not vertically elongated. Also, the juveniles of *E. macrospilos* do not have white spots. It may also be confused with *E. howlandi*, but that species differs in lacking dark spots on the belly, chest and underside of head, the posterior nostrils are not vertically elongated, and the juveniles do not have white spots.

2008 IUCN RED LIST STATUS Data Deficient

Epinephelus corallicola is a western Pacific species that is found in Thailand, Hong Kong and Taiwan, extending east to the Solomon and Mariana Islands, including Indonesia, Singapore, Philippines, Christmas Island, Micronesia, Papua New Guinea and Palau, and south to northern Australia.

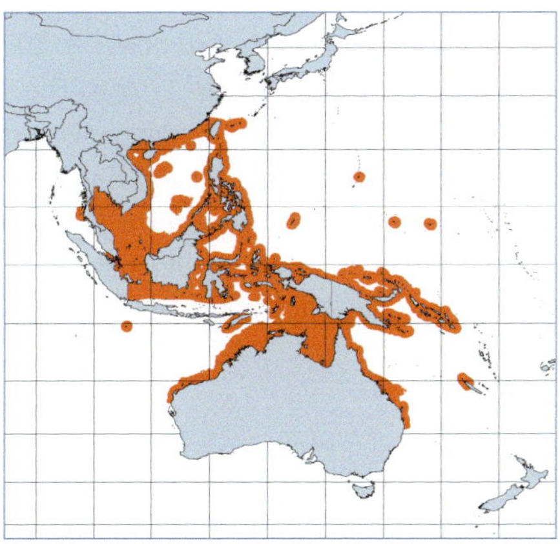

Authors: *PC Heemstra, M Kulbicki, DA Pollard, KL Rhodes, BC Russell and MA Samoilys*

Epinephelus costae (Steindachner 1878)

goldblotch grouper (E), mérou badèche (F), falso abadejo (S)

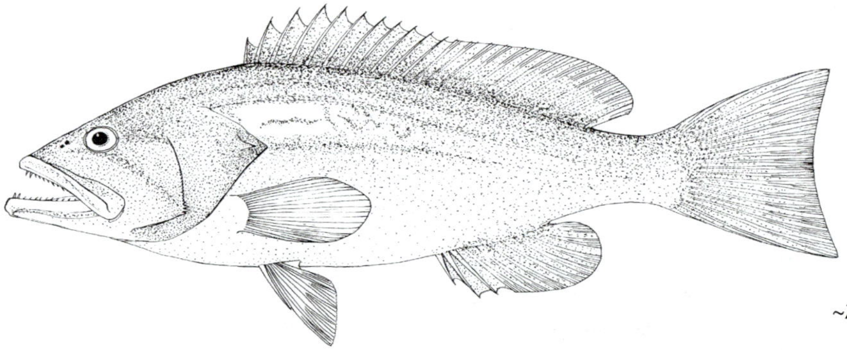

~70 cm TL
© FAO

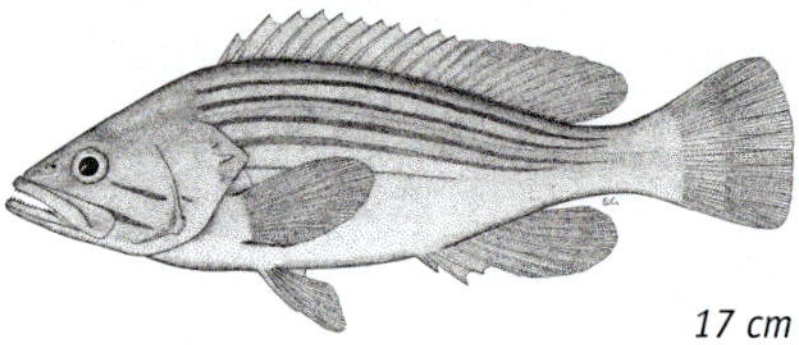

17 cm TL (Angola)
© SAIAB

Body depth less than head length, depth contained 3.0–3.4 times in standard length. Head length 2.5–2.7 times in standard length; interorbital area convex; preopercle angular, with 2 or 3 greatly enlarged serrae at the angle; in adults larger than 40 cm SL, the preopercle angle is produced into a rounded lobe, with an indentation immediately above the lobe; middle and lower opercular spines flat but distinct, upper spine not apparent; upper edge of operculum straight or slightly convex; posterior and anterior nostrils subequal in specimens less than 30 cm SL, posterior nostril diameter about twice that of anterior nostril in fish of 40–50 cm SL; maxilla usually reaching a vertical at rear edge of eye, ventral edge of maxilla with a low step, no scales on maxilla; midlateral part of lower jaw with 2 rows of teeth. Gill rakers 8–10 on upper limb, 16–18 on lower limb, 24–27 total. Dorsal fin with 11 spines and 15–17 rays, the 3rd or 4th spine longest, subequal to longest dorsal-fin ray, the interspinous membranes distinctly incised; anal fin with 3 spines and 8 soft rays; pectoral-fin rays 18 or 19; pectoral fins usually longer than pelvic fins, contained 1.6–2.1 times in head length; caudal fin truncate or slightly convex in juveniles, becoming concave or lunate in adults larger than 40 cm SL. Lateral-body scales ctenoid, with auxiliary scales in adults; lateral-line scales 70–73; lateral-scale series 113–130. Pyloric caeca 17.

MAXIMUM RECORDED SIZE 140 cm TL

POPULATION **General** *E. costae* was reported as common in the Mediterranean and tropical west coast of Africa in the 1970s (Seret 1981), but current population trends are unknown.

POPULATION TREND Unknown

MAJOR THREATS No major threats to this species are currently known.

HABITAT AND ECOLOGY **General** *E. costae* is found on sand, mud or rock bottoms in depths of 20–80 m. Juveniles form small groups in shallow water. **Feeding** Feeds on fishes and invertebrates.

COLOUR IN LIFE Head and body brownish, the fins darker. Juveniles less than 15 cm SL with 3–5 narrow dark stripes parallelling the lateral line on the dorsal part of the body (2 stripes above and 1–3 stripes below the lateral line). Two dark lines on the head, one from the lower edge of the eye to the ventral rear edge of the interopercle, the 2nd from the dark maxillary streak to the lower edge of the preopercle. Adults brown or greyish brown, often with a large, distinct golden yellow blotch (vaguely defined at periphery) on the body below the spinous dorsal fin.

COLOUR IN MARKET Maintains similar colouration in death although the body colour may be paler.

© Patrick Louisy

Juvenile *E. costae* (Kaş, Turkey)

© Thorke Østergaard

Subadult *E. costae* (Turkey)

Subadult *E. costae* (Ibiza, Balearic Islands, Mediterranean)

© Robert Patzner

Adult *E. costae* (Mallorca, Balearic Islands, Mediterranean)

© Eric Clua

Epinephelus costae is found in the Mediterranean (Malta, Greece, Lebanon, Israel and Egypt) and in the eastern Atlantic from Portugal south to the central coastal region of Angola, including Cape Verde and the Canary Islands (Spain).

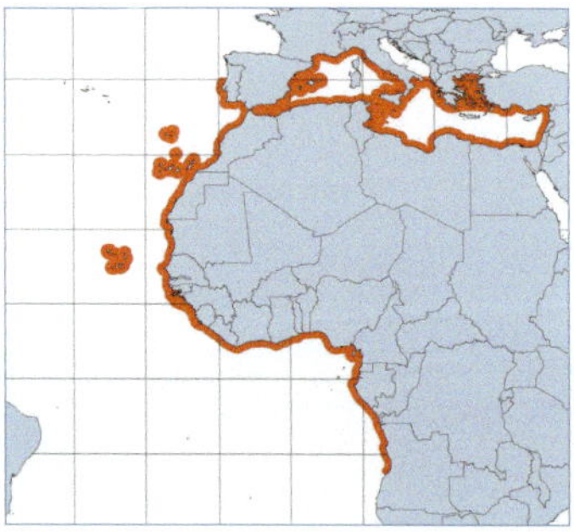

COMMENTS *E. costae* is similar to *E. goreensis* but these two species can be distinguished by their colour patterns. Juveniles of *E. costae* have 3–5 narrow dark stripes parallelling the lateral line (no dark stripes in *E. goreensis*), and live (or freshly dead) adults of *E. costae* often have a large golden yellow blotch on the body below the spinous dorsal fin. *E. costae* does not have the sub-vertical dark bars or 'saddles' on the dorsal part of the body that are characteristic of *E. goreensis*.
2008 IUCN RED LIST STATUS Data Deficient

Authors: *AA Bertoncini, JH Choat, MT Craig, BP Ferreira, PC Heemstra, DA Pollard and LA Rocha*

Epinephelus cyanopodus (Richardson 1846)

speckledblue grouper (E), mérou bleu (F), mero azul (S)

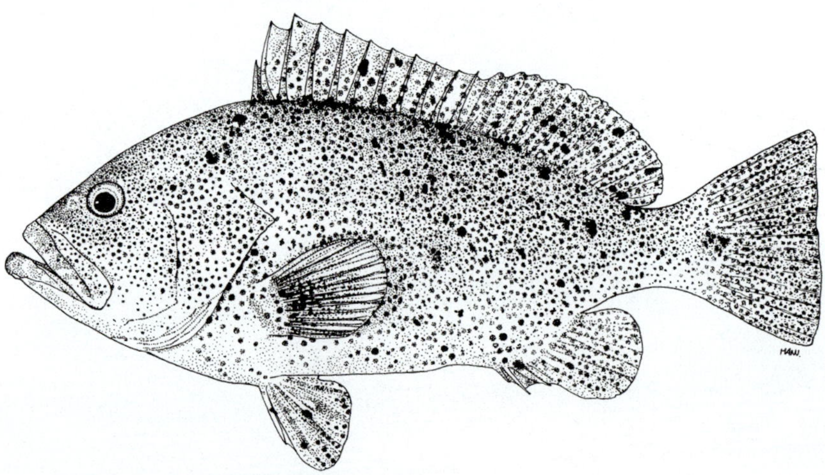

32 cm SL
© FAO

Body deep and compressed, the depth subequal to head length, contained 2.4–2.7 times in standard length; body width 1.9–2.8 times in the depth. Dorsal head profile steep, the interorbital area distinctly convex; preopercle subangular, finely serrate, serrae at rounded corner slightly enlarged; opercular spines inconspicuous; upper edge of operculum straight; posterior nostrils of adults 2 or 3 times larger than anterior nostrils; maxilla reaches about to vertical at rear edge of eye; midlateral part of lower jaw with 2–4 rows of teeth. Gill rakers

9 or 10 on upper limb, 15–17 on lower limb, 24–27 total. Dorsal fin with 11 spines and 16 or 17 rays, the 3rd or 4th spine longest, contained 2.2–2.8 times in head length, the interspinous membranes not or only slightly incised; anal fin with 3 spines and 8 rays; pectoral fins thin (not fleshy), with 18–20 rays; pectoral fins subequal to pelvic fins, their length 1.7–2.0 times in head length; caudal fin truncate or slightly emarginate. Lateral-body scales distinctly ctenoid, with auxiliary scales; lateral-line scales 63–75; lateral-scale series 128–147.

MAXIMUM RECORDED SIZE 120 cm TL

COLOUR IN LIFE Adults usually pale bluish grey, covered with black dots and a few scattered, irregular black spots; broad black submarginal band on the caudal fin of some large juveniles; pelvic fins usually black-tipped. Juveniles with yellow fins; body of small juveniles (to 12 cm SL) mainly yellow, the head and front part of the body with a wash of bluish grey and faint dark dots.

COLOUR IN MARKET Bluish grey body colour fades to a uniform grey or tan, the black dots and spots becoming more prominent, sometimes coalescing into blotches. Large individuals (>70 cm TL) often have no markings on the body and are a solid drab grey.

© Gerald R Allen

Juvenile *E. cyanopodus* (Milne Bay, Papua New Guinea)

© Matthew T Craig

Adult *E. cyanopodus* (Okinawa Fish Market, Japan)

© Robert F Myers

Adult *E. cyanopodus* (Long Beach Aquarium, California, USA)

POPULATION GENERAL *Epinephelus cyanopodus* is relatively uncommon throughout most of its range, but may be locally abundant (e.g. in New Caledonia). FISHERY INDEPENDENT Fishery-independent data for *E. cyanopodus* are restricted to New Caledonia. There, recruitment and spawning aggregation size have been shown to be variable and perhaps linked to large-scale oceanographic changes such as El Niño Southern Oscillations (M Kulbicki pers. obs.).

Epinephelus cyanopodus is a western Pacific species, occurring from southern Japan to southern Queensland (Australia), the Gulf of Thailand, and extending east to Fiji, the islands of Micronesia (except for the Mariana Islands) and to Tonga.

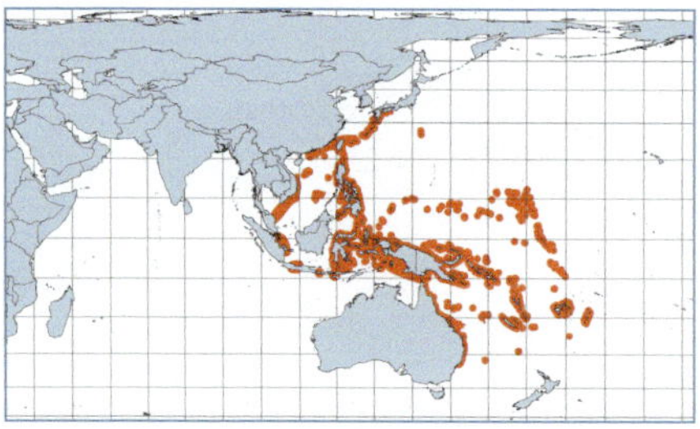

POPULATION TREND Unknown

MAJOR THREATS No major threats are known to this species although it has undergone declines in Fiji according to fisher interviews (YJ Sadovy de Mitcheson pers. comm.).

HABITAT AND ECOLOGY GENERAL *E. cyanopodus* is usually found around isolated coral heads in lagoons or bays, but is also known from outer reef areas in depths of 2–150 m. According to Myers (1999) the species usually swims several metres above the bottom and can be considered semi-pelagic. FEEDING The species feeds mainly on sand-dwelling fishes and crustaceans such as snake eels and box crabs (Thollot 1996, Kulbicki et al. 2005b). REPRODUCTION In New Caledonia and Fiji, *E. cyanopodus* is an aggregate spawner (C Chauvet pers. comm.). Aggregations in New Caledonia are known to form in November and December in mixed species aggregations. In New Caledonia, mature males have been observed as small as 52 cm TL, while mature females have been observed as small as 31.5 cm TL.

2008 IUCN RED LIST STATUS Least Concern

Authors: *AS Cabanban, M Kulbicki, DA Pollard, BC Russell and BM Yeeting*

Epinephelus daemelii (Günther 1876)

saddletail grouper (E), mérou troussequin (F), mero montado (S)

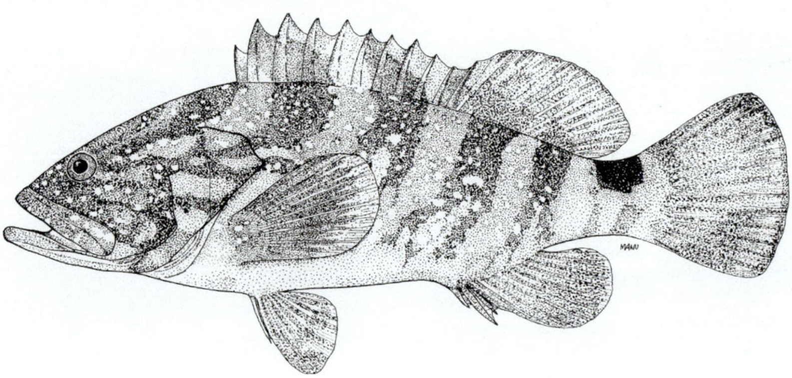

24 cm SL
© FAO

Body depth contained 2.9–3.3 times in standard length. Head length 2.3–2.5 times in standard length; interorbital area flat to slightly convex; dorsal head profile almost straight; preopercle rounded, finely serrate; opercular spines inconspicuous; upper edge of operculum distinctly convex; posterior nostrils of adults enlarged, 2–4 times larger than anterior nostrils; maxilla reaches past vertical at rear edge of eye; canines at front of jaws well developed, particularly the inner depressible teeth at symphysis of upper jaw; midlateral part of lower jaw with 2 rows of large teeth. Gill rakers 9–12 on upper limb, 15–19 on lower limb, 25–28 total; longest gill raker shorter than longest gill filaments. Dorsal fin with 11 spines and 14 rays, the 3rd to last spines subequal and shorter than longest dorsal-fin rays and the interspinous membranes incised; anal fin with 3 spines and 8 rays; pectoral fins large and fleshy, with 17–19 rays; pectoral-fin length contained 1.6–2.1 times in head length; pelvic fins not reaching anus, their length 2.1–2.5 times in head length; caudal fin rounded. Lateral-body scales smooth (except for area covered by pectoral fins), with auxiliary scales; lateral-line scales 63–71; lateral-scale series 111–126.

MAXIMUM RECORDED SIZE 122 cm TL

COLOUR IN LIFE Head and body greyish brown or dark greenish brown with small irregular pale yellow or whitish spots; black saddle blotch on the peduncle. Juveniles and subadults with 4 irregular oblique dark brown bars on the body and a dark band from the nape to the eye; median and pelvic fins with narrow white margins on some fish. Large adults often uniformly dark brown or black, the underside of the head pale and a few pale spots on the cheeks.

COLOUR IN MARKET Body may fade to generally tan to grey and yellow-white spots become less visible. The black saddle blotch usually remains prominent in death.

© Rudie Kuiter

Adult *E. daemelii* (Manly, Sydney, Australia)

Epinephelus daemelii occurs in the warm temperate and subtropical waters of the South Pacific Ocean along the south-eastern coast of Australia, extending east to Lord Howe, Norfolk and Kermadec islands, and to northern New Zealand.

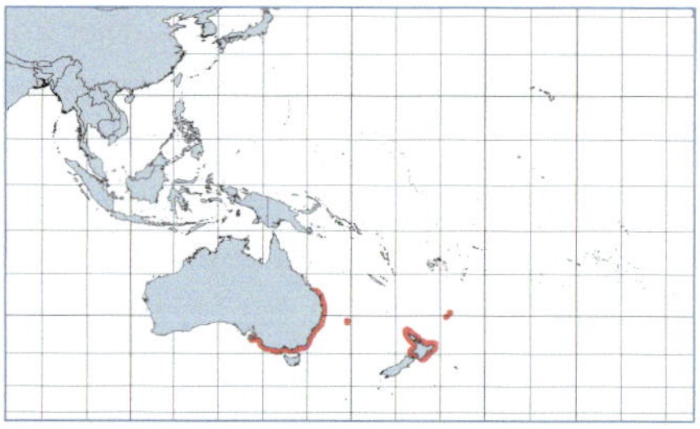

POPULATION **GENERAL** *E. daemelii* is now considered an uncommon species throughout most its range. **FISHERY DEPENDENT** No explicit fishery-dependent data are available; however, grey literature suggests catches have decreased considerably in both numbers and average size of fish in many places. The species is particularly susceptible to overfishing by line and spearfishers, and numbers have been much reduced by recreational and commercial fishing in Australia (Pogonoski et al. 2002).

POPULATION TREND Unknown

MAJOR THREATS Fishing pressure is the major threat to this species.

HABITAT AND ECOLOGY **GENERAL** *E. daemelii* is a territorial species known to occur in caves and channels of rocky reefs from nearshore shallow waters to depths of at least 100 m. Small adults are also known from estuaries, and juveniles may be found in tide pools. **REPRODUCTION** *E. daemelii* is a protogynous hermaphrodite, with sex change occurring at around 100–110 cm TL (Paulin and Roberts 1992).

2006 IUCN RED LIST STATUS Near Threatened

Authors: *W-C Ng, DA Pollard and C-M Chan*

Epinephelus diacanthus (Valenciennes 1828)

spinycheek grouper (E), mérou épineux (F), mero espinudo (S)

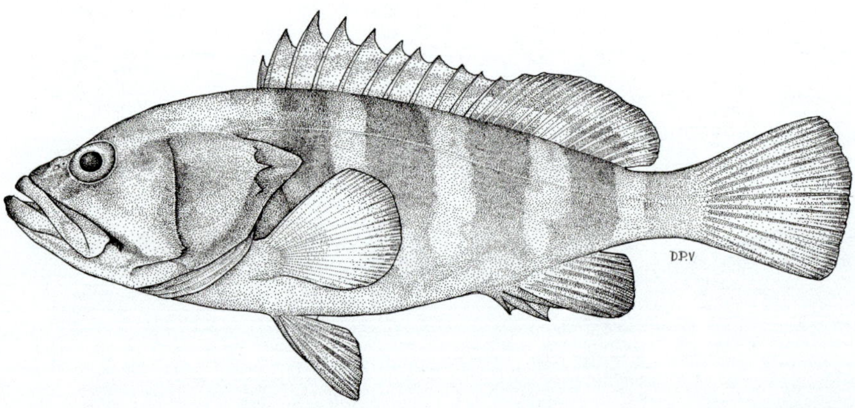

33 cm SL (Cochin, India)

Body depth contained 2.8–3.5 times in standard length (for fish 10–34 cm SL). Head large, its length 2.2–2.4 times in standard length; interorbital region flat or slightly convex, the dorsal profile convex; preopercle with 1–5 prominent spines at the angle; upper edge of operculum straight or slightly convex; posterior and anterior nostrils subequal; anterior nostril tubular, the margin usually with a large, bilobed flap of skin; maxilla reaches to or almost to vertical at rear edge of eye, the lower edge smoothly curved; midlateral part of lower jaw with 2 rows of short, subequal teeth. Gill rakers 8–10 on upper limb, 15–17 on lower limb; numerous bony plates on sides of gill arches. Dorsal fin with 11 spines and 15–17 rays, the 3rd or 4th spine longest, its length contained 2.8–3.6 times in head length and longer than longest ray, the interspinous membranes incised; anal fin with 3 spines and 8 rays; pectoral fins with 17–20 rays; pectoral-fin length contained 1.7–2.1 times in head length; pelvic fins end well short of anus, their length 2.0–2.6 times in head length; caudal fin usually rounded or convex; caudal-peduncle depth 3.7–4.7 times in head length. Lateral-body scales ctenoid, with auxiliary scales in adults; lateral-line scales 52–60; lateral-scale series 103–121. Pyloric caeca 7 or 8.
MAXIMUM RECORDED SIZE 55 cm TL

COLOUR IN LIFE Body pale greyish brown, usually with 5 dark vertical bars broader than the interspaces, 4 occurring below the dorsal fin and the 5th (faintest) on the peduncle; ventral part of the head and body often pink or reddish; dark maxillary streak continues faintly to the lower edge of the preopercle; fins dusky grey without spots.
COLOUR IN MARKET Body pale brown to grey with faded dark bars on the posterior body.

© GS Govindaraju and P Jayasankar

Adult *E. diacanthus* (India)

© Jean-Francois Helias

Adult *E. diacanthus* (Gulf of Aden, Yemen)

POPULATION **GENERAL** *E. diacanthus* is a moderately common species that is heavily fished throughout its range. **FISHERY DEPENDENT** *E. diacanthus* is an important component of the trawl fisheries of the northern Indian Ocean. Declines are inferred over much of its range from the trawl fisheries, which take many juveniles, expanding into deeper water and from the apparent overexploitation of juveniles and adults in shallower water (IUCN Red List Assessment 2008). Exploitation of juveniles and the associated overfishing will lead to depletion of the stock of *E. diacanthus* from the Dakshina Kannada coast in India (Zacharia et al. 1995). Catch rates during the late 1980s to early 1990s increased greatly in India (Zacharia et al. 1995). A concomitant decrease in the mean size of fish was noted indicating significant impact on the standing stock of this species. The species is also vulnerable to net fisheries because of its grouping behaviour (De Bruin et al. 1994). **FISHERY INDEPENDENT** *E. diacanthus* is moderately common in north-western India in visual surveys (Sluka and Lazarus 2010).

Epinephelus diacanthus is an Indian Ocean species found on the continental shelf of the northern Indian Ocean from the Gulf of Aden to Sri Lanka and Madras (India). It is not known from the Persian Gulf or the Red Sea.

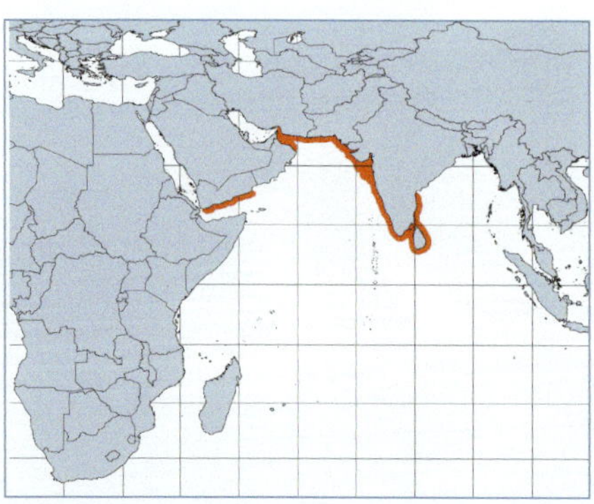

POPULATION TREND Decreasing

MAJOR THREATS The major threat to *E. diacanthus* is fishing pressure on both juveniles and adults and the absence of effective fishing management.

HABITAT AND ECOLOGY **GENERAL** *E. diacanthus* occurs on mud or sand bottoms in depths of 10 m to over 120 m. **FEEDING** Feeds on crustaceans (mainly crabs and small prawns) and fishes (mainly *Ambassis* spp. and *Leiognathus* spp.) (Zacharia et al. 1995). **AGE AND GROWTH** The von Bertalanffy growth equation for *E. diacanthus* was $L = 502(1 - e^{0.61(t_1 - t_0)})$, where L is total length in mm (Chakraborty 1994). **REPRODUCTION** *E. diacanthus* is a protogynous hermaphrodite which changes sex at 2–3 years of age in Taiwan: intersexual fish were found within the 2- to 6-year-old classes and were absent during spawning, and males became dominant after an age of 4 years (Chen et al. 1980b). Based on specimens collected in 2000–2001 in Oman, 87.2% and 71.7% of the fish sampled were mature females and males, respectively: sizes of first maturity for female and male *E. diacanthus* were found to be 39.2 mm FL and 42.5 mm FL, respectively (McIlwain et al. 2003). Off western India the smallest transitional fish was found to be 27.5 cm TL and the largest 31.4 cm TL (Rao and Drishnan 2007).

2008 IUCN RED LIST STATUS Near Threatened

Authors: *PC Heemstra, DA Pollard, BC Russell and YJ Sadovy de Mitcheson*

Epinephelus drummondhayi

Goode and Bean 1879

speckled hind (E), mérou grivelé (F), mero pintaroja (S)

~40 cm SL (composite)
© SAIAB

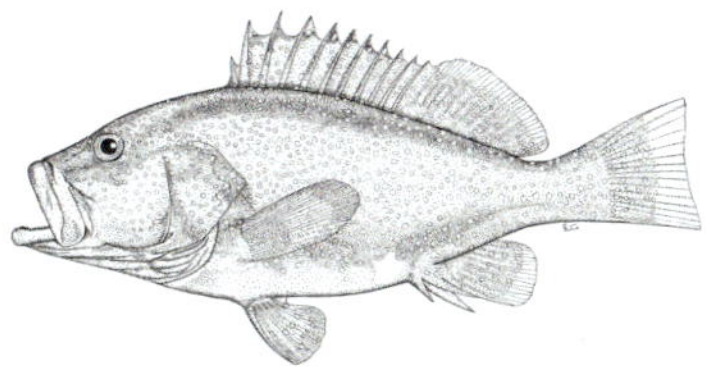

~15 cm SL
© SAIAB

Body depth subequal to head length, depth contained 2.4–2.6 times in standard length. Interorbital area distinctly convex; preopercle evenly serrate, without salient angle; interopercle and subopercle finely serrate, but the serrae covered by skin; posterior and anterior nostrils subequal. Gill rakers 9 or 10 on upper limb, 17 or 18 on lower limb, 26–28 total. Dorsal fin with 11 spines and 15 or 16 rays, the membranes slightly incised between the anterior spines; anal fin with 3 spines and 9 rays; pectoral-fin rays 18; caudal fin truncate or slightly emarginate, the corners acute. Scales strongly ctenoid, with numerous auxiliary scales; lateral-line scales 72–76; lateral-scale series about 125.

MAXIMUM RECORDED SIZE 110 cm TL

POPULATION GENERAL This once-common species has been depleted throughout much of its historical range, particularly in shallow waters. **FISHERY DEPENDENT** Landings in the USA increased steadily from 1980 (0.7 t) to 2004 (47.4 t) (NMFS database). This increase was likely facilitated by the development of offshore, deep-water fishing. It is unclear how long this fishery will remain sustainable given the absence of this species from shallow waters where it was once common. *E. drummondhayi* is an important species in the recreational and commercial fisheries of the south-eastern USA and is considered overfished by the South Atlantic Fisheries Management Council. The species is mainly caught with hook and line, but some might also be taken in traps, trawls and on bottom-set longlines. Despite management, there continues to be killing of the species in bycatch which is unlikely to be reported. It appears to have largely disappeared from Bermuda, and Musick et al. (2000) consider the species to be endangered in USA waters.

COLOUR IN LIFE Adults (greater than 30 cm SL) dark reddish brown to black, densely covered with small pearly white spots; pectoral fins yellowish distally. Some adults pale grey, with white spots joining to form irregular wavy streaks; margin of median fins blue. Juveniles (15–27 cm SL) are bright yellow covered with small bluish white spots. **COLOUR IN MARKET** Body a dark grey with densely scattered pearly spots.

© Matthew T Craig

Adult *E. drummondhayi* (Florida, USA)

© D Ross Robertson

Adult *E. drummondhayi* (Georgia Aquarium, Atlanta, USA)

POPULATION TREND Unknown
MAJOR THREATS Fishing pressure is the major threat to this species. The species is also threatened by release mortality as a result of fishing or bycatch releases due to barotraumas since it is a deep-water species.

Epinephelus drummondhayi occurs at Bermuda, along the coast of the USA from North Carolina to the Florida Keys, and in the northern and eastern Gulf of Mexico.

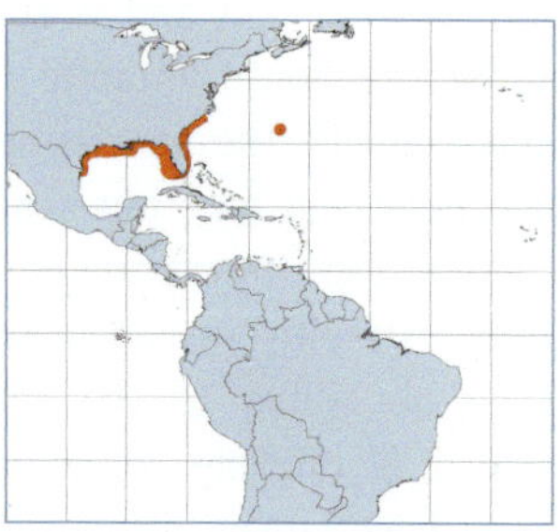

HABITAT AND ECOLOGY GENERAL *E. drummondhayi* is a reef-associated species occurring at depths of 10 m to at least 180 m. REPRODUCTION *E. drummondhayi* is a protogynous hermaphrodite (Brulé et al. 2000); females mature at 4 or 5 years of age (45–60 cm TL) and the transition into males has been estimated to occur at ages 7–14 years (Brulé et al. 2000). Spawning occurs from April to May and from July to September (Brulé et al. 2000). No direct observation of spawning has been reported; spawning may occur at >70 m depth based on the catch of ripe individuals (Gilmore and Jones 1992, Brulé et al. 2000).
2008 IUCN RED LIST STATUS Critically Endangered (A2d+3d)

Authors: *GR Huntsman and W-C Ng*

Epinephelus epistictus

(Temminck and Schlegel 1842)

dotted grouper (E), mérou pale (F), mero pálido (S)

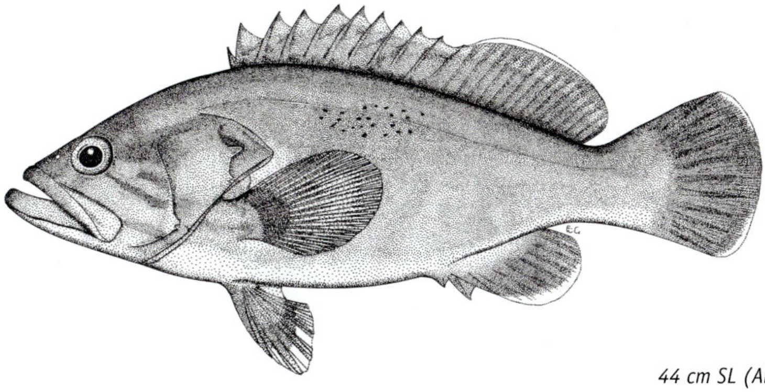

44 cm SL (Ambon, Indonesia)
© SAIAB

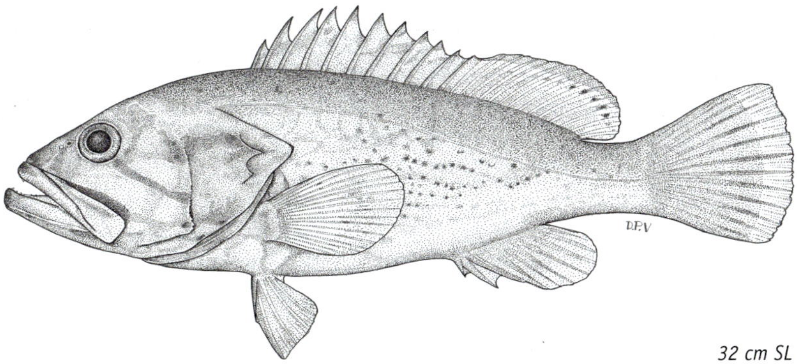

32 cm SL (West India)
© SAIAB

Body depth contained 3.0–3.3 times in standard length. Head length 2.2–2.5 times in standard length; interorbital area and dorsal head profile slightly convex; preopercle angle with 3–5 distinctly enlarged serrae; upper edge of operculum straight or slightly convex; anterior and posterior nostrils of adults subequal, or posterior nostrils about twice as large as anterior ones; maxilla reaches to, almost to, or slightly past, vertical at rear edge of eye; adults with a step or abrupt bend on ventral edge of maxilla; midlateral part of lower jaw with 2 rows of teeth, the inner ones slightly larger. Gill rakers 7–10 on upper limb, 15–19 on lower limb; length of longest gill raker subequal to longest gill filaments. Dorsal fin with 11 spines and 14 or 15 rays, the 3rd or 4th spine longest, usually shorter than longest ray, the interspinous membranes distinctly incised; anal fin with 3 spines and 8 rays; pectoral-fin rays 17–19; pectoral-fin length contained 1.6–2.1 times in head length; pelvic fins end well short of anus, their length 2.0–2.6 times in head length; caudal fin slightly to moderately rounded. Lateral-body scales ctenoid, with a few auxiliary scales in adults; lateral-line scales 57–70; lateral-scale series 105–127. Pyloric caeca 7–10.

MAXIMUM RECORDED SIZE 80 cm TL

COLOUR IN LIFE In the Pacific Ocean, head and body are pale brownish or greenish grey, with conspicuous small brownish black spots on the dorsolateral part of the body, the rear part of the head and the median fins; some specimens with a faint dark band from the eye to the end of the operculum, another band from the eye to the notch of preopercle, and a 3rd band a continuation of the maxillary streak; pectoral-fin rays brownish, the membranes clear. Juveniles with dark spots on the head and body arranged in 3 longitudinal rows.

In the Indian Ocean, the head, body and fins are brown or olive; faint brownish black dots are usually visible on the dorsolateral part of the body and sometimes on the postorbital part of the head; a dark brown maxillary streak is present; some specimens with a row of faint dark spots along the middle of the soft dorsal fin, and a faint pale submarginal band or row of faint pale spots immediately distal to these dark spots; adults from South Africa are dark brown, with a pale or white edge on the soft dorsal and anal fins and on the corners of the caudal fin. Juveniles with 3 dark bands on the head: one from the eye to the rear end of the operculum; a second band from the eye to the preopercle notch and a third band a continuation of the maxillary streak; pectoral-fin rays brown or greyish yellow, the membranes clear.

COLOUR IN MARKET Similar to life colour but the spots are less conspicuous.

© Phillip C Heemstra

Adult *E. epistictus* (Taiwan)

Epinephelus epistictus is a widespread Indo-West Pacific species, occurring from South Africa and the Red Sea, extending east to Papua New Guinea, north to southern Japan, and south to the Gulf of Carpentaria and north-western Australia.

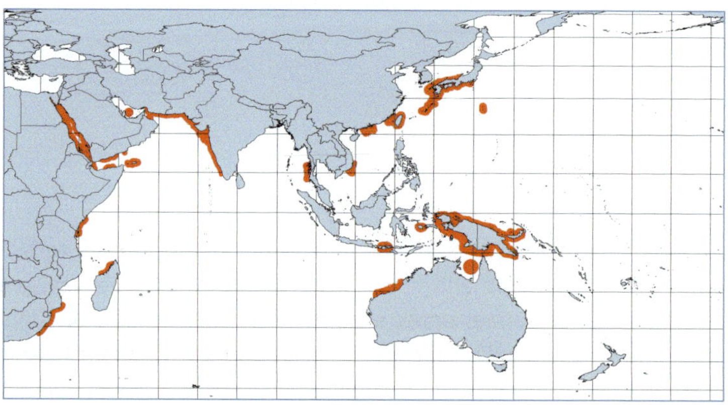

POPULATION **GENERAL** *E. epistictus* is generally rare throughout its range. **FISHERY DEPENDENT** Fishery-dependent data are scarce for this species but indicate that the species is not directly targeted, although highly marketable when caught. **FISHERY INDEPENDENT** Few fishery-independent data are currently available for this species. One market survey (W-L To unpublished data) counted few specimens in 2004 and 2006 in Hong Kong where it was once common.

POPULATION TREND Unknown

MAJOR THREATS Fishing pressure is potentially a threat to this species.

HABITAT AND ECOLOGY **GENERAL** *E. epistictus* is a deep-water species that occurs on the continental shelf over soft and rocky substrate in depths of 71 m to over 200 m.

COMMENTS *E. heniochus* is similar to *E. epistictus*. However, it has fewer lateral-scale series (89–100), the dark spots on the body (if present) are faint and smaller than the nostrils and the fins are unspotted. The juveniles do not have 3 longitudinal rows of dark spots on the head and body.
2008 IUCN RED LIST STATUS Data Deficient

Authors: *DA Pollard and W-L To*

Epinephelus erythrurus (Valenciennes 1828)

cloudy grouper (E), mérou nebuleux (F), mero nublado (S)

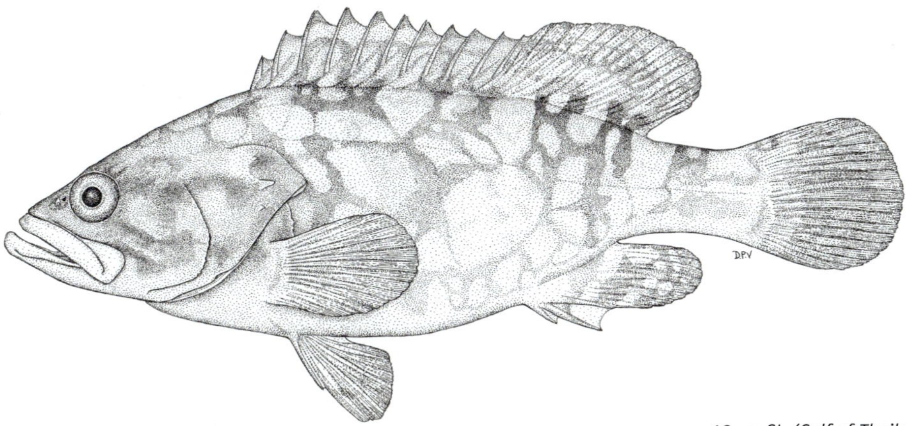

18 cm SL (Gulf of Thailand)

Body depth contained 2.8–3.2 times in standard length. Head length 2.4–2.7 times in standard length; interorbital area flat, the dorsal head profile almost straight; preopercle rounded, finely serrate, the lower serrae slightly enlarged; upper edge of operculum straight, except for bend at upper end; posterior nostril about twice size of anterior nostril; maxilla reaches to or beyond vertical at rear edge of eye; midlateral part of lower jaw with 2 or 3 rows of teeth; canines at front of jaws inconspicuous. Gill rakers 8 or 9 on upper limb, 14–17 on lower limb; gill rakers shorter than gill filaments. Dorsal fin with 11 spines and 15–17 rays, the 3rd to 9th spines subequal, the longest contained 3.0–3.75 times in head length and distinctly shorter than longest ray; interspinous membranes of dorsal fin not incised or moderately incised; anal fin with 3 spines and 8 rays; pectoral fins with 17–19 rays; pectoral-fin length contained 1.7–2.2 times in head length; pelvic fins not reaching anus, their length 2.3–2.5 times in head length; caudal fin well rounded. Lateral-body scales distinctly ctenoid, with auxiliary scales; lateral-line scales 53–62; lateral-scale series 92–107. Pyloric caeca about 6.
MAXIMUM RECORDED SIZE 43 cm TL

COLOUR IN LIFE Olive to reddish brown, usually with irregular pale spots and blotches that join randomly to form an irregular dark reticulum of the background colour; 3 dark streaks across the operculum, the 1st (broadest and darkest) from the eye to the lower edge of the operculum, the 2nd from the lower edge of the eye to the subopercle, and the 3rd (faintest) from the dark maxillary streak to the lower edge of the preopercle; median and pelvic fins mottled; pectoral fins uniform. Some specimens, especially the larger ones, nearly uniform brown or with the pale blotches on the body only faintly visible.

COLOUR IN MARKET Generally drab brown with few distinguishing marks, especially in larger individuals. Streaks across the face, if present, are faint. A blotchy pattern of tan spots may also be present where a fish is in contact with the deck of a boat or the fish market floor.

© Allen W-L To

Adult *E. erythrurus* (Ap Lei Chow Market, Hong Kong)

© Roger Steene

Adult *E. erythrurus* (Sabah, Malaysia)

Epinephelus erythrurus is an Indo-West Pacific species ranging from Pakistan to India, including the Laccadive Islands (India) and Sri Lanka, through to the larger islands of the East Indies and the Gulf of Thailand.

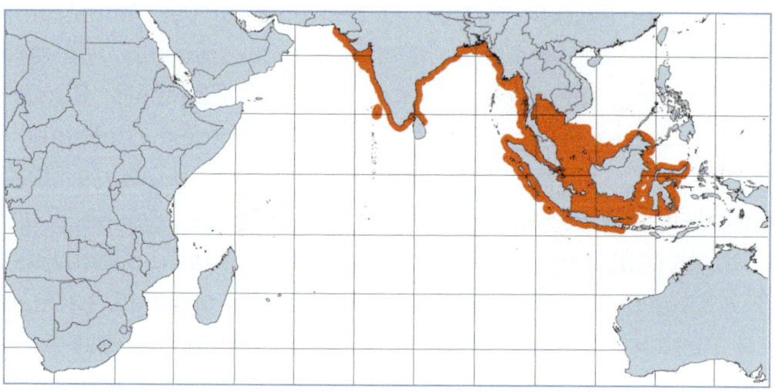

POPULATION **GENERAL** *E. erythrurus* is a rare species throughout its range.
POPULATION TREND Unknown
MAJOR THREATS No specific major threats have been identified for this species. However, the species occurs in shallow, often heavily fished, poorly managed and under-studied areas where it is taken by a range of fishing gears in multi-species fisheries.

HABITAT AND ECOLOGY **GENERAL** *E. erythrurus* inhabits areas with mud or silty sand bottoms such as harbours and estuaries. **REPRODUCTION** Heemstra and Randall (1993) report a 16 cm individual with hydrated ova.
2008 IUCN RED LIST STATUS Data Deficient

Authors: *AA Bertoncini, JH Choat, MT Craig, BP Ferreira, PC Heemstra and LA Rocha*

Epinephelus fasciatomaculosus

(Peters 1866)

rock grouper (E), mérou rocaille (F), mero de las piedras (S)

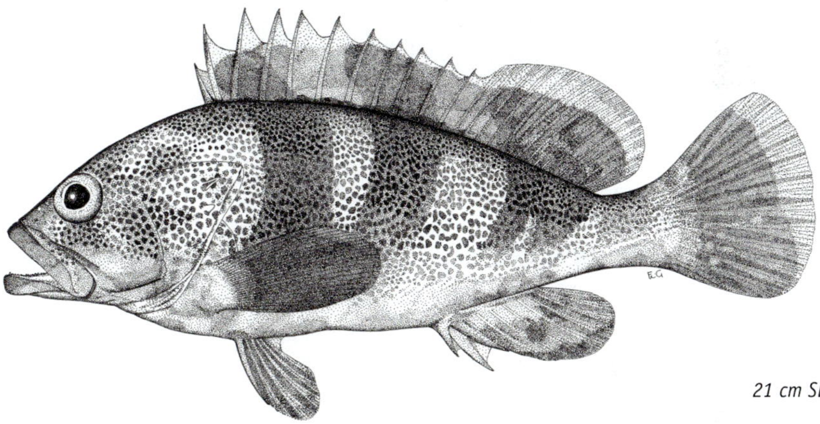

21 cm SL (Taiwan)
© SAIAB

Body depth contained 2.8–3.3 times in standard length. Head length 2.4–2.6 times in standard length; interorbital area and dorsal head profile slightly convex; preopercle angular, the serrae at the angle slightly enlarged; upper edge of operculum straight; posterior and anterior nostrils subequal; maxilla reaches about to below rear third of eye or a little past eye; midlateral part of lower jaw with 2 or 3 rows of subequal teeth. Gill rakers 7 or 8 on upper limb, 14–16 on lower limb. Dorsal fin with 11 spines and 15–17 rays, the 3rd or 4th spine longest, contained 2.5–3.1 times in head length, and a little shorter than longest dorsal-fin rays, the interspinous membranes incised; anal fin with 3 spines and 8 rays; pectoral-fin rays 17–19; pectoral-fin length 1.5–1.8 times in head length; pelvic fins not reaching past anus, their length 1.9–2.3 times in head length; caudal fin rounded. Lateral-body scales ctenoid, with auxiliary scales; lateral-line scales 48–52; lateral-scale series 92–106.
MAXIMUM RECORDED SIZE 30 cm TL

COLOUR IN LIFE Head and body pale greyish brown, with numerous small dark brown, brownish yellow or russet spots; 3–5 slightly oblique broad dark bars usually visible on the body, the last covering the peduncle, the first 4 extending into the dorsal fin; the first 2 bars are darker dorsally and extend to the margin of the spinous dorsal fin; ventral parts of the head and body whitish or pinkish; soft dorsal, caudal and anal fins with faint pale spots and streaks; a minute gold tag is often visible at the tips of some of the dorsal-fin spines.

COLOUR IN MARKET General body colour remains similar with small spots remaining prominent. Body bars fade, particularly towards the lower half of the fish's body. The pectoral fin may become pale at the base with an orange to red colour.

© Allen W-L To

Adult *E. fasciatomaculosus* (Trio Island, Hong Kong)

POPULATION GENERAL Little is known about *E. fasciatomaculosus*. It was once known to be common in Hong Kong, but now is only moderately common in markets (Sadovy and Cornish 2000).

POPULATION TREND Unknown

Epinephelus fasciatomaculosus is a western Pacific species ranging from southern Japan to Taiwan, including mainland China, Philippines, Vietnam, Hong Kong, Hainan and Sarawak.

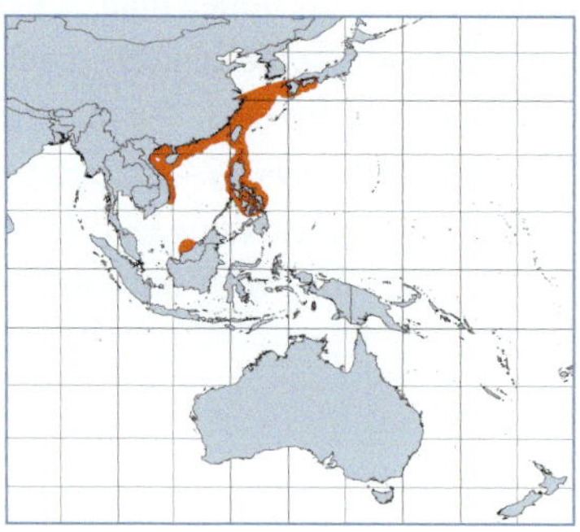

MAJOR THREATS No major threats have been identified for this species, although as a shallow-water species it may be threatened by coastal development. In general, as larger groupers become overfished, fishing pressure on smaller groupers increases, particularly within the limited geographic range of *E. fasciatomaculosus*. This effect is compounded by the absence of effective management in these areas.

HABITAT AND ECOLOGY GENERAL *E. fasciatomaculosus* occurs in shallow rocky areas of bedrock and boulders in depths of at least 20 m. **FEEDING** Feeds on fishes, prawns, crabs, worms and gastropods (Chan 1968). **AGE AND GROWTH** The von Bertalanffy growth equation was given by Chen et al. (1980a) as $L = 228(1 - e^{-0.3571(t + 0.6993)})$ where L is the standard length in mm. **REPRODUCTION** This species is protogynous; females are mature at 12.5 cm SL, and most sexual transition occurs in fish 2 or 3 years old (14–16 cm SL) (Chen et al. 1980b).

COMMENTS *E. fasciatomaculosus* is similar to the sympatric *E. akaara*. The dark bars on the body of *E. akaara* are fainter and they extend only onto the base of the dorsal fin. The prominent spots on the head and body are red, orange or gold, and there are dusky yellow or orange spots on the spinous dorsal fin.

E. fasciatomaculosus has sometimes been reported as *E. diacanthus* in the literature.

2008 IUCN RED LIST STATUS Data Deficient

Author: *MA Samoilys*

Epinephelus fasciatus (Forsskål 1775)

blacktip grouper (E), mérou oriflamme (F), mero banderilla (S)

17 cm SL (Tanzania)
© SAIAB

7 cm SL (northern Mozambique)
© SAIAB

Body depth contained 2.8–3.3 times in standard length. Head length 2.3–2.6 times in standard length; interorbital area flat, but dorsal head profile convex; snout length contained 4.3–5.1 times in head length; preopercle rounded, rear edge finely serrate, with lowermost serrae slightly enlarged; upper edge of operculum straight; posterior and anterior nostrils subequal; maxilla reaches to below rear third of eye or a little past eye; midlateral part of lower jaw with 2–4 rows of teeth. Gill rakers 6–8 on upper limb, 15–17 on lower limb. Dorsal fin with 11 spines and 15–17 rays, the 3rd to 11th spines subequal and slightly shorter than longest dorsal-fin ray; interspinous membranes of dorsal fin distinctly incised; anal fin with 3 spines and 8 rays; pectoral-fin rays 18–20; pectoral-fin length contained 1.5–2.0 times in head length; pelvic fins not reaching past anus, their length 2.0–2.4 times in head length; caudal fin slightly to moderately rounded, central Pacific specimens often with truncate caudal fins. Lateral-body scales ctenoid, with numerous auxiliary scales; nape and dorsoposterior part of head densely covered with minute auxiliary scales; lateral-line scales 49–75; lateral-scale series 92–135. Pyloric caeca 10–16.

MAXIMUM RECORDED SIZE 40 cm TL

COLOUR IN LIFE Ground colour varying from pale greenish grey, to pale reddish yellow to scarlet; the body often with 5 or 6 faint dark bars, the last on the peduncle; body scales (except ventrally) with pale centre and dark rear margin, producing a faint checked pattern; dorsal part of the head and nape, including the upper jaw, dark red or reddish brown or with bands and blotches of similar colour; often a dark band from below the eye to the interopercle; rim of the orbit black and often bordered by a pale bluish line. Fins reddish orange, pale yellowish green, or greenish brown, the outer triangular part of the interspinous membranes of the dorsal fin black (dark red in fish from Western Australia and in some specimens from deep water), with a pale yellow or white spot behind the tip of each spine; soft dorsal, anal and caudal fins often with pale yellow, white or pale blue margins; pectoral fins may be yellowish distally. Pacific specimens usually with irregular pale or white blotches and spots on the body (often a midlateral series diminishing in size posteriorly from behind the pectoral-fin base to the caudal peduncle) and a blackish brown line along the crease of the dorsal-fin base.

COLOUR IN MARKET One of the few species to retain nearly identical colour patterns in death. The bars on the body may fade, but the black tips of the dorsal fin are prominent.

© Luiz A Rocha

Adult *E. fasciatus* (Seychelles)

© Matthew T Craig

Adult *E. fasciatus* (Okinawa Fish Market, Japan)

© Robert F Myers

Adult *E. fasciatus* (Bali, Indonesia)

Epinephelus fasciatus is an Indo-Pacific species ranging from the Red Sea to Port Alfred (South Africa) in the west, and throughout the western and central Pacific. Heemstra and Randall (1993) recognise 6 subpopulations based on colour patterns and scale counts: (1) Indian Ocean and Red Sea: lateral-line scales 49–56, lateral-scale series 91–123; (2) Western Australia: lateral-line scales 49–56, lateral-scale series 91–123; (3) western Pacific: lateral-line scales 49–54, lateral-scale series 87–98; (4) Japan: lateral-line scales 55–61, lateral-scale series 102–119; (5) Pacific islands: lateral-line scales 59–75, lateral-scale series 109–135; and (6) Marquesas Islands: lateral-line scales 59–75, lateral-scale series 99–112. Specimens from Western Australia differ notably in colour with the margin of the spinous dorsal fin being dark red rather than black, as in most of the other *E. fasciatus*. At the Marquesas Islands, the *E. fasciatus* has numerous irregular white spots scattered over the head and body.

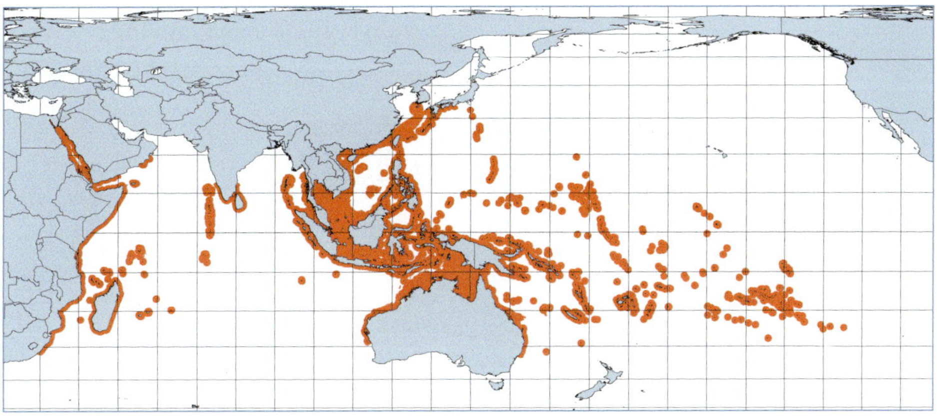

POPULATION **GENERAL** *E. fasciatus* is one of the most widespread and common of all grouper species. **FISHERY DEPENDENT** *E. fasciatus* is an important component of small-scale, multi-species fisheries throughout its range. Catch and catch-rate data are limited, mainly because they are seldom identified to species level in catch statistics, but are lumped with other groupers. Information on catch trends vary with location. There has been a steady decline in CPUE in southern Japan from 1989 to 2003, to less than one-third of that at the beginning of the time-series. There has been inconsistent variation in catches from Mozambique. In Réunion, total catch has dropped by 40% in comparison to that of 1992, with similar total effort levels, although data are not considered to be accurate. In the Pacific, catches have declined to about 25% of what they were in 1957 at one locality, but in other localities this trend is not apparent (IUCN Red List Assessment 2008). **FISHERY INDEPENDENT** *E. fasciatus* is one of the most commonly observed groupers in diver surveys. It may be locally rare, especially in areas with intense fishing pressure.

POPULATION TREND Decreasing

MAJOR THREATS *E. fasciatus* is strongly tied to areas with live coral cover, thus habitat loss is considered a major threat for this species.

HABITAT AND ECOLOGY **GENERAL** *E. fasciatus* is a reef-associated species that is common on outer reef slopes of 15–160 m depth. It also occurs to a lesser degree in protected bays and lagoons rich in live coral. **FEEDING** In Madagascar, *E. fasciatus* feeds on brachyuran crabs, fishes, shrimp and galatheid crabs. In Kenya, it feeds on crabs, stomatopods,

fishes, ophiuroids and octopus, and in the Red Sea, mostly fishes and some crustaceans (mainly crabs) are consumed (Morgans 1964, Harmelin-Vivien and Bouchon 1976). **AGE AND GROWTH** The maximum age reported for *E. fasciatus* is 19 years (Pears 2005). A length-weight relationship is given by Kulbicki et al. (2005a) of $W = 0.0138 \times L^{3.04066}$ where W is the weight in g and L is the fork length in cm. **REPRODUCTION** *E. fasciatus* can attain sexual maturity at 12 cm SL (Sadovy and Cornish 2000).
2008 IUCN RED LIST STATUS Least Concern

Authors: *AS Cabanban, JH Choat, ST Fennessy, M Kulbicki and RF Myers*

Epinephelus faveatus (Valenciennes 1828)

barred-chest grouper (E), mérou écharpe (F), mero bandeado (S)

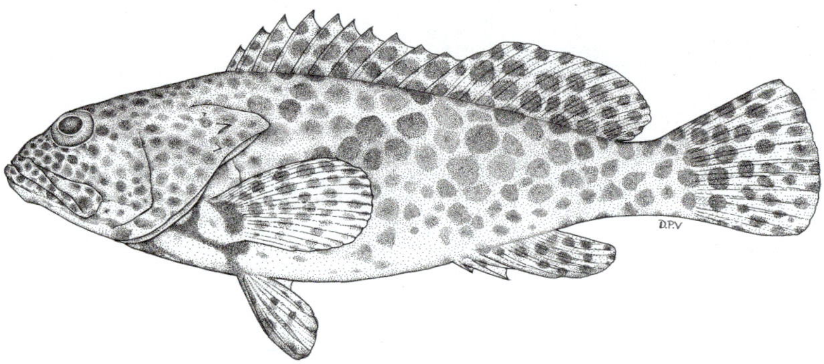

21 cm SL (Tuticorin, India)

Body depth contained 3.0–3.5 times in standard length. Head length 2.3–2.6 times in standard length; interorbital area flat or slightly concave, the dorsal head profile convex; preopercle rounded, with shallow indentation just above angle and slightly enlarged serrae at the angle; upper edge of operculum slightly convex; posterior and anterior nostrils subequal; maxilla reaches past vertical at rear edge of eye; midlateral part of lower jaw with 2 rows of teeth. Gill rakers 7–10 on upper limb, 14–16 on lower limb. Dorsal fin with 11 spines and 16–18 rays, the 3rd or 4th spine longest, its length contained 2.6–3.9 times in head length and shorter than longest dorsal-fin rays, the interspinous membranes incised; anal fin with 3 spines and 8 rays, the 2nd and 3rd spines subequal, their length 3.6–4.3 times in head length; pectoral-fin rays 17–19; pectoral-fin length 1.7–2.2 times in head length; pelvic fins not reaching anus, their length 2.2–2.6 times in head length; caudal-peduncle depth 3.8–4.3 times in head length; caudal fin rounded, length of middle rays 1.65–1.9 times in head length. Lateral-body scales smooth (except for area covered by pectoral fin), with numerous auxiliary scales; lateral-line scales 48–52; lateral-scale series 83–98. Pyloric caeca about 25.
MAXIMUM RECORDED SIZE 32 cm TL

COLOUR IN LIFE Head, body and fins pale, covered with close-set roundish brown spots of unequal size, most of those on the body larger than the pupil; four groups of 2 or 3 spots at the base of the dorsal fin darker than the other spots on the body; 2 oblique dark bands on the side of the chest; pectoral fins dusky, with a dark blotch at the base and obscure dark spots more distinct on the inner surface of the fins; spots on the median fins darker than those on the body.

COLOUR IN MARKET Maintains nearly identical colouration to life; the spots, however, may fade.

Adult *E. faveatus* (India)

POPULATION GENERAL Little information is available on the population status of *E. faveatus* across its range. It is very common in India.

POPULATION TREND Unknown

MAJOR THREATS No major threats have been identified for this species. However, very little is known of *E. faveatus* population trends as it is readily confused with similar-looking species.

HABITAT AND ECOLOGY GENERAL *E. faveatus* is a reef-associated species that occurs over coral reefs and rocky substrate in shallow waters to depths of 125 m.

COMMENTS *E. faveatus* is one of nine similar species known as the 'reticulated groupers' that have a rounded caudal fin and close-set dark brown spots with the pale interspaces forming a network on the body. The other eight species differ from *E. faveatus* as follows:

- *E. bilobatus* has ctenoid lateral-body scales, a series of 3 bilobed dark blotches or close-set pairs of dark brown to black spots along the base of the dorsal fin and adjacent body, 3rd or 4th dorsal-fin spine longest (its length contained 2.1–2.5 times in head length), and longest anal-fin spine 2.5–3.2 times in head length.
- *E. hexagonatus* has the dark spots mostly hexagonal and partially merging, leaving only a conspicuous white triangular dot at each corner of the dark spots; no dark bands on the chest; ctenoid lateral-body scales; 61–70 lateral-line scales; and 2nd anal-fin spine contained 2.1–2.5 times in head length.
- *E. macrospilos* has 15–17 dorsal-fin rays, chest and pectoral-fin base with dark spots (not bands), and pectoral fins mostly dusky with pale edge.
- *E. maculatus* has the dorsal-fin membranes not incised between the spines, pelvic-fin length contained 1.7–2.0 times in head length, lateral-body scales distinctly ctenoid, longest dorsal-fin spine 2.1–2.6 times in head length, and juveniles yellowish brown, with large irregular white blotches.
- *E. melanostigma* has a large black blotch on the body and dorsal fin at the base of

Epinephelus faveatus is found in the eastern Indian Ocean from southern India, Sri Lanka, southern Indonesia (Bali, Lombok) and the Cocos-Keeling Islands.

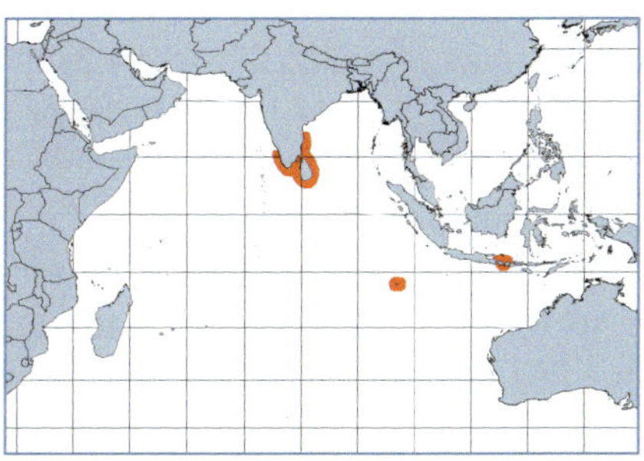

the last 3 spines, no dark bands on the chest, ctenoid lateral-body scales, 57–61 lateral-line scales, and 3–5 rows of teeth at the side of the lower jaw.

- *E. merra* has pectoral and median fins with prominent small dark spots (much smaller than the spots on the body), no dark bands on the chest, ctenoid lateral-body scales, and 98–114 lateral-scale series.
- *E. quoyanus* has the lateral-body scales distinctly ctenoid, larger pectoral fins (their length contained 1.2–1.7 times in head length) and length of middle caudal-fin rays 1.3–1.6 times in head length.
- *E. spilotoceps* has no dark bands on chest, spots on front of the head less than twice the size of the nostrils, ctenoid lateral-body scales, 59–69 lateral-line scales, and 14–16 dorsal-fin rays.

2008 IUCN RED LIST STATUS Data Deficient

Authors: *AS Cabanban, ST Fennessy, PC Heemstra, RF Myers, BC Russell and MA Samoilys*

Epinephelus flavocaeruleus

(Lacepède 1802)

blue-and-yellow grouper (E), mérou faraud (F), mero azul y amarillo (S)

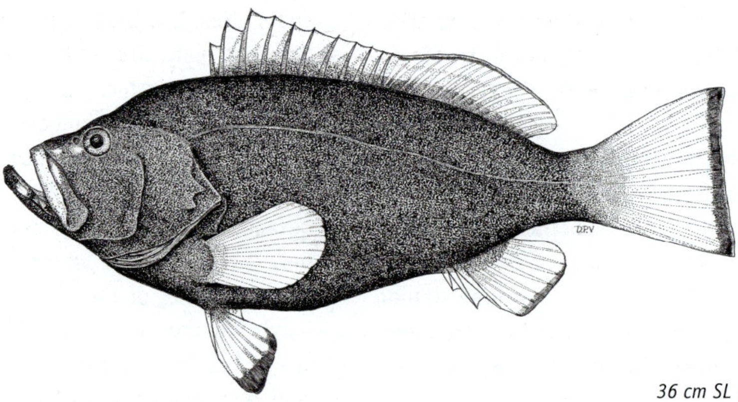

36 cm SL (South Africa)

Body deep and compressed, the depth contained 2.3–2.7 times in standard length, body width 2.0–2.8 times in depth. Head length 2.4–2.7 times in standard length; preorbital depth 6.8–8.8 times in head length for fish 36–66 cm SL; interorbital area convex; posterior nostrils of adults 4 or 5 times larger than anterior nostrils; preopercle subangular, the serrae at angle enlarged; upper edge of operculum straight or slightly convex; maxilla reaches to or slightly past vertical at rear edge of eye; midlateral part of lower jaw with 2–4 rows of teeth. Gill rakers 8–10 on upper limb, 15–17 on lower limb. Dorsal fin with 11 spines and 16 or 17 rays, the 3rd or 4th spine longest, its length contained 2.5–3.0 times in head length and subequal to longest rays, the interspinous membranes not incised; anal fin with 3 spines and 8 rays; pectoral-fin rays 18–20; pectoral-fin length 1.7–2.1 times in head length; pelvic fins not reaching anus (except in fish <15 cm SL) but often longer than pectoral fins, its length 1.7–2.0 times in head length; caudal fin truncate. Lateral-body scales ctenoid, with a few auxiliary scales in adults; lateral-line scales 61–74; lateral-scale series 129–148.

MAXIMUM RECORDED SIZE 90 cm TL

COLOUR IN LIFE Head and body dark bluish violet to dark greyish blue, some flecked with pale blue; fins and jaws bright yellow; in some fish the corners of the caudal fin, the margin of the soft dorsal and anal fins and the tips of the pelvic fins are blackish. The yellow of the fins and jaws diminishes with growth, and large adults are dark greyish, dark blue, violet, chestnut brown, or almost black.
COLOUR IN MARKET Generally the body colour remains similar depending on colour when taken (dark blue or greyish blue). Yellow fins remain conspicuous in smaller fish.

© Matthew T Craig

Adult *E. flavocaeruleus* (Sodwana Bay, South Africa)

© Robert F Myers

Subadult *E. flavocaeruleus* (Phuket Aquarium, Thailand)

POPULATION **GENERAL** *E. flavocaeruleus* is uncommon throughout its range. **FISHERY DEPENDENT** Few fishery-dependent data are available, and information reported is often based instead on the similar *E. multinotatus* which is more common. It is not clear whether the paucity of data reflects a decline in catches or naturally low levels of abundance, as fisheries statistics for most of the region where it occurs are poor. **FISHERY INDEPENDENT** Fishery-independent surveys are scarce but indicate the rarity of this species.

POPULATION TREND Unknown

MAJOR THREATS Given that it is naturally uncommon, fishing pressure and habitat loss are potential threats to this species.

HABITAT AND ECOLOGY **GENERAL** *E. flavocaeruleus* is a reef-associated species. Juveniles inhabit shallow reefs while adults are found in depths of 10–150 m. **FEEDING** *E. flavocaeruleus* feeds on a variety of organisms including fishes, crabs, shrimps, spiny lobsters, squid and small octopus (Morgans 1964). **REPRODUCTION** Reproductively active fish have been observed in January on the North Kenya Banks, in October off Mafia Island (Tanzania), as well as in August and November, the former month coinciding with very large catches (Morgans 1964, Nzioka 1979). Females mature at around 50 cm (Morgans 1982).

Epinephelus flavocaeruleus is found in the Indian Ocean from the Gulf of Aden south to Port Alfred (South Africa) and east to the north-west tip of Sumatra (Indonesia), including the islands of the western Indian Ocean. It is not known from the Red Sea or the Persian Gulf.

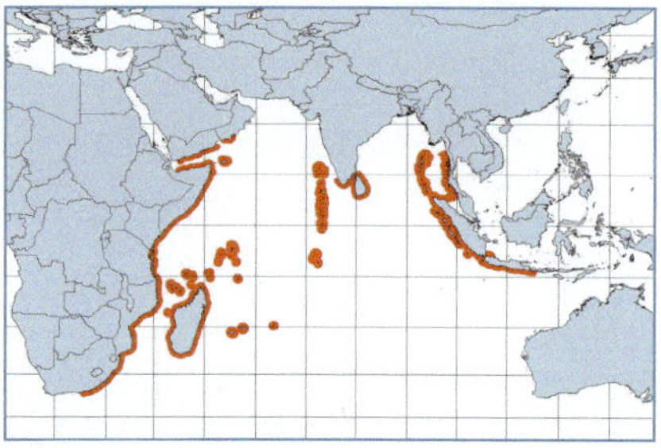

COMMENTS *E. flavocaeruleus* is similar in appearance and closely related to *E. multinotatus* of the northern and western Indian Ocean, and to *E. cyanopodus* of the western and central Pacific. The only significant differences for these three species are their colour patterns. *E. multinotatus* does not attain a deep blue hue and has numerous scattered white blotches and spots. *E. cyanopodus* is generally blue-grey with scattered black spots.

2008 IUCN RED LIST STATUS Least Concern

Authors: *ST Fennessy, RF Myers, DA Pollard, BC Russell and MA Samoilys*

Epinephelus fuscoguttatus (Forsskål 1775)

brownmarbled grouper (E), mérou marron (F), mero manchado (S)

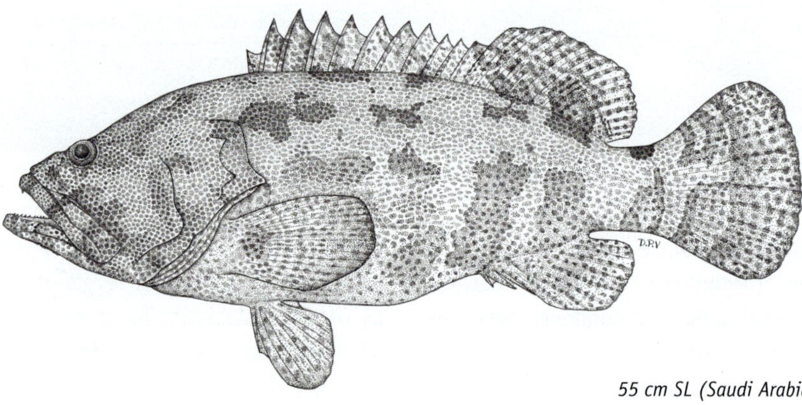

55 cm SL (Saudi Arabia, Red Sea)

Body depth contained 2.6–2.9 times in standard length. Head length 2.3–2.5 times in standard length; interorbital area flat or slightly concave; dorsal head profile of adults indented at eyes and distinctly convex from there to dorsal-fin origin; preopercle rounded, finely serrate; upper edge of operculum distinctly convex, descending almost vertically to rear end of operculum; anterior edge of preorbital bone deeply indented below nostrils; posterior nostrils triangular, 4–7 times larger than anterior nostrils in adults; maxilla extends well posterior to eye; midlateral part of lower jaw with 3 or 4 rows of teeth, the inner teeth about twice longer than outer teeth; canines inconspicuous; nostrils close together. Gill rakers 10–12 on upper limb, 17–21 on lower limb (but rudiments often difficult to count). Dorsal fin with 11 spines and 14 or 15 rays, the 3rd or 4th spine longest, contained 2.9–3.5 times in head length and distinctly shorter than longest dorsal-fin rays, the interspinous membranes distinctly incised; anal fin with 3 spines and 8 rays; pectoral-fin rays 18–20; pectoral-fin length 1.7–2.1 times in head length; pelvic fins not reaching anus, their length 2.0–2.5 times in head length; caudal fin rounded. Lateral-body scales of fish more than 10 cm SL smooth, with auxiliary scales; lateral-line scales 52–58; lateral-scale series 102–115.

MAXIMUM RECORDED SIZE 120 cm TL

COLOUR IN LIFE Pale yellowish brown, with 5 vertical series of dark brown blotches that form irregular bars; head, body and fins covered with close-set small brown spots, those on the dark blotches much darker than the spots in-between the blotches; small diffuse black saddle spot on the rear half of the peduncle; 2 or 3 faint dark bars at the side of the jaws.

COLOUR IN MARKET Generally similar to life colour but may take on a more blotchy appearance. Vertical series of dark brown blotches become less distinct and may be interrupted by large patches of pale brown.

© Stephen Newman

Adult *E. fuscoguttatus* (Pilbara coast, Western Australia)

© Robert F Myers

Adult *E. fuscoguttatus* (Blue Corner, Palau)

Epinephelus fuscoguttatus is an Indo-Pacific species, occurring in the Red Sea, along the east coast of Africa to Mozambique, extending east to Samoa and the Phoenix Islands, north to Japan, and south to Australia.

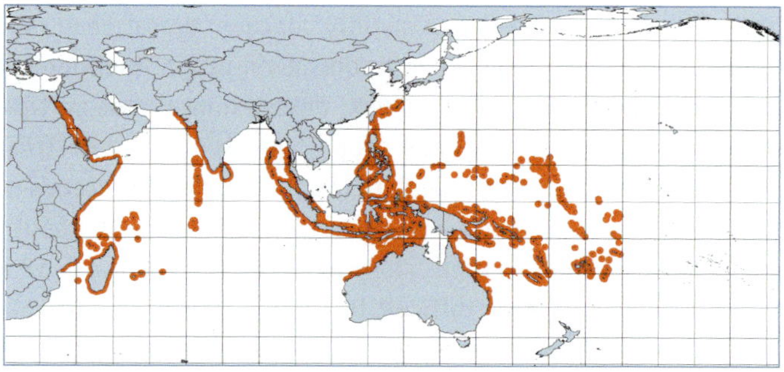

POPULATION GENERAL *E. fuscoguttatus* is generally considered common throughout its range. It is ciguatoxic in some areas and as a consequence is not readily taken. **FISHERY DEPENDENT** A common component of small-scale, multi-species fisheries, there are few reported fisheries data for this species. It is one of the more common species in the live reef food fish trade centred in South-East Asia. While juveniles are still harvested for the grow-out mariculture industry, *E. fuscoguttatus* is one of the relatively few groupers that can now be commercially produced in hatcheries (Liu and Sadovy de Mitcheson 2008). Peaks in catch rates on the Great Barrier Reef (Australia) have been noted during spawning periods (Pears et al. 2007).

POPULATION TREND Unknown

MAJOR THREATS *E. fuscogutattus* is threatened by fishing pressure, particularly as juveniles and on spawning aggregations.

HABITAT AND ECOLOGY GENERAL *E. fuscoguttatus* inhabits shallow coral reefs and rocky bottoms to depths of 60 m. Juveniles are often found in seagrass areas. **AGE AND GROWTH** *E. fuscoguttatus* reaches ages in excess of 42 years (Pears et al. 2006). **REPRODUCTION** *E. fuscoguttatus* is a protogynous hermaphrodite that forms large spawning aggregations (Pears et al. 2007). Females change sex at about 68 cm TL. Spawning on the Great Barrier Reef (Australia) occurs from November to January (Pears et al. 2007) and there are many records of 2–4 month spawning seasons and aggregations in the Indo-Pacific region (database on www.scrfa.org). Estimates of size and age at 50% effective maturity are 566 mm FL and about 9 years of age (Pears et al. 2006). Often the species aggregates in the same general areas as *Plectropomus areolatus* and *E. polyphekadion*.

COMMENTS *E. fuscoguttatus* is similar to *E. polyphekadion* which has fewer pectoral-fin rays (16 or 17), usually fewer lower gill rakers (16–18), smoothly convex dorsal head profile, and interspinous dorsal-fin membranes less deeply incised. *E. polyphekadion* also has a distinct black saddle on the caudal peduncle which is less pronounced (sometimes absent) in *E. fuscoguttatus*.

2004 IUCN RED LIST STATUS Near Threatened

Author: *AS Cornish*

Epinephelus gabriellae

Randall and Heemstra 1991

multi-spotted grouper (E), mérou passoire (F), mero pintitas (S)

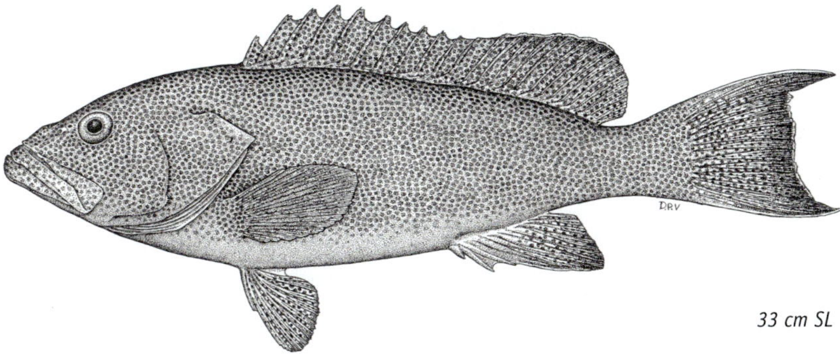

33 cm SL (Somalia)

Body depth contained 3.2–3.6 times in standard length. Head length 2.4–2.6 times in standard length; interorbital area flat to slightly convex, the dorsal head profile with slight indentation above nostrils; preopercle angular, with enlarged serrae at the angle and a shallow notch just above the angle; upper edge of operculum straight; posterior nostril slightly larger than anterior one; maxilla reaches vertical at rear edge of eye; ventral edge of maxilla with a low step-like bend one orbit diameter from rear end of bone; midlateral part of lower jaw with 2 rows of teeth. Gill rakers 10–12 on upper limb, 17–19 on lower limb, no rudiments. Dorsal fin with 11 spines and 14 or 15 rays, the 3rd or 4th spine longest, contained 2.7–3.6 times in head length, slightly shorter than longest rays; anal fin with 3 spines and 8 rays, the fin margin angular; pectoral-fin rays 17 or 18; pectoral-fin length 1.6–2.1 times in head length; pelvic fins not reaching anus, their length 2.0–2.4 times in head length; caudal-peduncle depth 3.5–3.9 times in head length, 10.5–11.3% of standard length; caudal fin emarginate to concave, the concavity 5–15 times in head length. Lateral-body scales ctenoid, with numerous auxiliary scales; lateral-line scales 52–54; lateral-scale series 106–126. Pyloric caeca about 76.

MAXIMUM RECORDED SIZE 70 cm TL

Epinephelus gabriellae is narrowly distributed from the coast of Oman (Muscat) to the eastern border of Yemen, as well as along the coasts of Socotra (Yemen) and Somalia (Sommer et al. 1996).

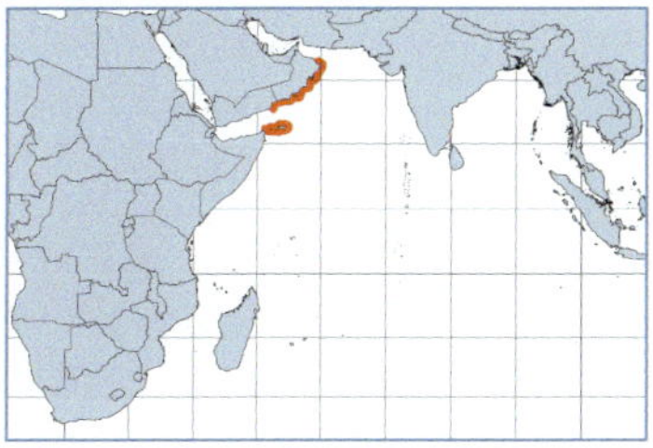

POPULATION GENERAL *E. gabriellae* is one of the most common species of groupers along the southern coast of Oman and is a major component of the grouper fishery there. It is comparatively less common in shallow waters. **FISHERY DEPENDENT** In Oman, *E. gabriellae* made up approximately 11% of the grouper catch in 2004/2005 (JL McIlwain pers. obs.). Fishers suggest that CPUE and grouper landings decreased by approximately 75% between 1991 and 2002 while aggregated 'grouper' landings data suggest a slight increase. This period coincides with a dramatic growth in the numbers of fishers over the past 20 years, coupled with heavy government subsidies for new vessels, outboard engines, etc. There are no regulations specific for *E. gabriellae* although some general controls on fishing effort were recently implemented. **Fishery independent** Fishery-independent surveys in shallow depths suggest that *E. gabriellae* is rare, but this is likely due to the species preference for deeper water inaccessible to visual census.

POPULATION TREND Decreasing

MAJOR THREATS *E. gabriellae* is threatened by fishing pressure which is increasing throughout its depth range and limited area of distribution. There are no protected areas within the range of this species and no regulations specific for *E. gabriellae*.

HABITAT AND ECOLOGY **General** *E. gabriellae* is a reef-associated species. Juveniles/subadults (<40 cm TL) are taken in shallow waters with high coral cover, and larger fish commonly inhabit depths of 40 m to over 80 m. **Feeding** Likely feeds on fishes and invertebrates. **Age and growth** *E. gabriellae* reaches ages of at least 26 years (IUCN Red List Assessment 2009).

COLOUR IN LIFE Head and body pale brownish grey, densely covered (except ventrally) with tiny close-set dark orange-brown spots; median and paired fins spotted like body; caudal, soft dorsal and anal fins with white edge and a submarginal blackish zone.

COLOUR IN MARKET Body generally pale grey, small spots remain distinctly visible or slightly muted. The white edges of the fins are often less distinct but the blackish zone remains dark.

© John E Randall

Adult *E. gabriellae* (Oman)

© Klaus Fiedler

Adult *E. gabriellae* (Oman)

COMMENTS *E. gabriellae* is similar in appearance to *E. chlorostigma*, which differs in having more dorsal-fin rays (16–18), deeper body (depth 2.8–3.3 times in standard length), deeper caudal peduncle (peduncle depth 3.0–3.6 times in head length), and shorter pelvic fins (length 1.7–2.3 times in head length). In addition, the spots are dark brown and slightly larger, and the outer surface of the pectoral fins is distinctly spotted. *E. polylepis* has a similar colour pattern and an emarginate caudal fin, but it has more scales (65–72 lateral-line scales, 126–142 lateral-scale series), 16 or 17 dorsal-fin rays, a deeper body (depth 2.6–3.3 times in standard length), and the caudal fin of adults is usually less concave (concavity 9–33 times in head length).

2008 IUCN RED LIST STATUS Vulnerable (B1ab(v))

Authors: *AA Bertoncini, JH Choat, MT Craig, BP Ferreira, JL McIlwain and LA Rocha*

Epinephelus goreensis (Valenciennes 1830)

dungat grouper (E), mérou de gorée (F), mero de gorea (S)

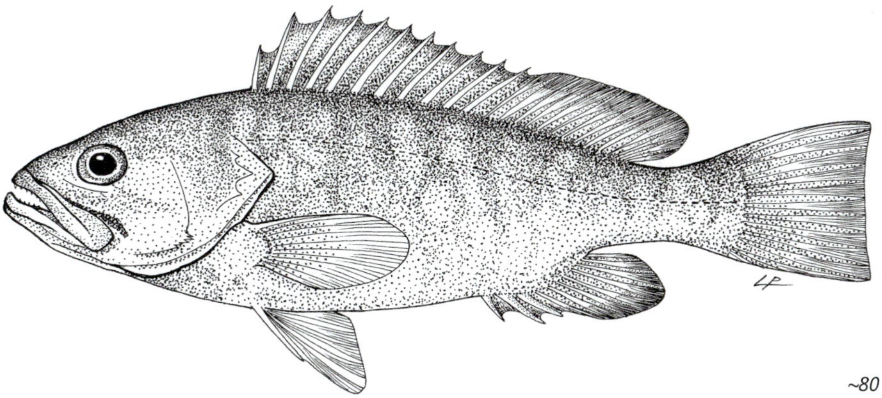

~80 cm TL
© FAO

Body depth less than head length, depth contained 2.9–3.2 times in standard length. Head length 2.5–2.7 times in standard length; interorbital area flat or slightly convex; preopercle angular, with 3 or 4 enlarged serrae at the angle, the lowermost directed ventrally; interopercle and subopercle serrate; middle and lower opercular spines well developed, the upper spine not apparent; upper edge of operculum approximately straight; maxilla reaches vertical at rear edge of eye; no step on ventral edge of maxilla; maxilla naked or with a few minute scales dorsally; midlateral part of lower jaw with 2 rows of teeth; posterior nostril about twice the size of anterior nostril. Gill rakers 8 or 9 on upper limb, 16 or 17 on lower limb, 24–26 total; longest gill raker longer than longest gill filaments. Dorsal fin with 11 spines and 16 rays, the 3rd or 4th spine longest but shorter than longest ray, the fin membrane incised between the spines; anal fin with 3 spines and 8 rays; pectoral-fin rays 17–19, pectoral-fin length contained 1.8–2.1 times in head length; pelvic fins shorter than or equal to pectoral fins; caudal fin concave, truncate or slightly convex. Lateral-body scales ctenoid, no auxiliary scales; lateral-line scales 68–74; lateral-scale series 120–129. Pyloric caeca 13, long and slender.

MAXIMUM RECORDED SIZE 140 cm TL

COLOUR IN LIFE Head and body brownish; 3 or 4 broad oblique dark bars on the dorsal part of the body and another on the dorsal half of the peduncle; 2 narrow faint dark bands extending posteriorly from the lower half of the eye; dark moustache streak present, but not extending past the rear end of the maxilla.
COLOUR IN MARKET Generally light brown. Dark body bars may fade to absence, especially on the lower portion. Dark bars radiating from the eye are faint. Moustache streak is prominent.

Adult *E. goreensis* (Cape Verde)

POPULATION GENERAL Little information is known regarding the population status of *E. goreensis*. **FISHERY DEPENDENT** The only fishery-dependent data available are from Sierra Leone where catches have fluctuated with no marked trends from 1994 to 2001 (FAO database).
POPULATION TREND Unknown
MAJOR THREATS Fishing pressure may threaten *E. goreensis*.
HABITAT AND ECOLOGY GENERAL *E. goreensis* is reported to occur in a variety of habitats (rock, mud and sand) in depths to 100 m.

Epinephelus goreensis is an eastern Atlantic species that is distributed from Mauritania and Senegal to southern Angola, including Cape Verde and the islands of the Gulf of Guinea.

COMMENTS *E. goreensis* is similar to *E. costae*. Juveniles of *E. costae* have 3–5 narrow dark stripes parallelling the lateral line which are absent in *E. goreensis*. Adults of *E. costae* often have a large golden yellow blotch on the body below the spinous dorsal fin.
2008 IUCN RED LIST STATUS Data Deficient

Authors: *AA Bertoncini, JH Choat, MT Craig, BP Ferreira, LA Rocha and CLS Sampaio*

Epinephelus guttatus (Linnaeus 1758)

red hind (E), mérou couronné (F), mero colorado (S)

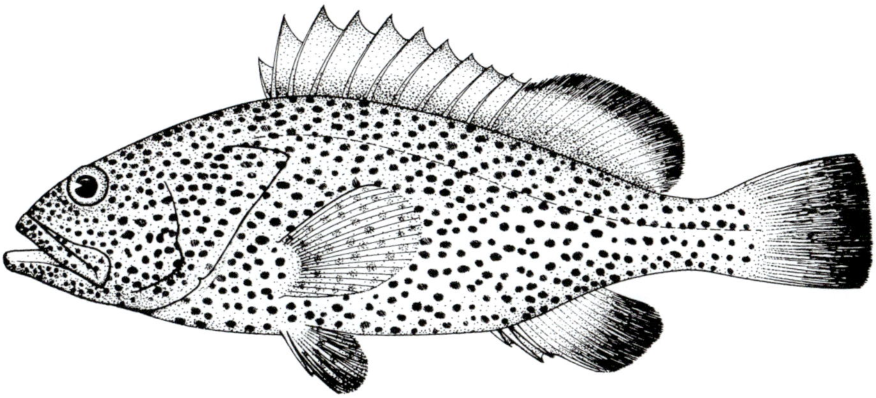

~24 cm TL

Body depth distinctly less than head length, depth contained 2.7–3.1 times in standard length (for fish 17–38 cm SL). Head length 2.3–2.4 times in standard length; preopercle evenly serrate, without salient angle; posterior nostril larger than anterior nostril. Gill rakers 8 or 9 on upper limb, 16–18 on lower limb, 24–26 total. Dorsal fin with 11 spines and 15 or 16 rays, the 3rd or 4th spine longest, the interspinous membrane incised and produced into a short flag behind tip of each spine; anal fin with 3 spines and 8 rays; pectoral-fin rays 16–18; rear edge of caudal fin rounded. Lateral-body scales ctenoid; lateral-scale series 92–104.

MAXIMUM RECORDED SIZE 76 cm TL

Epinephelus guttatus is a western Atlantic species that ranges from North Carolina (USA) to Venezuela.

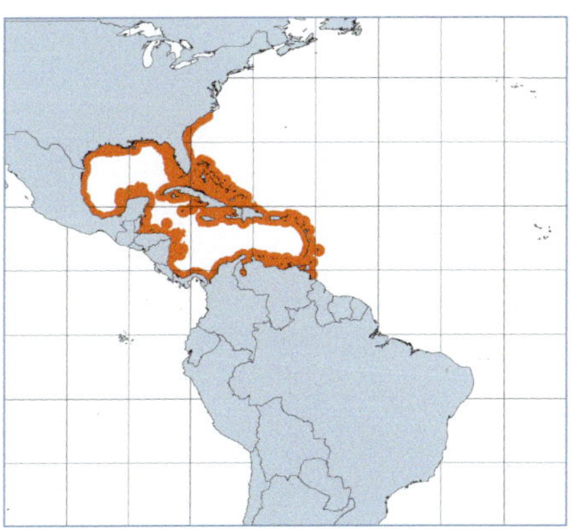

POPULATION **GENERAL** *E. guttatus* is a common species throughout its range and may be particularly abundant at some localities (e.g. the West Indies). **FISHERY DEPENDENT** The species is widely fished throughout most of its range. Spawning aggregations of *E. guttatus* are protected in St Thomas (US Virgin Islands), Puerto Rico, British Virgin Islands and Bermuda, and there are positive signs of recovery in the US Virgin Islands, including an increase in catches and in the average size of fish (Beets and Friedlander 1999, Nemeth 2005). *E. guttatus* contributed 67–99% of the total annual commercial finfish catch in the US Virgin Islands between 1987 and 1991, and over 30% of the grouper catch in Bermuda.

In Puerto Rico, *E. guttatus* has long been a very important grouper in local catches and the most important following the decline of *E. striatus* (Matos-Caraballo 2002). **FISHERY INDEPENDENT** Fishery-independent surveys in Puerto Rico (SEAMAP programme) showed a slight decline in CPUE for *E. guttatus* from 1989 through 2006.

POPULATION TREND Decreasing

MAJOR THREATS The major threat to the species is unmanaged fishing pressure, particularly on its annual spawning aggregations. Fishing pressure on medium-sized groupers like *E. guttatus* has increased following declines in larger grouper species from overfishing.

HABITAT AND ECOLOGY **GENERAL** *E. guttatus* is a reef-associated species found in shallow reefs and rocky bottoms to depths of at least 100 m. **FEEDING** Feeds mainly on crabs (*Calapa* and *Mithrax*) and other crustaceans (alpheid shrimps and scyllarid lobsters), fishes (labrids and haemulids) and octopus (Randall 1967). **AGE AND GROWTH** *E. guttatus* reaches at least 17 years with an exceptional maximum of about 22 years at 72 cm FL (Luckhurst et al. 1992). The von Bertalanffy growth parameters for Puerto Rico were: $L_\infty = 514.5$; $k = 0.1013$; $t_0 = -2.944$; $\log W = -5.21 + 3.142 \log L$, where W is the weight in g and L is the fork length in mm (Sadovy et al. 1992). Potts and Manooch (1995) provide the growth parameters

COLOUR IN LIFE Ground colour buff, greenish white or pale reddish brown, the head and body covered with bright red spots, the dorsal spots reddish brown; spinous dorsal fin olive with yellow flags at the tips of the spines; soft dorsal, caudal and anal fins olivaceous, with a broad blackish submarginal band and narrow pale edge; pectoral fins pale orange-red with darker red spots on the base; pelvic fins coloured like the body but darker distally and along the leading edge.

COLOUR IN MARKET One of the few grouper species to maintain nearly identical colouration in death.

© Thierry Brulé

Adult *E. guttatus* (Merida Fish Market, Mexico)

© Robert F Myers

Adult *E. guttatus* (San Salvador, Bahamas)

L_∞ = 471.4, k = 0.2, t = −2.397, and the weight–length relationship of $W = 1.8 \times 10^{-7} \times L^{2.614}$, where W is the weight in kg and L is the total length in mm, for populations in the Dry Tortugas, Florida (USA). **REPRODUCTION** *E. guttatus* is a protogynous hermaphrodite. Individuals are solitary outside of the spawning season and have home ranges that range from 112 m^2 to 5636 m^2 in Puerto Rico (Shapiro et al. 1994). The species forms spawning aggregations within which pair-spawning occurs in groups of a male and several females and reproduces exclusively within the aggregation period (Shapiro et al. 1993a, 1993b). *E. guttatus* seasonally aggregates to spawn from December to April in the Caribbean, and from May to July in Bermuda. Sexual transition occurs around 28 cm TL, and most fish larger than 40 cm are males; mean size of female sexual maturation is about 240 mm FL at about 3 years (Sadovy et al. 1992).

2008 IUCN RED LIST STATUS Least Concern

Authors: *AA Bertoncini, JH Choat, MT Craig, BP Ferreira, LA Rocha and YJ Sadovy de Mitcheson*

Epinephelus heniochus Fowler 1904

bridled grouper (E), mérou bride (F), mero embridado (S)

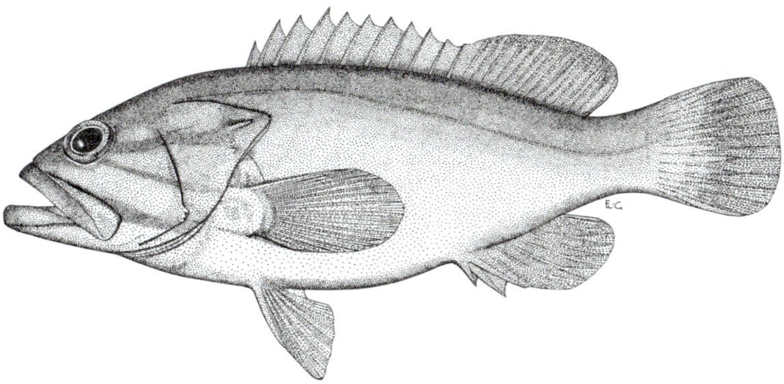

27 cm SL (north-western Australia)
© SAIAB

Body depth contained 2.7–3.2 times in standard length. Head length 2.2–2.4 times in standard length; interorbital area slightly convex, dorsal head profile distinctly convex; preopercle angular, with 2–4 large spines at the angle; upper edge of operculum approximately straight; posterior nostril about twice as large as anterior nostril; maxilla usually reaches to or slightly past a vertical at rear edge of eye; lower edge of maxilla with a step-like extension in adults; canines at front of jaws well developed, especially in upper jaw; midlateral part of lower jaw with 2 rows of teeth. Gill rakers 7–9 on upper limb, 14–16 on lower limb. Dorsal fin with 11 spines and 14 or 15 rays, the 3rd or 4th spine longest, its length contained 3.0–3.8 times in head length and shorter than longest dorsal-fin rays; anal fin with 3 spines and 8 rays; pectoral-fin rays 16–18; pectoral-fin length 1.6–1.9 times and pelvic-fin length 1.9–2.4 times in head length; caudal fin rounded. Lateral-body scales ctenoid, without auxiliary scales; lateral-line scales 54–60; lateral-scale series 89–100. Pyloric caeca 7 or 8.

MAXIMUM RECORDED SIZE 35 cm SL

COLOUR IN LIFE Head and body pale brown dorsally, shading to whitish or pale pink ventrally; some specimens with minute brownish black dots on the body and rear part of the head; faint dark brown stripe from the eye to the end of the operculum, another darker stripe from the lower edge of the eye to the subopercle and a 3rd stripe from the edge of the preorbital to the interopercle; pectoral fins hyaline greyish yellow; lower part of the caudal fin sometimes darker than the rest of the fin; margin of the interspinous dorsal-fin membranes yellow.

COLOUR IN MARKET Body pale brown with few distinguishing markings. Caudal fin retains a darker hue, and lines radiating from the eye are faint.

Adult *E. heniochus* (north-western Australia)

POPULATION **GENERAL** *E. heniochus* is a rare species throughout its range.

POPULATION TREND Unknown

MAJOR THREATS *E. heniochus* is likely threatened by trawl fisheries.

HABITAT AND ECOLOGY **GENERAL** *E. heniochus* inhabits soft substrates in depths of 40–235 m.

Epinephelus heniochus is a tropical western Pacific species distributed from Indonesia to Papua New Guinea, including the Philippines, Gulf of Thailand, Vietnam, northern Australia and New Britain (Papua New Guinea). It is distributed northwards to Japan and South Korea.

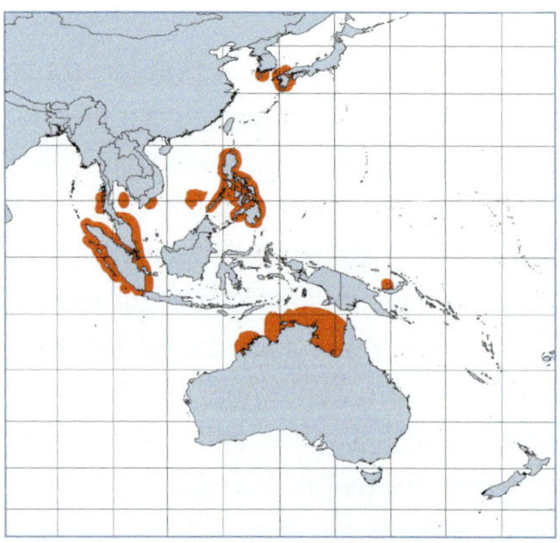

COMMENTS *E. heniochus* is similar to *E. epistictus*. The latter species differs from *E. heniochus* in having more numerous scales (57–70 lateral-line scales and 105–127 lateral-scale series). Also the dark brown or black dots on the body and the rear part of the head are conspicuous and larger than the exposed part of each scale and extend onto the dorsal and caudal fins; whereas the dark dots of *E. heniochus* (if they are present) are faint and smaller than the exposed parts of the scales, and the dorsal and caudal fins are usually unspotted.

2008 IUCN RED LIST STATUS Data Deficient

Authors: *DA Pollard and BC Russell*

Epinephelus hexagonatus (Forster 1801)

starspotted grouper (E), mérou mélifère (F), mero mielero (S)

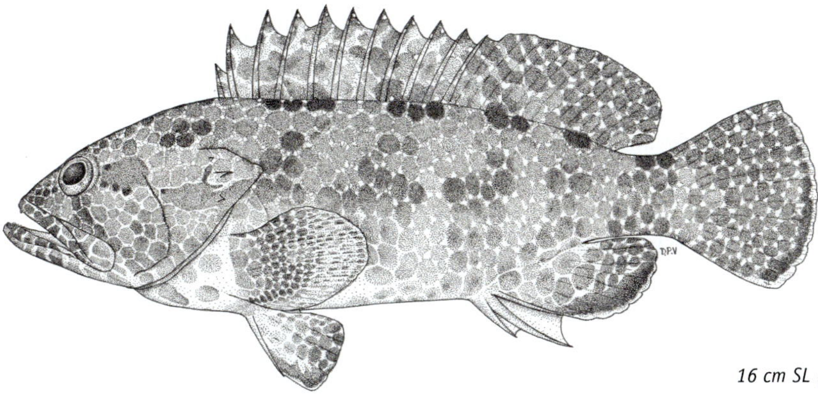

16 cm SL (Mauritius)
© SAIAB

Body depth contained 2.8–3.4 times in standard length. Head length 2.5–2.6 times in standard length; interorbital area flat, the dorsal head profile convex; preopercle rounded, the ventral serrae slightly enlarged; upper edge of operculum convex; posterior and anterior nostrils subequal; maxilla reaches to or past vertical at rear edge of eye; ventral edge of maxilla smoothly curved; midlateral part of lower jaw with 3–5 rows of teeth. Gill rakers 7–9 on upper limb, 17–19 on lower limb. Dorsal fin with 11 spines and 15–17 rays, the 5th to 9th spines subequal (their length contained 2.5–2.8 times in head length) and slightly shorter than longest rays, the interspinous membranes incised; anal fin with 3 spines and 8 rays, the 2nd spine 2.1–2.5 times in head length, distinctly longer than 3rd spine or depth of peduncle; pectoral-fin rays 17–19; pectoral-fin length 1.6–1.9 times in head length; pelvic fins not reaching past anus, their length 1.8–2.1 times in head length; caudal fin rounded. Lateral-body scales ctenoid, with auxiliary scales; lateral-line scales 61–70; lateral-scale series 93–114.

MAXIMUM RECORDED SIZE 27 cm TL

Epinephelus hexagonatus is a widely distributed Indo-Pacific species known from East Africa (Madagascar and the Mascarene Islands) east to Pitcairn Island, extending north to Ogasawara and Wake Atoll, and south to New Caledonia and Rapa Island. It is unknown from the Red Sea, Persian Gulf and most of the east coast of Africa except Kenya.

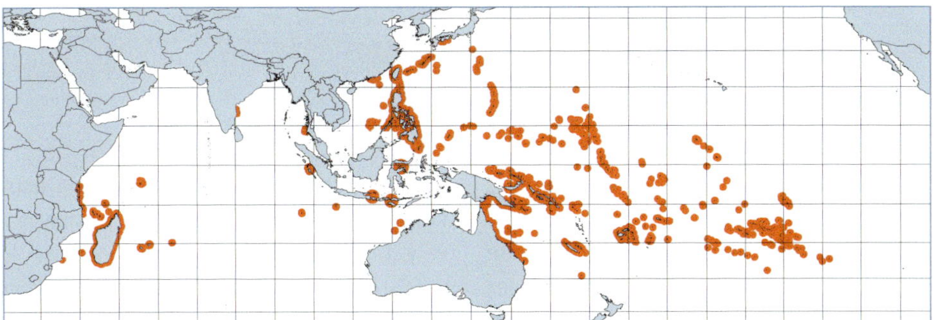

COLOUR IN LIFE Head and body covered with polygonal brown spots that tend to merge, leaving only triangular white dots at the corners of the polygons; spots on the belly and the ventral part of the head more rounded and separated and often reddish brown; 4 or 5 brownish black saddle blotches (formed by groups of darker spots) on the dorsal part of the body and caudal peduncle, the first 4 extending onto the base of the dorsal fin; irregular dark bar, formed by darker polygonal spots, on the lower part of the body below each saddle blotch; dark spots on the head progressively smaller anteriorly; large dark brown or olive spot just behind the eye, sometimes joined to a horizontally elongate spot of the same colour on the opercle; all fins with close-set dark brown or reddish brown spots and white dots, except the outer half of the pectoral fins which has faint dark spots and no white dots; leading edge of the pelvic fins and distal margin of anal fin with a pale edge and dark brown submarginal band; margin of interspinous dorsal-fin membranes with a blackish brown triangle and short white or pale yellow filament behind the tip of each spine. The conspicuous white dots on the body and fins often persist on preserved specimens.

COLOUR IN MARKET As with all members of the 'reticulated groupers' (see Comments below), this species retains a similar colour pattern in death.

© Kwang-Tsao Shao

Adult *E. hexagonatus* (Nanwan, Pintung County, Taiwan)

© Robert F Myers

Adult *E. hexagonatus* (Guam)

POPULATION GENERAL *E. hexagonatus* is a common species throughout its range. **FISHERY DEPENDENT** *E. hexagonatus* is a common component of small-scale, multi-species fisheries. It rarely accounts for more than 1% of total catches where statistics are known. **FISHERY INDEPENDENT** *E. hexagonatus* may reach very high densities at some localities (e.g. >2000 fish per km^2 in French Polynesia; M Kulbicki pers. obs.).

POPULATION TREND Stable

MAJOR THREATS *E. hexagonatus* is reliant upon clear waters and healthy reefs, thus habitat loss is the primary threat to this species.

HABITAT AND ECOLOGY **General** *E. hexagonatus* is an insular, shallow-water species that inhabits coral reefs and rubble bottoms exposed to oceanic influence to depths of up to 30 m. **Feeding** Feeds on fishes and crustaceans. **Age and growth** Available information suggests a rapid growth rate, probably reaching sexual maturity at 2–3 years (M Kulbicki pers. obs.). **Reproduction** Size at maturity is reached at 19 cm TL in New Caledonia (M Kulbicki pers. obs.).

COMMENTS *E. hexagonatus* is one of nine shallow-water coral reef species that have a rounded caudal fin and close-set dark brown spots, with the pale interspaces forming a network on the body. These 'reticulated groupers' have been much confused in the literature, and many museum specimens have been misidentified. These other species differ from *E. hexagonatus* as follows:

- *E. bilobatus* has 3 bilobed dark blotches or close-set pairs of dark brown spots on the body and the base of the dorsal fin, no white dots on the body, 17 or 18 dorsal-fin rays, 2nd and 3rd anal-fin spines subequal, and 48–52 lateral-line scales.
- *E. faveatus* has smooth lateral-body scales (except for the area covered by the pectoral fins), no white dots on the body, midlateral part of lower jaw with 2 rows of teeth, and 48–52 lateral-line scales.
- *E. macrospilos* has the lateral-body scales mostly smooth, 48–52 lateral-line scales, 14–17 lower gill rakers, no triangular white dots on the body, and pectoral fins dusky with a narrow white edge.
- *E. maculatus* has the dorsal-fin membranes not incised between the spines, 3rd or 4th dorsal-fin spine longest (its length contained 2.1–2.6 times in head length and usually longer than dorsal-fin rays), no white dots on the body, and juveniles have a few large white blotches on the body and dorsal fin.
- *E. melanostigma* has a single black blotch at the base of the last 4 dorsal-fin spines and no white dots on the body, 2nd anal-fin spine subequal to 3rd, its length 2.6–3.6 times in head length and not more than the depth of the peduncle.
- *E. merra* has the pectoral fins with small black spots largely confined to the rays, no white dots on the body, no black blotches at the base of the dorsal fin, and 48–53 lateral-lines scales.
- *E. quoyanus* has 48–52 lateral-line scales, 14–16 lower gill rakers, larger pectoral fins (its length 1.2–1.7 times in head length), 2nd and 3rd anal-fin spines subequal and not much longer than the depth of the peduncle, and no triangular white dots on the body.
- *E. spilotoceps* has distinct black dots on the snout, no triangular white dots on the body, no large olive-brown spot behind the eye, shorter 2nd anal-fin spine (its length 2.4–3.4 times in head length), and a distinct notch in the rear edge of the preopercle just above the angle.

2008 IUCN RED LIST STATUS Least Concern

Authors: *PC Heemstra, M Kulbicki, RF Myers and KL Rhodes*

Epinephelus howlandi (Günther 1873)

blacksaddle grouper (E), mérou à selle noir (F), mero montura negra (S)

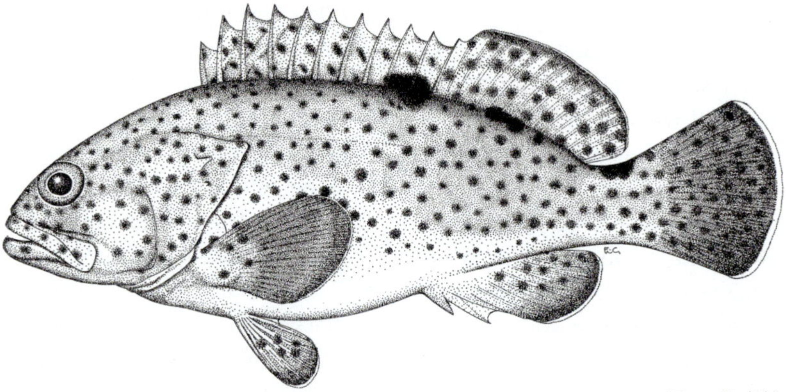

27 cm SL (Okinawa, Japan)
© SAIAB

Body depth contained 2.9–3.3 times in standard length. Head length 2.2–2.5 times in standard length; interorbital area flat, the dorsal head profile convex; preopercle rounded, the serrae at angle slightly enlarged; upper edge of operculum almost straight; posterior and anterior nostrils subequal; maxilla reaches past vertical at rear edge of eye; midlateral part of lower jaw with 2–4 rows of teeth. Gill rakers 7–9 on upper limb, 15–17 on lower limb, 23–26 total; gill raker at angle distinctly longer than adjacent rakers. Dorsal fin with 11 spines and 15–17 rays, the 3rd or 4th spine longest, subequal to longest rays and contained 2.7–3.6 times in head length, the interspinous membranes incised; anal fin with 3 spines and 8 rays, the 2nd and 3rd spines subequal; pectoral-fin rays 17–19; pectoral-fin length 1.6–2.1 times in head length; pelvic fins not reaching anus except in small specimens, their length 2.0–2.4 times in head length; caudal fin rounded. Lateral-body scales usually smooth (a few ctenoid scales may be found in area covered by pectoral fin), with numerous auxiliary scales; lateral-line scales 49–52; lateral-scale series 85–102. Pseudobranch pocket well developed; pyloric caeca 13.

MAXIMUM RECORDED SIZE 44 cm TL

Epinephelus howlandi occurs on the islands of the South Pacific and Micronesia from the Line Islands (Palmyra, Kingsman Reef, Jarvis Island) to the Mariana Islands, Papua New Guinea, Palau and the Great Barrier Reef (Australia). It is also present offshore in the China Sea, in the Ryukyu and Ogasawara island groups (Japan). The southern limit of this species is Lord Howe Island.

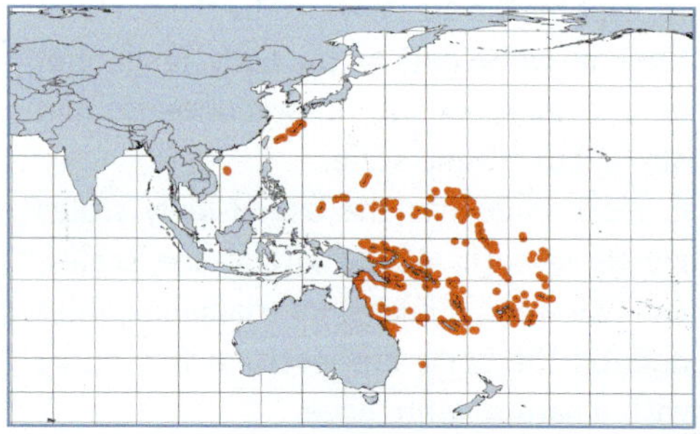

POPULATION GENERAL *E. howlandi* is an uncommon species across most of its range but may be locally abundant. **FISHERY DEPENDENT** *E. howlandi* represents about 0.3% of reported reef fish catch in the Commonwealth of the northern Mariana Islands and <0.01% in Guam (RF Myers pers. comm.). It is not reported in catches from American Samoa, and represents

COLOUR IN LIFE Head and body pale grey or brownish grey, covered with small brownish black spots with diffuse edges, separated from the adjacent spots by spaces equal to or greater than the width of the spots; black saddle blotch on the body and dorsal fin at the base of the last 3 spines and a smaller black saddle blotch on the caudal peduncle; no dark spots on the belly, chest or underside of the head; median and pelvic fins with dark spots as on the body, the margins of the soft dorsal, caudal and anal fins dusky, with a white line along the edge; pectoral fins dusky with a few dark spots basally.

COLOUR IN MARKET Body generally pale grey, spots remain prominent, especially black centres. Dark saddle blotches generally remain visible, especially on the caudal peduncle. White line on the edge of the median fins remains but fades to buff.

© Matthew T Craig

Adult *E. howlandi* (Okinawa Fish Market, Japan)

© Robert F Myers

Subadult *E. howlandi* (Guam)

© Robert F Myers

Adult *E. howlandi* (Coral Corner, Viti Levu, Fiji)

2.5% in numbers and 0.5% in weight of groupers caught in experimental fishing in New Caledonia. The species constitutes 12% of the grouper catch in Pohnpei (Rhodes and Tupper 2007). **FISHERY INDEPENDENT** Diver surveys indicate that this species may reach densities of ~200 fish per km^2.

POPULATION TREND Unknown

MAJOR THREATS No major threats have been identified for this species, although fishing pressure may cause local declines in abundance.

HABITAT AND ECOLOGY **GENERAL** *E. howlandi* is an insular species found on shallow back reefs in both well-developed and marginal coral habitats to depths of at least 37 m. The largest specimens are nearly always found on outer reefs with clear waters. **FEEDING** *E. howlandi* is an ambush predator. Its diet is comprised of fishes and crustaceans (shrimp and crab) with occasional molluscs or worms. **REPRODUCTION** Sexual maturity is reached at 25 cm TL in females in New Caledonia and there is some indication of a highly skewed sex ratio (95% females) (M Kulbicki unpublished data).

COMMENTS *E. howlandi* is often confused with *E. macrospilos* and *E. corallicola*. The colour pattern of *E. corallicola* is similar to *E. howlandi*, but the latter lacks dark spots on the belly, chest, underside of the head, and on the dorsal part of the eye (these areas are spotted in *E. corallicola*). *E. howlandi* also has a white line along the edge of the anal and caudal fins, which is absent in *E. corallicola*. The latter species differs from *E. howlandi* in the ctenoid lateral-body scales, 53–63 lateral-line scales, the absence of a pseudobranch pocket, the posterior nostrils are vertically elongate in adults, and the gill raker at the angle of the first gill arch is subequal in length to the adjacent rakers. *E. macrospilos* differs from *E. howlandi* in lacking a black blotch on the body at the base of the last 3 dorsal-fin spines, and in having a submarginal series of dark spots parallelling the rear edge of the caudal fin, usually fewer gill rakers (total 21–25), pectoral-fin rays 17–20 (usually 19), and a strongly projecting lower jaw.

2008 IUCN RED LIST STATUS Least Concern

Author: *M Kulbicki*

Epinephelus indistinctus

(Randall and Heemstra 1991)

Somali grouper (E), mérou Somali (F), mero Somalí (S)

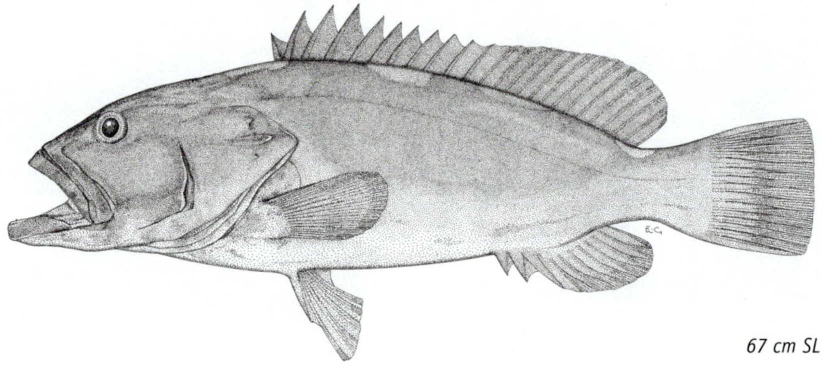

67 cm SL (Somalia)

Body depth contained 3.2 times in standard length (1 specimen, 672 mm SL). Head length 2.3 times in standard length; interorbital area convex, the dorsal head profile slightly convex; preopercle angular, with a shallow indentation just above the rounded angle, the serrae not noticeably enlarged at the angle; upper edge of operculum almost straight; posterior and anterior nostrils subequal; maxilla probably reaches vertical at rear edge of eye (holotype preserved with mouth wide open); midlateral part of lower jaw with 2 rows of subequal small teeth; small canines at front of jaws. Gill rakers 9 on upper limb, 14 on lower limb, the longest raker about two-thirds of length of longest gill filament; small bony platelets on gill arches. Dorsal fin with 11 spines and 14 rays, 4th spine longest, its length contained 4.2 times in head length and shorter than longest dorsal-fin rays, the interspinous membranes distinctly incised; anal fin with 3 spines and 8 rays; pectoral-fin rays 18; pectoral-fin length 2.4 times in head length; pelvic-fin length 2.7 times in head length; caudal-fin rear margin slightly convex. Lateral-body scales smooth (except for area covered by pectoral fins), with numerous auxiliary scales; lateral-line scales 64; lateral-scales series 114. Pyloric caeca more than 50.

MAXIMUM RECORDED SIZE 80 cm TL

COLOUR IN LIFE Known only from the preserved holotype, no live specimen is known.

COLOUR IN MARKET No fresh market specimens are known. In alcohol, the head, body and median fins are dark greyish brown, the exposed part of the dorsolateral-body scales are darker giving a faint finely dotted pattern; 4 faint pale blotches on the body at the base of the dorsal fin, the first at the origin of the fin and extending anterior to it, the last at the base of the last 3 or 4 rays and extending onto top of the front half of the peduncle.

Epinephelus indistinctus is known only from the holotype; however, anecdotal reports indicate its presence from the Puntland coast of Somalia and Oman.

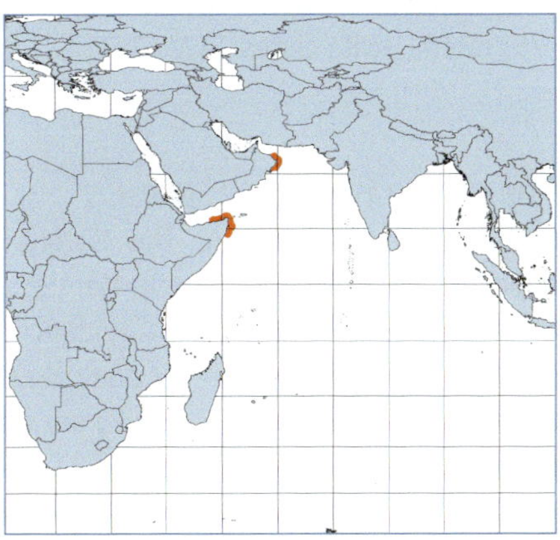

POPULATION GENERAL This species is known only from the holotype.

POPULATION TREND Unknown

MAJOR THREATS No major threats have been identified for this species.

HABITAT AND ECOLOGY GENERAL The holotype of *E. indistinctus* was taken off Somalia in 70–80 m with a trawl.

2008 IUCN RED LIST STATUS Data Deficient

Authors: *MT Craig and MA Samoilys*

Epinephelus irroratus (Forster 1801)

Marquesan grouper (E), mérou Marquises (F), mero marquesano (S)

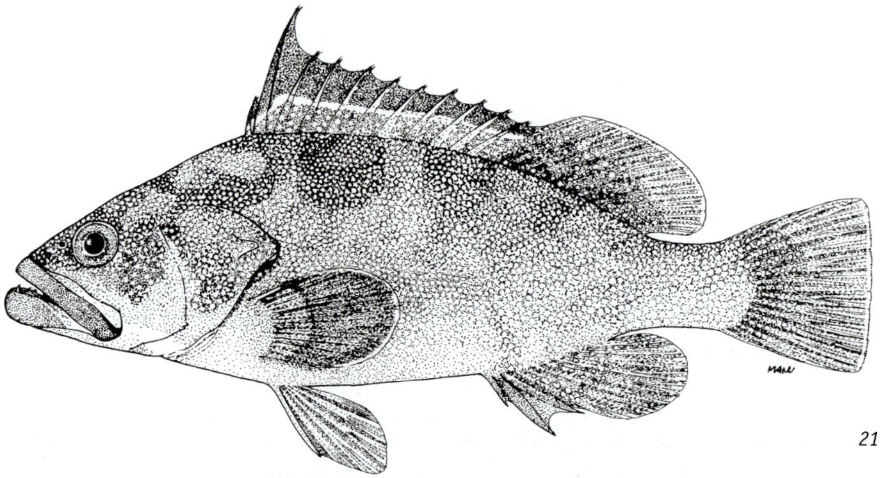

21 cm SL
© FAO

Body depth contained 2.7–3.3 times in standard length. Head length 2.4–2.6 times in standard length; interorbital area and dorsal head profile slightly convex; preopercle angle rounded and slightly indented, the lower serrae scarcely enlarged; upper edge of operculum almost straight; posterior nostrils about twice size of anterior nostrils; maxilla reaching to or just past vertical at rear edge of eye; midlateral part of lower jaw with 2–4 rows of teeth. Gill rakers 6–8 on upper limb, 13–16 on lower limb. Dorsal fin with 11 spines and 16 rays, the 2nd spine greatly elongated in adults, more than twice length of 3rd spine, the interspinous membranes not or only slightly incised; anal fin with 3 spines and 8 rays; pectoral-fin rays 18–20; pectoral-fin length contained 1.6–1.9 times in head length; pelvic fins not reaching past anus, their length 1.8–2.0 times in head length; caudal fin truncate to slightly rounded. Lateral-body scales ctenoid, with auxiliary scales; lateral-line scales 70–75; lateral-scale series 117–136.
MAXIMUM RECORDED SIZE 34 cm TL

Epinephelus irroratus is found only in the Marquesas Islands (French Polynesia) in the South Pacific.

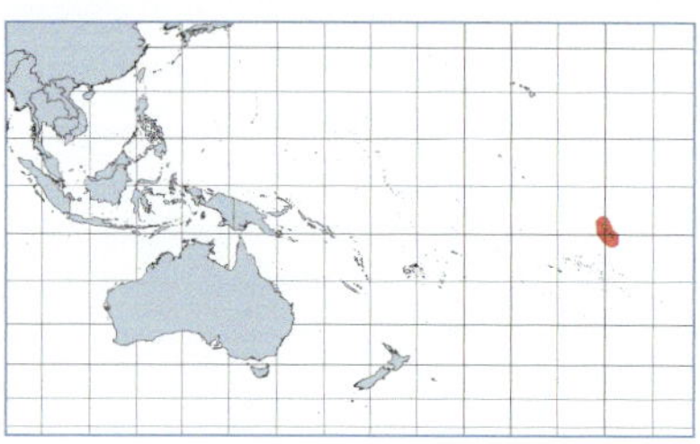

POPULATION GENERAL *E. irroratus* is a common species throughout the Marquesas archipelago. **FISHERY DEPENDENT** In the Marquesas, fishery-dependent data show no evidence of declines in CPUE (10–30 fish per boat day) or average weight of fish caught (1.8 kg per fish; C Ponsonnet, pers. comm., 2003).
POPULATION TREND Stable
MAJOR THREATS No major threats have been identified for this species.

COLOUR IN LIFE Reddish brown with a white dot on each scale giving a finely speckled appearance. Maxillary streak dark reddish brown; spinous dorsal fin with a prominent dark red margin; white line along the rear edges of the median and pectoral fins.

COLOUR IN MARKET Generally dark brown with a finely speckled appearance. Maxillary streak remains but is dull. The dark margin on the dorsal fin is not apparent, and the white edge of the median fins remains or fades from white to a translucent hue.

© Michelle Gaither

Adult *E. irroratus* (Marquesas Islands, French Polynesia)

© John E Randall

Adult *E. irroratus* (Marquesas Islands, French Polynesia)

HABITAT AND ECOLOGY **General** *E. irroratus* is a poorly known species that inhabits reefs in depths of at least 50 m.

2004 IUCN RED LIST STATUS Least Concern

Authors: *AS Cornish and Y-Y Situ*

Epinephelus itajara (Lichtenstein 1822)

Atlantic goliath grouper (formerly jewfish) (E), mérou goliath (F), mero guasa (S)

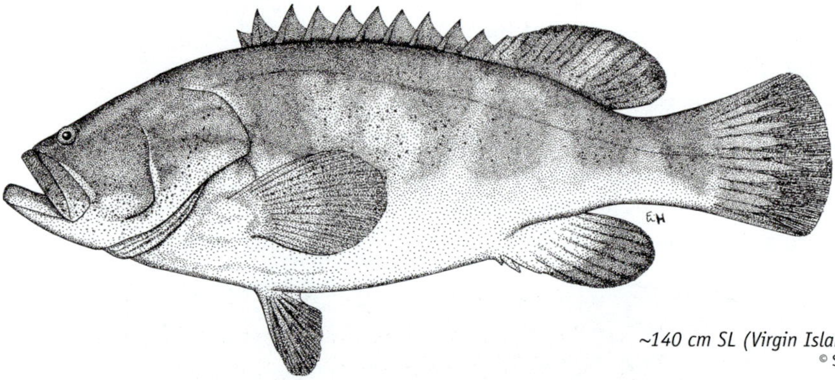

~140 cm SL (Virgin Islands)
© SAIAB

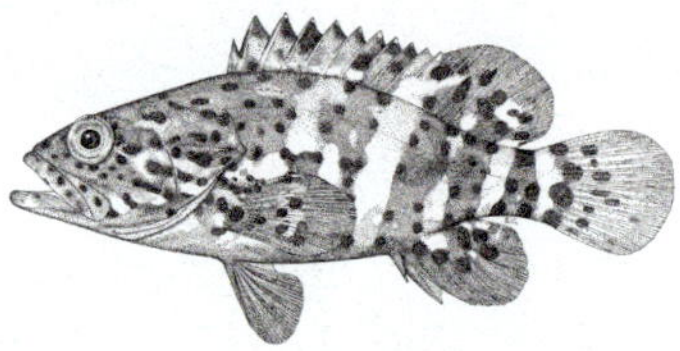

~7 cm SL (Gulf of Mexico)
© SAIAB

Body robust, elongate, the greatest width more than half body depth, which is distinctly less than head length (in fish 15–160 cm SL); body depth contained 2.7–3.4 times in standard length. Head length 2.3–2.9 times in standard length; head extremely broad; interorbital flat, the width equals eye diameter in fish 10–15 cm SL, distinctly greater than eye diameter in fish 18–25 cm SL, and 1.5–3.4 times greater than eye diameter in fish 30–160 cm SL; eye diameter contained 5–8 times in head length for fish 10–30 cm SL and 8–13 times in head length for fish 35–160 cm SL; preopercle rounded, finely serrate; nostrils round, subequal; maxilla scaly, reaching well past eye; midlateral part of lower jaw with 3–5 rows of subequal teeth; no canines at front of jaws. Gill rakers 8 or 9 on upper limb and 13–15 on lower limb, 21–24 total; gill arches covered with small bony plates. Dorsal fin with 11 spines and 15 or 16 rays, the spines short, 3rd to 11th subequal and shorter than the first ray, the membranes distinctly indented between the spines; anal fin with 3 spines and 8 rays; pectoral-fin rays 18 or 19; caudal fin rounded. Body scales strongly ctenoid; lateral-line scales 61–64, each with 4–6 radiating ridges; lateral-scale series 89–112.

MAXIMUM RECORDED SIZE 250 cm TL

POPULATION **GENERAL** *E. itajara* was once known to be common throughout its range. It has suffered severe population declines and is now rare throughout most of its range, with only localised pockets of abundance. **FISHERY DEPENDENT** Declines in landings of the species have been recorded throughout its geographic range, including Florida (USA), Belize, Mexico and Brazil, and attributed to a combination of overfishing and habitat degradation, especially the loss of mangrove habitat. Fishing methods include hook and line, traps and trawls; spear-fishing is particularly popular with sports divers. While populations historically supported

COLOUR IN LIFE Generally brownish yellow, grey or greenish; head and dorsal part of body and fins with small black spots, becoming smaller with growth. Fish of less than about 1 m show 3 or 4 faint irregular subvertical dark bars posteriorly on the body, and another covering the rear half of the caudal peduncle; large adults darker and more uniformly coloured than the juveniles.

COLOUR IN MARKET Uniform brown, sometimes with irregular blotches of lighter hue. A very indistinct grouper in death colour, this species is best recognised by its large size, wide body, proportionally small eyes and broad head.

© Maurício Hostim-Silva

Adult *E. itajara* (Brazil)

© Robert F Myers

Adult *E. itajara* (Boynton Beach, Florida, USA)

© Áthila Bertoncini Andrade

Adult *E. itajara* (Monoboia, Brazil)

both commercial and recreational fisheries, overfishing in shallow waters resulted in rapid declines nearly 50 years ago, and in deeper waters some 30 years ago led to declining landings and loss of, or severe declines in, spawning aggregations (Sadovy and Eklund 1999, Gehardingher et al. 2006). Detailed fishery-dependent data from recent years are lacking due to a moratorium on catches in countries that formerly had large landings (e.g. USA and Brazil). Prior to this moratorium, fishery-dependent data indicated massive declines in catch rates, particularly for larger fish (IUCN Red List Assessment 2006). **Fishery independent** Fishery-independent data indicate that the adult population is generally absent from appropriate habitat. Juvenile and young adult abundance has shown promising increases in areas where the species is protected (e.g. south-eastern USA) (e.g. Koenig et al. 2007). **Genetics** Population genetic data indicate some degree of partitioning throughout the eastern Atlantic range of the species. This may indicate that there is a substantial contribution to local populations by self-recruitment (Craig et al. 2009).

POPULATION TREND Unknown

MAJOR THREATS *E. itajara* is severely threatened by overfishing and habitat loss, particularly for habitat for juveniles.

HABITAT AND ECOLOGY **General** *E. itajara* inhabits a variety of habitats from shallow inshore waters to depths up to 100 m. The species shows an ontogenetic habitat shift with juveniles and young adults having an obligate tie to mangroves, bays and estuaries for 5–6 years, while adults inhabit offshore reefs with long-term site fidelity, as well as bays and harbours. Tagged individuals in Cuba have moved up to 168 km from a tagging site within a couple of years (Pina-Amargós and González-Sansón 2008). The species has a high tolerance for brackish water. **Feeding** *E. itajara* feeds at a rather low trophic level, preying on crustaceans (mudcrabs) and slow-moving, bottom-associated fishes (including stingrays, catfish, toadfish, burrfish and cowfish), but very few if any snappers or groupers. They may incidentally take octopus and young sea turtles (Randall 1967). **Age and growth** *E. itajara* grow fast relative to their maximum size. Growth rates for males and females are similar, averaging >10 cm per year through age 6 years, then slowing to about 3 cm per year by age 15, and finally declining to <10 mm per year after age 25. The von Bertalanffy growth model was found to be $L = 206(1 - e^{-0.126(\mathrm{Age} + 0.49)})$ where L is the total length in mm (Bullock et al. 1992). The oldest individual known was 37 years old (female) and 26 years (male) (Bullock et al. 1992) but greater longevity (possibly over 50 years old) is likely given the overfished state of the species at the time of the study. $W = 2.94 \times 10^{-5} \times L^{2.94}$ where W is the ungutted weight in g and L is the total length in mm, and TL = 32.45 + 1.18SL (in mm) (reviewed in Sadovy and Eklund 1999). **Reproduction** *E. itajara* is an aggregate spawner and is one of the few groupers that aggregate in relatively shallow (10–50 m) water (Coleman et al. 2002).

Epinephelus itajara is an amphi-Atlantic species. In the western Atlantic it is known from North Carolina (USA) to south-eastern Brazil, including the Caribbean and Gulf of Mexico, and in the eastern Atlantic from Senegal to Congo. It was previously thought to occur in the tropical eastern Pacific; however, populations there have recently been shown to be a distinct species (*E. quinquefasciatus*; Craig et al. 2009).

Its sexual pattern is unclear and it is most probably gonochoristic (Sadovy and Eklund 1999). The species has a short spawning period in the north-eastern Caribbean during July and August (Erdman 1976) and more protracted spawning in the Gulf of Mexico from June through December (Bullock et al. 1992). Females mature by 120 cm TL and males mature by 110 cm TL, and aggregations reported are relatively small, numbering 10s to maybe a hundred or so (Bullock et al. 1992, Sadovy and Eklund 1999).

COMMENTS *E. itajara* was listed as a candidate species on the US Endangered Species List, and from 1991 to 2006 it was considered a Species of Concern throughout its geographic range in US waters. Due to promising signs of recovery, particularly in the protected areas of the south-eastern USA, the species is no longer listed by US authorities in any special category. However, all harvesting of *E. itajara* in US federal waters of the south-eastern United States (including the Gulf of Mexico) is still, and has been, prohibited since 1990 by the South Atlantic Fisheries Management Council and the Gulf of Mexico Fisheries Management Council, and in the Caribbean since 1993 by the Caribbean Fishery Management Council. No other signs of recovery have been noted elsewhere in the range of this species.

2006 IUCN RED LIST STATUS Critically Endangered (A2d)

Authors: *T-C Chan and BP Ferreira*

Epinephelus labriformis (Jenyns 1843)

starry grouper (E), mérou étoile (F), cabrilla pedrera (S)

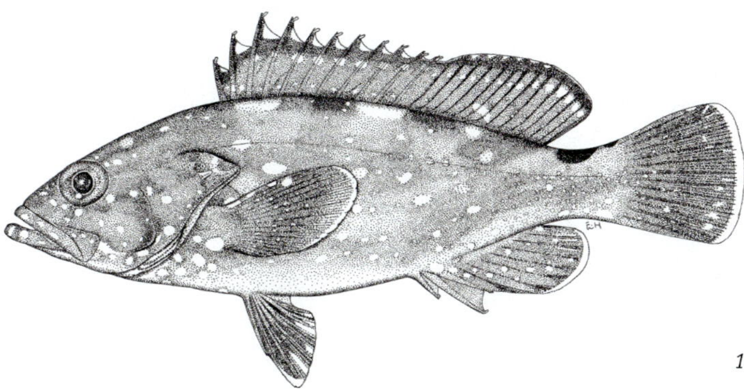

18 cm SL (Costa Rica)
© SAIAB

Body depth distinctly less than head length, depth contained 2.7–3.1 times in standard length. Head length 2.2–2.5 times in standard length; interorbital area flat; preopercle rounded, finely serrate, the ventral serrae slightly enlarged; subopercle and interopercle smooth; upper edge of operculum slightly convex; posterior and anterior nostrils subequal. Gill rakers 7–9 on upper limb, 15–17 on lower limb, 23–26 total. Dorsal fin with 11 spines and 16–18 rays, the 3rd or 4th spine longest (longer than longest dorsal-fin ray) and the interspinous membranes moderately incised; anal fin with 3 spines and 8 rays; pectoral-fin rays 18 or 19; pelvic fins much shorter than pectoral fins, their origin below or behind pectoral-fin base; caudal fin rounded. Lateral-body scales ctenoid, with a few auxiliary scales; lateral-line scales 48–51; lateral-scale series 84–100.

MAXIMUM RECORDED SIZE 52 cm TL

COLOUR IN LIFE Generally olive green to reddish brown with scattered irregular white and brown spots and blotches; white triangle at the margin of the interspinous dorsal-fin membranes and a bright white tag at the tip of each spine; juveniles with black spots on the top of the head and median fins reddish distally with a white edge; small black saddle on the peduncle; the inside of the mouth is red.

COLOUR IN MARKET Brown with pale ventral surface or whole body faded to pale and blotchy brown; white blotches less numerous and faded; reddish tint on the dorsal, pectoral and caudal fins; the black saddle on the caudal peduncle may fade.

© Matthew T Craig

Adult *E. labriformis* (Loreto, Mexico)

© Matthew T Craig

Adult *E. labriformis* (Loreto, Mexico)

© Octavio Aburto

Adult *E. labriformis* (Gulf of California, Mexico)

POPULATION **General** *E. labriformis* is a common species throughout its range. **Fishery dependent** Fishery-dependent data are scarce. However, casual observations of landings from 1998 to 2006 do not indicate declines in the Gulf of California (MT Craig and B Erisman pers. obs.). **Fishery independent** Diver surveys indicate abundance is locally stable, but declining in some parts of its range. **Genetics** Genetic data indicate strong population connectivity throughout the entire range of this species (Craig et al. 2006b).

POPULATION TREND Decreasing

MAJOR THREATS *E. labriformis* is taken by subsistence, artisanal and subnational fisheries, but fishing is currently not considered a major threat.

HABITAT AND ECOLOGY **General** *E. labriformis* inhabits rocky reefs to at least 30 m. **Feeding** *E. labriformis* is a solitary predator that feeds on fishes by day and crustaceans after dark. **Age and growth** *E. labriformis* reaches at least 30 years (Craig et al. 1999). **Reproduction** *E. labriformis* is a protogynous hermaphrodite that forms male-dominated harems. In the Gulf of California, it does not appear to form aggregations, and spawning occurs at dusk in mid- to late summer during full-moon phases (Erisman 2008).

COMMENTS *E. labriformis* is similar to *E. clippertonensis* Allen and Robertson 1999. The former is generally restricted to continental localities (but occurs at the Galápagos Islands), while the latter is more common at islands. *E. clippertonensis* lacks the distinct scattered white spots and olive-green body colour seen in *E. labriformis*.

2008 IUCN RED LIST STATUS Least Concern

Epinephelus labriformis is an eastern Pacific species that occurs from San Diego, California (USA) to Peru, including the offshore islands of Cocos, Revillagigedo, Malpelo and Galápagos.

Authors: *AA Bertoncini, JH Choat, MT Craig, BP Ferreira and LA Rocha*

Epinephelus lanceolatus (Bloch 1790)

giant grouper (E), mérou lancéolé (F), mero lanceolado (S)

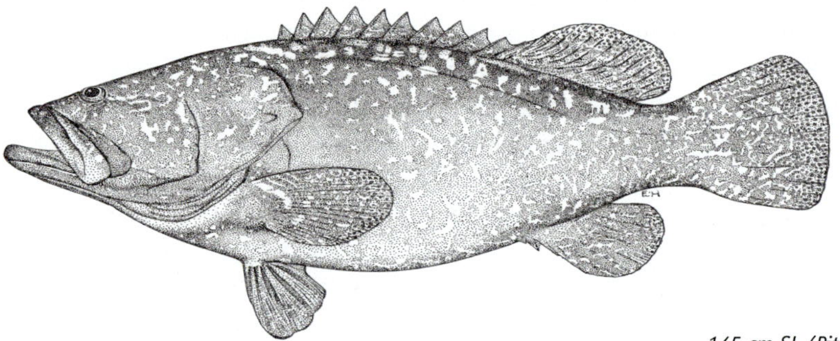

145 cm SL (Pitcairn Islands)
© SAIAB

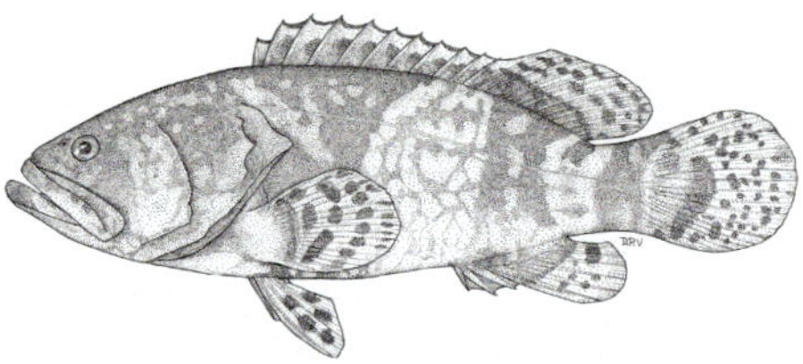

23 cm SL (South Africa)
© SAIAB

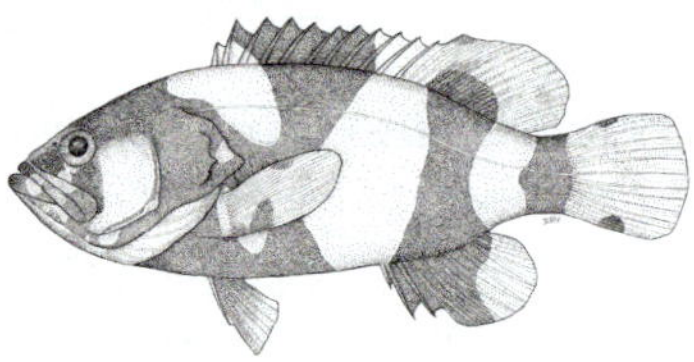

3 cm SL (South Africa)
© SAIAB

Body robust, the depth contained 2.4–3.4 times in standard length, the body width 1.5–1.8 times in the depth. Head length 2.2–2.7 times in standard length; interorbital width 3.3 (for fish 177 cm SL) to 6.2 (for fish 12 cm SL) times in head length; interorbital area flat to slightly convex, the dorsal head profile convex; preopercle subangular, finely serrate, the angle rounded; upper edge of operculum convex; eye diameter 5.8–14 in head length; posterior and anterior nostrils subequal; maxilla reaching past vertical at rear edge of eye; midlateral part of lower jaw with 2 or 3 rows of teeth (specimens of 20–25 cm SL) increasing to 15–16 rows in a fish of 177 cm SL; canine teeth at front of jaws small or absent. Gill rakers of juveniles 8–10 on upper limb, 14–17 on lower limb; rudiments in adults are difficult to distinguish from the bony platelets covering the gill arch. Dorsal fin with 11 spines and 14–16 rays, the 3rd to 11th spines subequal, their length contained 3.1–5.7 times in head length and much shorter than longest rays in adults; anal fin with 3 spines and 8 rays; pectoral-fin rays 18–20; pectoral-fin length 1.8–2.2 times in head length; pelvic fins not reaching anus, their length 2.1–2.6 in head length; caudal fin rounded. Lateral-body scales smooth, with auxiliary scales; lateral-line scales 54–62, the anterior scales with branched tubules (except small juveniles); lateral-scale series 95–105.

MAXIMUM RECORDED SIZE 270 cm TL

POPULATION **GENERAL** *E. lanceolatus* is a rare species throughout its range. **FISHERY DEPENDENT** No explicit fishery-dependent data are currently available. However, anecdotal reports indicate that natural populations have suffered severe declines. The species is now subject to a growing mariculture industry and can be successfully hatchery-produced, especially in Taiwan.

POPULATION TREND Decreasing

COLOUR IN LIFE Small juveniles (12 cm SL) yellow, with irregular broad black bars on the body, the 1st from the spinous dorsal fin to the belly and chest and extending onto the head, the 2nd from the base of the soft dorsal fin to the anal fin, and the last at the base of the caudal fin; small adults (20–50 cm SL) with irregular white or yellow spots on the black areas and fins with irregular black spots; adults (80–150 cm SL) dark brown with faint mottling, the fins with numerous small black spots; large adults (160–230 cm SL) dark brown, the fins darker.

COLOUR IN MARKET Generally dark brown or green, with irregular markings taking the form of indistinct bars. May also lose all markings and be completely dark brown. Small black spots are faint or absent.

© Matthew T Craig

Juvenile *E. lanceolatus* (Los Angeles fish store, California, USA)

© Allen W-L To

Adult *E. lanceolatus* (Cape d'Aguilar Marine Reserve, Hong Kong)

Epinephelus lanceolatus is a widespread grouper throughout the Indo-Pacific, including the Red Sea. It occurs from Algoa Bay (South Africa), eastward to the Pitcairn Islands, including the Hawaiian Islands (USA), where its occurrence is extremely rare, and Micronesia. It ranges northwards to southern Japan and southwards to Australia.

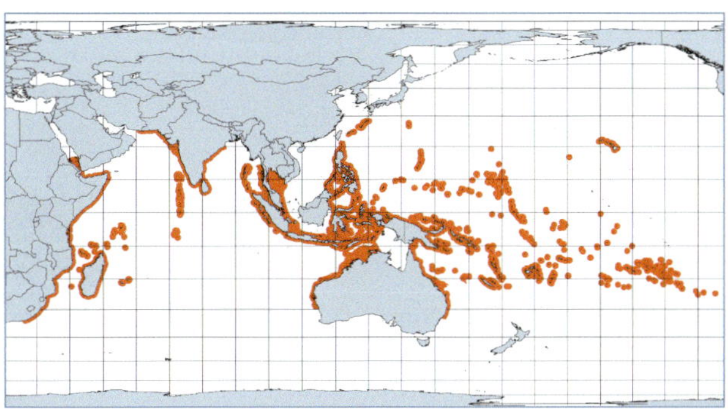

MAJOR THREATS Given its large size and desirability as a food fish, *E. lanceolatus* is considered to be threatened by fishing pressure.

HABITAT AND ECOLOGY **GENERAL** *E. lanceolatus* is the largest of all coral reef-dwelling bony fishes. It tends to be solitary and inhabits lagoons and seaward reefs to depths of at least 50 m. The species is also known to inhabit estuaries as adults and juveniles. **FEEDING** *E. lanceolatus* feeds on a variety of prey including spiny lobsters and a variety of fishes, including small sharks and batoids, and juvenile sea turtles (Heemstra and Randall 1993). In South African estuaries, the main prey item is the mud crab. **REPRODUCTION** It is not known whether *E. lanceolatus* forms spawning aggregations. However, seasonal increases in catch rates suggest that this may occur (Sadovy and Liu 2004, P Chan pers. comm.). Maturity size is thought to occur at approximately 129 cm TL (Lau and Li 2000).

COMMENTS *E. lanceolatus* is a common component of the live reef food fish trade in Asia where it is sold for large profit. Many individuals present in this trade are below the size at first maturity indicating removal of immature individuals. However, given that this species is successfully hatchery-produced at commercial levels, it is not known what proportion of these fish are from natural populations vs those from hatcheries. Large fish are still taken from the wild. It is a protected species in some parts of Australia.

2006 IUCN RED LIST STATUS Vulnerable (A2d)

Authors: *C-M Chan, ST Fennessy, W-C Ng and DA Pollard*

Epinephelus latifasciatus

(Temminck and Schlegel 1842)

striped grouper (E), mérou à bandes (F), mero abanderado (S)

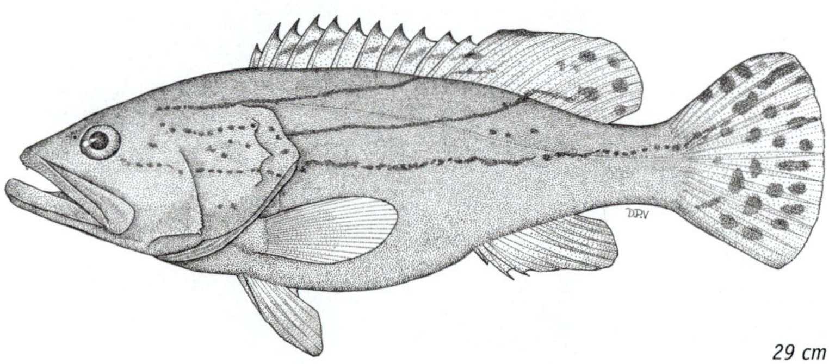

29 cm SL (West India)
© SAIAB

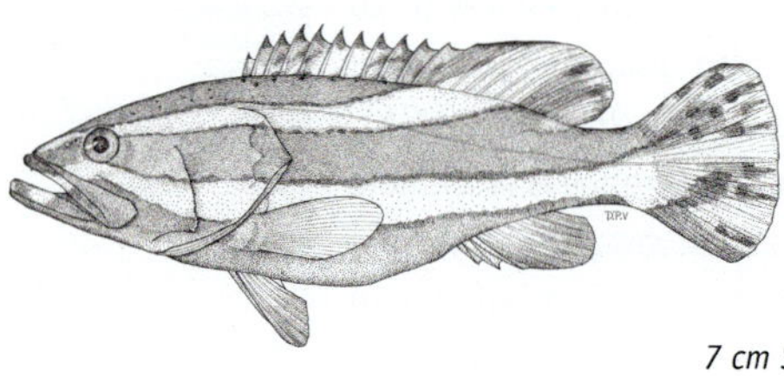

7 cm SL (Japan)
© SAIAB

Body depth contained 2.9–3.4 times in standard length. Head length 2.3–2.6 times in standard length; interorbital area convex, the dorsal head profile convex; preopercle angular, with 3–7 distinctly enlarged serrae at angle; upper edge of operculum distinctly convex; posterior and anterior nostrils subequal; maxilla reaches past vertical at rear edge of eye; midlateral part of lower jaw with 2 or 3 rows of teeth. Gill rakers 8–11 on upper limb, 15–18 on lower limb. Dorsal fin with 11 spines and 12–14 rays, the 3rd or 4th spine longest, contained 2.9–3.9 times in head length and not much shorter than longest rays, the interspinous membranes distinctly incised; anal fin with 3 spines and 8 rays; pectoral-fin rays 17–19; pectoral-fin length 1.8–2.2 times in head length; pelvic fins not reaching anus, their length 2.2–2.6 times in head length; caudal fin truncate in large adults, the rear margin convex in juveniles. Lateral-body scales smooth; lateral-line scales 56–65; lateral-scale series 91–106.

MAXIMUM RECORDED SIZE 157 cm TL

COLOUR IN LIFE Juveniles lavender-grey or pale brownish, shading to whitish ventrally; 2 black-edged white longitudinal bands, the upper band extending from above the eye to the anterior dorsal-fin rays, and the lower band from below the eye to the lower caudal-fin rays; dorsal and caudal fins with black spots and streaks; white bands disappearing on adults, the dark edges breaking into dashes and spots; head and body of large adults uniformly grey.

COLOUR IN MARKET Pale grey or brownish, longitudinal bands remain prominent in smaller specimens, or present as a series of dashes in larger specimens.

Adult *E. latifasciatus* (Jimbaran Fish Market, Bali, Indonesia)

Epinephelus latifasciatus is an Indo-West Pacific species distributed in the Red Sea and Persian Gulf, to north-western Australia. It is not known from the east coast of Africa, the Indian Ocean islands, Indonesia, Philippines, or Papua New Guinea.

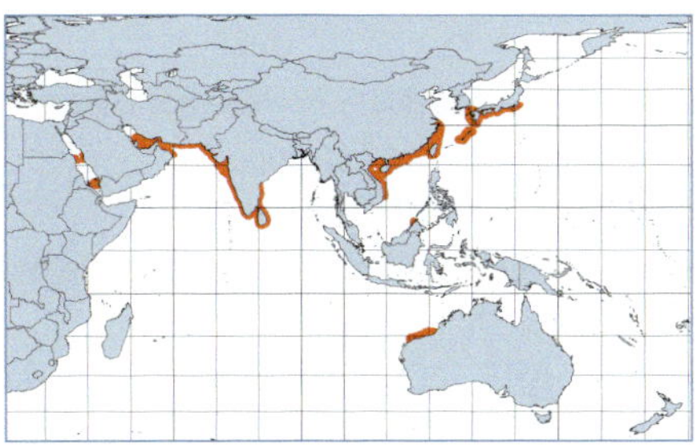

POPULATION **GENERAL** *E. latifasciatus* is commonly seen in Asian fish markets.

POPULATION TREND Unknown

MAJOR THREATS The major threat to *E. latifasciatus* is fishing pressure.

HABITAT AND ECOLOGY **GENERAL** *E. latifasciatus* is a continental species that inhabits areas of low relief in depths of 20–240 m.

2008 IUCN RED LIST STATUS Data Deficient

Authors: *PC Heemstra and BC Russell*

Epinephelus lebretonianus

(Hombron and Jacquinot 1853)

mystery grouper (E), mérou arcane (F), mero misterioso (S)

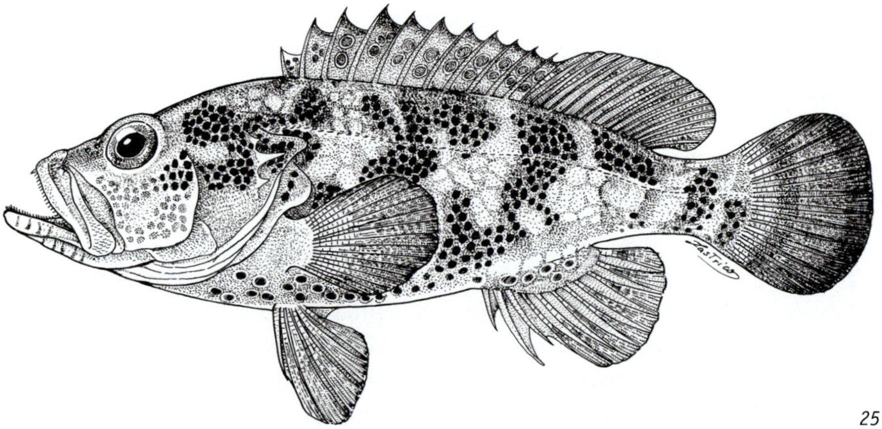

25 cm SL
© FAO

Body depth contained 3.4 times in standard length. Head length 2.3 times in standard length; interorbital area slightly convex; preopercle margin finely serrate, the lower rear edge mostly smooth; posterior nostril slightly larger than anterior nostril; maxilla reaches vertical at rear edge of orbit; midlateral part of lower jaw with 3 rows of teeth. Gill rakers 7 on upper limb, 17–18 on lower limb. Dorsal fin with 11 spines and 17 rays, the 4th and 5th spines longest, their length contained 3.5 times in head length and shorter than longest dorsal-fin ray, the interspinous membranes incised; anal fin with 3 spines and 8 rays; pectoral-fin rays 20; pectoral-fin length 1.9 times in head length; pelvic-fin length 2.3 times in head length; caudal fin rounded. Lateral-body scales ctenoid; lateral-line scales 73; lateral-scale series about 120.
MAXIMUM RECORDED SIZE 25 cm SL

COLOUR IN LIFE Body mottled with brown and covered with small close-set dark brown spots; median fins brown, with small blue-edged brown ocelli, the fin margin blackish; paired fins whitish proximally, brown distally. **COLOUR IN MARKET** No market specimens known.

Epinephelus lebretonianus probably has an Indo-Pacific distribution.

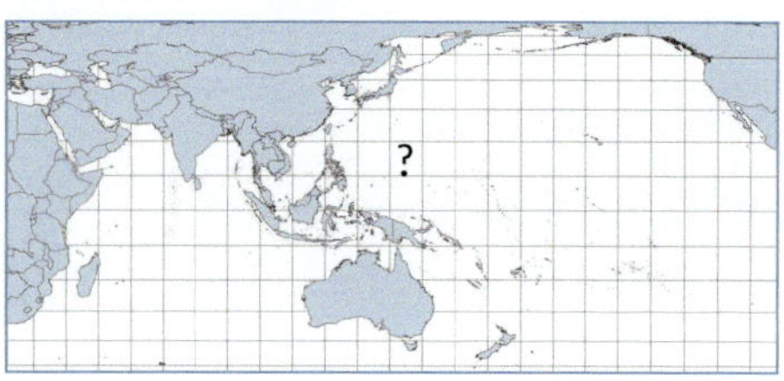

POPULATION GENERAL *E. lebretonianus* is known only from the holotype.
POPULATION TREND Unknown
MAJOR THREATS No major threats have been identified for this species.
HABITAT AND ECOLOGY GENERAL *E. lebretonianus* is known only from a single specimen and no information on its habitat and ecology is currently available.

COMMENTS Known only from a single specimen of uncertain locality although it is likely to be from the Indo-Pacific, collected during a circumglobal voyage of the corvettes *L'Astrolabe* and *La Zélée* (Hombron and Jacquinot 1853).

2008 IUCN RED LIST STATUS Data Deficient

Authors: *DA Pollard and MA Samoilys*

Epinephelus longispinis (Kner 1864)

longspine grouper (E), mérou longues épines (F), mero espigón (S)

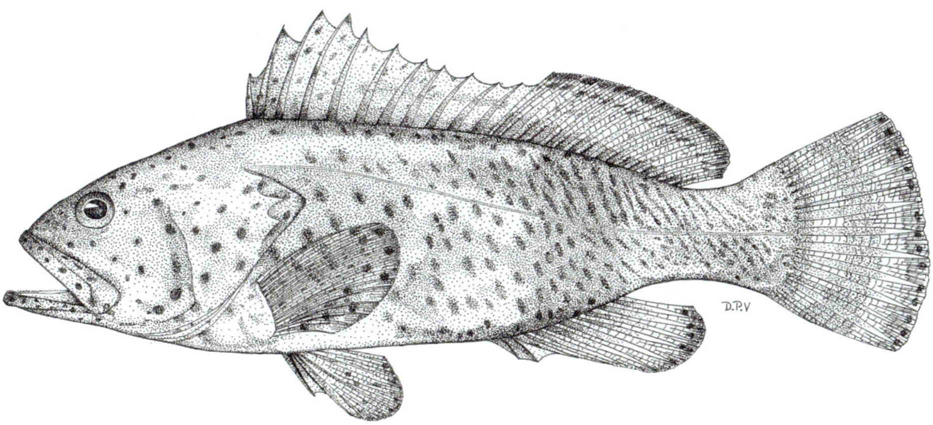

31 cm SL (South Africa)

Body depth contained 2.8–3.3 times in standard length. Head length 2.4–2.6 times in standard length; interorbital area flat to slightly convex, the dorsal head profile convex; preopercle angle with enlarged serrae and a shallow indentation just above the angle; upper edge of operculum straight or slightly convex; posterior nostril not much larger than anterior nostril; maxilla reaches to or past vertical at rear edge of eye, the ventral edge with a blunt hook-like process distally in fish larger than 35 cm SL; midlateral part of lower jaw with 2 rows of teeth. Gill rakers 8–11 on upper limb, 15–17 on lower limb. Dorsal fin with 11 spines and 16 or 17 rays, the 3rd or 4th spine longest, its length contained 2.1–2.6 times in head length and distinctly longer than longest dorsal rays, the interspinous membranes slightly incised between anterior spines and not incised between posterior spines; anal fin with 3 spines and 8 rays; pectoral-fin rays 17–19; pectoral-fin length 1.5–1.9 times in head length; pelvic fins not reaching anus, their length 1.8–2.2 times in head length; caudal fin convex. Lateral-body scales distinctly ctenoid, with numerous auxiliary scales; lateral-line scales 49–53; lateral-scale series 98–121.

MAXIMUM RECORDED SIZE 55 cm TL

COLOUR IN LIFE Head and body pale; greyish brown, covered with small dark reddish brown spots that are rounded and widely spaced on the head and front half of the body, but obliquely elongated and closer together and darker posteriorly; fins with similar dark brown spots; a row of dark spots along the distal margin of the soft dorsal and caudal fins. Juveniles with fewer dark spots on head and body, and 2–4 dark blotches on body at base of dorsal fin and extending onto fin; ventral part of head and body with irregular white spots; white blotch on upper part of the operculum.

COLOUR IN MARKET Generally pale grey to brown, dark spots remain prominent. White blotch on upper operculum fades but is sometimes visible.

© Patrice Marker

Adult *E. longispinis* (Komodo area, Indonesia)

POPULATION **GENERAL** *E. longispinis* is a rare species throughout its range. **FISHERY DEPENDENT** *E. longispinis* is only occasionally reported in fisheries statistics, most likely due to its natural rarity. **FISHERY INDEPENDENT** Fishery-independent data are scarce, but surveys in South Africa and Mozambique did not record this species (Chater et al. 1993, Robertson et al. 1996, Pereira 2003).

Epinephelus longispinis is an Indo-West Pacific species ranging from Kenya, through Tanzania and Mozambique to South Africa (32° S), and east to the Watubela Group of islands in the eastern Banda Sea.

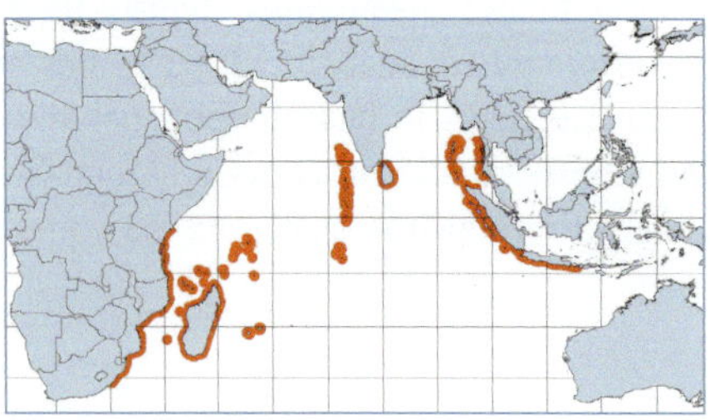

POPULATION TREND Unknown

MAJOR THREATS Fishing pressure is likely to pose a threat to this species.

HABITAT AND ECOLOGY **GENERAL** *E. longispinis* is a reef-associated species that inhabits a variety of habitats in depths to 70 m. **FEEDING** *E. longispinis* feeds mainly on crustaceans. **REPRODUCTION** Morgans (1964) reported a mature specimen weighing 1.59 kg (TL estimate of 44 cm). He also reported 'nearly-ripe' fish in August and September. A reproductively mature fish was obtained in January on the North Kenya Banks (Nzioka 1979).

COMMENTS *E. longispinis* is similar to *E. maculatus*. Adults of *E. maculatus* differ in having all the dark spots on head and body round (none elongated into oblique streaks) and close-set (not more crowded posteriorly), a large pale blotch on the middle of the dorsal fin, and no row of dark spots along the rear edge of the caudal fin.

2008 IUCN RED LIST STATUS Least Concern

Authors: *ST Fennessy, RF Myers and DA Pollard*

Epinephelus macrospilos (Bleeker 1855)

snubnose grouper (E), mérou tapis (F), mero alfombrado (S)

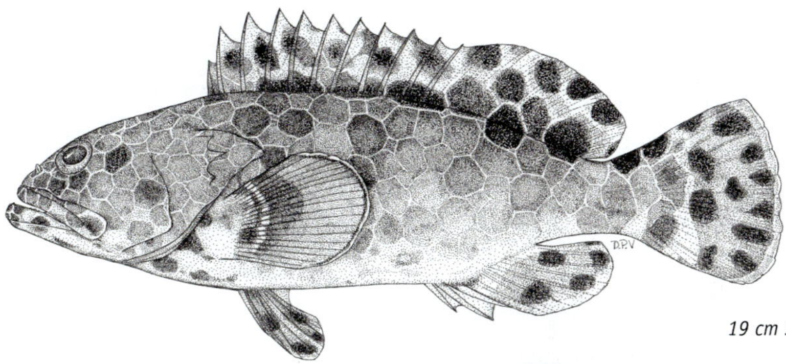

19 cm SL (Mauritius)
© SAIAB

6 cm SL (Taiwan)
© SAIAB

Body depth contained 3.0–3.6 times in standard length. Head length 2.3–2.6 times in standard length; interorbital area flat or slightly concave, the dorsal head profile of adults with a ventral bend at orbits; preopercle rounded, with minute serrae mostly covered by skin and a shallow indentation just above the angle; upper edge of operculum straight or slightly convex; posterior nostril usually distinctly larger than anterior nostril; maxilla reaches to or past vertical at rear edge of orbit, the ventral edge smoothly curved at distal expansion; lower jaw strongly projecting, the midlateral part with 2–4 rows of teeth. Gill rakers 7–9 on upper limb, 14–17 on lower limb, 21–26 total. Dorsal fin with 11 spines and 15–17 rays, the 3rd or 4th spine longest, its length contained 2.3–3.4 times in head length and shorter than longest dorsal-fin rays, the interspinous membranes moderately incised; anal fin with 3 spines and 8 rays, the 2nd and 3rd spines subequal, their length 2.8–4.2 times in head length; pectoral-fin rays 17–20; pectoral-fin length 1.5–2.0 times in head length; pelvic-fin length 1.8–2.5 times in head length; caudal-peduncle depth 3.2–3.7 times in head length; caudal fin rounded; length of middle caudal-fin rays 1.5–2.0 times in head length. Lateral-body scales smooth, except for area covered by pectoral fins; lateral-line scales 48–52; lateral-scale series 86–103. Pyloric caeca 25–27.

MAXIMUM RECORDED SIZE 52 cm TL

COLOUR IN LIFE This species has two colour patterns. Pacific Ocean: head and body pale greyish brown with dark brown spots (the centre of the spots darker than the diffuse edges) that are large and well separated in young, becoming relatively smaller, more numerous and closer together in adults; median and pelvic fins with similar dark spots; pectoral fins usually dusky, with a white line along the edge and usually a few faint dark spots; some specimens with a faint oblique dark line across the lower part of the chest; median fins with a narrow pale margin posteriorly; no dark spots on the underside of the lower jaw.

Indian Ocean: head and body with round to polygonal brown to dark brown spots, variable in size (some spots on juveniles may be as large as the eye) and close-set, the narrow interspaces forming a pale network pattern; median fins yellowish, with blackish brown spots like those on the body; soft dorsal, caudal and anal fins with pale margin; underside of the lower jaw usually with dark spots. Small juveniles (5–7 cm SL) with dark spots on the head and body fewer and much larger than in adults, those on the caudal fin represented by a broad black area covering two-thirds of the fin.

COLOUR IN MARKET Body colour remains similar to that in life with spots and reticulations remaining prominent. Pectoral fins retain dusky appearance in death.

© Robin Stobbs

Subadult *E. macrospilos* (KwaZulu-Natal, South Africa)

© Robert F Myers

Adult *E. macrospilos* (Kisiti National Park, Kenya)

© Phillip C Heemstra

Adult *E. macrospilos* (KwaZulu-Natal, South Africa)

© Robert F Myers

Adult *E. macrospilos* (West Pacific variant; Palau)

Epinephelus macrospilos is widespread, extending from South Africa, Mozambique, Comoros Islands, Madagascar, Mascarenes, Seychelles, Chagos and Sri Lanka to Indonesia, Australia and the Marquesas Islands (French Polynesia), northwards to the Ryukyu and Ogasawara island groups (Japan), and southwards to New Caledonia.

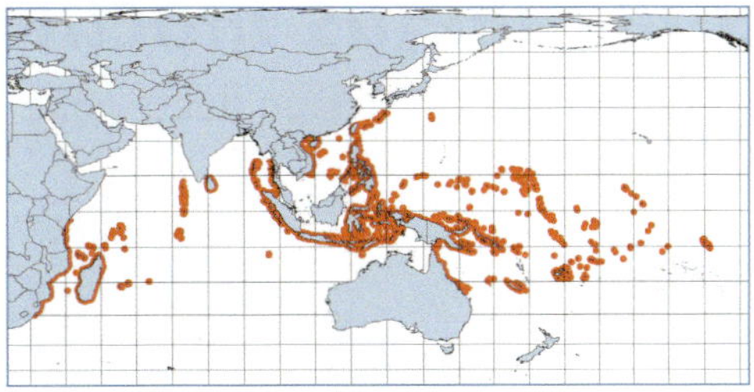

POPULATION **General** *E. macrospilos* is an uncommon species throughout its range. **Fishery dependent** *E. macrospilos* rarely represents more than 1% of grouper catches throughout its range, although it may be locally more common in small-scale, multi-species fisheries. **Fishery independent** Diver surveys indicate that this species may vary considerably in density throughout its range, from 8 to 138 fish per km^2 (M Kulbicki pers. obs.).

POPULATION TREND Unknown

MAJOR THREATS No major threats have been identified for this species.

HABITAT AND ECOLOGY **General** *E. macrospilos* occurs on coral reefs, including outer reef slopes to depths of 44 m. It is more abundant on oceanic reefs and rare on coastal reefs. **Feeding** Feeds mainly on crabs, and to a lesser degree on fishes. **Reproduction** Size at first maturity was found to be 34 cm TL (female) in New Caledonia (M Kulbicki unpublished data).

COMMENTS Western Indian Ocean specimens of *E. macrospilos* are often confused with the other 'reticulated groupers' (*E. bilobatus*, *E. faveatus*, *E. hexagonatus*, *E. maculatus*, *E. melanostigma*, *E. merra*, *E. quoyanus* and *E. spilotoceps*):

- *E. maculatus* has the dorsal-fin membranes not incised between the spines, ctenoid lateral-body scales and 102–120 lateral-scale series; juveniles are yellowish brown, with irregular white spots and blotches on the head, body and dorsal fin and a few small black spots on head and fins.
- *E. melanostigma* has ctenoid lateral-body scales, 56–68 lateral-line scales, and a large black blotch (which extends halfway to the dorsal-fin margin) on the body at the base of the last 4 dorsal-fin spines.
- *E. merra* has ctenoid lateral-body scales, and pectoral fins covered with distinct small black spots mainly confined to the rays.
- *E. quoyanus* has larger pectoral fins (their length contained 1.2–1.7 times in head length), lateral-body scales distinctly ctenoid, chest with 2 oblique dark brown bands (or large blotches linked by narrow bands), pectoral fins with semicircular reddish brown blotch covering most of the base, and small juveniles are coloured essentially like the adults.
- *E. spilotoceps* has ctenoid lateral-body scales, 59–69 lateral-line scales, black spots on the snout about the size of the nostrils, and caudal-peduncle depth contained 3.7–4.3 times in head length.

2008 IUCN RED LIST STATUS Least Concern

Authors: *PC Heemstra, M Kulbicki and RF Myers*

Epinephelus maculatus (Bloch 1890)

highfin grouper (E), mérou haute voile (F), mero aleta alta (S)

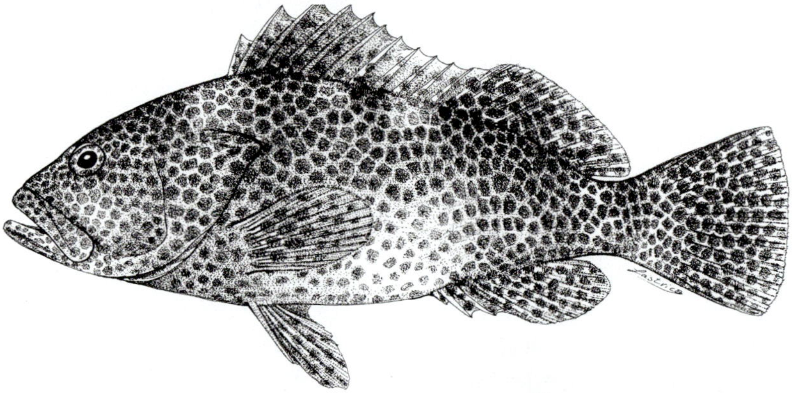

40 cm SL
© FAO

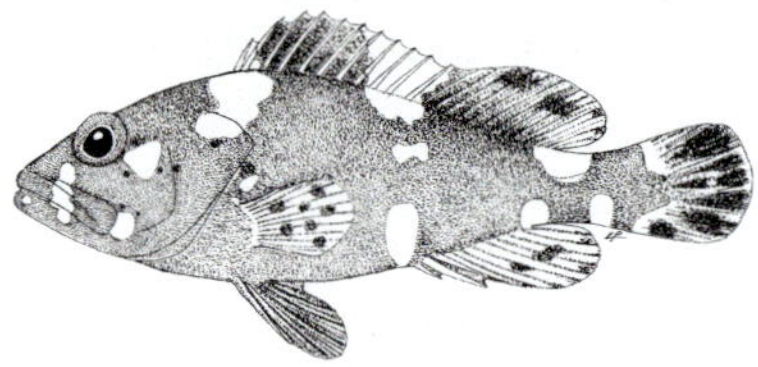

8 cm SL
© FAO

Body depth contained 2.8–3.1 times in standard length. Head length 2.4–2.6 times in standard length; interorbital area flat to slightly convex, the dorsal head profile convex; preopercle with a shallow indentation just above the enlarged serrae at the angle; upper edge of operculum straight or slightly convex, posterior and anterior nostrils subequal; maxilla reaches to or past vertical at rear edge of eye, the ventral edge with a blunt hook-like process distally in fish larger than 35 cm SL; midlateral part of lower jaw with 2 rows of teeth. Gill rakers 8–10 on upper limb, 15–17 on lower limb. Dorsal fin with 11 spines and 15–17 rays, the 3rd or 4th spine longest, its length contained 2.1–2.6 times in head length and distinctly longer than dorsal-fin rays, the interspinous membranes only slightly incised between anterior spines and not incised posteriorly; anal fin with 3 spines and 8 rays, the 3rd spine longest, its length 3.3–4.1 times in head length and equal to or less than depth of caudal peduncle; pectoral-fin rays 17–19; pectoral-fin length 1.5–1.9 times in head length; pelvic fins reaching to or near anus, their length 1.7–2.0 times in head length; caudal fin convex or rounded. Lateral-body scales distinctly ctenoid, with numerous auxiliary scales; lateral-line scales 49–52; lateral-scale series 102–120. Pyloric caeca 30–45.

MAXIMUM RECORDED SIZE 60.5 cm TL

Epinephelus maculatus is an Indo-Pacific species ranging from the Cocos-Keeling Islands eastwards to the Line Islands, extending north to southern Japan and south to south-eastern Australia and Lord Howe Island.

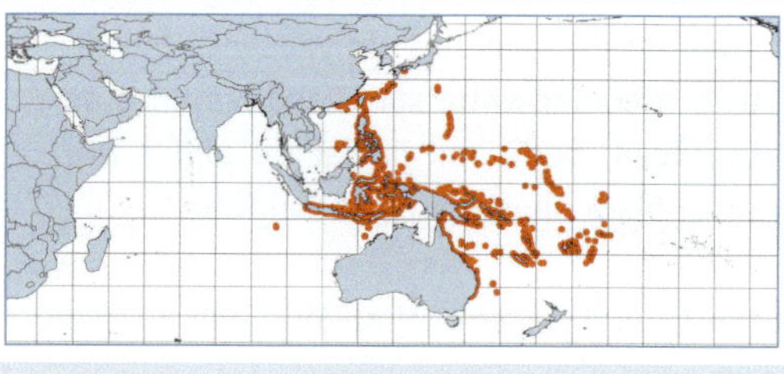

COLOUR IN LIFE Head, body and fins of adults pale brown, covered with small, round to hexagonal, close-set dark brown spots, extending onto the belly, chest and underside of the head; 2 large diffuse dusky areas on the dorsal part of the body and dorsal fin, the largest extending over the dorsal fin from the 2nd to 5th spines, the 2nd dark blotch from the 10th spine to the 2nd soft ray, the dorsal fin pale between these 2 dusky blotches, but with small dark spots. One resting colour phase is broadly mottled with large dark and pale areas; in the pale areas, the usual dark brown spots are mostly whitish, with small dark brown centres. Small juveniles (4–8 cm SL) yellowish brown, with well-separated small black spots (mainly on the head and fins) and prominent irregular white blotches and spots, the largest on the middle of the dorsal fin.

COLOUR IN MARKET Generally similar to colour in life, but the background body colour may fade to become very pale.

© Matthew T Craig

Adult *E. maculatus* (Okinawa Fish Market, Japan)

© Matthew T Craig

Adult *E. maculatus* (Okinawa Fish Market, Japan)

Subadult *E. maculatus* (Pulau Banta, Indonesia)

© Robert F Myers

Adult *E. maculatus* (Seraya, Bali, Indonesia)

© Robert F Myers

POPULATION **GENERAL** *E. maculatus* is a somewhat common species throughout its range. **FISHERY DEPENDENT** In New Caledonia, where *E. maculatus* is caught in experimental fishing, it makes up 24% by number and 16% by weight of all serranids caught (M Kulbicki pers. obs.). Catches in 1987 were approximately twice as high as in 2000, and average weight changed from 1.1 to 0.9 kg per fish for the same levels of fishing effort (M Kulbicki pers. obs.). **FISHERY INDEPENDENT** Diver surveys indicate that *E. maculatus* maintains relatively low population densities.

POPULATION TREND Decreasing

MAJOR THREATS *E. maculatus* may be threatened by fishing pressure.

HABITAT AND ECOLOGY **GENERAL** *E. maculatus* is a reef-associated species found at depths of 1–100 m. It tends to occur on more sheltered reef localities on the Great Barrier Reef (Australia) (Pears 2005) but in the Pacific it is associated with outer reef habitats (Kulbicki 2006). May be ciguatoxic in some areas (e.g. New Caledonia) (Myers 1999). Juveniles are found in shallow coral rubble. **FEEDING** *E. maculatus* feeds mainly on small fishes, crabs and octopus. **AGE AND GROWTH** *E. maculatus* reaches ages of at least 7 years (IUCN Red List Assessment 2009).

COMMENTS *E. maculatus* is one of nine 'reticulated groupers'. The other species in this group are *E. bilobatus*, *E. faveatus*, *E. hexagonatus*, *E. macrospilos*, *E. melanostigma*, *E. merra*, *E. quoyanus* and *E. spilotoceps*. This species differs from the other 'reticulated groupers' in having the dorsal-fin membranes not (or only slightly) incised between the fin spines, and in the colour pattern of small juveniles (4–8 cm SL).

2008 IUCN RED LIST STATUS Least Concern

Authors: *JH Choat, M Kulbicki and YJ Sadovy de Mitcheson*

Epinephelus magniscuttis

Postel, Fourmanoir & Guézé 1963

speckled grouper (E), mérou grandes écailles (F), mero bacalao (S)

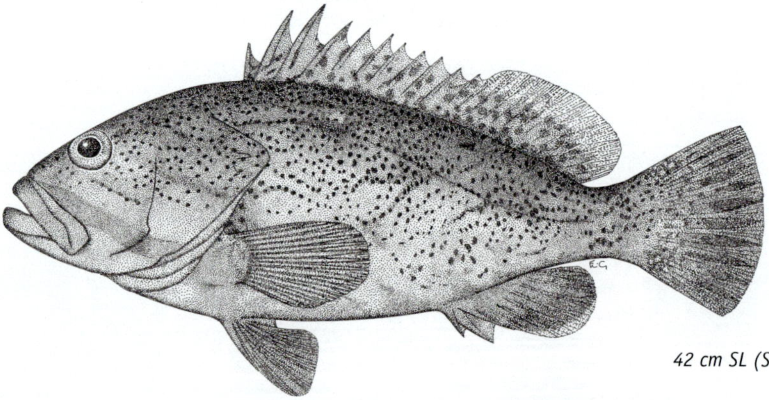

42 cm SL (South Africa)
© SAIAB

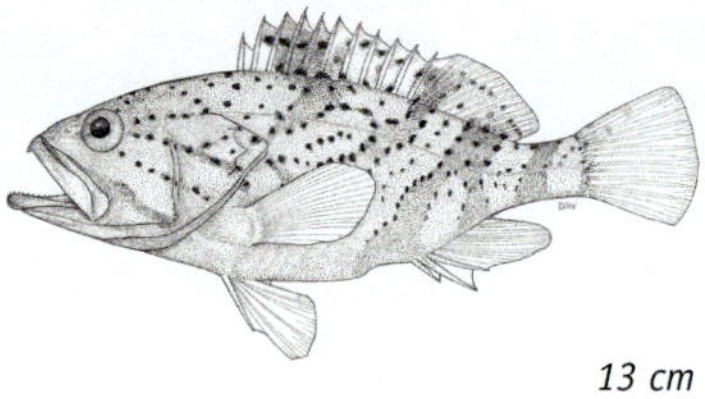

13 cm SL (Mauritius)
© SAIAB

Body depth contained 2.7–3.2 times in standard length. Head length 2.2–2.4 times in standard length; interorbital area flat to convex, the dorsal head profile almost straight; preopercle angular, with 2–4 distinctly enlarged serrae at angle; upper edge of operculum slightly convex; posterior nostril not much larger than anterior nostril; maxilla reaches to below rear half of eye, the ventral edge with a low step at the distal expansion; midlateral part of lower jaw with 2 rows of teeth, the inner teeth larger than outer teeth. Gill rakers 8 or 9 on upper limb, 16 or 17 on lower limb. Dorsal fin with 11 spines and 14 or 15 rays, the 3rd or 4th spine longest, its depth contained 2.5–3.0 times in head length and longer than longest dorsal-fin ray, the interspinous membranes deeply incised; anal fin with 3 spines and 8 rays; pectoral-fin rays 17–19; pectoral-fin length 1.7–2.0 times in head length; pelvic fins not reaching anus, their length 2.1–2.3 times in head length. Lateral-body scales distinctly ctenoid and without auxiliary scales; lateral-line scales 55–62; lateral-scale series 103–122. Pyloric caeca 7.

MAXIMUM RECORDED SIZE 150 cm TL

COLOUR IN LIFE Generally pale brown, with small dark brown to greyish green spots unevenly scattered on the dorsolateral parts of the head (posterior to the eyes) and body, dorsal fin and caudal fin; one row of close-set spots running posteriorly from the eye in a distinct line; no spots on the body below the level of the pectoral fins or on the anal and paired fins. According to Postel et al. (1963), juveniles have 6 or 7 dark lines running horizontally along the body and these disappear with age.

COLOUR IN MARKET Body pale brown, small dark spots remain prominent. The row of spots running in a line posteriorly from the eye is conspicuous in death.

© John E Randall

Adult *E. magniscuttis* (KwaZulu-Natal, South Africa)

Epinephelus magniscuttis is a widely distributed Indo-West Pacific species, with a disjunct distribution. It ranges from the KwaZulu-Natal coast of South Africa and southern coast of Mozambique to Réunion, Mauritius and Fiji, and was recently recorded from Tonga (Randall et al. 2003). There is one record from Australia (Myrmidon Reef, Great Barrier Reef; Kramer et al. 1994); it is unknown in the Red Sea and Persian Gulf.

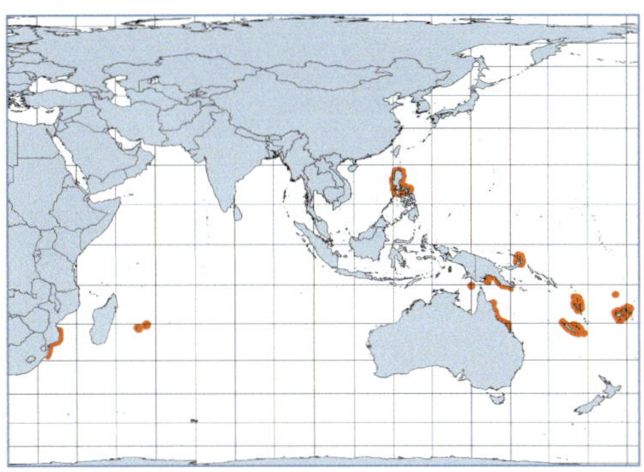

POPULATION **GENERAL** *E. magniscuttis* is a poorly known species. It is reportedly common at Réunion Island and off the coast of Mozambique. **FISHERY DEPENDENT** *E. magniscuttis* occurs regularly in the line and trap fisheries of Mozambique (Heemstra and Randall 1993, FRI unpublished data); it is of some importance to the fisheries at Réunion and Mozambique (FRI unpublished data), where it is fairly common.

POPULATION TREND Unknown

MAJOR THREATS *E. magniscuttis* may be threatened by commercial fishing pressure. It is not specifically managed anywhere.

HABITAT AND ECOLOGY **GENERAL** *E. magniscuttis* is a poorly known deep-water species that inhabits depths of 128–300 m.

COMMENTS *E. magniscuttis* is similar in appearance to *E. epistictus*. The latter species has fewer (and smaller) dark spots on the head and body, and in juveniles the dark spots are arranged in 3 longitudinal rows on the body.

2008 IUCN RED LIST STATUS Data Deficient

Authors: *ST Fennessy and BC Russell*

Epinephelus malabaricus

(Bloch & Schneider 1801)

Malabar grouper (E), mérou Malabare (F), mero Malabárico (S)

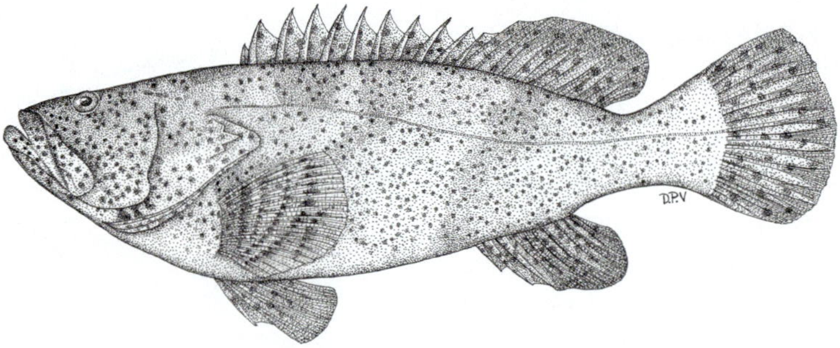

50 cm TL
© SAIAB

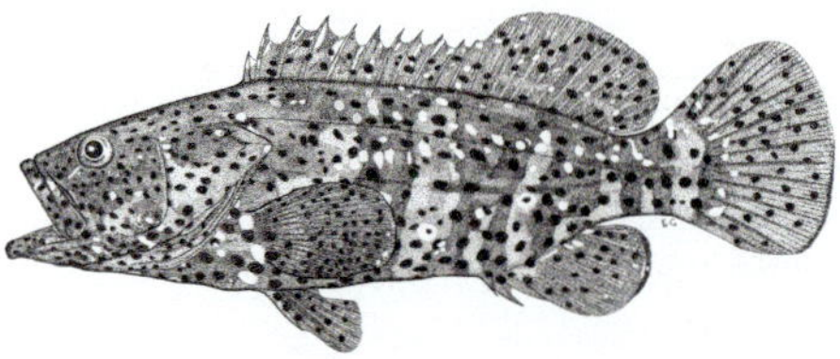

25 cm SL (Taiwan)
© SAIAB

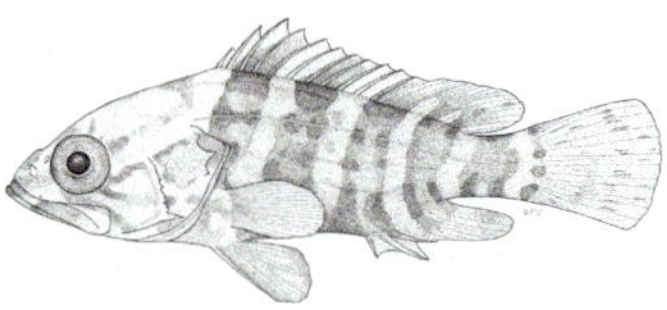

2 cm SL (South Africa)
© SAIAB

Body elongate, depth contained 3.0–3.7 times in standard length; body width 1.4–1.9 times in depth. Head length 2.3–2.6 times in standard length; snout length 1.7–2.0 times in upper jaw length; interorbital width 4.5–6.5 times in head length and 2.1–3.0 times in upper jaw length; interorbital area flat or slightly convex; preopercle subangular, with enlarged serrae at the angle; upper edge of operculum almost straight; posterior and anterior nostrils subequal, except in large adults which have the posterior nostrils slightly larger; maxilla extends past vertical at rear edge of orbit, maxilla width 4.5–6.5% of standard length; upper jaw length 17–22% of standard length, midlateral part of lower jaw with 2–5 rows of teeth. Gill rakers 8–11 on upper limb, 14–18 on lower limb, 23–27 total; rudiments difficult to distinguish from small bony platelets on outer face of first gill arch. Dorsal fin with 11 spines and 14–16 rays, the 3rd to

5th spines usually slightly longer than posterior spines, their length contained 3.1–4.0 times in head length and distinctly shorter than longest rays, the interspinous membranes incised; anal fin with 3 spines and 8 rays, the 3rd spine usually longest; pectoral-fin rays 18–20; pectoral-fin length 1.7–2.2 times in head length, pelvic-fin length 2.1–2.6 times in head length; caudal fin rounded. Lateral-body scales ctenoid, with auxiliary scales; lateral-line scales 54–64; anterior lateral-line tubes of large adults with 2–4 branches; lateral-scale series 101–117. Pyloric caeca numerous (more than 80 branches).

MAXIMUM RECORDED SIZE 150 cm TL

COLOUR IN LIFE Head and body brownish, covered with small well-separated blackish brown spots that extend onto the chest, lower jaw and gular area and roof of the mouth; head and body also with scattered white spots and blotches; 5 irregular oblique dark brown bars (more or less interrupted by pale spots) often visible on the body; fins with scattered small black spots. On preserved fish the blackish spots are conspicuous against the drab background.

COLOUR IN MARKET Generally brown or dark green, small spots remain prominent. The oblique bars are generally not visible. The spots on the fins are especially prominent.

© Stephen Newman

Adult *E. malabaricus* (Robe River, Western Australia)

© Stephen Newman

Adult *E. malabaricus* (Kimberley coast, Western Australia)

© Gerald R Allen

Juvenile *E. malabaricus* (Raja Ampat Islands, Indonesia)

© Matthew Pember

Juvenile *E. malabaricus* (Port Smith, Western Australia)

© Brad Erisman

Adult *E. malabaricus* (Fiji)

Epinephelus malabaricus is found in the Indo-Pacific from the Red Sea and East Africa to the Comoros Islands, Madagascar, Seychelles, India, Sri Lanka to Indonesia, Australia, Papua New Guinea and Tonga, extending north to Japan. Not known from the Persian Gulf.

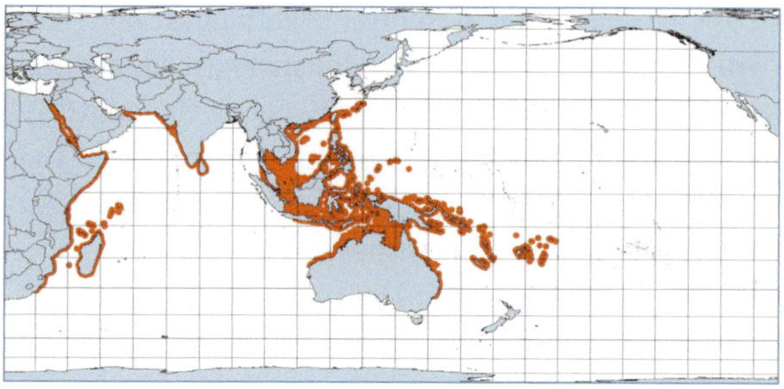

POPULATION **GENERAL** *E. malabaricus* is a common species throughout its range. **FISHERY DEPENDENT** Fishery-dependent data are difficult to interpret for this species given its confusion with similar species. It is one of the most important grouper species to fisheries in the Indo-Pacific. Landings data indicate a substantial decrease in catch for all members of this similar group and the species is taken with a wide range of gear types (Heemstra and Randall 1993). There has also been a growing mariculture industry, and *E. malabaricus* is now part of this enterprise which, for this species, is based on the grow-out of both hatchery-produced and wild-caught juveniles (Liu and Sadovy de Mitcheson 2008). Annual production from hatcheries in Thailand increased from 15 000 fish in 1991 to 265 200 in 1995 (Yashiro 1999) and larval rearing also occurs in Taiwan. Available information from New Caledonia and from the fry fisheries in the Philippines demonstrates that fishing can greatly reduce populations (IUCN Red List Assessment 2008). *E. malabaricus* is sold live in the Hong Kong live reef food fish market most commonly at 35–55 cm TL which is at or below the approximate size of sexual maturation in the wild, likely to be about 45–50 cm TL (Lau and Parry-Jones 1999). **FISHERY INDEPENDENT** Fishery-independent data suggest declines in abundance of *E. malabaricus*.

POPULATION TREND Decreasing

MAJOR THREATS *E. malabaricus* is primarily threatened by fishing pressure on both juveniles and adults. Habitat loss also poses a threat, as juveniles are common in estuaries and mangrove areas which may be damaged by coastal development. The species is nowhere specifically managed.

HABITAT AND ECOLOGY **GENERAL** *E. malabaricus* may be found in a variety of habitats in depths to 150 m. Juveniles are found nearshore and in estuaries. **FEEDING** Feeds on fishes and invertebrates. **REPRODUCTION** *E. malabaricus* is a protogynous hermaphrodite. Sex change occurs at approximately 114 cm TL (Lau and Li 2000).

COMMENTS *E. malabaricus* is similar in appearance to *E. coioides* and *E. tauvina*. However, *E. coioides* differs in having no white spots on the body and orange or reddish brown spots (compared with dark brown or black in *E. malabaricus*), and in having fewer pyloric caeca. *E. malabaricus* differs from *E. tauvina* in having roughly ctenoid scales (smooth in *E. tauvina*) and fewer lateral-line scales (63–74 in *E. tauvina*), and shorter jaws.

2006 IUCN RED LIST STATUS Near Threatened

Authors: *AS Cornish and DA Pollard*

Epinephelus marginatus (Lowe 1834)

dusky grouper (E), mérou noir (F), mero moreno (S)

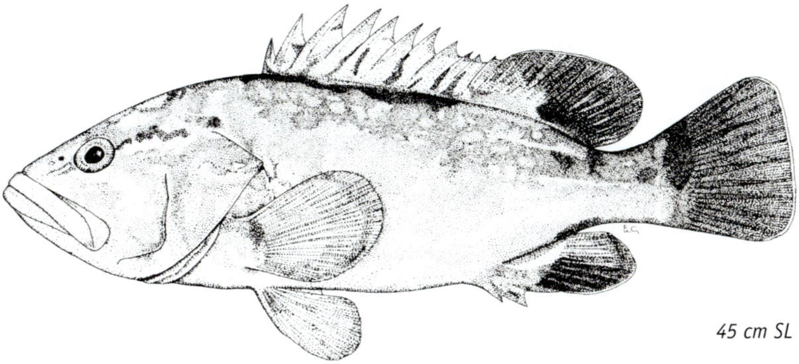

45 cm SL (South Africa)
© SAIAB

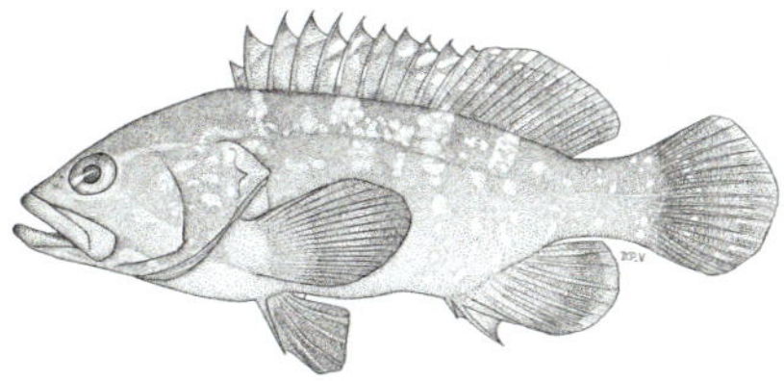

8 cm SL (South Africa)
© SAIAB

Body depth less than head length, depth contained 2.6–3.1 times in standard length (for fish 15–62 cm SL). Head length 2.3–2.5 times in standard length; interorbital area convex; preopercle rounded, finely serrate, the serrae at angle slightly enlarged; subopercle and interopercle smooth; eye diameter greater than or subequal to interorbital width in fish 10–30 cm SL, less than interorbital in fish over 40 cm SL; posterior and anterior nostrils subequal or posterior nostril slightly larger; maxilla naked, reaching to or slightly past vertical at rear edge of eye; midlateral part of lower jaw with 2–4 rows of subequal teeth. Gill rakers 7–10 on upper limb, 14–16 on lower limb, 22–25 total. Dorsal fin with 11 spines and 14–16 rays, 3rd or 4th spine longest, longer than longest dorsal-fin ray and contained 2.3–2.8 times in head length, the interspinous membrane distinctly incised; anal fin with 3 spines and 8 rays (one of 87 fish counted with 9 rays); pectoral-fin rays 17–19, the fin length 1.6–2.0 times in head length; pelvic fins distinctly shorter than pectoral fins, not reaching anus (except in some fish less than 20 cm SL), pelvic-fin length 1.8–2.4 times in head length; pelvic-fin origin below or slightly posterior to base of pectoral fins; caudal fin rounded (in juveniles) or truncate with rounded corners (large adults). Midlateral-body scales ctenoid (at least in area covered by pectoral fins); adults with numerous auxiliary scales; lateral-line scales 62–73; lateral-scale series 98–116. Pyloric caeca 26–50.

MAXIMUM RECORDED SIZE 143 cm TL

POPULATION **GENERAL** *E. marginatus* was once considered a common species throughout its range. It is highly likely that the two primary areas where *E. marginatus* occurs (in West Africa and Europe, and in South America), represent (at least) two separate subpopulations with minimal gene flow due to the very large distance between them. **FISHERY DEPENDENT** The species has experienced declines in abundance over the past few decades,

COLOUR IN LIFE Head and body dark reddish brown or greyish dorsally, usually yellowish gold ventrally; irregular white, pale greenish yellow or silvery grey blotches usually visible on the body and head and mostly arranged in vertical series; black maxillary streak more or less distinct; median fins dark brown; distal edge of the anal and caudal fins, and often the pectoral fins, narrowly white; pelvic fins blackish distally; pectoral fins dark reddish brown or grey; margin of the spinous dorsal fin and basal part of the paired fins often golden yellow.

COLOUR IN MARKET Body takes on a dark brown hue with pale yellowish underside. Blotches usually fade in death. Maxillary streak, if distinct in life, will be prominent.

© Matthew T Craig

Adult *E. marginatus* (Umkomaas, South Africa)

© Áthila Bertoncini Andrade

Juvenile *E. marginatus* (Arvoredo Island, Brazil)

Adult *E. marginatus* (Arvoredo Island, Brazil)

© Áthila Bertoncini Andrade

with localised recovery in certain areas. In the eastern Atlantic and Mediterranean Sea (France, Spain, Portugal, Greece, Italy, Turkey and UK), total landings of *E. marginatus* declined by 88% between 1990 and 2001 (FAO 2002a). Regional landings have dropped by up to 95% (Italy) in some cases. Estimates of effort during this time indicate that it had remained consistent, thus landings data are presumed to reflect population decreases. In South Africa, spawning biomass per recruit was estimated at 33%, and there has been a reduction in the mean size of fish landed since the mid-1980s (Fennessy 2000). In Brazil, *E. marginatus* is heavily fished during periods of aggregation behaviour. Protection in a few small MPAs has led to some promising recovery. **FISHERY INDEPENDENT** Along the French Mediterranean coast, diver observations have shown steady increases of *E. marginatus* since the mid-1980s with fish over 40 cm showing increases as well (Bodilis et al. 2003). This increase has been attributed to a ban on spearfishing for *E. marginatus.*

Epinephelus marginatus is an amphi-Atlantic and western Indian Ocean species. In the western Atlantic it is known from southern Brazil, Uruguay and Argentina. In the eastern Atlantic, it is known from throughout the Mediterranean Sea and southwards around the southern tip of Africa to southern Mozambique. There are also reports of this species from Oman (Randall 1995a) and India.

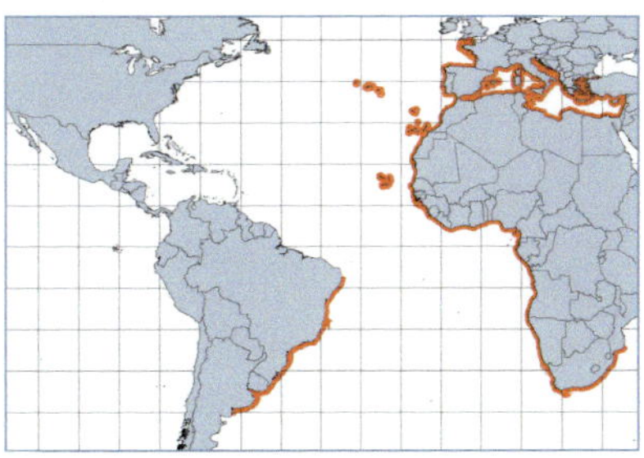

POPULATION TREND Decreasing

MAJOR THREATS Fishing pressure is the major threat to *E. marginatus*.

HABITAT AND ECOLOGY **GENERAL** *E. marginatus* inhabits rocky reefs in depths of 8–300 m. **FEEDING** Feeds on fishes and invertebrates. **AGE AND GROWTH** In the western Mediterranean, *E. marginatus* may reach ages in excess of 60 years; von Bertalanffy growth parameters were $L_\infty = 95.5$ cm, $k = 0.087$, $t_0 = -1.12$ years (Reñones et al. 2007). In south-eastern Africa, the von Bertalanffy growth parameters were $L_\infty = 1249$ mm TL, $k = 0.09$, $t_0 = -1.43$ years (Fennessy 2000). **REPRODUCTION** *E. marginatus* is a protogynous hermaphrodite that forms spawning aggregations. In Tunisia, females reach sexual maturity at age 5, and sex change to males takes place between ages 9 and 16 years (Chauvet 1988). In south-eastern Africa, spawning occurs during summer, female size-at-50% maturity was 622 mm TL, the species is a monandric protogynous hermaphrodite and the smallest male recorded was 800 mm TL (Fennessy 2006). In Brazil, 50% of females reach sexual maturity at about 470 mm TL; the species matures in spring and spawns in early summer with a peak in December (Andrade et al. 2003). In Spain, spawning aggregations have been noted from June to August. *E. marginatus* forms small spawning aggregations of a few tens of individuals. Aggregation sites also appear to exist in Brazil as spearfishermen have reported large catches (up to 400 kg) in a few days from the same areas (Andrade et al. 2003).

COMMENTS Craig and Hastings (2007) suggest that this species is best treated as a member of the genus *Mycteroperca*. It is retained here in *Epinephelus* pending further study.

2004 IUCN RED LIST STATUS Endangered (A2d)

Authors: *AS Cornish, ML Harmelin-Vivien and DA Pollard*

Epinephelus melanostigma Schutlz 1953

oneblotch grouper (E), mérou dossard (F), mero espaldarón (S)

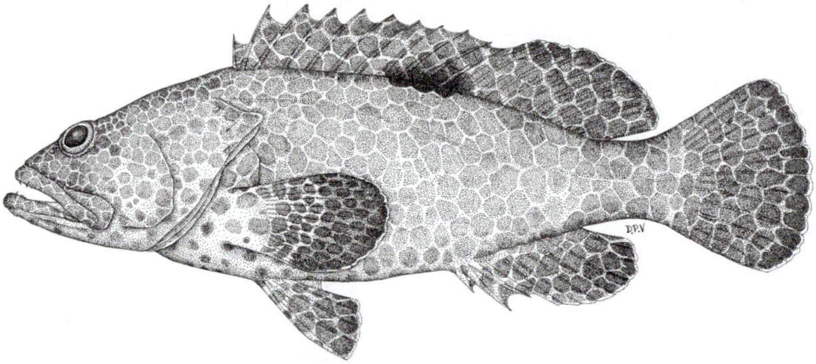

20 cm SL (South Africa)
© SAIAB

Body elongate, the depth contained 3.0–3.4 times in standard length. Head length 2.2–2.4 times in standard length; interorbital area flat or slightly concave; preopercle rounded, the serrae mostly covered by skin; upper edge of operculum straight or slightly convex; posterior and anterior nostrils subequal in size; maxilla reaches past vertical at rear edge of eye; midlateral part of lower jaw with 3–5 rows of teeth, the inner teeth about twice as long as the outer ones. Gill rakers 7–10 on upper limb, 16–19 on lower limb. Dorsal fin with 11 spines and 14–16 rays, the 3rd to 10th spines subequal, the 4th spine contained 2.8–3.8 times in head length, the interspinous membranes incised; anal fin with 3 spines and 8 rays, the 2nd spine less than or subequal to 3rd spine, length of 2nd spine 2.6–3.6 times in head length and usually less than depth of caudal peduncle; pectoral-fin rays 17–19; pectoral-fin length 1.6–2.2 times in head length; pelvic-fin length 2.0–2.6 times in head length; caudal-peduncle depth 3.2–3.8 times in head length; caudal fin rounded. Lateral-body scales ctenoid, with numerous auxiliary scales; lateral-line scales 56–68; lateral-scale series 83–99.

MAXIMUM RECORDED SIZE 35 cm TL

Epinephelus melanostigma is an Indo-Pacific species found from KwaZulu-Natal (South Africa) to Mauritius and the central Pacific. It is not known from the Red Sea or the Persian Gulf.

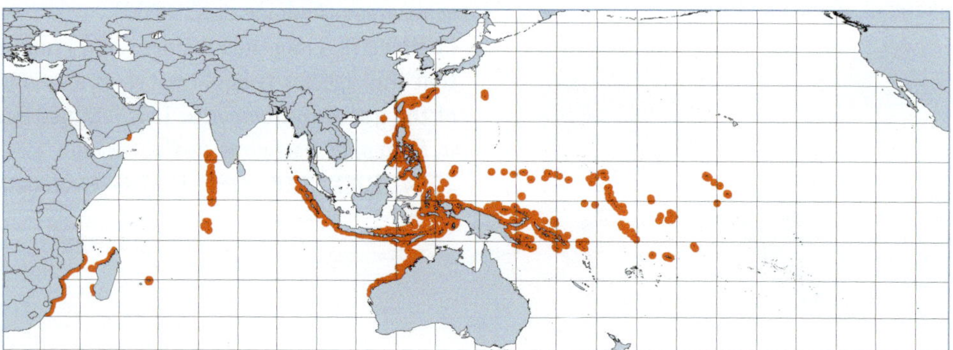

COLOUR IN LIFE Head, body and fins covered with close-set dark brown spots, the pale interspaces forming a network pattern; spots on the ventral parts of the head and body are reddish brown and not so closely set; spots on the head are progressively smaller anteriorly; a prominent black blotch on the body at the base of the last 4 dorsal-fin spines, extending onto the fin at least halfway to its margin; no other black blotches at the base of the dorsal fin or on the caudal peduncle; median and pectoral fins with narrow white edge.

COLOUR IN MARKET Body colour remains generally the same although it may fade to take on a pale hue. Network pattern is clearly visible; blotch on the body at the base of the dorsal fin remains prominent.

© Matthew T Craig

Adult *E. melanostigma* (Okinawa Fish Market, Japan)

© Robert F Myers

Adult *E. melanostigma* (Maldives)

POPULATION GENERAL *E. melanostigma* is an uncommon species throughout its range. **FISHERY DEPENDENT** *E. melanostigma* is found in subsistence fisheries but, because of its small size and apparent rarity, it has been of relatively little commercial importance. In Pohnpei, *E. melanostigma* constituted less than 2% of catch in reef fish fisheries in 2006 (Rhodes and Tupper 2007).

POPULATION TREND Unknown

MAJOR THREATS No major threats have been identified for this species.

HABITAT AND ECOLOGY GENERAL *E. melanostigma* is a coral reef-associated species occurring in outer reef flats and shallow lagoons in depths of up to 30 m.

COMMENTS *E. melanostigma* is one of the 'reticulated groupers'. These species are often confused in the literature.

- *E. bilobatus* has 3 large bilobed dark blotches or close-set pairs of dark brown spots on the body and base of the dorsal fin, 17 or 18 dorsal-fin rays, 48–52 lateral-line scales, and 14–16 lower gill rakers.
- *E. faveatus* has the lateral-body scales smooth (except for area covered by pectoral fins), 16–18 dorsal-fin rays, length of 2nd anal-fin spine contained 3.6–4.3 times in head length, and sides of lower jaw with 2 rows of teeth.
- *E. hexagonatus* has conspicuous white dots on the body between the dark spots, 5th to 9th dorsal-fin spines subequal (length contained 2.5–2.8 times in head length), head length 2.5–2.6 times in standard length, length of 2nd anal-fin spine 2.1–2.5 times in head length and distinctly longer than 3rd spine or depth of caudal peduncle.
- *E. macrospilos* has the lateral-body scales mostly smooth, 48–52 lateral-line scales and no large black blotch at the middle of the dorsal-fin base.
- *E. maculatus* has the dorsal-fin membranes not incised between the spines; 3rd or 4th dorsal-fin spine longest, its length contained 2.1–2.6 times in head length and usually longer than the dorsal-fin rays.
- *E. merra* has pectoral fins covered with small black spots largely confined to the rays, 3rd to last dorsal-fin spines subequal (the longest contained 2.6–2.9 times in head length), and 2nd anal-fin spine subequal to 3rd and much longer than the depth of the caudal peduncle.
- *E. quoyanus* has the pectoral fins with indistinct dark brown spots, the base with a large semicircular dark brown spot edged posteriorly with white, 2 dark brown bands or blotches linked by bands on sides of chest, 16–18 dorsal-fin rays, and longest dorsal-fin spine contained 2.4–3.0 times in head length.
- *E. spilotoceps* has 3 black saddle blotches at the base of the dorsal fin and tiny (nostril-size) black spots on the snout; caudal-peduncle depth contained 3.5–4.1 times in head length and usually less than the length of the 2nd anal-fin spine, shallow notch in the rear edge of the preopercle just above the angle, and 59–69 lateral-line scales.
- *E. tauvina* is similar to *E. melanostigma* in having a black blotch at middle of dorsal-fin base, but the dark spots on the head and body are poorly defined and well separated (distance between spots about equal to their diameter); *E. tauvina* also has 63–74 lateral-line scales and 95–112 lateral-scale series.

2008 IUCN RED LIST STATUS Data Deficient

Authors: *PC Heemstra, M Kulbicki, RF Myers, KL Rhodes and BC Russell*

Epinephelus merra (Bloch 1793)

honeycomb grouper (E), mérou gâteau de cire (F), mero panal (S)

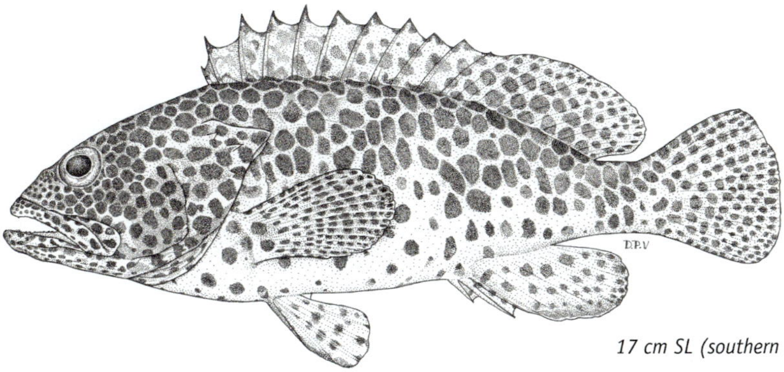

17 cm SL (southern Mozambique)
© SAIAB

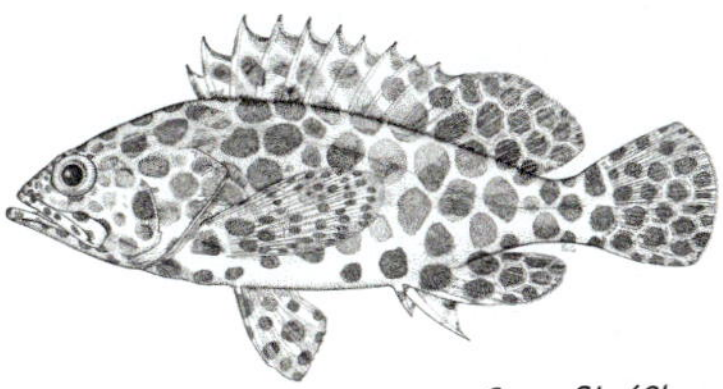

6 cm SL (Chagos Archipelago)
© SAIAB

Body depth contained 2.8–3.3 times in standard length. Head length 2.3–2.6 times in standard length; interorbital area flat, the dorsal head profile convex; preopercle rounded or subangular, the serrae at angle enlarged; upper edge of operculum almost straight; posterior and anterior nostrils subequal or posterior nostrils larger; maxilla reaches past vertical at rear edge of eye; midlateral part of lower jaw with 2–4 rows of teeth, the inner teeth about twice length of outer teeth. Gill rakers 6–9 on upper limb, 14–17 on lower limb. Dorsal fin with 11 spines and 15–17 rays, the 3rd to last spines subequal, the longest 2.4–3.2 times in head length; anal fin with 3 spines and 8 rays, the 2nd spine subequal to the 3rd, its length 2.1–3.0 times in head length and much longer than peduncle depth; pectoral-fin rays 16–18; pectoral-fin length 1.5–1.8 times in head length; pelvic-fin length 1.8–2.2 times in head length; caudal-peduncle depth 3.2–4.1 times in head length; caudal fin rounded. Lateral-body scales ctenoid, with auxiliary scales; lateral-line scales 48–54; lateral-scale series 98–114. Pyloric caeca 8.

MAXIMUM RECORDED SIZE 32 cm TL

Epinephelus merra is one of the most widely distributed and common small groupers in the Indo-Pacific, occurring from South Africa to Pitcairn Island, excluding the Middle East, coastal India and Hawaii. It occurs north to southern Japan and south to Lord Howe Island.

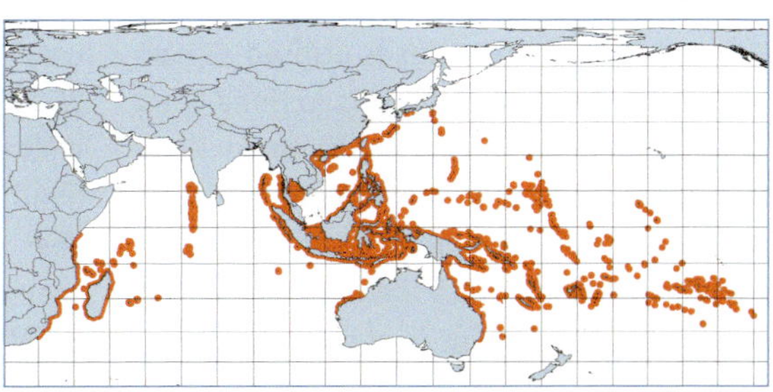

COLOUR IN LIFE Head, body and fins pale, covered with close-set dark brown or reddish brown spots, the interspaces forming an irregular pale reticulum; spots on the ventral part of the body paler, more widely spaced and diffuse in outline; some midlateral spots are often joined to form horizontal bands; 5 irregular dark bars can be displayed by differential darkening of some of the body spots, but black dorsal blotches are never present; dark spots on median fins become smaller towards the fin margin; pectoral fins covered with distinct small black spots, largely confined to the rays; tips of interspinous dorsal-fin membranes white or pale yellow, with a small submarginal black spot.

COLOUR IN MARKET One of the few grouper species that maintains nearly identical colouration in death.

© Matthew T Craig

Adult *E. merra* (Okinawa Fish Market, Japan)

© Robert F Myers

Adult *E. merra* (Fiji)

POPULATION GENERAL *E. merra* is very abundant throughout its range. FISHERY DEPENDENT Fishery-dependent data are scarce for this species. The species is utilised as part of the live reef-fish food markets in Hong Kong but rarely in large numbers. In Hong Kong wet markets, 95% of *E. merra* were immature sized (W-L To pers. comm.). FISHERY INDEPENDENT Fishery-independent data are also scarce, but indicate that the species is abundant.

POPULATION TREND Stable

MAJOR THREATS The major threat to *E. merra* is fishing pressure. Wild-caught juveniles are used for grow-out in Vietnam (Tuan et al. 2000), and adults are present in small-scale, multi-species fisheries and occasionally in the live fish trade. Although utilised heavily, this does not appear to be a major threat to this species.

HABITAT AND ECOLOGY GENERAL *E. merra* is

a reef-associated species inhabiting shallow depths to at least 50 m but mainly <20 m (Heemstra and Randall 1993). Juveniles are common in thickets of staghorn corals of the genus *Acropora*. Home-range size in Hong Kong was 47.7 ± 11 m^2 for fish between 134–205 mm TL (To 2009). **FEEDING** Feeds on fishes and small invertebrates. **AGE AND GROWTH** The TL–SL relationship was reported as SL = 0.75TL + 26.84; growth rate is relatively high, with *k* ranging from 0.43 to 0.8, and natural mortality was reported to be 0.9–1.67 (Pothin et al. 2004). **REPRODUCTION** *E. merra* is a protogynous hermaphrodite with sex change occurring at 3–5 years (Pothin et al. 2004). Spawning season varies by locality, and has been reported as January through April in the Society Islands, between May and August in Okinawa (Japan), March and April in Chuuk (Federated States of Micronesia), and September–February in New Caledonia (IUCN Red List Assessment 2009). Around Sesoko Island, Okinawa, mature *E. merra* disappeared from their inhabited fringing reefs for a few days after the full moon and appeared again in the same areas; it was proposed that *E. merra* formed a spawning aggregation outside of their inhabited reefs (Soyano et al. 2003). See also the SCRFA database (www.scrfa.org).

COMMENTS *E. merra* was intentionally introduced into the Hawaiian Islands in the 1950s; however, the species did not successfully colonise the area and is no longer present.

E. merra is one of the 'reticulated groupers' and can be distinguished from the other reticulated groupers by its pectoral-fin pattern of conspicuous black dots that are largely confined to the rays of the fin. In addition, *E. merra* differs in the following:

- *E. bilobatus* has 3 bilobed dark blotches or close-set pairs of dark brown spots on the body and base of the dorsal fin, dark spots on median fins not much smaller distally, and 17 or 18 dorsal-fin rays.
- *E. faveatus* has the lateral-body scales smooth (except for area covered by pectoral fins), longest anal-fin spine contained 3.6–4.3 times in head length, dark bands on the chest, and 83–98 lateral-scale series.
- *E. hexagonatus* has conspicuous white dots on the body between the dark spots, 4 or 5 brownish black blotches on the body at the base of the dorsal fin, a large olive-brown spot or band just behind the eye, and 61–70 lateral-line scales.
- *E. macrospilos* has the lateral-body scales mostly smooth and the pectoral fins dusky with narrow white edge.
- *E. maculatus* has the interspinous dorsal-fin membranes not incised, 3rd or 4th dorsal-fin spine longest (its length contained 2.1–2.6 times in head length and longer than the dorsal-fin rays); and juveniles are yellowish brown with irregular white blotches on the body.
- *E. melanostigma* has a large black blotch at the base of the last 4 dorsal-fin spines, length of the 2nd anal-fin spine is not longer than the caudal-peduncle depth, and 56–68 lateral-line scales.
- *E. quoyanus* has the pectoral fins with indistinct dark brown spots and the base with a large semicircular dark brown spot edged posteriorly with white, 2 oblique dark brown bands or blotches linked by bands on the sides of the chest, 16–18 dorsal-fin rays, and 3rd to 5th dorsal-fin spines longest (length contained 2.4–3.0 times in head length).
- *E. spilotoceps* has very small dark spots on the snout (about the size of the nostrils), 3 black blotches at the base of the dorsal fin and another on top of the caudal peduncle, and 59–69 lateral-line scales.

2008 IUCN RED LIST STATUS Least Concern

Authors: *M Liu and BM Yeeting*

Epinephelus miliaris (Valenciennes 1830)

netfin grouper (E), mérou abeille (F), mero colmenar (S)

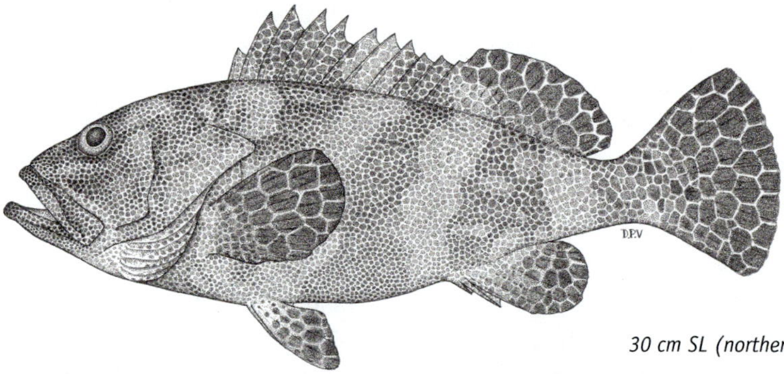

30 cm SL (northern Mozambique)
© SAIAB

3 cm SL (Zanzibar, Tanzania)
© SAIAB

Body depth contained 2.8–3.2 times in standard length. Head length 2.4–2.6 times in standard length; interorbital convex, but dorsal head profile slightly concave at eyes; preopercle subangular, with a shallow notch just above the angle and the serrae at the angle enlarged in juveniles; upper edge of operculum straight; diameter of posterior nostrils about twice that of anterior nostrils; maxilla reaches vertical at rear edge of eye or thereabouts; adults larger than 25 cm SL with a low step on lower edge of maxilla at distal expansion; midlateral part of lower jaw with 2–4 rows of teeth. Gill rakers 8 or 9 on upper limb, 14–16 on lower limb. Dorsal fin with 11 spines and 16 or 17 rays, the 3rd or 4th spine longest, its length contained 2.2–2.7 times in head length and distinctly longer than the longest rays, the interspinous membranes distinctly incised; anal fin with 3 spines and 8 rays, the 3rd spine usually slightly longer than 2nd, length of 2nd spine 3.2–4.0 times in head length; pectoral-fin rays 17 or 18; pectoral-fin length 1.6–1.9 times in head length; pelvic fins reach slightly past anus or not so far, their length 1.7–2.1 times in head length; caudal fin rounded. Lateral-body scales distinctly ctenoid, with a few auxiliary scales; lateral-line scales 48–52; lateral-scale series 92–108. Pyloric caeca 30.

MAXIMUM RECORDED SIZE 43 cm TL

Epinephelus miliaris is primarily an insular Indo-Central Pacific species ranging from East Africa (excluding the Red Sea and Persian Gulf) to the Gilbert Islands and Samoa, extending north to the Ryukyu Islands (Japan).

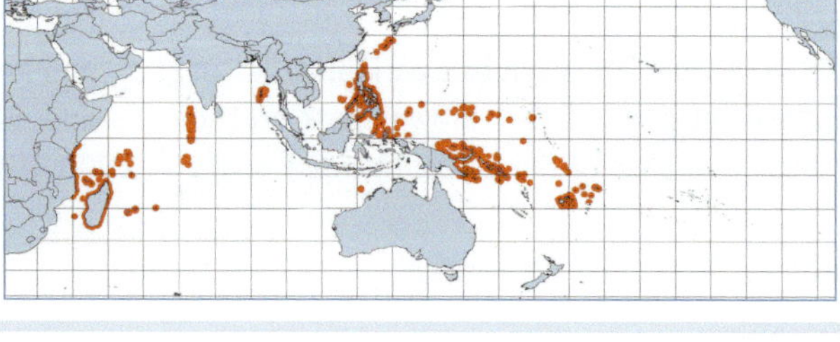

COLOUR IN LIFE Head and body pale, covered with small close-set polygonal dark brown, reddish brown or yellowish brown spots, the ground colour forming a pale irregular small-mesh network; fins with pattern similar to the body, but the dark spots are darker and distinctly larger than on the body. Some spots on the body and dorsal fin are darker than others, forming 4 or 5 broad oblique dark bars (1 on the nape, 3 extending into the dorsal fin and 1 at the base of the caudal fin).

COLOUR IN MARKET Body colour remains nearly identical to that in life; however, the dark bars are generally not prominent.

© Robert F Myers

Adult *E. miliaris* (Guam)

© Dieter Eichler

Adult *E. miliaris* (locality unknown)

POPULATION **General** *E. miliaris* is typically found in low densities throughout its range.

POPULATION TREND Unknown

MAJOR THREATS Fishing pressure may pose a threat to this species. Juvenile habitat loss is also of particular concern.

HABITAT AND ECOLOGY **General** *E. miliaris* is a reef-associated species found in a variety of habitats. Juveniles (8–21 cm SL) have been taken from mud bottoms, seagrass beds, mangrove swamps, and from coral reefs in depths of 1–16 m; adults in coral reefs at depths of 18–180 m.

2008 IUCN RED LIST STATUS Least Concern

Authors: *AS Cabanban and YJ Sadovy de Mitcheson*

Epinephelus morio (Valenciennes 1828)

red grouper (E), mérou rouge (F), mero americano (S)

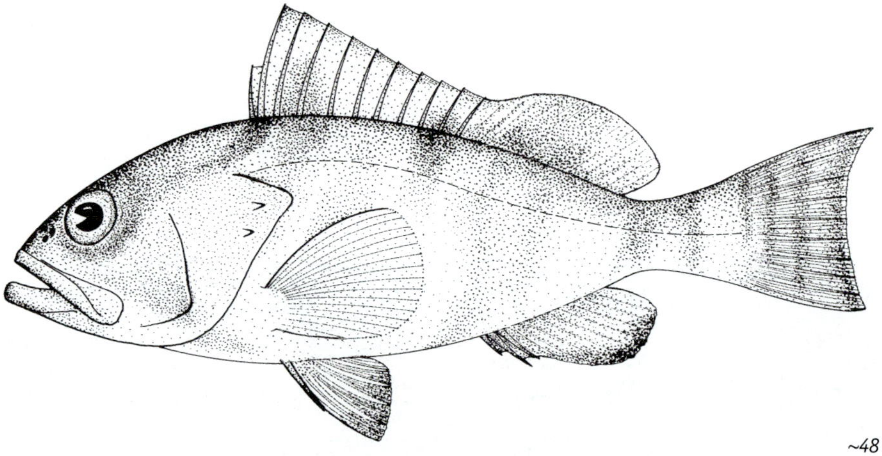

~48 cm TL
© FAO

Body depth distinctly less than head length, depth contained 2.6–3.0 times in standard length. Head length 2.3–2.5 times in standard length; interorbital convex; preopercle subangular, the serrae at angle slightly enlarged; upper edge of operculum straight; posterior and anterior nostrils subequal in size. Gill rakers 8 or 9 on upper limb and 15 or 16 on lower limb, 23–25 total. Dorsal fin with 11 spines and 16 or 17 rays, the membrane not incised between the spines and the 2nd spine longest, giving the fin a triangular sail-like aspect; anal fin with 3 spines and 8–10 rays; pectoral-fin rays 16–18; caudal fin convex in fish less than 15 cm SL, truncate or slightly concave in larger fish. Lateral-body scales ctenoid, with auxiliary scales; lateral-line scales 60–68; lateral-scale series 112–128.

MAXIMUM RECORDED SIZE 120 cm TL

Epinephelus morio is found in the western Atlantic, from North Carolina (USA) to southern Brazil, including the Gulf of Mexico, the Caribbean and Bermuda; it strays north to Massachusetts (USA).

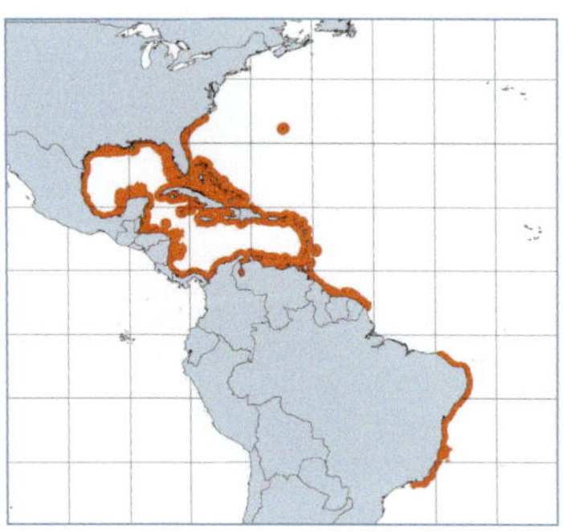

POPULATION **GENERAL** *E. morio* is a common species throughout the central portion of its range. **FISHERY DEPENDENT** *E. morio* is a continental species only rarely caught in the West Indies, with major fisheries in the eastern Gulf of Mexico (Moe 1969, Smith et al. 1975), Campeche Bank (Solis-Ramírez 1970), Isla Margarita (Cervigon 1966) and off Brazil (Stevenson 1981). It is one of the most abundant groupers in west Florida commercial catches (Bullock and Smith 1991). Fishery-dependent data show steady declines in catches under similar levels of effort since the 1950s. Most quantitative estimates in the USA and Mexico indicate that this species is overfished and, as a result, quotas, minimum

COLOUR IN LIFE Head and body dark reddish brown, shading to pink or reddish below; soft dorsal, caudal and anal fins dark distally with a narrow white edge; a few dark dots on the snout and/or cheeks; body often with irregular white spots and/or large pale blotches.
COLOUR IN MARKET Generally reddish brown, with irregular blotches and a pale underbelly.

© Phillip C Heemstra

Adult *E. morio* (Gulf of Mexico)

© Robert F Myers

Adult *E. morio* (Fort Lauderdale, Florida, USA)

sizes, closed seasons and sales controls have been variously introduced.
POPULATION TREND Decreasing
MAJOR THREATS Fishing pressure is the major threat to this species.
HABITAT AND ECOLOGY **General** Juveniles of 3–20 cm SL are occasionally found on shallow seagrass beds and inshore reefs. Larger juveniles (20–40 cm SL) are commonly found in crevices and under ledges on rocky reef bottoms in depths of 5-25 m. At 40–50 cm SL, adults migrate to deeper water (50–300 m) where they also occur over sandy or mud bottoms. **Feeding** Feeds on fishes and crustaceans. **Age and growth** *E. morio* may reach at least 25 years (Moe 1969, IUCN Red List Assessment 2009). **Reproduction** *E. morio* is a protogynous hermaphrodite with sexual transition variously reported at 45 cm SL and 50% transition at 7.2 years and 690 mm TL (Moe 1969). Burgos et al. (2007) note 50% maturation at 2.4 years and 487 mm TL with a sex ratio of 1 male to 6.6 females; spawning in females occurs from mid-February to mid-June off North and South Carolina (USA) at between 33 and 90 m. Although catchability increases during the reproductive season, the species is not thought to form concentrated spawning aggregations (Moe 1969, Arreguin-Sanchez and Pitcher 1999, Burgos et al. 2007).
2004 IUCN RED LIST STATUS Near Threatened

Authors: *AM Eklund and G García-Moliner*

Epinephelus morrhua (Valenciennes 1833)

comet grouper (E), mérou comète (F), mero cometa (S)

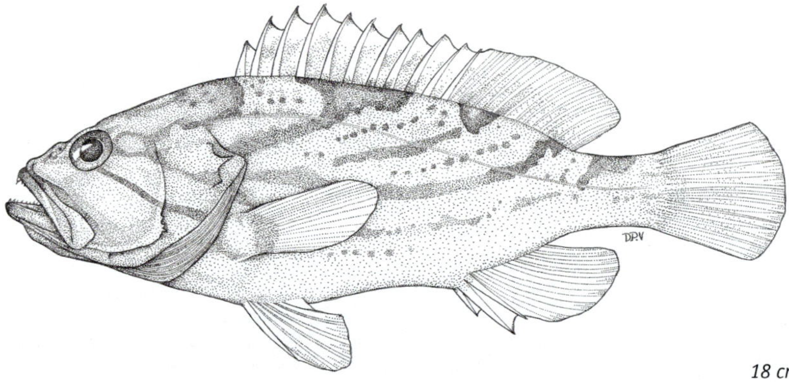

18 cm SL (Kenya)
© SAIAB

Body depth contained 2.8–3.1 times in standard length. Head length 2.3–2.5 times in standard length; interorbital area flat to moderately convex, the dorsal head profile slightly convex; preopercle with a shallow indentation just above the enlarged serrae at the angle; upper edge of operculum almost straight; adults with posterior nostril diameter 2 or 3 times larger than anterior nostril; maxilla reaches to or past vertical at rear edge of eye; midlateral part of lower jaw with 2 rows of teeth. Gill rakers 8–10 on upper limb, 15–18 on lower limb, the longest gill rakers shorter than longest gill filaments. Dorsal fin with 11 spines and 14 or 15 rays, the 3rd or 4th spine longest, its length contained 2.6–3.3 times in head length and subequal to longest dorsal-fin ray; anal fin with 3 spines and 7 or 8 rays, the 2nd and 3rd spines subequal; pectoral-fin rays 17 or 18; pectoral-fin length 1.8–2.2 times in head length; pelvic fins not reaching anus, their length 2.0–2.7 times in head length; caudal fin convex to moderately rounded. Lateral-body scales distinctly ctenoid, occasionally with a few auxiliary scales; lateral-line scales 55–64; lateral-scale series 108–125.

MAXIMUM RECORDED SIZE 100 cm TL

Epinephelus morrhua ranges from the Red Sea to South Africa in the west, and to the Tuamotu Archipelago in the east.

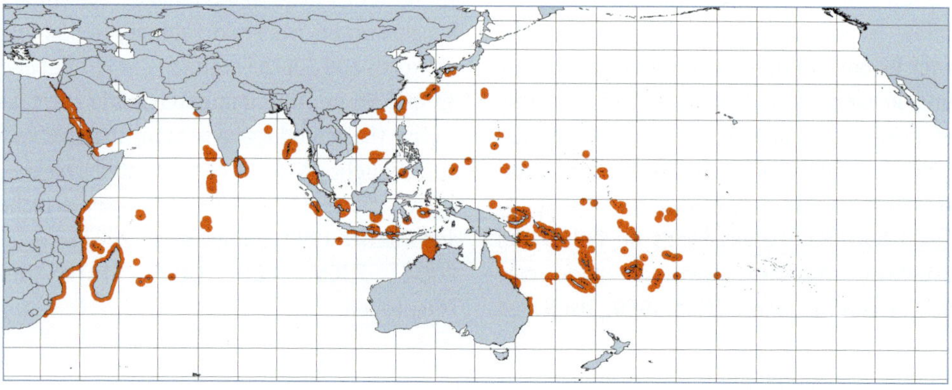

COLOUR IN LIFE Head and body tan, with dark brown bands including: a bifurcate band that begins at the rear edge of the eye, the upper branch extending to a dark brown saddle blotch on the nape just in front of the dorsal fin, the lower branch running to the lower opercular spine and continuing on the body as a midlateral band that bifurcates above the pectoral fin, the upper branch of this band running to a dark blotch at the base of the 3rd to 7th dorsal-fin rays, the lower branch curving up to the base of the last 4 dorsal-fin rays; another dark band from the upper edge of the operculum to the base of the 5th to 9th dorsal-fin spines; a narrow band from the lower edge of the eye to the pectoral-fin base, continued as a broken band along the lower part of the body and curving up to the dorsal part of the peduncle; and a broad band from the maxillary groove to the posterior end of the interopercle. Small dark brown spots often present in the pale areas between the bands and usually arranged in series parallelling the bands; fins generally unmarked; pectoral fins hyaline yellow. In juveniles, the dark bands appear first as series of dark blotches. Dark bands fade with growth.

COLOUR IN MARKET Body colour buff brown, banding remains prominent except on large adults. May have light-coloured blotches in irregular patterns, particularly at the midbody.

© Matthew T Craig

Adult *E. morrhua* (Tsukiji Fish Market, Tokyo, Japan)

POPULATION **General** *E. morrhua* is a common member of the deep-water fish community throughout its range. **Fishery dependent** *E. morrhua* is one of the major species of bottom handline or longline fishing on deep-water reef slopes in the Pacific for both local and export markets; it is the most common grouper species in the catch of bottom line-fishing on reefs in Melanesia, Micronesia and Polynesia. It is commercially fished in the Red Sea, but it does not seem to be a major component of the deep-reef bottom linefishing in the western Indian Ocean. In Mozambique, *E. morrhua* makes up 20% of groupers caught in the commercial trap fishery and is regularly taken by line. Although the species can reach 10 kg, most catches are of much smaller fish, averaging 2.5 kg in the South Pacific and 0.9–2 kg in the West (Papua New Guinea, Solomon Islands, Palau). In small isolated islands (Tuvalu, Tonga seamounts, Cook, French Polynesia, Kiribati), the fish average 2–5 kg. Seasonal variations in the catch suggest that there could be migrations (IUCN Red List Assessment 2008).

POPULATION TREND Decreasing

MAJOR THREATS No major threats have been identified for this species except where fishing on deep-water reefs is intensive.

HABITAT AND ECOLOGY **General** *E. morrhua* is found in deep waters from 80 to 370 m on the slopes of islands, sea mounts or continental shelves. **Feeding** *E. morrhua* feeds on benthic fishes and large invertebrates.

COMMENTS *E. morrhua* may be confused with *E. poecilonotus*, *E. radiatus* and *E. tuamotuensis*. Juveniles of *E. poecilonotus* have a large dark brown or black saddle blotch on the body at the base of the spinous dorsal fin; this blotch is isolated from the other dark bands on the body and extends over the front half of the spinous dorsal fin; in adults this blotch breaks up into small dark spots, as do the dark bands on the body, and in large adults most of the dark spots and bands have disappeared; juveniles have a dark band from the eye to the lower opercular spine, continued as a dark curving midlateral stripe or series of spots to a dark saddle blotch on the peduncle. On adults, the triangular interspinous dorsal-fin margins are brownish yellow or gold. Juveniles of *E. radiatus* have 5 irregular solid dark brown bands (with age only the edges remain dark) that run down and forward from the dorsal edge of the body, the 1st from the nape to the eye, the 2nd from the base of the middle dorsal-fin spines to the upper end of the gill opening, the 3rd and 4th dark bands from the anterior and posterior dorsal-fin rays, both branching as they pass ventrally, and the last dark band on the caudal peduncle; with growth, the dark bands break into spots and disappear ventrally on adults; soft dorsal fin and dorsal part of the caudal fin densely spotted. *E. tuamotuensis* has the dark bands on the head and body forming a coarse reticulum that does not extend to the ventral part of the body.

2008 IUCN RED LIST STATUS Least Concern

Authors: *M Kulbicki and BM Yeeting*

Epinephelus multinotatus (Peters 1876)

whiteblotched grouper (E), mérou plate grise (F), mero de lunares (S)

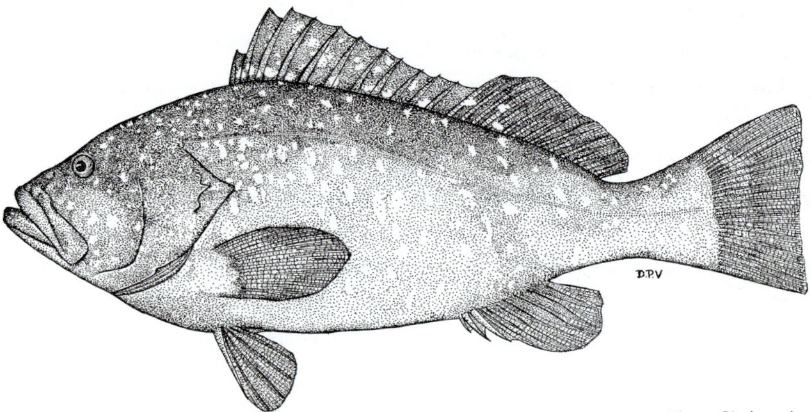

52 cm SL (southern Mozambique)
© SAIAB

Body depth contained 2.6–2.9 times in standard length. Head length 2.4–2.7 times in standard length; interorbital area distinctly convex, the dorsal head profile slightly convex; preopercle subangular, with a shallow notch (more distinct in adults) just above the angle, the serrae at the angle not or only slightly enlarged; upper edge of operculum straight or nearly so; posterior nostrils of adults about twice the size of anterior nostrils; maxilla reaches vertical at rear edge of eye or thereabouts; ventral edge of maxilla of adults with distinct step distally; midlateral part of lower jaw with 2 rows of teeth. Gill rakers 9–11 on upper limb, 15–17 on lower limb. Dorsal fin with 11 spines and 15–17 rays, the 3rd or 4th spine longest, contained 2.3–2.9 times in head length, the interspinous membranes not incised, the anterior rays distinctly longer than last spine; anal fin with 3 spines and 8 rays, the 3rd spine clearly longer than the 2nd, 3rd spine 3.4–4.6 times in head length and less than or subequal to peduncle depth; pectoral-fin rays 18–20; pectoral fins subequal to pelvic fins, pectoral-fin length 1.7–2.3 times in head length; pelvic fins not reaching anus; caudal fin truncate to slightly emarginate. Lateral-body scales distinctly ctenoid, with numerous auxiliary scales; lateral-line scales 64–81; lateral-scale series 130–162.

MAXIMUM RECORDED SIZE 100 cm TL

COLOUR IN LIFE Head and body dark purplish grey, with scattered irregular whitish spots and blotches; pelvic fins greyish black; rear margins of median fins with narrow white edge. Specimens from the western Indian Ocean (except the Persian Gulf and Gulf of Oman) usually have numerous small dark reddish brown spots over the ventral parts of the head and body. Juveniles (4–7 cm SL) are dark greyish blue with caudal fin, peduncle, soft dorsal fin and rear part of the anal fin yellow (which may be retained by adults in some cases); pelvic fins and ventral margin of the anal fin blackish.

COLOUR IN MARKET Generally drab grey to light brown; scattered spots and blotches often fade.

© Stephen Newman

Adult *E. multinotatus* (Kimberley coast, Western Australia)

© Robert F Myers

Adult *E. multinotatus* (Hong Kong Market)

Epinephelus multitnotatus is a western Indian Ocean species that is distributed along mainland East Africa from East London (South Africa) to Mozambique and in the island states (Maldives, Mauritius, Seychelles, Madagascar, Réunion, Rodrigues and Chagos), Persian Gulf and Gulf of Oman. The population along the coast of Western Australia may represent a distinct species.

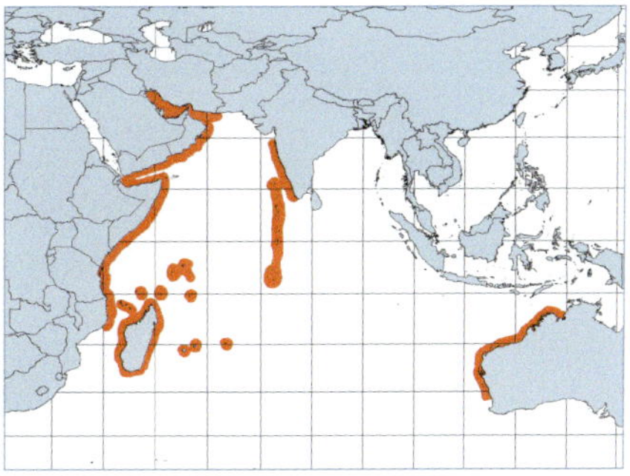

POPULATION GENERAL *E. multinotatus* is a common species throughout its range. There are three subpopulations based on meristic differences: western Indian Ocean, Persian Gulf and Gulf of Oman, and Western Australia (Heemstra and Randall 1993).

FISHERY DEPENDENT *E. multinotatus* is recorded as the most common of the three groupers taken in the demersal handline fisheries of the Seychelles (Mees 1992, MRAG 1996) although the same Creole common name is often used for *E. multinotatus* and *E. flavocaeruleus*, leading to uncertainty in fisher reports. Records of fishing spawning aggregations of *E. multinotatus* in the Seychelles date to the 1940s (Wheeler and Ommanney 1953, Robinson et al. 2004), and populations at many localities there are now overfished (Robinson et al. 2007). *E. multinotatus* is commonly taken in the fisheries of Réunion Island where declines in landings were noted from the late 1990s to the early 2000s (D Miossec, IFREMER, pers. comm.). It is also reported as the fifth most commonly caught grouper in the artisanal fisheries of southern Kenya, making up 4–5% of the total grouper catch (Agembe et al. 2010). **FISHERY INDEPENDENT** Fishery-independent data indicate that *E. multinotatus* may exceed densities of 0.5 per 1000 m^2 in waters <20 m in the Amirantes Isles (Pears 2005). **GENETICS** The populations in Shark Bay, Ningaloo, Pilbara and Broome (Western Australia) are recognised as separate stocks based on genetic studies (Stephenson et al. 2001).

POPULATION TREND Decreasing

MAJOR THREATS Fishing pressure is the major threat to this species.

HABITAT AND ECOLOGY **GENERAL** *E. multinotatus* is a reef-associated species found on rocky reefs at depths of up to 110 m. **AGE AND GROWTH** Individuals reach ages of at least 27 years (IUCN Red List Assessment 2009). **REPRODUCTION** *E. multinotatus* is an aggregate spawner: reproductive seasonality spans the entire year in the Seychelles, with peak activity from August to October when aggregations form (Robinson et al. 2007).

2008 IUCN RED LIST STATUS Least Concern

Authors: *JH Choat, LA Rocha and MA Samoilys*

Epinephelus ongus (Bloch 1790)

whitestreaked grouper (E), mérou à flocons (F), mero nubífero (S)

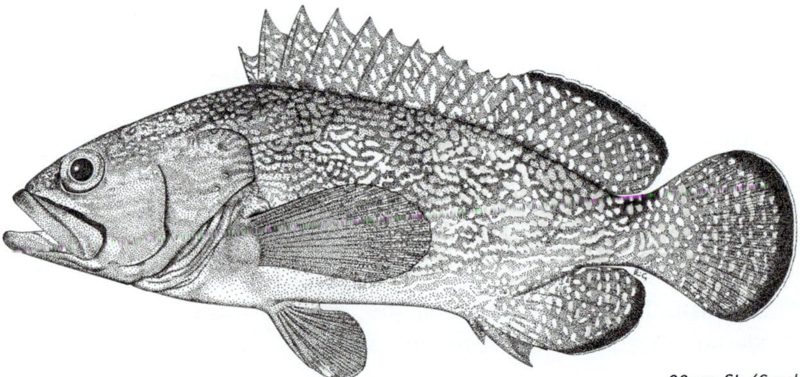

22 cm SL (Caroline Islands)
© SAIAB

11 cm SL (Kenya)
© SAIAB

Body depth contained 2.7–3.2 times in standard length; body width 1.7–2.2 times in the depth. Head length 2.3–2.5 times in standard length; head pointed, the interorbital area flat, the dorsal head profile slightly convex; preopercle rounded, the serrae small and mostly covered by skin; upper edge of operculum strongly convex, the rear edge almost vertical, the upper spine extending posterior to lower spine; posterior nostril diameter about twice that of anterior nostril; maxilla reaches to or slightly past vertical at rear edge of eye; small, embedded scales on maxilla; teeth small, 2–4 rows at midlateral part of lower jaw; canines at front of jaws small or absent. Gill rakers 8–10 on upper limb, 15–18 on lower limb (including 6–8 rudiments on each limb). Dorsal fin with 11 spines and 14–16 rays, the 3rd or 4th spine longest, its length contained 2.6–3.4 times in head length, the interspinous membranes incised; anal fin with 3 spines and 8 rays; pectoral fins large and fleshy, with 15–17 rays; pectoral-fin length 1.4–1.7 times in head length; pelvic fins end well short of anus, their length 2.0–2.3 times in head length; caudal fin rounded. Lateral-body scales ctenoid, with auxiliary scales; lateral-line scales 48–53; lateral-scale series 95–109.

MAXIMUM RECORDED SIZE 40 cm TL

POPULATION **GENERAL** *E. ongus* is a common species throughout its range. **FISHERY DEPENDENT** *E. ongus* is very commonly seen in fisheries catches. In 2001 it was the most numerically dominant grouper species in the Naha market in Okinawa (Craig 2007). It represented 0.3% of experimental grouper catch in New Caledonia (M Kulbicki pers. obs.). The species is sometimes targeted within spawning aggregations. **FISHERY INDEPENDENT** Diver surveys indicate that abundance of this species may vary by location; lowest densities recorded are 0–15 fish per 50 km^2 with highest densities on fringing reefs.

POPULATION TREND Unknown

Epinephelus ongus is an Indo-West Pacific species distributed from East Africa to Madagascar and the Seychelles, and to Indonesia, the Ryukyu Islands (Japan), Marshall Islands, Fiji, New Caledonia and northern Australia.

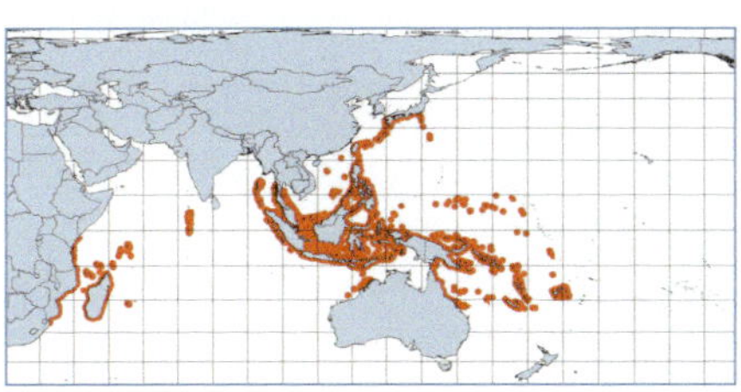

MAJOR THREATS Fishing pressure is the major threat to this species.

HABITAT AND ECOLOGY **GENERAL** *E. ongus* is a cryptic reef-associated species that inhabits inner coastal and lagoon reefs, but is also found in brackish waters where it occurs under ledges and in caves. Adults are usually found in 20–60 m depths. **FEEDING** Feeds on crustaceans and small fishes (Craig 2007). **AGE AND GROWTH** *E. ongus* may reach ages of at least 20 years (Craig 2007). **REPRODUCTION** There is limited evidence to suggest that *E. ongus* may be a protogynous hermaphrodite (Craig 2007). *E. ongus* is known to form spawning aggregations from October through January in some parts of its range, with more extended seasons in others (Hamilton and Kama 2004; database on www.scrfa.org). Craig (2007) reports that *E. ongus* is capable of reproducing by 3 years old and 16 cm SL.

COMMENTS *E. ongus* is similar to *E. summana*, which is endemic to the Red Sea. That species differs from *E. ongus* in having shorter pectoral fins (length contained 1.65–2.1 times in head length), shorter pelvic fins (length contained 2.2–2.7 times in head length), and the posterior nostrils of adults are vertically elongated (diameter of anterior nostrils contained 2–4 times in their length). The white-edged black margin on the median

COLOUR IN LIFE Body grey to brown, with numerous small white spots which, in fish larger than 10 cm SL, are horizontally elongate and, in adults, tend to form wavy white lines; several round or irregular pale blotches (eye-sized or larger) usually superimposed over small white spots; head brown, with numerous small white spots dorsally behind the eyes; black maxillary streak usually hidden by the maxilla; median fins with small white spots and streaks, the posterior margin blackish with a white edge; paired fins greyish brown. Juveniles of about 6 cm SL are brown, covered with small, dark-edged white spots which are round, except on the front part of the dorsal fin where they are elongated; the white spots on the paired fins become fewer and fainter with growth, and are absent in adults.

COLOUR IN MARKET Body brown to grey with white spots prominent. Patches of pale skin are often present in irregular patterns. White blotches are usually more prominent than spots in death.

© Matthew T Craig

Adult and subadult *E. ongus* (Okinawa Fish Market, Japan)

© Athila Bertoncini Andrade

Juvenile *E. ongus*
(Dumaguete, Philippines)

© Robert F Myers

Adult *E. ongus* (Tukangbesi, Indonesia)

fins of *E. ongus* is poorly developed or absent in *E. summana*.

The sympatric *E. coeruleopunctatus* has a similar colour pattern, but the caudal and anal fins have only a few white spots (confined mainly to proximal part of these fins), and the paired fins of juveniles are mostly unspotted. It also has 17–19 pectoral-fin rays, 51–61 lateral-line scales, upper edge of operculum sinuous or slightly convex, and the posterior nostrils of adults are vertically elongated, their length 5 or 6 times diameter of anterior nostrils. This species also attains a larger size than *E. ongus* (at least 60 cm TL).

2008 IUCN RED LIST STATUS Least Concern

Authors: *AS Cabanban, ST Fennessy, PC Heemstra, M Kulbicki, DA Pollard and BC Russell*

Epinephelus poecilonotus

(Temminck and Schlegel 1842)

dot-dash grouper (E), mérou morse (F), mero punto y linea (S)

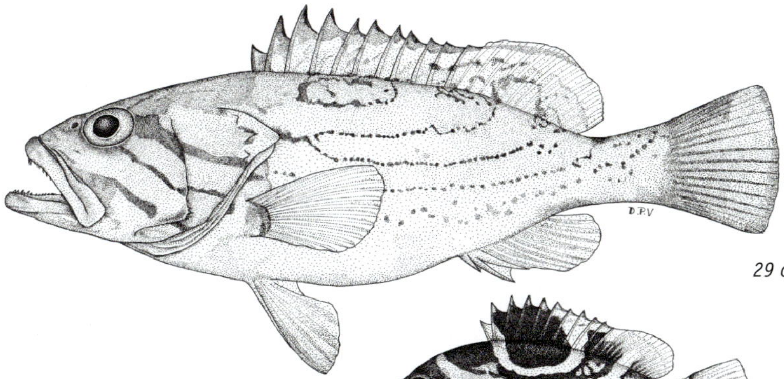

29 cm SL (Kenya)
© SAIAB

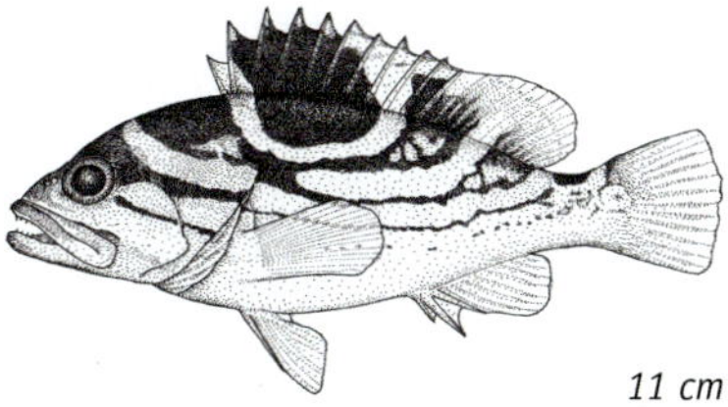

11 cm SL (South Africa)
© SAIAB

Body depth contained 2.6–3.1 times in standard length. Head length 2.3–2.5 times in standard length; interorbital area slightly convex; preopercle angle with 2–5 enlarged serrae; upper edge of operculum slightly convex; maxilla reaches about to vertical at rear edge of eye; adults with a step on ventral edge of maxilla; posterior nostrils of adults larger than anterior ones; midlateral part of lower jaw with 2 rows of teeth. Gill rakers 8–10 on upper limb, 15–18 on lower limb; longest gill raker subequal to longest gill filaments. Dorsal fin with 11 spines and 14 or 15 rays, the 3rd or 4th spine longest, its length contained 2.6–3.1 times in head length and subequal to longest dorsal-fin ray, the interspinous membranes deeply incised (adults with membrane-free length at front of 3rd and 4th spines for half or more of spine length); anal fin with 3 spines and 8 rays; pectoral fins not fleshy, with 17–19 rays; pectoral-fin length 1.7–2.1 times and pelvic-fin length 2.0–2.6 times in head length; caudal fin convex. Lateral-body scales ctenoid (with auxiliary scales on adults); lateral-line scales 54–63; lateral-scale series 110–121 (one specimen from Sri Lanka with 104 and one from Kenya with 136). Pyloric caeca 8 or 9.

MAXIMUM RECORDED SIZE 65 cm TL

COLOUR IN LIFE Small juveniles (5–12 cm SL) pale yellowish grey, with oval black blotch on the body between the bases of the 3rd to 9th dorsal-fin spines and extending to the edge of the fin between the 1st and 7th spines; a semicircular pale band passes ventrally around the oval black blotch and isolates it from a dark brown band that begins broadly on the nape and bifurcates just behind the operculum, the upper branch curving dorsally and expanding broadly over the basal half of the dorsal fin between the 9th spine and 4th soft ray, the lower branch also curving dorsally and expanding at the base of the last 4 dorsal-fin rays. A 2nd curved brown band, parallel to the one above, runs from the interorbital area and rear edge of the eye to a black saddle spot on the caudal peduncle; a 3rd (and narrowest) dark brown band runs from the lower edge of the eye to the subopercle and continues as a series of dark dots along the ventral part of the body to the base of the caudal fin. Fins pale yellow, except where dark markings occur on the dorsal fin. On larger juveniles (15–25 cm SL) the oval black blotch at the base of the dorsal-fin spines and the dark brown bands on the body break into series of black spots, a faint dark band runs from the dark maxillary streak to the angle of the preopercle. On small adults (40–50 cm SL) most of the dark spots on the body are faint or have disappeared completely and the dark bands on the head are fading too; at this stage the fins are yellowish brown, the triangular interspinous dorsal-fin margins abruptly orange-yellow or brownish yellow; soft dorsal, anal and caudal fins shading to blackish distally with a bluish white edge.

COLOUR IN MARKET Generally pale brown to tan. Markings on the body are nearly identical to life colour but may be interrupted by pale blotches if the specimen is left in contact with the ground or the deck of a boat.

© Stephen Newman

Adult *E. poecilonotus* (Scott Reef, Western Australia)

© Phillip C Heemstra

Juvenile *E. poecilonotus* (aquarium, Port Alfred, South Africa)

Epinephelus poecilonotus has a widespread but disjunct distribution within the Indo-West Pacific, ranging from the east coast of Africa to Japan, Korea, the South China Sea, Vietnam and Fiji. It is unknown from the Red Sea and Persian Gulf.

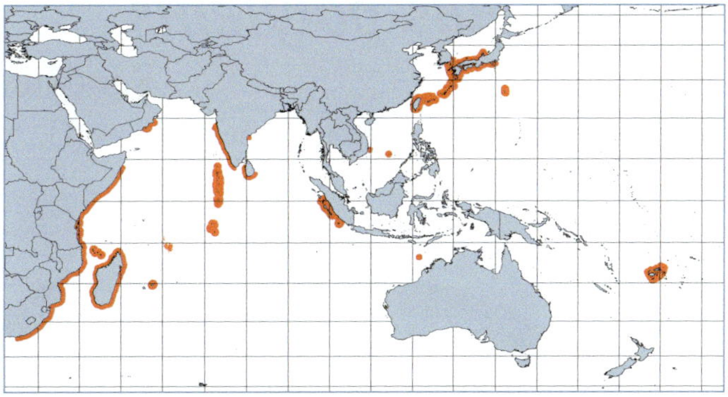

POPULATION **General** *E. poecilonotus* is a deep-water species and there is little information available on its population. **Fishery dependent** *E. poecilonotus* is taken in fisheries but is not known to be taken in appreciable numbers; however, data on landings are not species-specific. This does not appear to be a common species and is found in deep habitat where fishing and habitat destruction are not generally major concerns.

POPULATION TREND Unknown

MAJOR THREATS Fishing pressure may pose a threat to this species.

HABITAT AND ECOLOGY **General** *E. poecilonotus* is a reef-associated species that inhabits depths of 45–375 m, shallower in lower latitudes. **Feeding** Feeds on fishes and crustaceans. **Reproduction** Morgans (1964) reports 'nearly ripe' fish caught in November and December on North Kenya Banks, and that females mature at around 41 cm SL (Morgans 1982).

COMMENTS *E. poecilonotus* is similar in appearance to *E. morrhua*, *E. radiatus* and *E. tuamotuensis*. Juveniles of these species are readily identified by their colour patterns, but in large adults these distinctive markings are less obvious. *E. morrhua* has a dark blotch at the base of the 5th to 9th dorsal-fin spines connected by a dark band to the upper edge of the operculum. Juveniles of *E. radiatus* have 5 irregular solid dark brown bands (with age, only the edges remain dark) that run down and forward from the dorsal edge of the body, the 1st from the nape to the eye, the 2nd from the base of the middle dorsal-fin spines to the upper end of the gill opening, the 3rd and 4th dark bands from the anterior and posterior dorsal-fin rays, both branching as they pass ventrally, and the last dark band on the caudal peduncle; with growth, the dark bands break into spots and disappear ventrally on adults; soft dorsal fin and dorsal part of caudal fin densely spotted. *E. tuamotuensis* has the dark bands on the head and body forming a coarse reticulum that does not extend to the ventral part of the body.

2008 IUCN RED LIST STATUS Least Concern

Authors: *ST Fennessy, RF Myers and MA Samoilys*

Epinephelus polylepis

Randall and Heemstra 1991

smallscaled grouper (E), mérou petites écailles (F), mero escamoso (S)

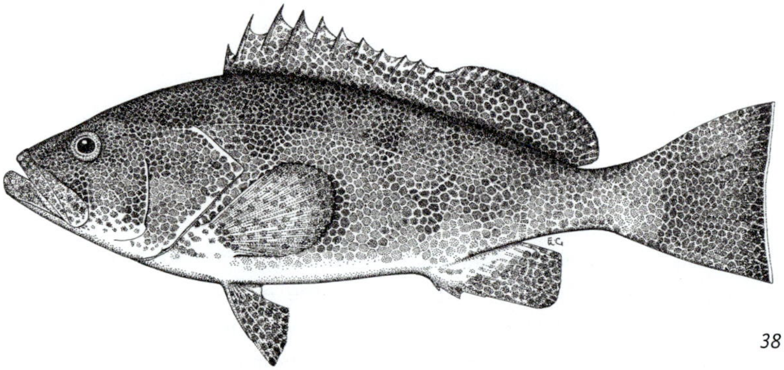

38 cm SL (India)
© SAIAB

Body depth contained 2.6–3.3 times in standard length; body width 1.8–2.4 times in the depth. Head length 2.5–2.8 times in standard length; interorbital slightly convex, the dorsal head profile almost straight; preopercle angular, with an indentation above the angle and enlarged serrae below the indentation; upper edge of operculum straight; posterior nostrils slightly larger than anterior ones; maxilla reaches to below rear half of eye; a pair of small fixed canines at front of jaws; midlateral part of lower jaw with 2 rows of teeth, the inner ones twice length of outer teeth. Gill rakers 9 or 10 on upper limb, 17 or 18 on lower limb, 25–28 total. Dorsal fin with 11 spines and 17 (rarely 16) rays, the 3rd or 4th spine longest, its length contained 2.5–2.9 times in head length, the anterior interspinous membranes moderately incised; anal fin rounded or slightly angular, with 3 spines and 8 rays, the 2nd or 3rd ray longest, its length 2.1–2.5 times in head length; pectoral fins not fleshy, with 18 or 19 rays; pectoral-fin length 1.6–2.0 times, pelvic-fin length 1.9–2.4 times in head length; caudal-peduncle depth 3.2–3.7 times in head length; caudal fin truncate or slightly emarginate. Lateral-body scales ctenoid, with auxiliary scales; lateral-line scales 65–72; lateral-scale series 126–137. Pyloric caeca 28–34.

MAXIMUM RECORDED SIZE 75 cm TL

COLOUR IN LIFE Head, body and fins pale, covered (except the ventral parts of the head and body) with numerous small close-set dark brown spots; spots on the fins and dorsal parts of the head and body smaller and closer together than those on the sides and ventrally. Spots may coalesce to form five bars on the body, the first under the longest dorsal-fin spines, the last over the caudal peduncle. Rear edge of the caudal fin with a white line and a row of blackish brown spots. Dark maxillary streak present.

COLOUR IN MARKET Body pale with close-set spots remaining prominent. Blotches formed by coalescing spots are faded and most obvious on the upper part of the midbody.

© John E Randall

Adult *E. polylepis* (Bahrain)

POPULATION **General** The population status of *E. polylepis* is not well known through its range, in part due to misidentification as *E. chlorostigma*. **Fishery dependent** *E. polylepis* has been heavily fished in many parts of its range. It is threatened by fishing pressure in the Arabian Sea coast of Oman which may have truncated the age structure of the population compared with the lighter fishing pressure in the Gulf of Oman where fish are older (JL McIlwain, pers. comm., 2007). *E. polylepis* is caught in large numbers in the Arabian Sea by both traditional and artisanal fishermen and by the industrial trawlers that operate in >50 m water (JL McIlwain, pers. comm., 2007). It is one of the most commonly taken groupers in the Arabian Sea. The species has been estimated to make up 14% of the total grouper catch in central Oman (JL McIlwain unpublished data). **Fishery independent** Fishery-independent surveys indicate the species is common in Muscat.

POPULATION TREND Decreasing

MAJOR THREATS Fishing pressure is the primary threat to this species and it is nowhere specifically managed.

Epinephelus polylepis is known from the west coast of India to the Yemen coast, but its range may be greater.

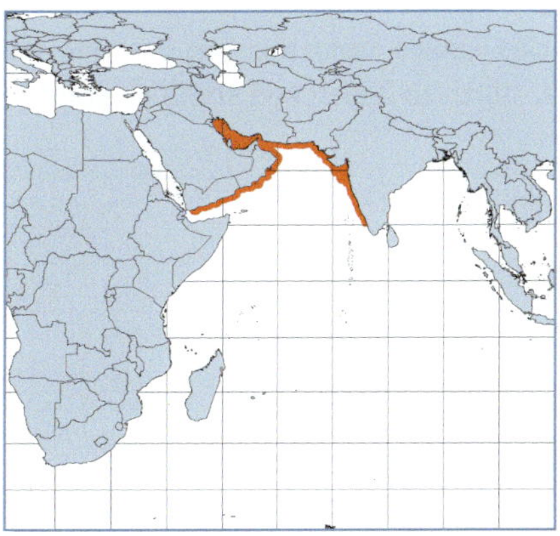

HABITAT AND ECOLOGY **General** *E. polylepis* is known from rocky areas adjacent to soft bottoms in depths of 10–155 m. **Reproduction** *E. polylepis* may be a diandric protogynous hermaphrodite (JL McIlwain unpublished data).

COMMENTS *E. polylepis* is similar in appearance to *E. chlorostigma* and *E. gabriellae*. The former differs in having fewer scales (lateral-line scales 48–53, lateral-scale series 96–122) and a more pointed anal fin in adults (4th or 5th ray longest, its length contained 1.9–2.3 times in head length). *E. gabriellae* differs from *E. polylepis* in having fewer scales (lateral-line scales 52–54, lateral-scale series 106–126), fewer dorsal-fin rays (14 or 15), more pyloric caeca (about 76), more elongate body (depth contained 3.2–3.6 times in standard length), and the caudal fin is more concave in adults (concavity 5–15 times in head length).

2008 IUCN RED LIST STATUS Near Threatened

Authors: *AA Bertoncini, JH Choat, MT Craig, BP Ferreira, LA Rocha and MA Samoilys*

Epinephelus polyphekadion (Bleeker 1849)

camouflage grouper (E), mérou camouflage (F), mero disfrazado (S)

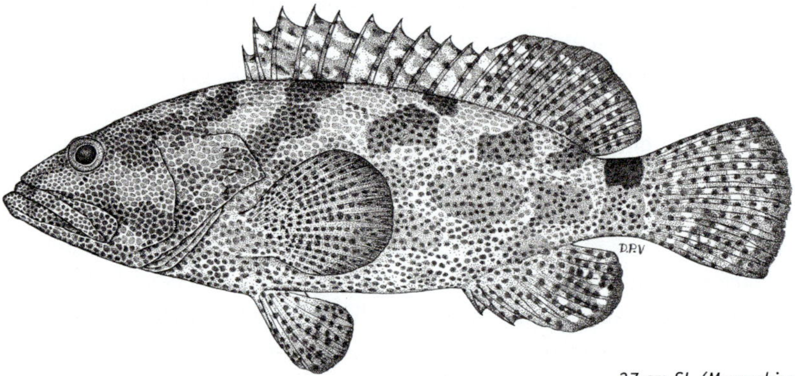

37 cm SL (Mozambique Channel)
© SAIAB

Body depth contained 2.7–3.1 times in standard length. Head length 2.3–2.5 times in standard length; interorbital area flat, the dorsal head profile evenly convex; preopercle rounded, the serrae at the angle slightly enlarged; upper edge of operculum very convex; posterior nostril diameter about twice diameter of anterior nostrils; maxilla reaches to or beyond vertical at rear edge of eye; midlateral part of lower jaw with 2 or 3 rows of teeth; fixed canines at front of jaws inconspicuous. Gill rakers 8–10 on upper limb, 15–17 on lower limb, 24–27 total. Dorsal fin with 11 spines and 14 or 15 rays, the 3rd or 4th spine longest, its length contained 2.7–3.3 times in head length and distinctly shorter than longest dorsal-fin ray, the interspinous membranes moderately incised; anal fin with 3 spines and 8 rays; pectoral-fin rays 16–18; pectoral-fin length 1.7–2.1 times, pelvic-fin length 1.9–2.4 times in head length; caudal fin rounded. Lateral-body scales ctenoid, with auxiliary scales; lateral-line scales 47–52; lateral-scale series 95–113.

MAXIMUM RECORDED SIZE 90 cm SL

Epinephelus polyphekadion is an Indo-Pacific species known from the Red Sea and east coast of Africa, extending east to French Polynesia, north to Japan and south to southern Queensland and Lord Howe Island (Australia).

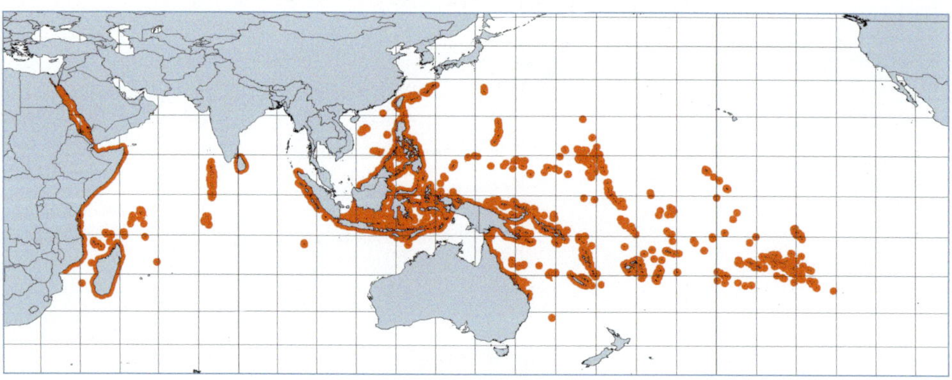

COLOUR IN LIFE Head, body and fins pale brown, covered with numerous small dark brown spots; head and body with irregular dark blotches superimposed over the dark spots; prominent black saddle blotch on the caudal peduncle; dark spots extend all over the head, including the lower jaw, lips, branchiostegal membranes, gular area and inside of the mouth; numerous small white spots on the fins and a few on the head and body. Juveniles with a pair of blackish spots on each side of the snout and a black spot at the margin of the 2nd and 3rd interspinous dorsal-fin membranes. Temporary colour changes, such as lightening of background body colour, occur during spawning activity.

COLOUR IN MARKET Generally grey to pale grey. Small spots are prominent, and blotches may fade. Saddle blotch on the caudal peduncle remains prominent.

© Matthew T Craig

Adult *E. polyphekadion* (Okinawa Fish Market, Japan)

© Robert F Myers

Adult *E. polyphekadion* (Chuuk, Federated States of Micronesia)

POPULATION **GENERAL** *E. polyphekadion* is a common species throughout its range. **FISHERY DEPENDENT** *E. polyphekadion* is heavily targeted by fisheries throughout much of its range. Known fishing stocks in many areas have been depleted by as much as 25%, particularly in South-East Asia and Micronesia. The species is commonly seen in the international live reef food fish trade (Sadovy et al. 2003). **FISHERY INDEPENDENT** Fishery-independent surveys indicate this species has variable population sizes depending on level of fishing pressure. **GENETICS** Genetic studies have shown discrete populations, both in terms of geography and sample year, in Pohnpei (Rhodes et al. 2003).

POPULATION TREND Decreasing

MAJOR THREATS *E. polyphekadion* is threatened by fishing pressure and lack of effective management, particularly during spawning periods when aggregations are targeted.

HABITAT AND ECOLOGY **GENERAL** *E. polyphekadion* is an insular species usually found in coral-rich areas of lagoons and outer reefs in depths of 2–46 m. **FEEDING** Feeds on fishes and crustaceans. **REPRODUCTION** *E. polyphekadion* is probably a gonochoric species that forms spawning aggregations of hundreds to thousands of individuals, often with *E. fuscoguttatus* and *Plectropomus areolatus*. Sexual maturity occurs at about 30 cm TL in Palau (Johannes et al. 1999). In Pohnpei, spawning occurs at night, for several nights during the full-moon periods between February and April (Rhodes and Sadovy 2002a, 2002b), and sexual maturity occurs at 27 cm SL. In Fiji, Palau and French Polynesia, spawning is also known to occur around the full moon.

COMMENTS *E. polyphekadion* is similar in appearance to *E. fuscoguttatus*. The latter species, however, has 18–20 pectoral-fin rays, 28–31 gill rakers, adults with distinct indentation in the dorsal head profile above the eyes, the posterior nostril diameter of adults is 4–8 times anterior nostrils, and it lacks a dark saddle on the caudal peduncle.

2006 IUCN RED LIST STATUS Near Threatened

Authors: *AS Cornish, BC Russell and Y-Y Situ*

Epinephelus polystigma (Bleeker 1853)

whitedotted grouper (E), mérou points blancs (F), mero punteado (S)

26 cm SL (Port Moresby, Papua New Guinea)
© SAIAB

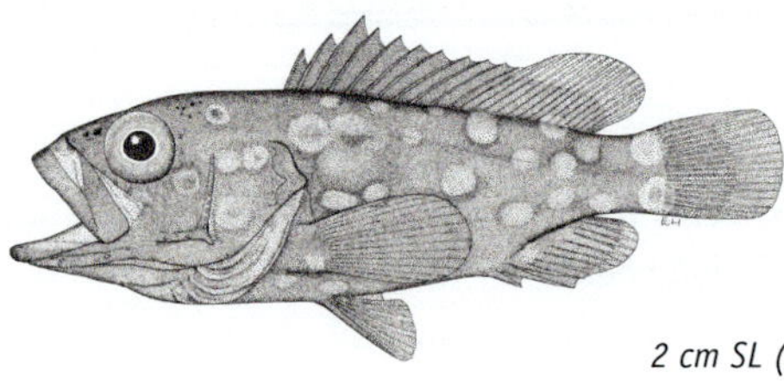

2 cm SL (Indonesia)
© SAIAB

Body depth contained 2.6–2.9 times in standard length (for fish 10–38 cm SL). Head length 2.3–2.7 times in standard length; interorbital area flat; snout short, its length subequal to eye diameter; preorbital depth less than greatest width of upper lip; preopercle rounded, finely serrate, the serrae at angle slightly enlarged; upper edge of operculum mostly straight; anterior nostrils funnel-shaped, the rear margin expanded as a flap reaching posterior nostril; posterior nostril diameter 2–4 times larger than anterior nostril diameter; maxilla reaches past vertical at rear edge of eye, the lower edge with a rounded step at front of widest part; midlateral part of lower jaw with 2 or 3 rows of small subequal teeth; 2 short stout canines at front of both jaws. Gill rakers shorter than gill filaments, 8 or 9 on upper limb, 13–16 on lower limb, 21–24 total. Dorsal fin with 11 spines and 15 or 16 rays, the 3rd or 4th spine longest, its length contained 2.7–3.3 times in head length and distinctly shorter than longest dorsal-fin rays, the interspinous membranes distinctly incised; anal fin with 3 spines and 8 rays; pectoral-fin rays 16–18; pectoral-fin length 1.6–1.8 times, pelvic-fin length 2.0–2.2 times in head length; caudal fin well rounded. Lateral-body scales ctenoid, with auxiliary scales. Lateral-line scales 49–55; lateral-scale series 81–91. Pyloric caeca 8.

MAXIMUM RECORDED SIZE 48 cm TL

COLOUR IN LIFE Head, body and fins dark brown, covered (except on the pelvic fins) with white or pale yellow dots; soft dorsal, anal, caudal and pectoral fins with a narrow orange-red margin. Juveniles (8 or 9 cm SL) with some larger, dark-edged, round or horizontally elongate, white spots on the body (in addition to the white dots); juveniles have dark-edged white spots, as large or larger than the pupil, in about 3 irregular rows on the body.

COLOUR IN MARKET Brown to pale brown, spots on the body may fade but are generally present.

Adult *E. polystigma* (Ambon, Indonesia)

Epinephelus polystigma is distributed from the Philippines, south through eastern Indonesia, Papua New Guinea and the Solomon Islands, to the northern tip of Cape York in Queensland, Australia.

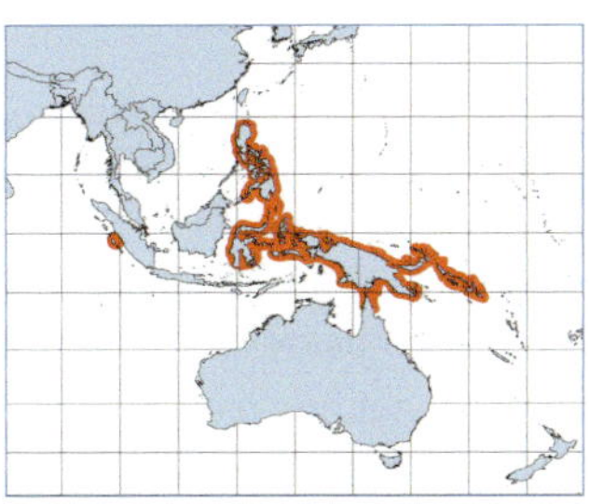

POPULATION **GENERAL** *E. polystigma* is a poorly known species throughout its range.

POPULATION TREND Unknown

MAJOR THREATS No major threats have been identified for *E. polystigma*, although coastal development may threaten this species, especially in mangrove and estuarine habitats.

HABITAT AND ECOLOGY **GENERAL** *E. polystigma* inhabits shallow brackish or freshwater areas, particularly in mangrove areas. **REPRODUCTION** From local fisher knowledge *E. polystigma* forms large nocturnal aggregations in shallow waters of river mouths and brackish mangrove regions throughout the year, prior to new moon. Sites were identified in several regions in Melanesia using maps during interviews. Field observations confirmed the local knowledge reports (Hamilton and Potuku 2007).

2008 IUCN RED LIST STATUS Data Deficient

Authors: *JH Choat and MA Samoilys*

Epinephelus posteli Fourmanoir & Crosnier 1964

stripedfin grouper (E), mérou aile zébrée (F), mero aleta listada (S)

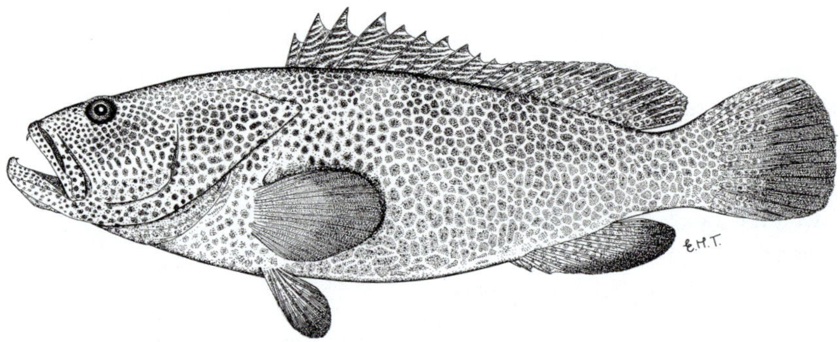

67 cm SL (South Africa)
© SAIAB

Body depth contained 3.1–3.5 times in standard length; body width 1.7–2.0 times in the depth. Head length 2.3–2.4 times in standard length; interorbital width 5.9–6.8 times in head length; interorbital area concave; preopercle rounded, finely serrate, with a shallow indentation just above the angle; upper edge of operculum almost straight; opercular spines inconspicuous; posterior and anterior nostrils subequal in size; maxilla not reaching past eye; a pair of large, curved, fixed canines (their length equal to half eye diameter) at front of upper jaw and a pair of similar but slightly smaller canines at front of lower jaw; lateral part of both jaws with an outer row of 8 or 9 slightly smaller fixed canines and an inner row of smaller, more slender, depressible teeth. Gill rakers 7 or 8 on upper limb (of which 7 are rudiments); 10–15 on lower limb (including 6–10 rudiments). Dorsal fin with 11 spines and 15 or 16 rays, the 3rd to 5th spines usually longest, their length contained 3.3–3.6 times in head length and shorter than the longest dorsal-fin rays, the interspinous membranes deeply incised; anal fin with 3 spines and 9 rays; pectoral-fin rays 18; pectoral-fin length 1.9–2.2 times, pelvic-fin length 2.6–2.8 times in head length; caudal fin rounded. Lateral-body scales smooth, mostly embedded, and covered with tiny auxiliary scales; lateral-line scales difficult to count (about 59–64); lateral-scales series also difficult to count (about 98–108).

MAXIMUM RECORDED SIZE 100 cm TL

POPULATION GENERAL *E. posteli* is a rare species throughout its range based on a limited number of diving and fishery surveys (IUCN Red List Assessment 2008). **FISHERY DEPENDENT** *E. posteli* is only rarely reported in fisheries catch data. **FISHERY INDEPENDENT** *E. posteli* is not commonly observed by divers.

POPULATION TREND Unknown

COLOUR IN LIFE Body colour dark reddish brown. Head and body covered with small, irregular, close-set, dark red or reddish brown spots, the pale interspaces forming an irregular pale network pattern; dark spots run in rows parallelling the fin rays of the soft dorsal and caudal fins; interspinous membranes of the dorsal fin with several horizontal dark streaks; live fish with 4 narrow, irregular, oblique, pale bars running down and forward on body; dark spots of some rows on the distal part of the caudal and soft dorsal fins of some fish merge to form dark streaks running along or between the fin rays; soft dorsal, caudal, anal and paired fins generally darker than body.
COLOUR IN MARKET Generally dark reddish brown and network of spots remains prominent. May have pale blotches if the specimen is left in contact with the ground or the deck of a boat.

Adult *E. posteli* (Sodwana Bay, South Africa)

Epinephelus posteli is found only in the western Indian Ocean and is known only from the south-west coast of Madagascar, KwaZulu-Natal (South Africa), and southern Mozambique.

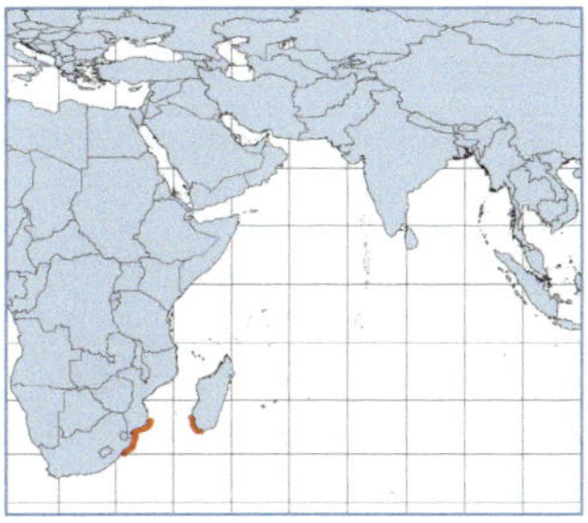

MAJOR THREATS No major threats have been identified for this species.
HABITAT AND ECOLOGY **General** *E. posteli* is a coral reef-associated species inhabiting depths of 20–50 m. **Feeding** Likely feeds on fishes and invertebrates.
2008 IUCN RED LIST STATUS Data Deficient

Author: *ST Fennessy*

Epinephelus quinquefasciatus

(Bocourt 1868)

Pacific goliath grouper (formerly jewfish) (E), mérou goliath (F), mero guasa (S)

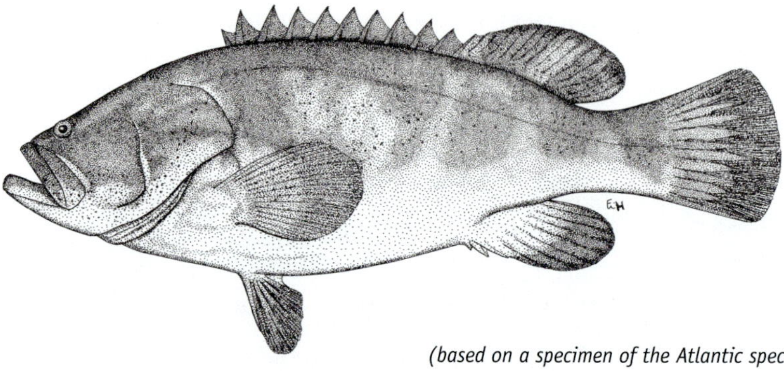

~140 cm SL
(based on a specimen of the Atlantic species, E. itajara*)*
© SAIAB

Body robust, elongate, the greatest width more than half of body depth, which is distinctly less than head length (in fish 15–160 cm SL); body depth contained 2.7–3.4 times in standard length. Head length 2.3–2.9 times in standard length; head extremely broad; interorbital flat, the width equals eye diameter in fish 10–15 cm SL, distinctly greater than eye diameter in fish 18–25 cm SL, and 1.5–3.4 times greater than eye diameter in fish 30–160 cm SL; eye diameter 5–8 times in head length for fish 10–30 cm SL and 8–13 times in head length for fish 35–160 cm SL; preopercle rounded, finely serrate; posterior and anterior nostrils round and subequal; maxilla scaly, reaching well past eye; midlateral part of lower jaw with 3–5 rows of subequal teeth; no canines at front of jaws. Gill rakers 8 or 9 on upper limb and 13–15 on lower limb, 21–24 total; gill arches covered with small bony plates. Dorsal fin with 11 spines and 15 or 16 rays, the spines short, 3rd to 11th subequal and shorter than the 1st ray, the membranes distinctly indented between the spines; anal fin with 3 spines and 8 rays; pectoral-fin rays 18 or 19; caudal fin rounded. Body scales strongly ctenoid; lateral-line scales 61–64, each with 4–6 radiating ridges; lateral-scale series 89–112.

MAXIMUM RECORDED SIZE 250 cm TL

COLOUR IN LIFE Generally brownish yellow, grey or greenish; head, dorsal part of body and fins with small black spots, becoming smaller with growth. Fish of less than about 100 cm show 3 or 4 faint, irregular, subvertical dark bars posteriorly on the body, and another covering the rear half of the caudal peduncle; large adults darker and more uniformly coloured than juveniles.

COLOUR IN MARKET Uniform brown, sometimes with irregular blotches of lighter hue, and with black spots less prominent. A very nondescript grouper after death, this species is best recognised by its large size, proportionally small eye, wide interorbital and short dorsal-fin spines.

© Matthew T Craig

Adult *E. quinquefasciatus* (Panama City Fish Market, Central America)

Epinephelus quinquefasciatus is an eastern Pacific species ranging from the Gulf of California to Peru. Formerly considered conspecific with populations in the tropical western Atlantic. Those populations, however, have been shown to be a distinct species (*E. itajara*; Craig et al. 2009).

POPULATION **GENERAL** *E. quinquefasciatus* was once known to be common throughout its range. Fisher interviews indicate severe population declines over the past few decades, and the species is now rare throughout most of its range, with only localised pockets of abundance (MO Aburto, B Erisman pers. comm.).

POPULATION TREND Unknown

MAJOR THREATS *E. quinquefasciatus* is severely threatened by overfishing and habitat loss, particularly for juveniles that inhabit mangroves and estuaries.

HABITAT AND ECOLOGY **GENERAL** *E. quinquefasciatus* inhabits a variety of habitats in depths of up to 100 m.

COMMENTS Although we are unable to find any morphological differences between the Atlantic and Pacific goliath grouper, the genetic data clearly support the existence of a transisthmian geminate species pair.

2011 IUCN RED LIST STATUS Data Deficient

Author: *MT Craig*

Epinephelus quoyanus (Valenciennes 1830)

longfin grouper (E), mérou longues ailes (F), mero aleta larga (S)

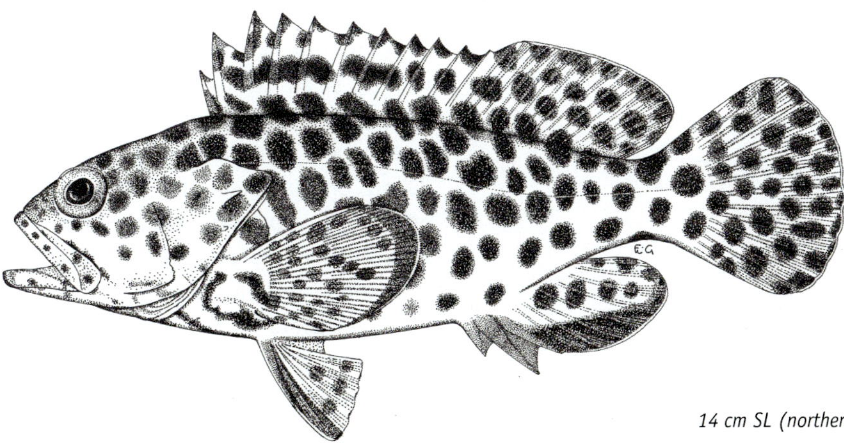

14 cm SL (northern Australia)

Body depth contained 2.8–3.2 times in standard length. Head length 2.3–2.6 times in standard length; dorsal head profile evenly curved; snout subequal to eye diameter, snout length 4.6–5.3 times in head length; preopercle rounded or subangular; upper edge of operculum almost straight; posterior nostril diameter about twice that of anterior nostrils; maxilla reaches to or past vertical at rear edge of eye; midlateral part of lower jaw with 2 or 3 rows of teeth; lower jaw barely projecting in front of upper jaw. Gill rakers 6–8 on upper limb, 14–16 on lower limb. Dorsal fin with 11 spines and 16–18 rays, the 4th spine usually longest, its length contained 2.3–3.0 times in head length and shorter than the longest dorsal-fin ray; anal fin with 3 spines and 8 rays, the 2nd and 3rd spines subequal, their length 2.8–3.8 times in head length and less than or subequal to depth of peduncle; pectoral-fin rays 17–19; pectoral-fin length 26–35% of standard length, 1.2–1.7 times in head length; pelvic-fin length 1.6–2.1 times in head length; caudal-peduncle depth 3.1–3.6 times in head length; caudal fin rounded; length of middle caudal-fin rays 1.35–2.0 times in head length. Lateral-body scales ctenoid, auxiliary scales present; lateral-line scales 48–52; lateral-scale series 80–96. Pyloric caeca 21.

MAXIMUM RECORDED SIZE 40 cm TL

POPULATION **GENERAL** *E. quoyanus* is abundant throughout its range. **FISHERY DEPENDENT** There are few fishery-related data for the species which is typically lumped in with other groupers in landings information. Small juveniles of 2–5 cm have been collected in Hong Kong in the summer and larger fish are taken by gillnets, handlines and traps (Sadovy and Cornish 2000). Wild-caught juveniles were grown out in floating cages in small-scale fisheries in South-East Asia. In Hong Kong markets, more than 30% of fish sold are immature (To and Sadovy de Mitcheson 2009). **FISHERY INDEPENDENT** *E. quoyanus* was one of the most common groupers in shallow and intertidal eastern waters of Hong Kong in the 1960s; however, it is infrequently observed today (Sadovy and Cornish 2000).

Epinephelus quoyanus is widespread within the Indo-Malay region and occurs from the Andaman Islands to northern New South Wales and Western Australia (Shark Bay), extending north to Japan and Korea.

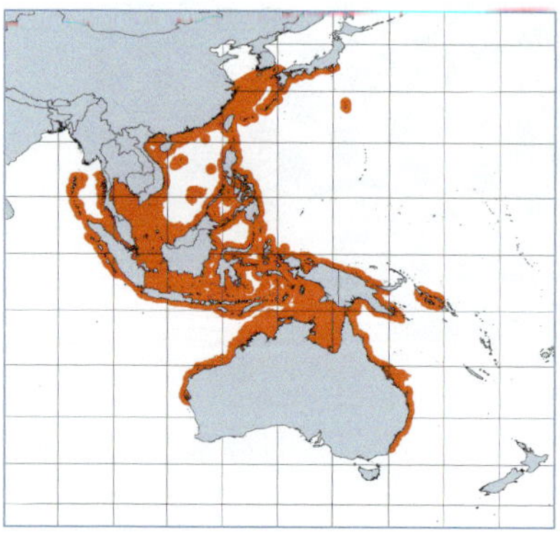

POPULATION TREND Decreasing

MAJOR THREATS The major threat to *E. quoyanus* is fishing pressure. Wild-caught juveniles are used for grow-out in Vietnam and elsewhere in South-East Asia (Tuan et al. 2000, Liu and Sadovy de Mitcheson 2008).

HABITAT AND ECOLOGY **GENERAL** *E. quoyanus* is a sedentary species and lives in shallow, often silty, inshore waters with a maximum depth of 50 m. **FEEDING** Feeds on shrimps, worms, small fishes (e.g. Apogonidae and Blenniidae species) and crabs (Connell 1998, Sadovy and Cornish 2000). **AGE AND GROWTH** Based on numbers of rings in sectioned otoliths, *E. quoyanus* female 50% maturity is 1.8 years, the youngest recorded male is 2.1 years, and the maximum recorded age is 14 years (To 2009). **REPRODUCTION** In Hong Kong, the spawning season is between May and September; 50% of females mature at about 18.9 cm SL and the smallest mature male was 19.8 cm SL (To 2009). *E. quoyanus* is a protogynous hermaphrodite (To 2009).

COLOUR IN LIFE Head and body pale, mostly covered with large, close-set, hexagonal to roundish dark brown spots (some spots reddish brown or black) — dorsally the spots are so close together that the pale interspaces form a reticulum, but ventrally the spots are more separated and their margins more diffuse; similar dark spots on the median fins; chest with 2 dark brown bands that are confluent below the pectoral-fin bases and nearly joined anteriorly at the gill opening, thus isolating a large pale area on the ventral part of the chest and another pale area between the upper dark band and dark blotch that usually covers most of the pectoral-fin base; isthmus and ventral surface of the chest with irregular dark brown bands and pale blotches; dark spots on the head smaller anteriorly, but always 3 or 4 times larger than the nostrils; a squarish unspotted white or pale area on the cheek at the rear end of the maxilla; ventral edge of the anal and caudal fins and leading edge of the pelvic fins with a white line and broad blackish submarginal band; pectoral fins mostly dusky with indistinct dark spots (more distinct in specimens from Australia).

COLOUR IN MARKET This species maintains nearly identical colouration in death.

© Phillip C Heemstra

Adult *E. quoyanus* (Northern Territory, Australia)

© Robert F Myers

Adult *E. quoyanus* (Bali, Indonesia)

COMMENTS *E. quoyanus* was intentionally introduced to the Hawaiian Islands in the 1950s, but the populations did not successfully colonise the area, and the species is no longer present.

E. quoyanus is a member of the 'reticulated groupers' and is very similar in appearance to several other species (see description of *E. merra*).

2008 IUCN RED LIST STATUS Least Concern

Authors: *M Liu, DA Pollard and BC Russell*

Epinephelus radiatus (Day 1867)

oblique-banded grouper (E), mérou zebra (F), mero acebrado (S)

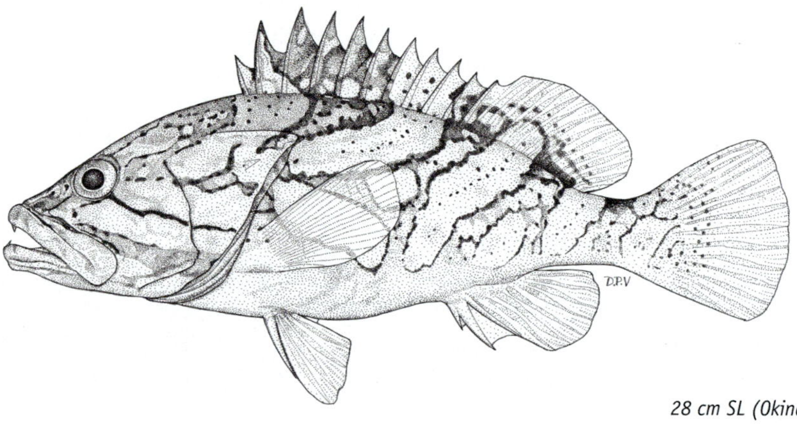

28 cm SL (Okinawa, Japan)

Body depth contained 2.6–3.0 times in standard length. Head length 2.1–2.3 times in standard length; interorbital area nearly flat, the dorsal head profile slightly convex; preopercle angle with 2–5 distinctly enlarged serrae; upper edge of operculum almost straight; adults with posterior nostril diameter 2 or 3 times that of anterior nostrils; maxilla reaches to or past vertical at rear edge of eye; midlateral part of lower jaw with 2 rows of teeth. Gill rakers 8 or 9 on upper limb, 16–18 on lower limb, the longest gill raker distinctly shorter than the longest gill filaments. Dorsal fin with 11 spines and 13–15 rays, the 3rd spine longest, its length contained 2.6–3.0 times in head length and longer than the longest dorsal-fin ray, the interspinous membranes deeply incised; anal fin with 3 spines and 8 rays, the 2nd and 3rd spines subequal; pectoral-fin rays 17 or 18, the fin almost transparent; pectoral-fin length 1.7–2.2 times, pelvic-fin length 2.1–2.8 times in head length; caudal fin convex to moderately rounded. Lateral-body scales distinctly ctenoid, without auxiliary scales; lateral-line scales 52–66; lateral-scale series 102–120.

MAXIMUM RECORDED SIZE 70 cm TL

Epinephelus radiatus is an Indo-West Pacific species known from the east coast of Africa to Papua New Guinea, including the Red Sea, Japan and northern Australia.

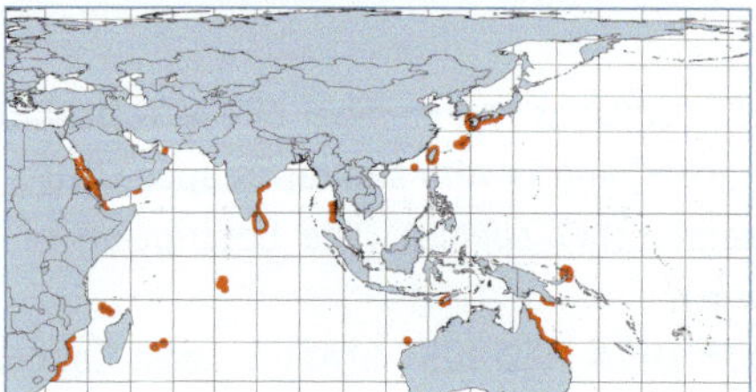

POPULATION **General** *E. radiatus* is patchily distributed throughout its range and may be locally abundant. **Fishery dependent** The species is caught commercially throughout parts of its range although there are few species-specific data and no reports of fisheries depletion. A possible exception may be in Sri Lanka and southern India, where groups of this species are trawled over offshore banks (e.g. Wadge and Pedro Banks off India). The species is 'of major interest to fisheries as large schools are caught by trawls on the limited trawling grounds off the Island [Sri Lanka]' (De Bruin

COLOUR IN LIFE Small adults (20–40 cm SL) tan with 5 oblique dark-edged pale bands: 1st band from the upper half of the eye, curving and broadening on the nape; 2nd band branching from the 1st band just behind the eye, bifurcating on the operculum, the upper branch continuing dorsally, broadening on the body and extending even more broadly over the middle of the spinous dorsal fin; 3rd band continued as lower branch of 2nd band, curving dorsally from the end of the operculum and expanding at the base of the last 2 spines and first 3 or 4 rays of dorsal fin (this band with a ventral extension from an included pale circle covered by the tip of the pectoral fin); 4th band runs from the rear end of the dorsal fin, branching at midside, with one branch going towards the anal-fin origin, the other to the rear end of the anal-fin base; 5th band on the caudal peduncle (sometimes divided into 2 short branches at the lower edge of the peduncle); dark bands with scattered small black spots and pale blotches, especially dorsally; pale interspaces (between the dark bands) with small dark brown spots, mainly arranged in series along the middle of the interspaces; dark brown line from the lower edge of the eye to the edge of the subopercle; faint dark band along the maxillary groove and continuing to the edge of the interopercle. Large adults (40–50 cm SL) with dark-edged bands replaced by series of dark spots (except for the dark line running posteriorly from the lower edge of the eye); no spots on the ventral third of the body; dorsal fin and most of the caudal fin covered with small dark spots. Juveniles (10–20 cm SL) dark brown with black-edged pale brown bands (equivalent to the white markings on smaller fish) enclosing numerous small black spots. Small juveniles (4–7 cm SL) are mostly dark greenish brown, with dark-edged immaculate white markings that represent the pale interspaces on larger specimens; fins translucent white, except for spinous dorsal fin which is coloured like the body.

COLOUR IN MARKET Generally tan. One of the few grouper species to retain nearly identical colouration in death.

© Matthew T Craig

Adult *E. radiatus* (Okinawa Fish Market, Japan)

et al. 1994). In Sri Lanka, it is quite common in deep-water longline and handline fisheries around the island (De Bruin et al. 1994). **FISHERY INDEPENDENT** Fishery-independent data indicate that this species is not abundant in shallow diving depths.

POPULATION TREND Unknown

MAJOR THREATS No major threats have been identified for this species.

HABITAT AND ECOLOGY GENERAL Adult *E. radiatus* are found over hard substrates at depths between 80 and 383 m. Juveniles are often found in 18–20 m.

COMMENTS *E. radiatus* is similar in appearance to *E. morrhua*. In the latter species, the oblique bands are wider and at a steeper angle than in *E. radiatus*.

2008 IUCN RED LIST STATUS Least Concern

Authors: *AS Cabanban, DA Pollard and BC Russell*

Epinephelus retouti Bleeker 1868

redtipped grouper (E), mérou à bout rouge (F), mero punteado (S)

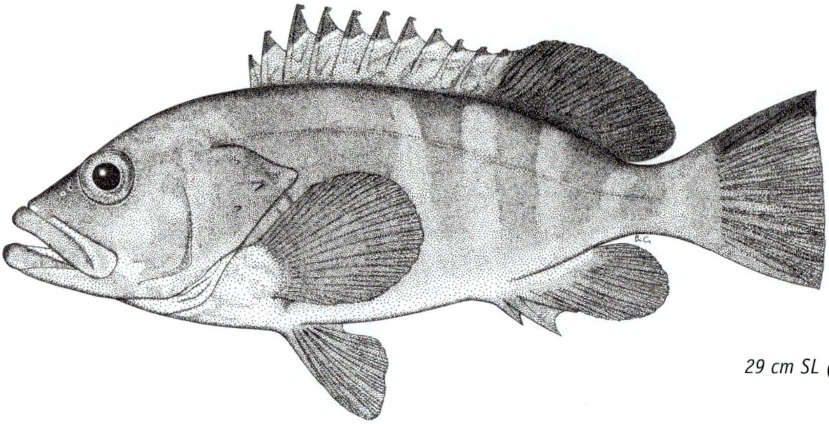

29 cm SL (Taiwan)
© SAIAB

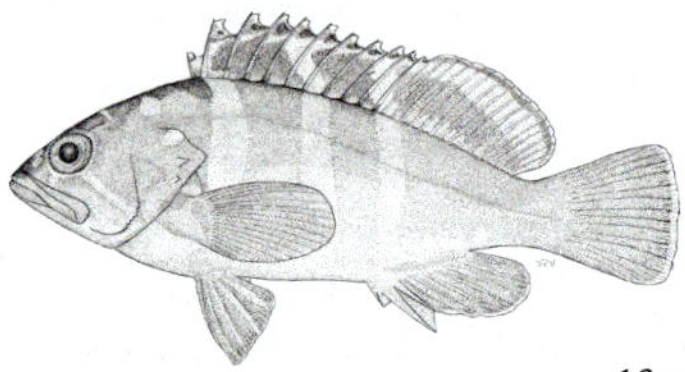

10 cm SL (Mauritius)
© SAIAB

Body depth contained 2.5–3.1 times in standard length. Head length 2.4–2.6 times in standard length; interorbital area flat to slightly convex, the dorsal head profile slightly convex; preopercle rounded, finely serrate; upper edge of operculum sinuous; snout length 3.5–3.9 times in head length; posterior and anterior nostrils subequal in size; maxilla reaches below rear half of eye; midlateral part of lower jaw with 3 or 4 rows of teeth; a pair of short fixed canines at front of both jaws. Gill rakers 6–8 on upper limb, 15–17 on lower limb, 21–24 total. Dorsal fin with 11 spines and 16 or 17 rays, the 3rd to 5th spines longest, their length contained 2.4–3.1 times in head length; anal fin with 3 spines and 8 rays; pectoral fins fleshy, with 19 or 20 rays; pectoral-fin length 1.5–1.8 times, pelvic-fin length 1.8–2.2 times in head length; caudal fin truncate to slightly convex. Lateral-body scales ctenoid, with numerous auxiliary scales; lateral-line scales 64–76; lateral-scale series 120–141. Pyloric caeca numerous (more than 40).

MAXIMUM RECORDED SIZE 50 cm TL

Epinephelus retouti is a widespread, primarily insular species known from Madagascar in the west to French Polynesia in the east, extending northwards to southern Japan. It is also known from KwaZulu-Natal (South Africa) and the United Arab Emirates.

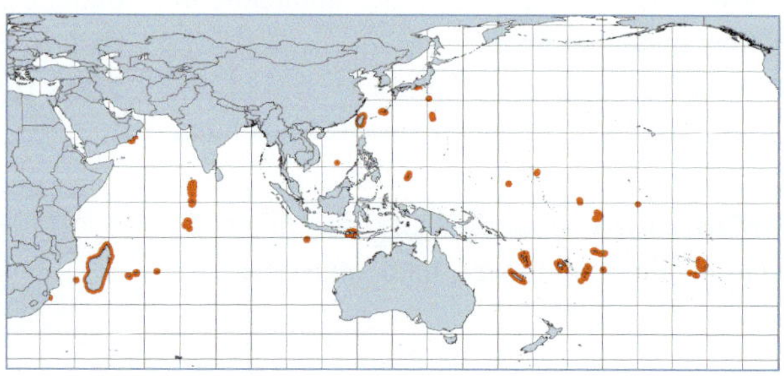

COLOUR IN LIFE Adults dull yellowish orange to brownish red, each scale on the dorsolateral part of the body with a dark greenish grey spot; usually 5 faint dark bars on the body, the 2nd and 3rd bars extend onto the spinous dorsal fin where they are dark brown; dorsal fin greenish brown, the soft-rayed part darker than the rest of the fin, the outer triangular part of the interspinous membranes dark red, set off from the rest of the fin by an orange-yellow stripe; dorsal-fin margin of the caudal fin dark greenish brown; dark red or brown line along the base of the dorsal fin; orbit narrowly edged, except anteriorly, with dark red; pale blue line adjacent to the red rim of the orbit and completely surrounding the eye. Juveniles with the dorsal part of the first 3 dark bars on body black, the 2nd and 3rd bars extending into the dorsal fin; dorsal part of the head also black, with 4 irregular transverse whitish bands across the dorsal surface.

COLOUR IN MARKET Generally dull orange to red; dark bars on the body are faint, more distinct directly below the dorsal fin. Tips of the spinous dorsal fin retain a darker hue.

Adult *E. retouti* (Okinawa Fish Market, Japan)

Juvenile *E. retouti* (locality unknown)

POPULATION **General** *E. retouti* is a rare species throughout its range. **Fishery dependent** *E. retouti* occurs in some deep-water handline fisheries but never constitutes a significant proportion of grouper catches.

POPULATION TREND Unknown

MAJOR THREATS No major threats have been identified for this species.

HABITAT AND ECOLOGY **General** *E. retouti* is a deep-water, mainly insular and reef-associated species. Juveniles occur mainly at depths between 20 and 40 m, and adults from 70 to 220 m on outer reef slopes.

COMMENTS *E. retouti* is similar in appearance to *E. fasciatus*, *E. rivulatus* and *E. irroratus*. *E. fasciatus* differs from *E. retouti* in having a shorter snout (length contained 4.3–5.1 times in head length), fewer lateral-line scales (except for *E. fasciatus* from Oceania, which have high counts like *E. retouti*), the tips of the interspinous dorsal-fin membranes are black (not dark red, except in specimens from Western Australia), the soft dorsal fin and upper caudal-fin rays are not darker than the rest of these fins, and the caudal-fin margin is convex or moderately rounded (except in central Pacific specimens). In *E. rivulatus*, the conspicuous body scale spots are white or pale blue, there is a semicircular red or reddish brown blotch at the base of the pectoral-fin rays and a similar spot anteriorly on the isthmus, and the snout is shorter (length 4.0–5.0 times in head length). In *E. irroratus*, there are no dark bars on the body, the scale spots are white, and the 2nd dorsal-fin spine is distinctly elongated in adults.

2008 IUCN RED LIST STATUS Data Deficient

Authors: *RF Myers, DA Pollard and BM Yeeting*

Epinephelus rivulatus

(Valenciennes in Cuv. and Val. 1830)

halfmoon grouper (E), mérou demi-lune (F), mero medialuna (S)

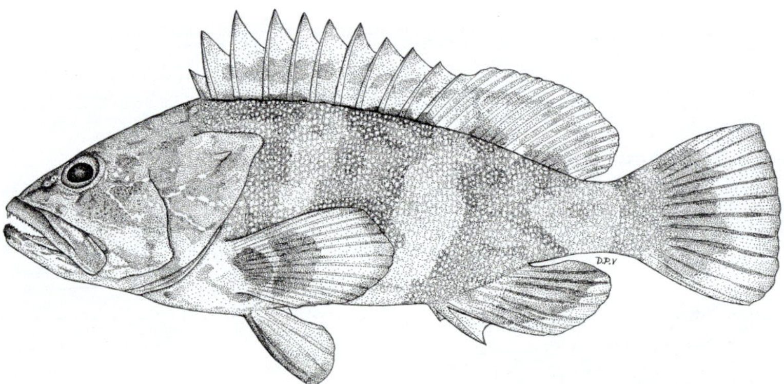

22 cm SL (KwaZulu-Natal, South Africa)

Body depth contained 2.7–3.2 times in standard length. Head length 2.3–2.6 times in standard length; interorbital area flat to slightly convex, the dorsal head profile convex; snout length 4.0–5.0 times in head length; preopercle angular, with enlarged serrae at angle; upper edge of operculum straight, almost horizontal; posterior and anterior nostrils subequal in size; maxilla reaches to or past vertical at rear edge of eye, the ventral edge with a slight step at distal expansion; midlateral part of lower jaw with 2 or 3 rows of teeth. Gill rakers 6–8 on upper limb, 14–16 on lower limb, 20–24 total. Dorsal fin with 11 spines and 16–18 rays, the 3rd or 4th spine longest, its length contained 2.4–3.2 times in head length and subequal to the longest dorsal-fin ray; the interspinous membranes distinctly incised; anal fin with 3 spines and 8 rays; pectoral fins fleshy, with 17–19 rays; pectoral-fin length 1.5–1.9 times, pelvic-fin length 1.9–2.4 times in head length; caudal fin rounded. Lateral-body scales ctenoid, with numerous auxiliary scales; nape and dorsoposterior part of head densely covered with minute auxiliary scales; lateral-line scales 48–53; lateral-scale series 86–98. Pyloric caeca 26.

MAXIMUM RECORDED SIZE 47 cm TL

COLOUR IN LIFE Body generally reddish to greenish brown, with a small white or pale blue spot (about the size of the nostrils) on each scale; 4 irregular dark bars usually visible on the body and extending onto the base of the dorsal fin, and a 5th dark bar on the peduncle, the 3rd and 4th bars run from the soft dorsal fin to the anal fin and are usually joined midlaterally; pectoral fins dusky, with a dark red or reddish brown semicircular blotch on the base of the rays; 2 dark red or reddish brown bands on the thorax and a dark brown spot on the front of the isthmus. Head mostly dark brown or pale reddish, with irregular pale blotches and pale blue or violet vermiculations; 4 white or pale spots along the lower part of the head (2 on the side of the lower jaw and upper lip, 1 just behind the end of the maxilla and the last on the interopercle); fins greenish yellow or greyish brown, the dorsal fin usually with a dark brown to black streak along the base (may be restricted to the base of the spinous part); margin of the interspinous dorsal-fin membranes yellow to reddish next to the spine, becoming transparent posteriorly, with a pale stripe below the triangular part of the fin. Some fish with white dots on the operculum like those on the body. Dark maxillary streak.

COLOUR IN MARKET Body pale tan to brown; small spots remain prominent, and bars fade and are only visible directly beneath the dorsal fin. Maxillary streak present.

Adult *E. rivulatus* (Umkomaas, South Africa)

Adult *E. rivulatus* (Western Australia)

Adult *E. rivulatus* (Okinawa Fish Market, Japan)

Adult *E. rivulatus* (Umkomaas, South Africa)

Epinephelus rivulatus is an Indo-West Pacific species that is distributed from Yemen, Kenya, Zanzibar, Mozambique, South Africa, Comoros Islands, Madagascar, Réunion, Mauritius and Chagos to Indonesia, Philippines, Taiwan, Japan, Australia, New Caledonia and New Zealand. It is not known from the Red Sea, Persian Gulf or Fiji.

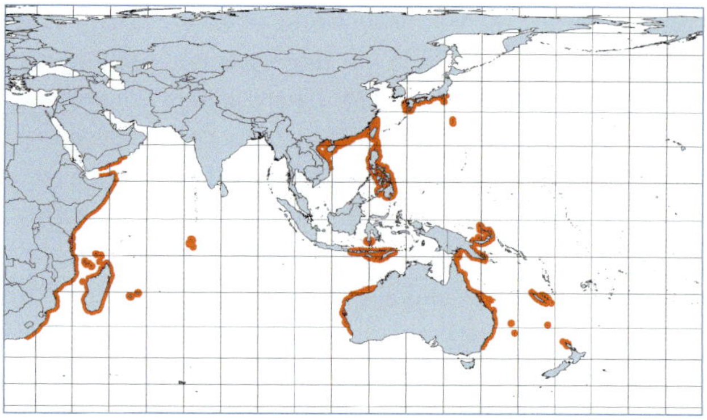

POPULATION **GENERAL** *E. rivulatus* is a widespread and often very common small grouper. **FISHERY DEPENDENT** Experimental catch decreased by about 50% in New Caledonia between 1987 and 2000 (M Kulbicki pers. obs.) and comprised 5% of the grouper catch. Recreational fishing off northern KwaZulu-Natal (South Africa) showed a 40% decline in catch rate over a 15-year period; in Mozambique, catches have been extremely variable over a 12- to 13-year period (M Kulbicki pers. obs.). Despite these reductions, spawner biomass per recruit levels in South Africa remained relatively high (Fennessy 2000).

POPULATION TREND Decreasing

MAJOR THREATS Fishing pressure and habitat loss are the main threats to this species.

HABITAT AND ECOLOGY **GENERAL** *E. rivulatus* is a sedentary reef-associated species that occurs on coral reefs, areas with rocky substrata, algal flats and seagrass beds in depths of up to 30 m. **FEEDING** Feeds on fishes and crustaceans. **AGE AND GROWTH** *E. rivulatus* may reach at least 24 years (Mackie 1998). The von Bertalanffy growth parameters vary depending on region: L_∞ is about 330 mm TL, k is around 0.45 per year, and t_0 is about −0.4 (Mackie 1998, Fennessy 2000). **REPRODUCTION** *E. rivulatus* is a monandric, protogynous hermaphrodite that does not form spawning aggregations. Sex change and maturation appear to vary by location. In Australia, *E. rivulatus* matures at a small size/age, with the smallest mature female recorded at 144 mm TL (one year old) (Mackie 1998); the smallest mature male was 221 mm TL (two years old) (Mackie 1998). In Western Australia, size at sex change occurred between 263 and 315 mm TL and is socially controlled (Mackie 2000, 2003).

COMMENTS *E. rivulatus* is similar in appearance to *E. fasciatus*, *E. irroratus* and *E. retouti,* but it is distinguished from the other members of this complex by the dark reddish blotch at the base of the pectoral fins, the dark bands on the thorax, and the margin of the interspinous dorsal-fin membranes is brown or gold (rather than dark red or black).

2008 IUCN RED LIST STATUS Least Concern

Authors: *ST Fennessy, M Kulbicki, RF Myers and DA Pollard*

Epinephelus sexfasciatus (Valenciennes 1828)

sixbar grouper (E), mérou six raies (F), mero de seis bandas (S)

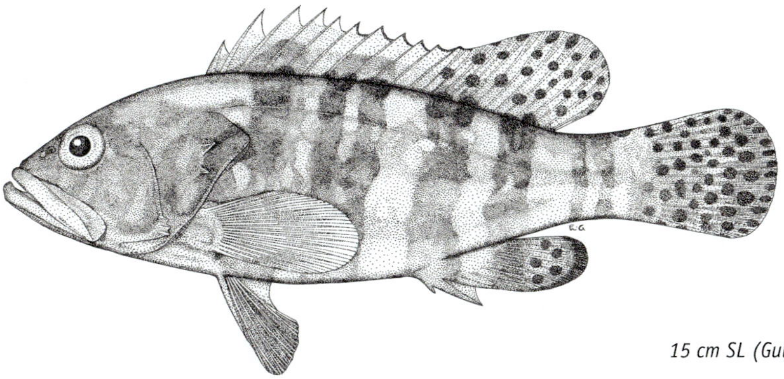

15 cm SL (Gulf of Thailand)
© SAIAB

Body depth contained 2.7–3.2 times in standard length. Head length 2.4–2.6 times in standard length; interorbital area flat or slightly convex, the dorsal head profile convex; preopercle with 2–4 greatly enlarged serrae at the angle; upper edge of operculum straight; posterior and anterior nostrils subequal in size; maxilla reaches to or slightly past vertical at rear edge of eye; midlateral part of lower jaw with 2 rows of subequal teeth. Gill rakers 7 or 8 on upper limb, 13–15 on lower limb, 20–23 total; longest gill raker shorter than longest gill filaments. Dorsal fin with 11 spines and 14–16 rays, the 3rd or 4th spine longest, its length contained 2.3–2.7 times in head length and about equal in length to longest dorsal-fin ray, the interspinous membranes distinctly incised; anal fin with 3 spines and 8 rays; pectoral fins not fleshy, with 17–19 rays; pectoral-fin length 1.5–1.7 times, pelvic-fin length 1.8–2.2 times in head length; caudal fin rounded; caudal-peduncle depth 2.6–3.4 times in head length. Lateral-body scales ctenoid, with a few auxiliary scales; lateral-line scales 46–51; lateral-scale series 82–96. Pyloric caeca 7 or 8.
MAXIMUM RECORDED SIZE 30 cm SL

Epinephelus sexfasciatus is a western Pacific species occurring in Indonesia, Singapore, Malaysia, Spratley Islands, Thailand, Vietnam, Philippines, Papua New Guinea, Solomon Islands, the Arafura Sea and Australia.

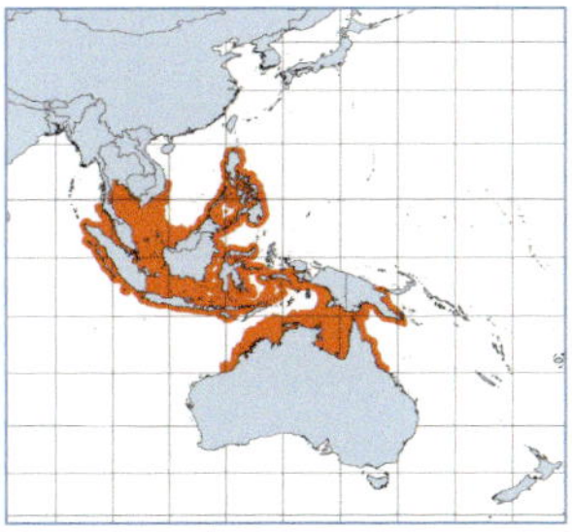

POPULATION **General** *E. sexfasciatus* is a common species usually found over the continental shelf.

POPULATION TREND Unknown

MAJOR THREATS No major threats have been identified for this species.

HABITAT AND ECOLOGY **General** *E. sexfasciatus* is found on silty, soft sand or mud bottoms in depths of 10–80 m. **Feeding** Feeds on fishes and crustaceans. **Reproduction** Little information is available regarding the reproductive biology of *E. sexfasciatus,* but females are thought to mature by 13 cm SL (Heemstra and Randall 1993).

COLOUR IN LIFE Head and body pale greyish brown; 5 dark brown bars on the body and 1 on the nape (dark bars usually more or less divided vertically by a narrow pale bar); scattered pale spots may be present on the body, and some faint small brown spots are often visible on the edges of the dark bars; soft dorsal, caudal and pelvic fins dusky grey, the pectoral fins greyish or dusky orange-red; jaws and ventral parts of the head sometimes pale reddish brown.

COLOUR IN MARKET Body dusky brown to grey. Dark bars usually remain prominent but may be distorted if the specimen is left on the deck of a boat or on the floor of a market.

© Matthew T Craig

Adult *E. sexfasciatus* (Jimbaran Fish Market, Bali, Indonesia)

© Thorke Østergaard

Juvenile *E. sexfasciatus* (Kuala Rompin, Malaysia)

COMMENTS *E. sexfasciatus* is similar in appearance to *E. diacanthus*. The latter species differs from *E. sexfasciatus* in having no spots on the median fins, more scales (lateral-line 52–60; lateral-scale series 103–121), a larger head (length contained 2.2–2.4 times in standard length), and a narrower caudal peduncle (peduncle depth 3.7–4.7 times in head length).

2008 IUCN RED LIST STATUS Data Deficient

Authors: *AS Cabanban, DA Pollard and BC Russell*

Epinephelus socialis (Günther 1873)

surge grouper (E), mérou houleux (F), mero oleado (S)

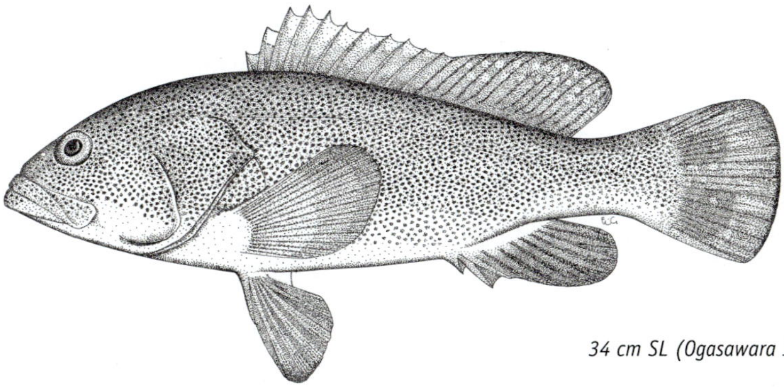

34 cm SL (Ogasawara Islands, Japan)
© SAIAB

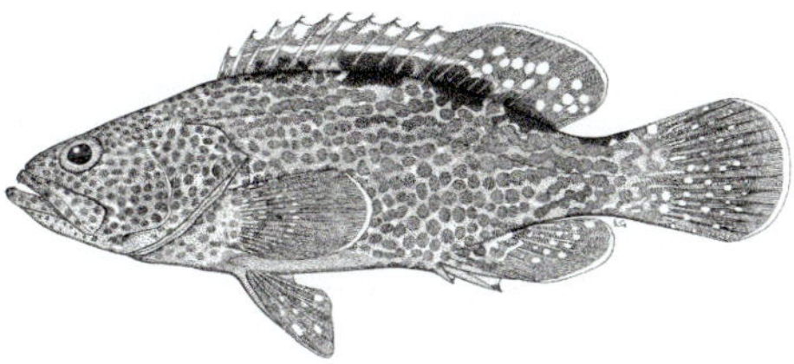

16 cm SL (Society Islands, French Polynesia)
© SAIAB

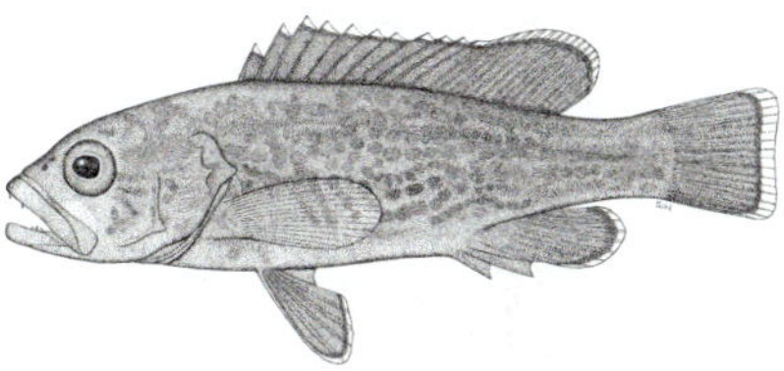

3 cm SL (Rapa, Austral Islands, French Polynesia)
© SAIAB

Body depth contained 2.9–3.4 times in standard length (for fish 10–34 cm SL). Head length 2.4–2.6 times in standard length; interorbital area flat or slightly convex, the dorsal head profile convex; preopercle rounded, finely serrate; upper edge of operculum sinuous; posterior and anterior nostrils subequal in size; maxilla reaches past vertical at rear edge of eye; midlateral part of lower jaw with 3 or 4 rows of teeth. Gill rakers 8–10 on upper limb, 16–20 on lower limb, 25–28 total. Dorsal fin with 11 spines and 14–16 rays, the 3rd or 4th spine longest, its length contained 2.6–3.6 times in head length, the interspinous membranes moderately incised; anal fin with 3 spines and 8 rays; pectoral fins fleshy, with 18 or 19 rays; pectoral-fin length 1.6–1.9 times and pelvic-fin length 1.8–2.2 times in head length; caudal fin rounded. Lateral-body scales of juveniles (less than 20 cm SL) ctenoid; lateral-body scales of subadults (20–30 cm SL) weakly ctenoid, with auxiliary scales; in adults (greater than 30 cm SL) the body scales are smooth (mostly embedded) with numerous auxiliary scales; lateral-line scales 64–70; lateral-scale series 97–111. Pyloric caeca about 20.

MAXIMUM RECORDED SIZE 52 cm TL

POPULATION GENERAL *E. socialis* is an uncommon species throughout its range. There is little information on population size but available information points to low natural abundances and restricted biotope. Populations can be considered as very fragmented. Local extinction is likely around islands, with a strong potential for recolonisation from nearby less populated or disturbed islands.

POPULATION TREND Decreasing

MAJOR THREATS *E. socialis* is threatened by fishing pressure because it is highly accessible, not common and not managed.

COLOUR IN LIFE Head and body whitish, covered (except the ventral part of the head, chest and abdomen) with small close-set blackish brown spots, those on the rear part of the body often coalesced to form irregular longitudinal bands; the dark spots become relatively smaller with growth, and on a fish of 21 cm TL they are black dots about the size of the posterior nostrils; 4 large blackish blotches usually visible on the body at the base of the dorsal fin and a 5th black saddle blotch on the peduncle. Median fins coloured like body basally, becoming dark greyish brown distally, the soft dorsal, caudal and anal fins with small white spots, the margin of these fins white, usually with a black submarginal band; pectoral fins dark greyish brown with a white margin posteriorly and small whitish spots; pelvic fins dark greyish brown, with a white leading edge.

COLOUR IN MARKET Generally white to pale grey body covered with small black spots that remain prominent in death. Organisation of spots into bands may be more prominent than in life.

Adult *E. socialis* (Pagan Island, northern Mariana Islands)

HABITAT AND ECOLOGY **GENERAL** *E. socialis* is found in depths of less than 10 m, mainly in waters of less than 3 m. *E. socialis* is found in a very specific habitat: the shallow part of reef flats exposed to the wave action from the ocean. It is not dependent on coral cover. **FISHERY DEPENDENT** This species is never abundant and it is a favoured target for subsistence fishing over most of its range. The habitat of this species is not very large in most of the islands where it is found and it is highly accessible to fishing due to its shallow occurrence. In parts of its range it is probably not eaten because of risks of ciguatera.

2008 IUCN RED LIST STATUS Near Threatened

Epinephelus socialis is found on coral reef islands from Japan to the Pitcairn Islands.

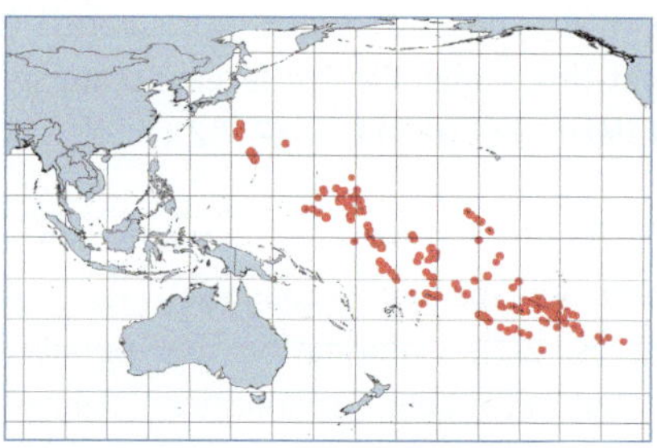

Authors: *M Kulbicki and YJ Sadovy de Mitcheson*

Epinephelus spilotoceps Schultz 1953

foursaddle grouper (E), mérou quatre selles (F), mero cuatro monturas (S)

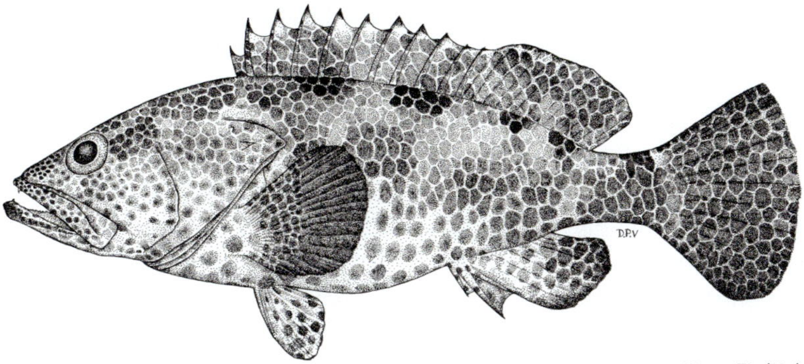

35 cm TL (Mahe, Seychelles)

Body depth contained 3.1–3.6 times in standard length. Head length 2.2–2.6 times in standard length; interorbital area flat or concave; dorsal head profile convex; preopercle rounded, with a shallow notch, below which the serrae are enlarged but covered with skin; upper edge of operculum almost straight; posterior and anterior nostrils subequal in size; maxilla reaches to or beyond vertical at rear edge of eye; midlateral part of lower jaw with 2–4 rows of teeth, lower jaw projecting in front of upper jaw. Gill rakers 6–9 on upper limb, 16–18 on lower limb. Dorsal fin with 11 spines and 14–16 rays, the 3rd or 4th spine longest, its length contained 2.8–3.6 times in head length and distinctly shorter than longest dorsal-fin ray; anal fin with 3 spines and 8 rays, the 2nd spine usually longest, its length 2.4–3.7 times in head length and usually more than peduncle depth; pectoral-fin rays 17–19; pectoral-fin length 21–24% of standard length, 1.7–2.1 times in head length; pelvic-fin length 1.9–2.4 times in head length; caudal-peduncle depth 3.5–4.1 times in head length; caudal fin rounded, middle caudal-fin rays 1.7–2.0 times in head length. Lateral-body scales ctenoid, with auxiliary scales; lateral-line scales 59–69; lateral-scale series 86–100.

MAXIMUM RECORDED SIZE 35 cm TL

Epinephelus spilotoceps has a widespread, primarily insular, tropical Indo-West Pacific distribution, from Zanzibar to Mozambique, Comoros Islands, Seychelles, Mauritius, Chagos and the Maldives, extending east to the Line Islands in the central Pacific.

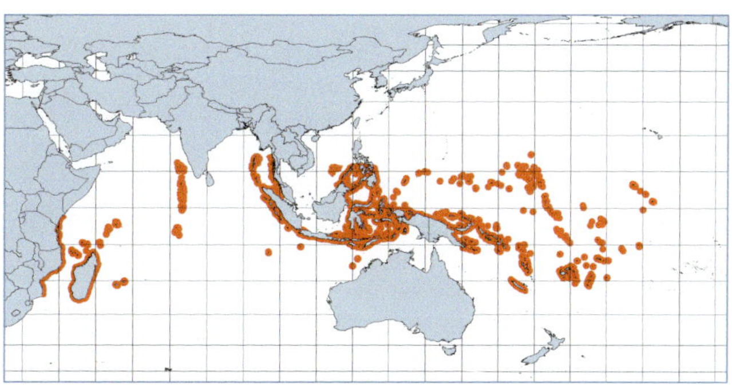

POPULATION GENERAL *E. spilotoceps* is relatively common throughout most of its range. FISHERY DEPENDENT This is one of nine species of related shallow-water coral reef-associated 'reticulated groupers' which are often confused with each other and species-specific population data are unavailable. The species does not appear to be a particular target of fishing, although it

COLOUR IN LIFE Head, body and median fins pale, mostly covered with close-set dark brown, olive or reddish brown spots, those on the dorsolateral parts of the head and body and on the median fins polygonal and close-set, the pale interspaces forming a white reticulum; spots on the ventral part of the head and body more rounded, more separated and often more reddish; large dark brown to black saddle blotch on the body at the base of the last 4 dorsal-fin spines and extending onto the basal part of the fin; 2 similar but smaller dark blotches at the base of the soft dorsal-fin rays and a third blotch on the rear end of the peduncle; these dark saddle blotches uniformly pigmented (on juveniles) or comprising a group of extra dark body spots separated by the white reticulum (on adults); spots on the head progressively smaller and darker anteriorly, those on the snout blackish brown and about the size of the nostrils, with 3 or 4 irregular rows along the front of the upper lip; pectoral fins with close-set reddish brown spots, larger and more distinct towards the base of the pectoral fins, the distal part dull yellowish green; tips of dorsal-fin spines blackish, with a short white filament.

COLOUR IN MARKET Body generally pale or light brown. Pattern of spots remains prominent in death with large dark blotches remaining.

Adult *E. spilotoceps* (Diego Garcia, Chagos Archipelago)

Adult *E. spilotoceps* (Sonsorol, Palau)

is common in many small-scale, multi-species fisheries throughout its range. In 2006, *E. spilotoceps* was the third most common grouper taken in Pohnpei fisheries, and represented 13.1% of the total grouper catch (Rhodes and Tupper 2007). In the Maldives, it is captured for the international live food fish trade (trade centre is Hong Kong), in which it is a common but not particularly important species. In this fishery, it has a (nominal) annual export quota of 8000 individuals (Shakeel and Ahmed 1997, Sattar and Adam 2005).

POPULATION TREND Unknown

MAJOR THREATS No major threats have been identified for this species.

HABITAT AND ECOLOGY **General** *E. spilotoceps* is an insular species inhabiting shallow coral reefs, patch reefs in lagoons, upper slopes of channels and reef margins, to depths of 30 m.

COMMENTS *E. spilotoceps* is part of the reticulated grouper complex. The other species differ from *E. spilotoceps* as follows:

- *E. bilobatus* has 3 bilobed dark brown blotches or close-set pairs of dark spots on body at the base of the dorsal fin but no dark saddle on the peduncle, 17 or 18 dorsal-fin rays, and 48–52 lateral-line scales.
- *E. faveatus* has the lateral-body scales smooth (except for area covered by pectoral fins), 16–18 dorsal-fin rays, 2 oblique dark bands on sides of chest, 2 rows of teeth at sides of lower jaw, and 48–52 lateral-line scales.
- *E. hexagonatus* has conspicuous white dots on the body between the dark spots (white dots not apparent on some preserved specimens) and a large, irregular dark brown or olive spot behind the eye; 2nd anal-fin spine distinctly longer than 3rd, length of 2nd anal-fin spine contained 2.1–2.5 times in head length; length of 5th dorsal-fin spine 2.5–2.8 times in head length; and pelvic-fin length 1.8–2.1 times in head length.
- *E. macrospilos* has the lateral-body scales mostly smooth, pectoral fins dusky with narrow white edge, no large black blotch at the base of the last 4 dorsal-fin spines (although 2 or 3 adjacent spots here may be darker than other brown spots on the body), dark spots on the snout much larger than the nostrils, and 48–52 lateral-line scales.
- *E. maculatus* has the 3rd or 4th dorsal-fin spine longest, its length contained 2.1–2.6 times in head length and usually longer than the dorsal-fin rays, the interspinous dorsal-fin membranes not or only slightly incised; and small juveniles are yellowish brown, with well-separated small black spots (mainly on the head and fins) and a few irregular white spots and blotches on the body and dorsal fin.
- *E. melanostigma* has a single black blotch at the base of the last 4 dorsal-fin spines, dark spots on the snout distinctly larger than the nostrils, and caudal-peduncle depth contained 3.2–3.8 times in head length and usually more than the length of the 2nd anal-fin spine.
- *E. merra* has pectoral fins covered with conspicuous, small black spots largely confined to the rays, no dark saddle blotches at the base of the dorsal fin, and 48–54 lateral-line scales.
- *E. quoyanus* has no black blotches at the base of the dorsal fin (although some brown body spots may be darker here), sides of the chest with 2 oblique dark brown bands, 16–18 dorsal-fin rays, body depth contained 2.8–3.2 times in standard length, caudal-peduncle depth 3.1–3.6 times in head length, and pectoral-fin length 1.2–1.7 times in head length.

2008 IUCN RED LIST STATUS Least Concern

Authors: *AA Bertoncini, JH Choat, MT Craig and DA Pollard*

Epinephelus stictus Randall and Allen 1987

blackdotted grouper (E), mérou points noirs (F), mero punteado negro (S)

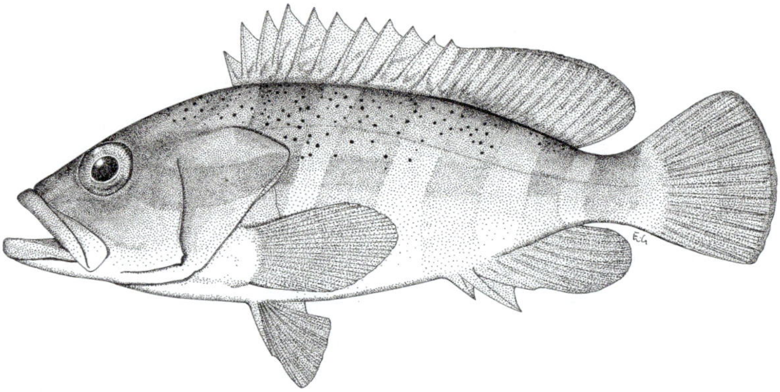

33 cm SL (Western Australia)
© SAIAB

Body depth contained 2.9–3.4 times in standard length. Head large, its length 2.2–2.4 times in standard length; interorbital region flat, the dorsal head profile slightly convex; preopercle with 1–4 distinctly enlarged serrae at the angle; upper edge of operculum straight; posterior and anterior nostrils subequal in size; maxilla reaches to or slightly past vertical at rear edge of eye; midlateral part of lower jaw with 2 rows of teeth. Gill rakers 7 or 8 on upper limb, 14–16 on lower limb. Dorsal fin with 11 spines and 15 or 16 rays, the 3rd or 4th spine longest, its length contained 2.7–3.4 times in head length and shorter than longest rays, the interspinous membranes incised; anal fin with 3 spines and 8 rays; pectoral fins not fleshy, with 18–20 rays; pectoral-fin length 1.7–1.9 times and pelvic-fin length 2.1–2.6 times in head length; caudal fin rounded; caudal-peduncle depth 3.8–4.2 times in head length. Lateral-body scales ctenoid, without auxiliary scales; lateral-line scales 48–51; lateral-scale series 84–96. Pyloric caeca 12.

MAXIMUM RECORDED SIZE 33 cm SL

Epinephelus stictus is known only from southern Japan, Hong Kong, Hainan Island (China), Vietnam, the South China Sea, Java and north-western Australia.

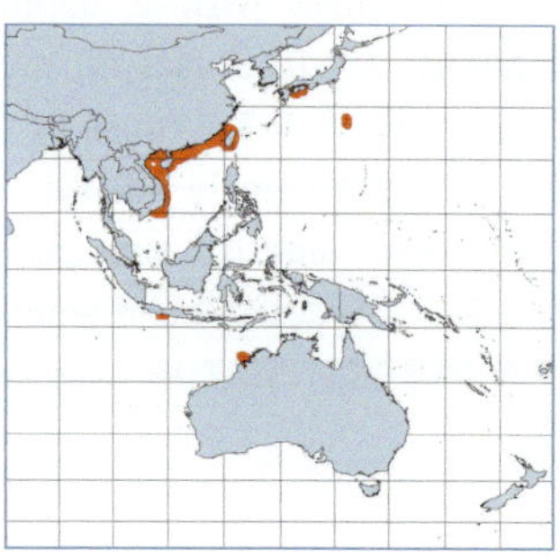

POPULATION **GENERAL** *E. stictus* was not an uncommon species throughout its relatively restricted range historically, but little is known of its current population status. **FISHERY DEPENDENT** Chan (1968) reports that *E. stictus* (reported as *E. diacanthus*) was common in trawl fisheries in Hong Kong but not of much commercial importance at the time. Today, the species is rarely seen in Hong Kong local markets (To 2009) despite the fact that small groupers are now commonly marketed and intensive trawling continues in the area.

POPULATION TREND Unknown

COLOUR IN LIFE Body and head yellowish brown dorsally; belly and rear part of the body whitish ventrally; chest and ventral parts of the head pale reddish orange; body with 5 faint oblique dark bars; midlateral part of each bar may be darker, representing a series of squarish dark blotches which lengthen the dark band running from the eye to the end of the operculum; numerous dark brown to black dots on the dorsolateral parts of the head and body, some concentrated within the dark bars and bands. Fins pale, the rays darker than the membranes; soft dorsal and caudal fins with narrow dark brown margin; pelvic and anal fins often darker than the other fins; anal fin may also have a narrow dark margin; dark line at the base of the dorsal fin (indistinct anteriorly).

COLOUR IN MARKET Upper portion of the body yellowish brown, underside of the body mottled white. Small dark dots are visible on the sides of the body, particularly on the upper, anterior portion. Pale blotches may be present on the entire body.

© Allen W-L To

Adult *E. stictus* (Tai Po Market, Hong Kong)

© Stephen Newman

Adult *E. stictus* (Kimberley coast, Western Australia)

MAJOR THREATS No major threats have been identified for this species.

HABITAT AND ECOLOGY **GENERAL** *E. stictus* is found on mud or sand bottoms in depths of 60–142 m.

COMMENTS *E. stictus* is similar in appearance to *E. diacanthus* of the northern Indian Ocean. *E. diacanthus* differs from *E. stictus* in lacking black dots on the body and in having 52–60 lateral-line scales, 103–121 lateral-scale series, and more gill rakers (8–10 on the upper limb, 15–17 on the lower limb).

2008 IUCN RED LIST STATUS Least Concern

Authors: *KE Carpenter, PC Heemstra, DA Pollard and BC Russell*

Epinephelus stoliczkae (Day 1875)

epaulette grouper (E), mérou epaulette (F), mero hombrero (S)

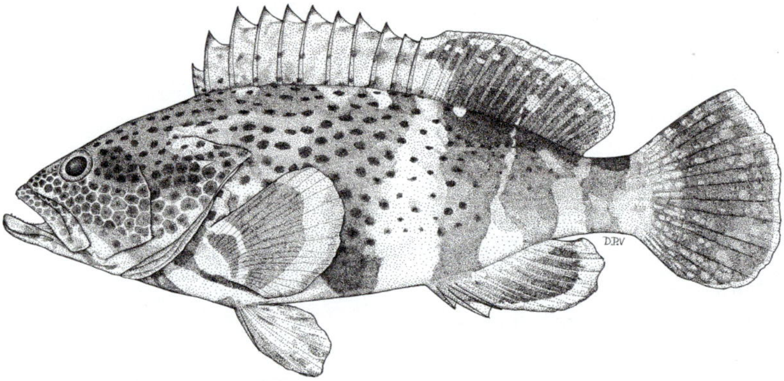

20 cm SL (Oman)

Body robust, the depth contained 2.8–3.3 times in standard length. Head length 2.3–2.6 times in standard length; interorbital slightly to moderately convex; serrae at angle of preopercle moderately enlarged; upper edge of operculum almost straight; nostrils subequal or posterior nostrils slightly larger; maxilla reaches past vertical at rear edge of eye; midlateral part of lower jaw with 2 rows of teeth. Gill rakers 6–8 on upper limb, 13–15 on lower limb, 20–23 total. Dorsal fin with 11 spines and 16–18 rays, the 4th spine usually longest, its length contained 2.4–3.1 times in head length and subequal to longest ray, the interspinous membranes incised; anal fin with 3 spines and 8 rays, the 2nd and 3rd spines subequal; pectoral-fin rays 17–19; pectoral-fin length 1.6–2.0 times and pelvic-fin length 2.0–2.5 times in head length; caudal fin rounded. Body scales smooth except for area covered by pectoral fin; lateral-line scales 48–53; lateral-scale series 93–106.

MAXIMUM RECORDED SIZE 38 cm TL

Epinephelus stoliczkae is known from the Red Sea (including the Gulf of Suez) and the north-western Indian Ocean from Yemen to the coast of Pakistan.

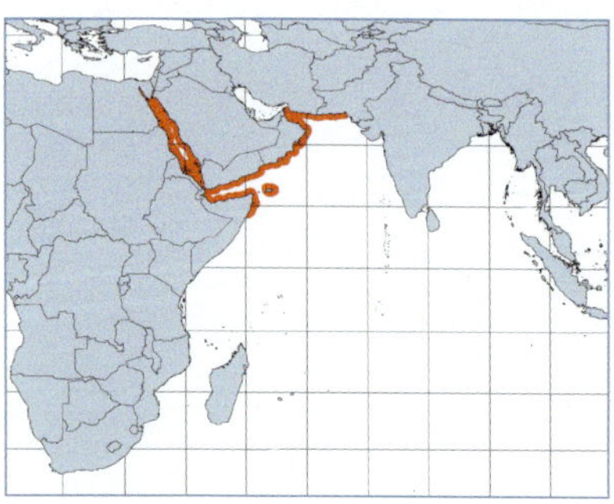

POPULATION GENERAL *E. stoliczkae* is a common species throughout its range.

POPULATION TREND Unknown

MAJOR THREATS No major threats have been identified for this species.

HABITAT AND ECOLOGY GENERAL *E. stoliczkae* inhabits coral reefs in depths of 5–50 m. It was very common during dive surveys carried out on inshore reefs (few corals, mainly dominated by red algae, ascidians and sponges) in north-eastern Somalia (Indian Ocean) in 1998 (Mann and Fielding 2000).

2008 IUCN RED LIST STATUS Data Deficient

COLOUR IN LIFE Head and body yellowish grey, with dark orange-red or reddish brown spots except ventrally and posteriorly; a dark grey bar below the posterior dorsal-fin spines, 2 more dark bars below the soft dorsal fin and another on the caudal peduncle (forming a black saddle dorsally); pectoral-fin base pale, with a dark oval or semicircular blotch across the base, central part of the fin dark brownish grey, distal third of the fin yellowish; chest pale, with dark bands; spinous dorsal fin yellowish, with a row of dark red spots along the base and 2 faint dark longitudinal bands; other fins dark yellowish grey-brown; median fins with a broad yellowish margin posteriorly.

COLOUR IN MARKET Body generally pale tan to grey. Spots on the anterior portion of the body may fade but remain visible. Bars on the posterior portion of the body are prominent. There is often a distinct whitish division between the spotted portion and non-spotted portion of the body in death.

© John E Randall

Adult *E. stoliczkae* (Oman)

© Klaus Fiedler

Adult *E. stoliczkae* (Oman)

Authors: *JH Choat, JL McIlwain, DA Pollard and MA Samoilys*

Epinephelus striatus (Bloch 1792)

Nassau grouper (E), mérou rayé (F), cherna criolla (S)

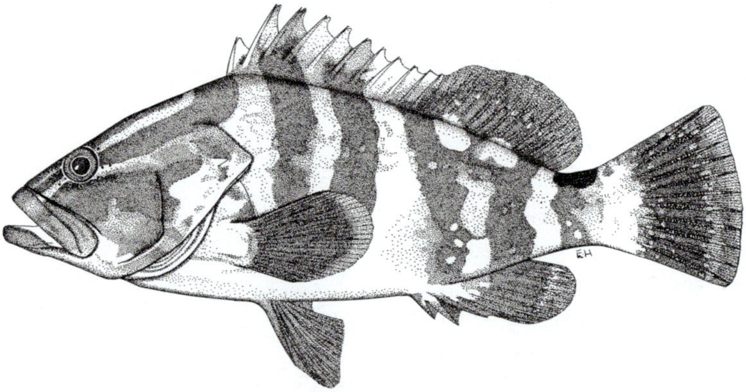

'Normal' barred pattern, 35 cm SL (Virgin Islands)
© SAIAB

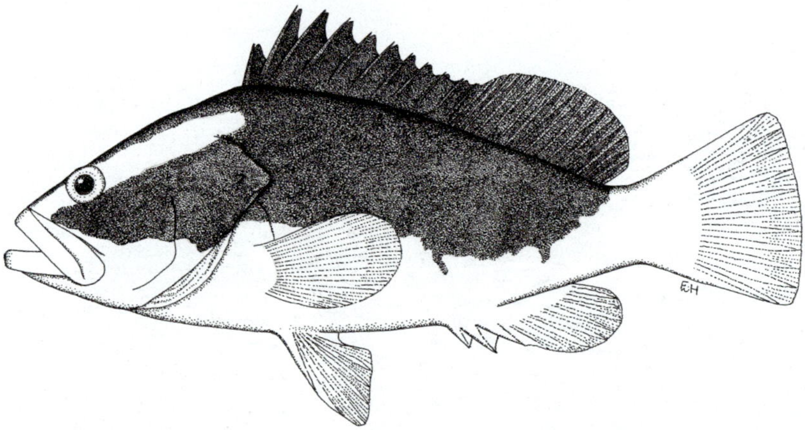

Bicoloured pattern, 35 cm SL (Virgin Islands)
© SAIAB

Body depth distinctly less than head length, depth contained 2.6–2.9 times in standard length (for fish 16–33 cm SL). Head length 2.4–2.6 times in standard length; interorbital convex; preopercle evenly serrate, without salient angle; posterior nostrils somewhat enlarged and elongated or comma-shaped in large adults. Gill rakers 8 or 9 on upper limb, 15–17 on lower limb, 23–26 total. Dorsal fin with 11 spines and 16–18 rays, the 3rd or 4th spine longest and the interspinous membranes distinctly incised; anal fin with 3 spines and 8 rays; pectoral-fin rays 17–19; caudal fin rounded in juveniles, convex in adults. Scales ctenoid, about 50 lateral-line scales and 98–108 lateral-scale series.

MAXIMUM RECORDED SIZE 122 cm TL

COLOUR IN LIFE Ground colour generally buff, with 5 dark brown vertical bars and a large black saddle blotch on top of the caudal peduncle; a row of black dots below and behind the eye. Distinctive dark tuning-fork mark beginning at the front of the upper jaw, extending dorsally along the interorbital region, and bifurcating on top of the head behind the eyes; another dark band from the tip of the snout through the eye and then curving upward to meet its fellow just before the dorsal-fin origin. Some fish have irregular pale spots and blotches all over the head and body; specimens from deep water are somewhat pinkish or reddish ventrally. May also display a bicoloured spawning pattern, with the dorsolateral parts of the head dark brown or blackish, the ventral parts, caudal peduncle and all fins except the dorsal are white, and there is an oblique white band from the tip of the snout through the eye, continued posteriorly towards the dorsal fin. A temporary all-dark colour phase occurs in females during the spawning aggregation and several other temporary colour phases have been described (Sadovy and Eklund 1999).

COLOUR IN MARKET *E. striatus* maintains very similar colour patterns in death as in life. The black saddle on the caudal peduncle remains prominent.

© Matthew T Craig

Adult *E. striatus* (Dry Tortugas, Florida, USA)

© Robert F Myers

Adult *E. striatus* (Hogsty Reef, Bahamas)

POPULATION **General** *E. striatus* was once considered a common species throughout its range. It has, however, shown substantial declines in abundance of up to 60% over the past three decades throughout most of its distribution, possible exceptions being the Turks and Caicos Islands and parts of the Bahamas (IUCN Red List Assessment 2003). **Fishery dependent** The species was once the most important commercial fish in the islands of the Caribbean. It is fished commercially and recreationally by handlines, longlines, fish traps, spearguns and gillnets. Spawning aggregations are heavily targeted, often yield most of the annual catches and are mainly exploited by handlines or fish traps, although gillnets are also used in Mexico and Cuba (IUCN Red List Assessment 2003). Fishery-dependent data in the USA showed marked declines and severe fluctuations in catch rates through the 1980s until a ban on capture of this species was enacted in 1990 (NMFS database). *E. striatus* is variously protected in Belize, the Bahamas, Bermuda, Cayman Islands, Cuba, Puerto Rico, United States Virgin Islands and Mexico following major declines in fisheries and significant losses of spawning aggregations in most places; however, enforcement is wanting in most locations (Sadovy and Eklund 1999, Sala et al. 2001, Bush et al. 2006, Gibson 2007, Aguilar-Perera et al. 2008, Sadovy de Mitcheson et al. 2008, Claro et al. 2009, www.scrfa.org database). There is a need for both national and regional management for recovery of this threatened species (Sadovy and Cornish 2006, García-Moliner and Sadovy 2007). **Fishery independent** The first scientific report of aggregating in this species estimated between 30 000 and 100 000 animals present at one Bahamas site based on direct observation (Smith 1972). The population size of *E. striatus* has been estimated at less than 10 000 mature individuals (NMFS database). **Genetics** Genetic data indicate high gene flow between samples taken in Florida, Cuba, Belize and the Bahamas (Sedberry et al. 1996). It is thus presumed that *E. striatus* maintains a single, genetically homogeneous population.

Epinephelus striatus is found from Bermuda and Florida (USA), throughout the Bahamas and Caribbean Sea. The previous report of *E. striatus* from the Brazilian coast south of the equator (Heemstra and Randall 1993, distribution map p 237) is unsubstantiated.

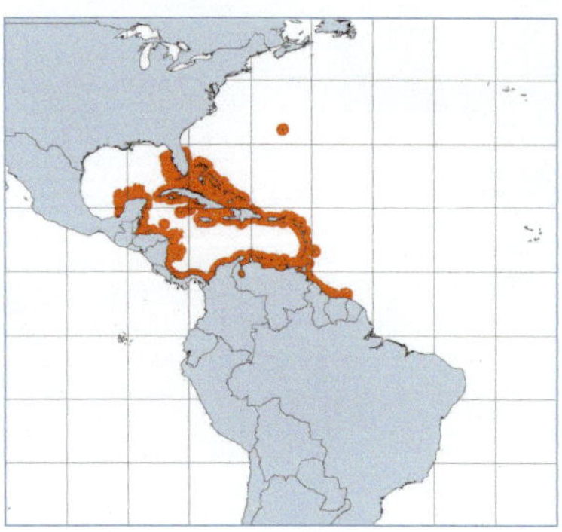

POPULATION TREND Decreasing

MAJOR THREATS Fishing pressure is the greatest threat to this species, particularly on spawning aggregations, while management is absent or not fully effective.

HABITAT AND ECOLOGY **General** *E. striatus* occurs in clear water around high-relief coral reefs or over rocky substrate to depths of at least 130 m and juveniles have been reported to settle inshore in macroalgae (Dahlgren and Eggleston 2001). **Feeding** *E. striatus* feeds on a wide range of fishes and invertebrates (summarised in Sadovy and Eklund 1999). **Age and growth** *E. striatus* may reach least 29 years (Bush et al. 2006); von Bertalanffy growth parameters are: L_∞ = 97.4 mm SL; k ranges from 0.063 to 0.185; log W = –4.67 + 3.03 log L, where W is the ungutted weight in g and L is the standard length in mm; TL (mm) = 2.24 + 1.11 SL (mm) (reviewed in Sadovy and Eklund 1999). **Reproduction** *E. striatus* is a gonochoric species that spawns in large (as much as tens of thousands) aggregations; it is possible that very large females change sex to males and that the species can exhibit diandric protogyny, but few large fish evidently remain due to overfishing; sexual maturity is reached between 40 and 45 cm TL at about 4 years of age (Sadovy and Colin 1995, Sadovy and Eklund 1999). Spawning occurs for a few months each year around the time of the full moon, usually between December and March (Colin 1992), although in Bermuda spawning occurs in the northern summer period (Sadovy and Eklund 1999). Individual fish are known to travel 220 km to spawn (Bolden 2000).

2003 IUCN RED LIST STATUS Endangered (A2ad)

Authors: *AS Cornish and AM Eklund*

Epinephelus suborbitalis

Amaoka and Randall 1990

seamount grouper (E), mérou guyot (F), mero de minami-koho (S)

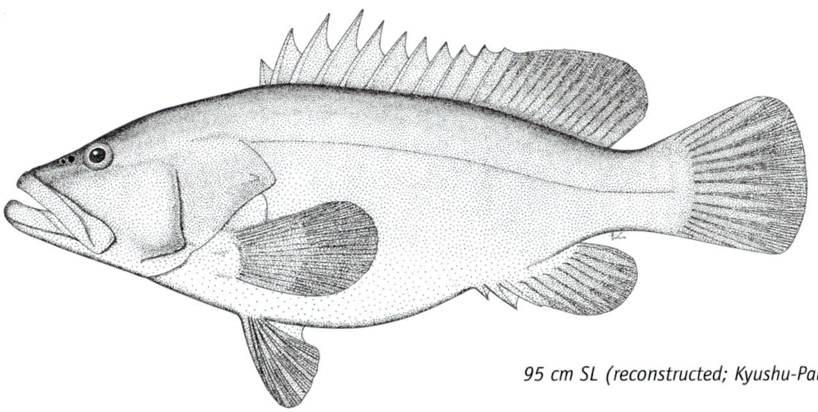

95 cm SL (reconstructed; Kyushu-Palau Ridge)
© SAIAB

Body robust, the depth contained 2.7 times in standard length; body width 2.25 times in the depth. Head length 2.4 times in standard length; interorbital area slightly convex, the width 4.4 times in head length; upper edge of operculum convex; posterior nostril of large adults broadly oval, its diameter about 3 times that of anterior nostril; ventral edge of maxilla with a step-like expansion distally; midlateral part of lower jaw with 3 rows of teeth; inner symphyseal teeth of upper jaw not enlarged. Gill rakers 8 on upper limb, 15 on lower limb. Dorsal fin with 11 spines and 14 rays; the 4th spine longest, its length contained 3.1 times in head length, the interspinous membranes deeply incised; anal fin with 3 spines and 8 rays; pectoral-fin rays 19; caudal-peduncle depth 3.3 times in head length; caudal fin slightly rounded. Lateral-body scales ctenoid, no auxiliary scales on body; lateral-line scales 63; lateral-scale series 122.

MAXIMUM RECORDED SIZE 118 cm TL

COLOUR IN LIFE Known only from a single specimen caught in a trawl. No life colouration is known.

COLOUR IN MARKET When taken from the trawl, the body was uniform greyish brown, the fins darker.

Epinephelus suborbitalis is known only from Kyushu Palau Ridge south of Japan.

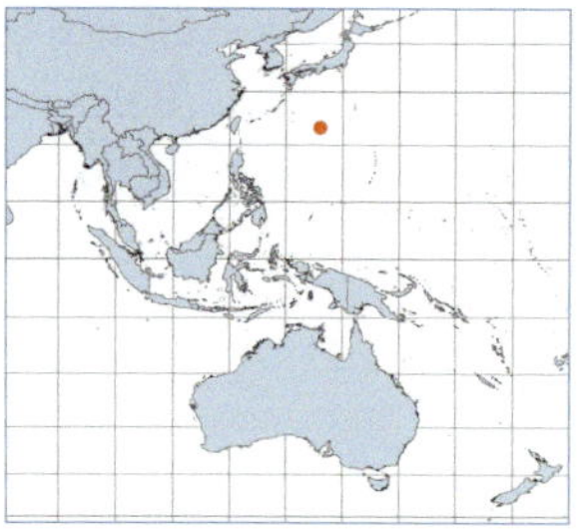

POPULATION General No population data are currently available.
POPULATION TREND Unknown
MAJOR THREATS No major threats have been identified for this species.
HABITAT AND ECOLOGY General *E. suborbitalis* is known only from a single specimen collected in a trawl from Kyushu Palau Ridge south of Japan at a depth of 360 to 570 m.
2008 IUCN RED LIST STATUS Data Deficient

Authors: *PC Heemstra and BC Russell*

Epinephelus summana (Forsskål 1775)

Summan grouper (E), mérou Summan (F), mero sumán (S)

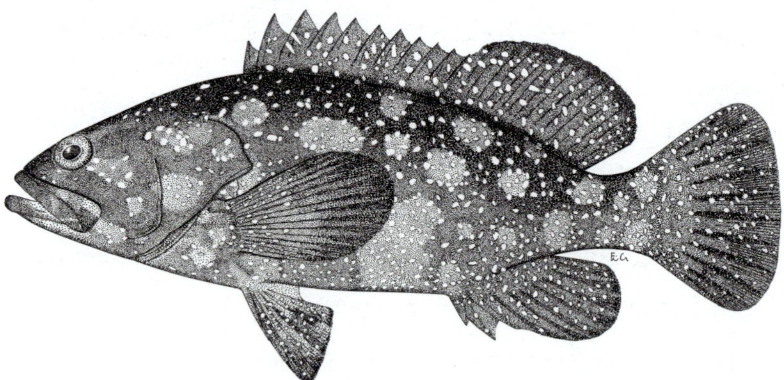

33 cm SL (Sudan)
© SAIAB

Body depth contained 2.7–3.1 times in standard length; body width 1.8–2.3 times in the depth. Head length 2.2–2.6 times in standard length; interorbital area flat or slightly concave; serrae at angle of preopercle slightly enlarged; upper edge of operculum strongly convex; posterior nostril of adults vertically elongate, its length 2–4 times diameter of anterior nostril; maxilla extends to vertical at rear edge of orbit; midlateral part of lower jaw with 2–4 rows of subequal teeth. Gill rakers 8–10 on upper limb, 14–17 on lower limb. Dorsal fin with 11 spines and 14–16 rays, the 3rd or 4th spine longest, its length contained 2.7–3.2 times in head length and subequal to longest dorsal-fin rays; interspinous dorsal-fin membranes incised; anal fin with 3 spines and 8 (rarely 9) rays; pectoral-fin rays 16–18; pectoral-fin length 1.65–2.1 times and pelvic-fin length 2.15–2.7 times in head length; caudal fin rounded. Lateral-body scales ctenoid, with auxiliary scales; lateral-line scales 49–54; lateral-scale series 95–110.
MAXIMUM RECORDED SIZE 52 cm TL

COLOUR IN LIFE Dark olive-brown to dark brownish grey; head and body with large pale botches (most larger than the eyes) and numerous small white spots overlying this pattern; fins covered with small white spots, except for the pectoral fins where the white spots are usually confined to the base; black maxillary streak present; blackish brown blotches are sometimes visible at the base of the dorsal fin and on top of the peduncle. Juveniles dark grey, with large dark-edged white spots of variable size on the head, body and fins; juveniles less than 4 cm SL with irregular black bands across pectoral fins.

COLOUR IN MARKET Generally same as in life but large pale blotches may obscure normal colour pattern.

© John E Randall

Adult *E. summana* (Sudan)

© Klaus Fiedler

Adult *E. summana* (Red Sea)

POPULATION GENERAL *E. summana* is an uncommon species throughout its limited range. FISHERY DEPENDENT FAO statistics report fluctuating annual catch between 18 and 54 t from Saudi Arabia between 2000 and 2004. Generally, the species does not appear to be an important fishery target but information is sparse, often lumped with other grouper catch, and groupers are generally desirable species.

POPULATION TREND Unknown

MAJOR THREATS Fishing pressure is potentially the main threat to this species.

HABITAT AND ECOLOGY GENERAL *E. summana* is a reef-associated species that occurs in shallow, protected coral reefs, shallow lagoons, seaward reef slopes or brackish-water environments in depths of 1–30 m. FEEDING Feeds on fishes and crustaceans.

COMMENTS *E. summana* is similar in appearance to *E. ongus* of the Indo-West Pacific. *E. ongus* has longer pectoral fins (length contained 1.4–1.7 times in head length), longer pelvic fins (length 2.0–2.3 times in head length) and adults do not have the rear nostrils vertically elongated. Also,

Epinephelus summana is a western Indian Ocean species known only from the Red Sea, Gulf of Aden, Socotra and Yemen.

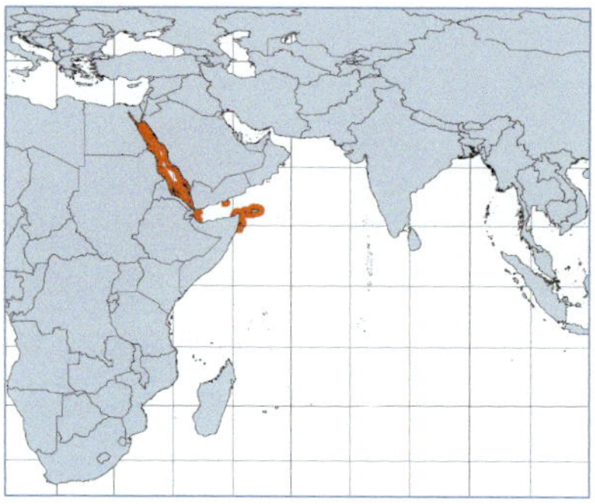

E. ongus has a narrow white margin and broad blackish submarginal band posteriorly on the median fins (these markings are absent or poorly developed in *E. summana*), and the white spots on adult *E. ongus* tend to coalesce posteriorly to form narrow wavy longitudinal stripes.

E. coeruleopunctatus, which does not occur in the Red Sea, is also similar to *E. summana*. It differs in having the upper edge of the operculum straight, sinuous or slightly convex, 17–19 pectoral-fin rays, and it has only a few white spots on the fins.

2008 IUCN RED LIST STATUS Data Deficient

Author: *KL Rhodes*

Epinephelus tauvina (Forsskål 1775)

greasy grouper (E), mérou loutre (F), mero lutra (S)

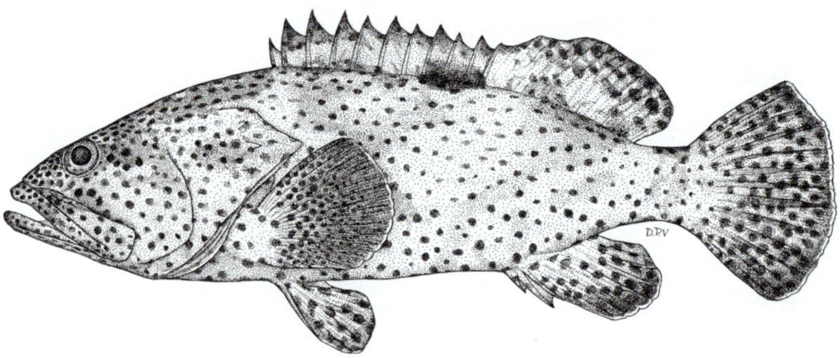

41 cm SL (Sudan)
© SAIAB

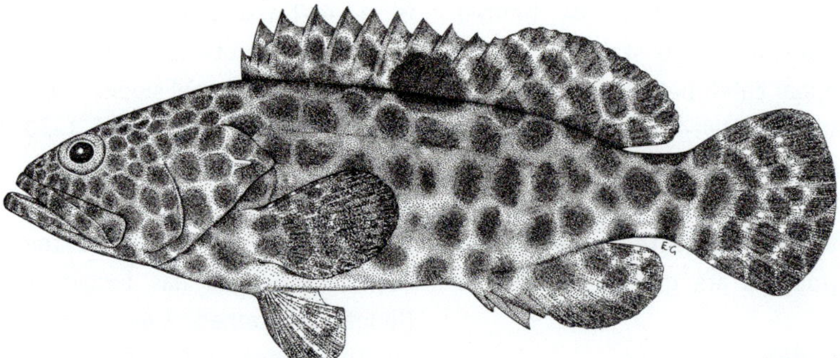

12 cm SL (South Africa)
© SAIAB

Body elongate, the depth contained 3.0–3.6 times in standard length. Head large, its length 2.1–2.4 times in standard length; snout length 2.0–2.4 times in upper-jaw length; interorbital area narrow, flat to slightly concave, interorbital width 6.8–8.1 times in head length and 3.1–4.0 times in upper-jaw length; preopercle broadly rounded (not angular), serrae at angle of preopercle slightly enlarged; upper edge of operculum almost straight; posterior nostrils distinctly larger than anterior nostrils; maxilla reaching well past eye, the greatest width about twice suborbital depth (least distance from eye to maxilla), maxilla width 6.8–8.1% of standard length; upper-jaw length 21–24% of standard length, midlateral part of lower jaw with 2–5 rows of teeth; inner teeth at symphysis of upper jaw are longer than the fixed canines at front of jaw. Gill rakers 8–10 on upper limb, 17–20 on lower limb; no bony platelets on side of gill arch. Dorsal fin with 11 spines and 13–16 rays, the 3rd to 5th spines longest, their length contained 3.1–4.7 times in head length and distinctly shorter than longest dorsal-fin rays; interspinous dorsal-fin membranes incised; anal fin with 3 spines and 8 rays; pectoral-fin rays 18–19; pectoral-fin length 1.7–2.4 times and pelvic-fin length 2.2–2.8 times in head length; caudal fin rounded. Midlateral-body scales ctenoid in juveniles; scales smooth in adults, except for small patch covered by pectoral fin; lateral-line scales 63–74; lateral-scale series 95–112. Pyloric caeca 16–18.

MAXIMUM RECORDED SIZE 75 cm TL

Epinephelus tauvina is known from the Red Sea and the East African coast (South Africa to Kenya), Comoros Islands, Madagascar, Mascarene Islands, Seychelles, Chagos, Maldives to Australia, extending east to the Pitcairn group, north to the Ryukyu Islands (Japan), and south to Lord Howe Island (Australia).

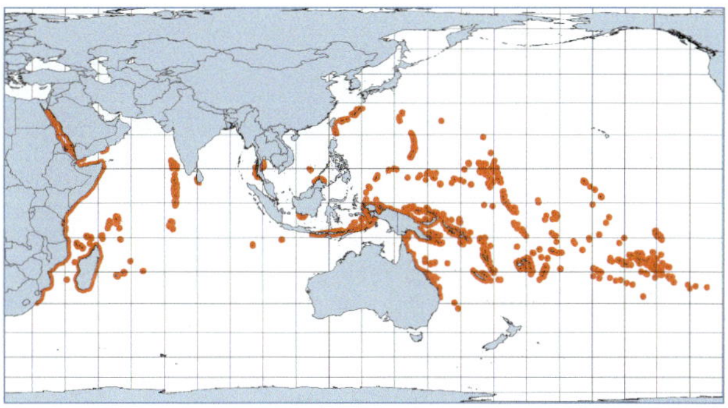

POPULATION **General** *E. tauvina* is considered common in some parts of its range. However, due to confusion with similar species, few explicit data are available on population and catch data. **Fishery dependent** *E. tauvina* is important to artisanal and recreational fisheries. It is occasionally found in the live reef food fish trade in Hong Kong (Lee and Sadovy 1998). In southern Mozambique, *E. tauvina* was one of the commercially important fishes caught by hook and line in 1993–2005 (ST Fennessy pers. comm.).

POPULATION TREND Unknown

MAJOR THREATS Fishing pressure in the primary threat to this species.

HABITAT AND ECOLOGY **General** *E. tauvina* prefers clear water areas on coral reefs in depths to 50 m. Juveniles can be found in reef flats and in tide pools. **Feeding** Feeds primarily on fishes and occasional crustaceans.

COMMENTS *E. tauvina* is similar to *E. coioides* and *E. malabaricus*. Juveniles of *E. tauvina* are also sometimes mistaken for the various species of the 'reticulated groupers'.

2008 IUCN RED LIST STATUS Data Deficient

COLOUR IN LIFE Head and body pale greenish grey or brown, covered with roundish dark spots that vary from dull orange-red to dark brown, the centres darker than the edges; spots on the head progressively smaller anteriorly; a large black blotch (or group of black spots) often visible on the body at the base of the last 4 dorsal-fin spines and extending onto the lower part of the fin; 5 faint subvertical dark bars may be present on the body, 4 below dorsal fin and the 5th on the peduncle (these bars may be represented by dusky blotches at the base of the dorsal fin and a dark saddle blotch on the peduncle); fins also covered with dark spots, those on the pectoral fins becoming smaller and less distinct on the distal part of the fin; posterior margin of the caudal, anal and pectoral fins often with a white edge; dark spots on the soft dorsal, caudal and anal fins of juveniles are so close-set that the pale interspaces form a white reticulum.

COLOUR IN MARKET This species maintains nearly identical colouration after death.

© Matthew T Craig

Adult *E. tauvina* (Okinawa Fish Market, Japan)

© Robert F Myers

Adult *E. tauvina* (Red Sea)

Authors: *M Kulbicki, DA Pollard, KL Rhodes and BC Russell*

Epinephelus timorensis Randall and Allen 1987

yellowspotted grouper (E), mérou taches jaunes (F), mero de pintas amarillas (S)

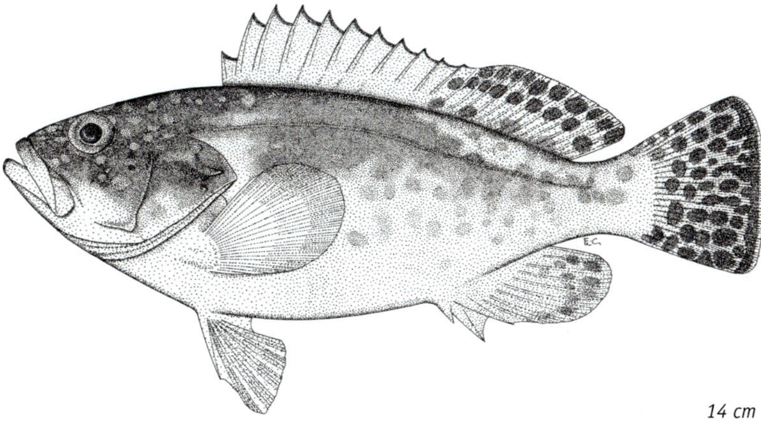

14 cm SL (Timor Sea)
© SAIAB

Body depth contained 2.7–3.4 times in standard length (for fish 14–27 cm SL). Head length 2.4–2.7 times in standard length; interorbital area flat to convex; preopercle angle with 2–4 distinctly enlarged serrae; upper edge of operculum straight or nearly so; posterior and anterior nostrils small, subequal; maxilla extends to below rear half of eye; midlateral part of lower jaw with 2 rows of teeth. Gill rakers 8 or 9 on upper limb, 14–16 on lower limb. Dorsal fin with 11 spines and 16 or 17 rays, the 4th spine longest, its length contained 2.7–3.0 times in head length and subequal to longest dorsal-fin rays, the interspinous membranes incised; anal fin with 3 spines and 8 rays; pectoral-fin rays 17 or 18; pectoral-fin length 1.6–1.8 times in head length; pelvic-fin length 1.8–2.1 times in head length; caudal fin of adults truncate or emarginate, but slightly rounded in juveniles. Lateral-body scales distinctly ctenoid, with auxiliary scales; lateral-line scales 49–53; lateral-scale series 97–108. Pyloric caeca 13 or 14.
MAXIMUM RECORDED SIZE 32 cm TL

Epinephelus timorensis is broadly, but sparsely, distributed across the eastern Indian Ocean and the eastern central Pacific. It is known from American Samoa (Tuitila), the Phoenix Islands (Canton Island), Kiribati, Fiji and limited areas of north-western Australia.

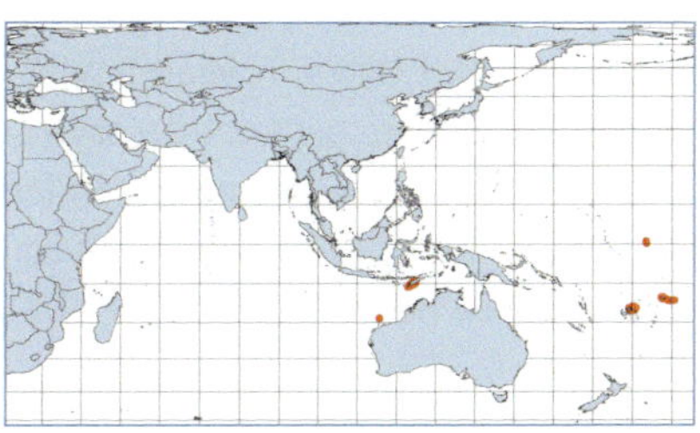

POPULATION **General** *E. timorensis* is considered rare across its range; however, there is a paucity of information likely due to its deep-water habitat. **Fishery dependent** Very little data on the fishery of this species is available. Annual reported catch in American Samoa peaked in the late 1980s at about 200 kg, but dropped to 13 kg by 1996 (PROCFISH database).

POPULATION TREND Unknown

MAJOR THREATS No major threats have been identified for this species.

HABITAT AND ECOLOGY **General** *E. timorensis* is found at depths of 73–210 m.

2008 IUCN RED LIST STATUS Data Deficient

COLOUR IN LIFE Head and body pale brownish grey; yellow or golden brown spots on the head, nape and chest; spots on the snout smaller than those on the operculum; a row of dark yellowish brown spots along the margin of the soft dorsal and anal fins and rear edge of the caudal fin; caudal fin of juveniles covered with close-set dark spots, the pale interspaces forming a white reticulum; adults with 5 faint, irregular, dark bars or series of dark blotches on the body, the first 4 bars extending into the dorsal fin; pectoral-fin rays pale yellowish, the membranes transparent; pelvic fins streaked with yellowish brown and whitish, with a few yellowish brown spots distally.

COLOUR IN MARKET Head and body generally pale tan to grey; spots on the head remain prominent, as does the spotting pattern on the rear edge of the caudal fin. Body bars may fade but usually remain faintly visible and do not disappear altogether.

© Richard C Wass

Adult *E. timorensis* (Samoa Market)

Authors: *PC Heemstra and BC Russell*

Epinephelus trimaculatus (Valenciennes 1828)

threespot grouper (E), mérou trois taches (F), mero de tres manchas (S)

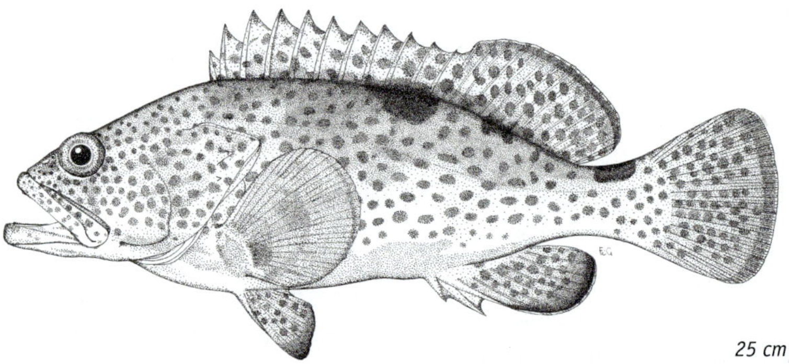

25 cm SL (Taiwan)
© SAIAB

Body depth contained 2.7–3.3 times in standard length. Head length 2.4–2.6 times in standard length; interorbital area slightly convex; preopercle rounded, the serrae mostly hidden by skin; upper edge of operculum straight; posterior and anterior nostrils small, subequal; maxilla reaches to or past vertical at rear edge of eye; midlateral part of lower jaw with 2 rows of teeth. Gill rakers 7–9 on upper limb, 14–16 on lower limb. Dorsal fin with 11 spines and 16 or 17 rays, the 3rd or 4th spine longest, its length contained 2.7–3.5 times in head length and shorter than longest dorsal-fin rays; anal fin with 3 spines and 8 rays; pectoral-fin rays 17 or 18; pectoral-fin length 1.7–2.0 times in head length; pelvic fins not reaching anus, their length 2.0–2.5 times in head length; caudal fin rounded. Midlateral-body scales of adults (28 cm SL) smooth; juveniles with patch of ctenoid scales on area covered by pectoral-fin tip; lateral-line scales 48–52; lateral-scale series 88–100. Pyloric caeca 27–36.

MAXIMUM RECORDED SIZE 50 cm TL

COLOUR IN LIFE Head and body pale brown, covered (except ventrally) with small red or reddish brown spots; median fins darker, but also covered with dark reddish brown spots and often with a narrow white edge; a black blotch larger than the eye occurs on the body at the base of the last 3 dorsal-fin spines, a smaller blotch occurs at the base of the middle dorsal-fin rays and another on the top of the caudal peduncle; paired fins yellowish, with a few small faint orange spots.
COLOUR IN MARKET Body generally pale brown to a light tan; reddish or reddish brown spots remain prominent; median fins are only slightly darker than general body colour. Blotches at the base of the soft dorsal fin remain especially visible.

© Andrew S Cornish

Adult *E. trimaculatus* (Hong Kong)

POPULATION **GENERAL** No explicit information is currently available regarding the population status of this species. **FISHERY DEPENDENT** This species is caught commercially throughout its range using vertical longlines, handlines and fish traps. *E. trimaculatus* is commonly sold in fish markets in Hong Kong and Japan, both as freshly caught fish and as part of the live reef food fish trade. In Hong Kong, over 60% of this species on sale alive are below 30 cm TL and likely to be sexually immature (To and Sadovy de Mitcheson 2009). No information on catch rates or landings is currently available. In South-East Asia, it is also caught extensively as juveniles from the wild and grown-out to market size and can also be produced in hatcheries through full-cycle aquaculture. **FISHERY INDEPENDENT** The species is moderately abundant in clear eastern waters of Hong Kong and outer islands (Sadovy and Cornish 2000).
POPULATION TREND Decreasing
MAJOR THREATS No major threats have been identified for this species.
HABITAT AND ECOLOGY **GENERAL** *E. trimaculatus* is generally reef-associated, with juveniles being common in tide pools and in shallow clear water around rocks and coral reefs; adults are generally found in deeper water to at least 30 m. **REPRODUCTION** Sex reversal and artificial spawning have been experimentally induced by Kuo et al. (1988), but no information on its reproductive biology in the wild is currently available.
2008 IUCN RED LIST STATUS Least Concern

Epinephelus trimaculatus is a subtropical species occurring in coastal East Asian seas, from southern China to southern South Korea and southern Japan.

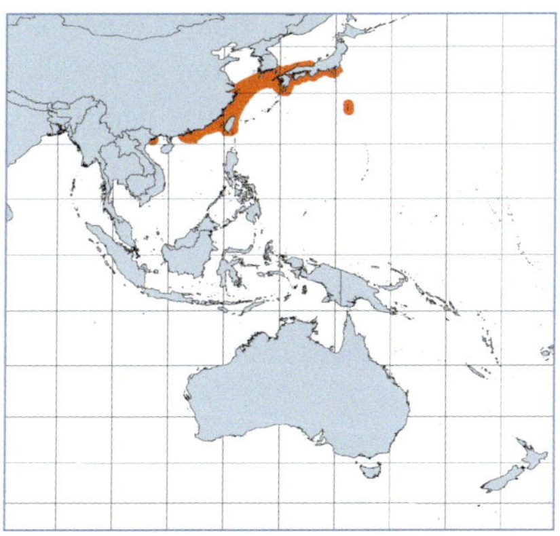

Authors: *JH Choat, DA Pollard, YJ Sadovy de Mitcheson and W-L To*

Epinephelus trophis Randall and Allen 1987

plump grouper (E), mérou rondelet (F), mero rollizo (S)

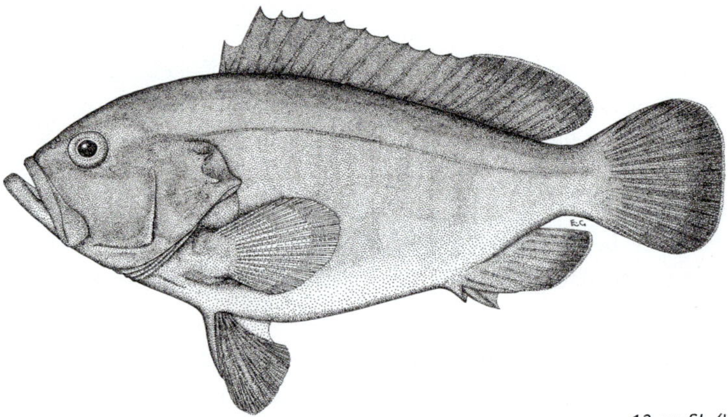

13 cm SL (Western Australia)
© SAIAB

Body depth subequal to head length, depth contained 2.4 times in standard length. Interorbital area convex; preopercle angle with 2 or 3 enlarged serrae; upper edge of operculum convex; posterior and anterior nostrils small, subequal; maxilla reaches vertical at rear edge of orbit; midlateral part of lower jaw with 2 rows of teeth. Gill rakers 10 on upper limb, 15 or 16 on lower limb. Dorsal fin with 11 spines and 16 or 17 rays, the 4th spine longest, its length contained 2.8–2.9 times in head length and shorter than longest dorsal-fin rays, the interspinous membranes slightly incised; anal fin with 3 spines and 8 rays; pectoral-fin rays 17 or 18; pelvic fins not reaching anus; pectoral fins subequal to pelvic fins, pectoral-fin length 1.3–2.0 times in head length; caudal fin convex. Lateral-body scales ctenoid, no auxiliary scales; lateral-line scales 67–69; lateral-scale series 143–145.

MAXIMUM RECORDED SIZE 15 cm TL

Epinephelus trophis is known only from Dillon Shoals (~11° S, ~125° E) in the Timor Sea off north-western Western Australia.

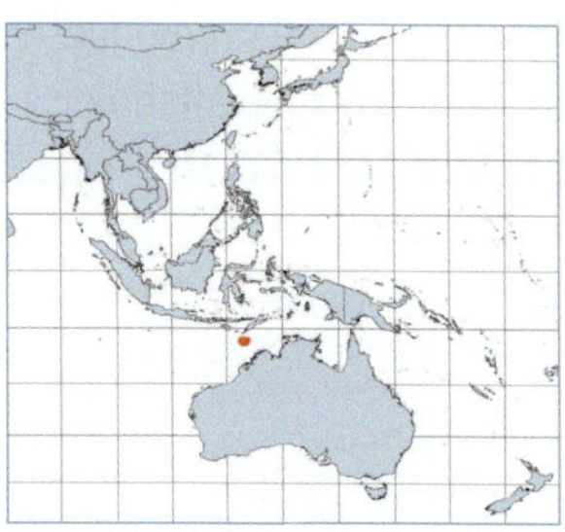

POPULATION GENERAL *E. trophis* is only known from two specimens collected at the base of an oil platform in 130 m water depth.

POPULATION TREND Unknown

MAJOR THREATS No major threats have been identified for this species.

HABITAT AND ECOLOGY GENERAL Known only from two specimens collected in ~130 m depth.

COMMENTS *E. trophis* is very similar to *E. flavocaeruleus*. Comparing *E. flavocaeruleus* with the type specimens of *E. trophis*, it appears that the caudal fin is more truncate in *E. flavocaeruleus* (although it is convex when spread open in a 112 mm fish), the caudal peduncle is slightly longer (length 17–20% of standard length vs 16% of standard length in *E. trophis*) and the pelvic fins reach to or within 3 mm of the anus (8 mm from anus in a 126 mm *E. trophis*). *E. trophis* and *E. flavocaeruleus* differ notably in colour; juveniles of *E. flavocaeruleus* have the fins and jaws bright yellow.

2008 IUCN RED LIST STATUS Data Deficient

COLOUR IN LIFE Known only from two specimens and no life colour is known.

COLOUR IN MARKET Head, body and median fins dark charcoal grey; pectoral fins pale; pelvic fins blackish.

Authors: *AA Bertoncini, JH Choat, MT Craig and DA Pollard*

Epinephelus tuamotuensis Fourmanoir 1971

reticulate grouper (E), mérou réseau (F), mero reticular (S)

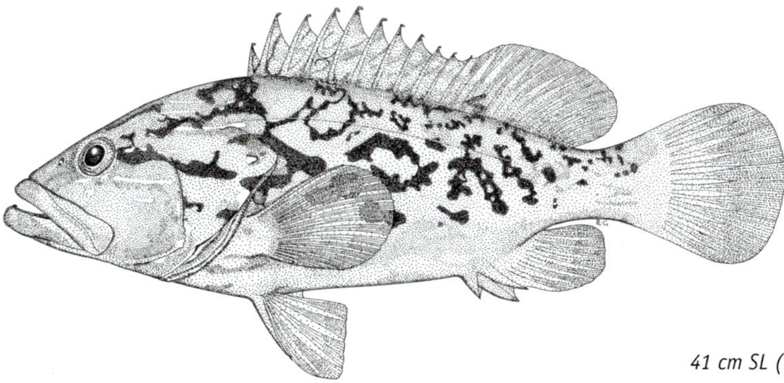

41 cm SL (Pitcairn Island)
© SAIAB

Body depth contained 2.7–2.9 times in standard length. Head length 2.2–2.4 times in standard length; interorbital area convex; preopercle subangular, with 3 or 4 distinctly enlarged serrae at the angle; upper edge of operculum straight or nearly so; posterior nostrils larger than anterior ones; maxilla reaches almost to or slightly past vertical at rear edge of eye; midlateral part of lower jaw with 2 rows of teeth, the inner teeth larger; canine teeth at front of jaws well developed. Gill rakers 9–11 on upper limb, 16 or 17 on lower limb. Dorsal fin with 11 spines and 15 rays, the 4th or 5th spine longest, its length contained 3.0–3.3 times in head length and almost equal to longest dorsal-fin ray; interspinous dorsal-fin membranes deeply incised; anal fin with 3 spines and 8 rays; pectoral-fin rays 17, the fins not fleshy; pectoral-fin length 1.9–2.1 times in head length; pelvic fins not reaching anus, their length 2.2–2.4 times in head length; caudal fin rounded. Lateral-body scales ctenoid, a few auxiliary scales on adults (40 cm SL); lateral-line scales 60–64; lateral-scale series 143–145.

MAXIMUM RECORDED SIZE 66 cm SL

COLOUR IN LIFE Pale yellowish brown with an irregular coarse reticulum of dark brown on the body and rear part of the head (some sections of the network are broken, and there may be small pale spots with the dark bands); fins unmarked.

COLOUR IN MARKET Generally pale brown to a light tan; reticulated pattern remains prominent, although it may be interrupted by patches of light brown if the fish has been left on the deck of a boat or on the floor of a fish market.

Adult *E. tuamotuensis* (Tuamotu Archipelago, French Polynesia)

Epinephelus tuamotuensis is known only from French Polynesia, the Pitcairn group and Raratonga in the Cook Islands.

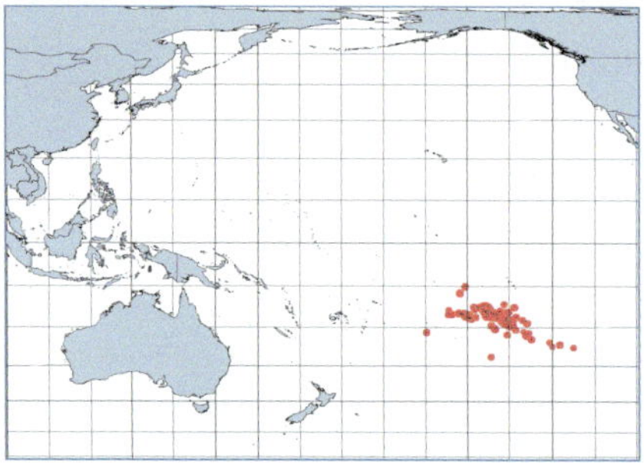

POPULATION GENERAL No information on the population of *E. tuamotuensis* is currently available.

POPULATION TREND Unknown

MAJOR THREATS No major threats have been identified for this species.

HABITAT AND ECOLOGY GENERAL *E. tuamotuensis* inhabits the outer slopes of coral reefs in depths of 120–250 m.

COMMENTS *E. tuamotuensis* is similar in appearance to *E. morrhua*, *E. poecilonotus* and *E. radiatus*. All four species may be distinguished by their colour patterns.

2008 IUCN RED LIST STATUS Least Concern

Authors: *PC Heemstra, M Kulbicki and KL Rhodes*

Epinephelus tukula Morgans 1959

potato grouper (E), mérou patate (F), mero patata (S)

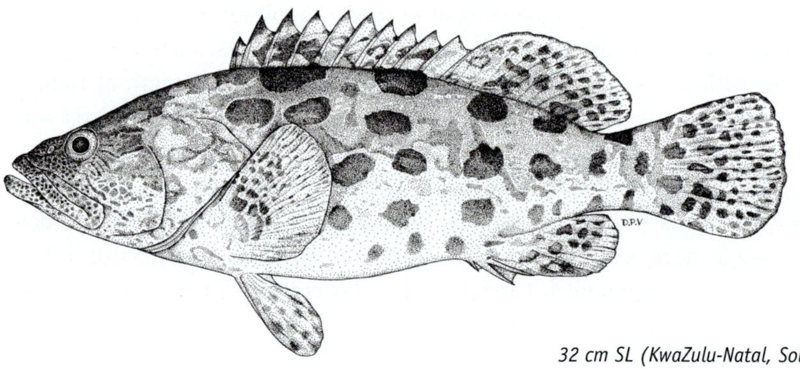

32 cm SL (KwaZulu-Natal, South Africa)
© SAIAB

Body depth contained 2.9–3.5 times in standard length. Head length 2.3–2.6 times in standard length; interorbital area slightly convex; the dorsal head profile straight; preopercle rounded or subangular, the serrae at the angle slightly enlarged; upper edge of operculum almost straight; posterior and anterior nostrils subequal in size; maxilla reaches to or beyond vertical at rear edge of eye; midlateral part of lower jaw with 2–6 rows of teeth. Gill rakers 8–10 on upper limb, 15–18 on lower limb. Dorsal fin with 11 spines and 14 or 15 rays, the 3rd or 4th spine longest, its length contained 3.1–3.8 times in head length and distinctly shorter than longest dorsal-fin rays; interspinous dorsal-fin membranes distinctly incised; anal fin with 3 spines and 8 rays; pectoral-fin rays 18–20; pectoral-fin length 1.6–2.1 times in head length; pelvic fins usually not reaching anus, their length 1.9–2.4 times in head length; caudal fin rounded. Lateral-body scales ctenoid, with auxiliary scales in adults; lateral-line scales 62–70; lateral-scale series 113–130.

MAXIMUM RECORDED SIZE 200 cm TL

COLOUR IN LIFE Body pale brownish grey with several large, dark brown to black, widely spaced blotches, mostly larger than the eye and varying in shape from round to oval or dumbbell-shaped; head with smaller dark brown spots and streaks (many radiating from the eye, especially posteriorly); dark spots on the fins, smaller distally. Large adults may be nearly black.

COLOUR IN MARKET Maintains nearly identical colouration in death, but the blotches eventually fade and become almost invisible.

© Matthew T Craig

Adult *E. tukula* (Okinawa Fish Market, Japan)

© Phillip C Heemstra

Juvenile *E. tukula* (Scottburgh, South Africa)

© Robert F Myers

Adult *E. tukula* (Great Barrier Reef, Australia)

Epinephelus tukula is a widely (but sparsely) distributed Indo-Pacific species that can be found from the Red Sea and the Gulf of Aden to Kenya, Tanzania, Mozambique, South Africa, Comoros Islands, the Seychelles, Oman, Pakistan, India, Sri Lanka, Maldives and Laccadives, as well as from northern Australia (Western Australia and Queensland), Taiwan, Japan and Papua New Guinea.

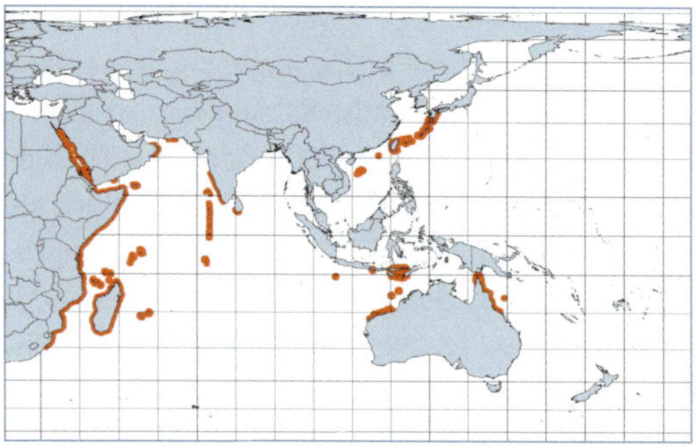

POPULATION **GENERAL** *E. tukula* appears to be a naturally uncommon species throughout its range. **FISHERY DEPENDENT** *E. tukula* is taken opportunistically by fishermen although never in large numbers, and is occasionally seen in the live reef food fish trade in Hong Kong (Lee and Sadovy 1998). **FISHERY INDEPENDENT** Observations by divers indicate that this species is usually solitary and does not reach high abundances.

POPULATION TREND Unknown

MAJOR THREATS Fishing pressure is the primary threat to this species.

HABITAT AND ECOLOGY **GENERAL** *E. tukula* is a territorial, reef-associated species mainly found in deep reef channels, seamounts, and other current-prone areas, in depths of 10–400 m. Juveniles may be found in tide pools. **FEEDING** Feeds on a variety of fishes, including skates, and invertebrates. **AGE AND GROWTH** The maximum age is 26 years (Grandcourt 2005). The length-weight relationship is given as $W = 1.0^{-5} L^{3.07}$ where W is the weight in kg and L is the total length in cm (Grandcourt 2005). *E. tukula* is a slow-growing species: the von Bertalanffy parameters are $k = 0.13$ y^{-1}, $L_\infty = 114.9$ cm TL (t_0 assumed to be zero), natural mortality = 0.16 y^{-1} (Grandcourt 2005). **REPRODUCTION** Yeh et al. (2003) report inducing protogynous sex change in captive specimens, but natural reproductive mode is unknown. *E. tukula* matures at about 90 cm SL (Morgans 1982, Yeh et al. 2003). In East Africa, sexually mature fish have been observed in February, October and December (Nzioka 1977). Localised small aggregations of large fish have been reported anecdotally by recreational SCUBA divers in numerous locations, particularly in areas where fishing effort is low, such as MPAs in northern KwaZulu-Natal, 'Bass City' in southern Mozambique (Pereira 2003), Bassas d'India in the Mozambique channel (Oceanographic Research Institute, South Africa, unpublished data), Aldabra (Grandcourt 2005) and isolated localities in the Seychelles (Robinson et al. 2004). Such aggregations may be for spawning.

2008 IUCN RED LIST STATUS Least Concern

Authors: *ST Fennessy, RF Myers and DA Pollard*

Epinephelus undulatostriatus (Peters 1867)

Maori grouper (E), mérou māori (F), mero maorí (S)

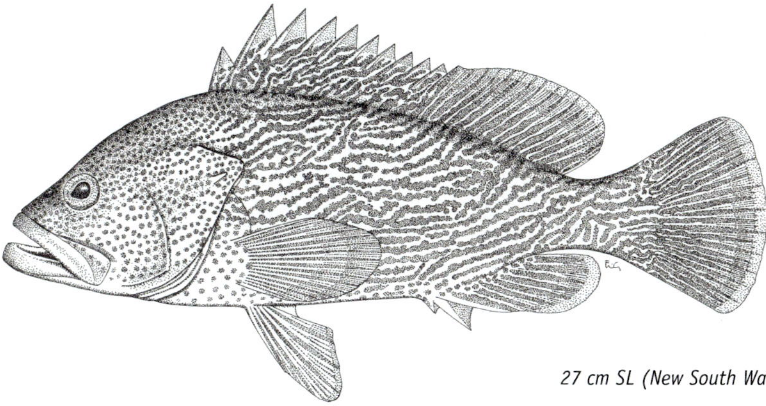

27 cm SL (New South Wales, Australia)
© SAIAB

Body depth contained 2.5–2.9 times in standard length. Head length 2.3–2.5 times in standard length; interorbital area convex, the dorsal head profile almost straight; preopercle subangular, the serrae at angle slightly enlarged; upper edge of operculum almost straight; posterior and anterior nostrils subequal in size; maxilla reaches to or slightly past vertical at rear edge of eye; midlateral part of lower jaw with 2 rows of teeth. Gill rakers 9 or 10 on upper limb, 15 or 16 on lower limb. Dorsal fin with 11 spines and 15–17 rays, the 3rd or 4th spine longest, its length contained 2.2–2.8 times in head length and longer than longest dorsal-fin rays; interspinous dorsal-fin membranes distinctly incised; anal fin with 3 spines and 8 rays; pectoral fins fleshy, with 17–19 rays; pectoral-fin length 1.6–2.0 times in head length; pelvic fins not reaching anus (except in small juveniles), pelvic-fin length 1.8–2.0 times in head length; caudal fin rounded. Lateral-body scales ctenoid, with auxiliary scales; lateral-line scales 48–55; lateral-scale series 98–108. Pyloric caeca 11.

MAXIMUM RECORDED SIZE 65 cm TL

COLOUR IN LIFE Body buff, covered with close-set wavy longitudinal brownish orange, golden brown or reddish brown stripes; some stripes broken into series of small spots; head pale greyish brown, covered with small spots like those on the body; maxillary streak orange-brown; fin rays pale, with small brown spots, the inter-radial membranes dark brown; median fins with narrow yellow margin; pectoral fins diffusely yellow distally.

COLOUR IN MARKET Body colour generally matches that at time of death; close-set wavy lines remain especially prominent. Fins generally lighter in colour than the background body colour.

© John E Randall

Adult *E. undulatostriatus* (Great Barrier Reef, Australia)

Epinephelus undulatostriatus is endemic to Australia, from southern Queensland to Bateman's Bay, New South Wales.

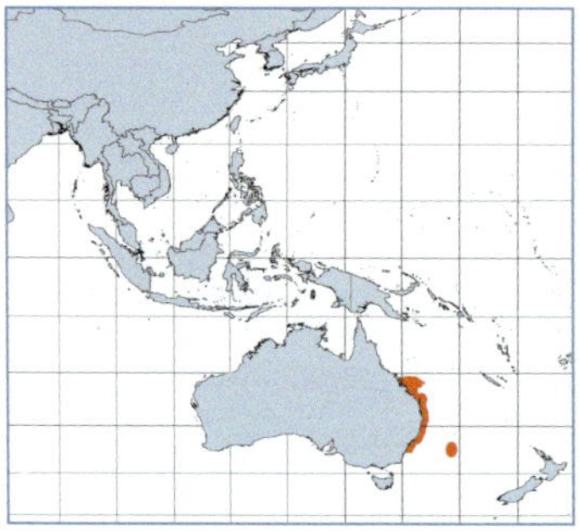

POPULATION GENERAL *E. undulatostriatus* is a common species throughout its range and is a popular sport fish species.

POPULATION TREND Unknown

MAJOR THREATS No major threats have been identified for this species.

HABITAT AND ECOLOGY GENERAL *E. undulatostriatus* is a reef-associated species that inhabits coral reefs and rocky areas in depths of 27–80 m.

2008 IUCN RED LIST STATUS Data Deficient

Authors: *DA Pollard, KL Rhodes and BC Russell*

Epinephelus undulosus

(Quoy and Gaimard 1824)

wavy-lined grouper (E), mérou nodule (F), mero ondulado (S)

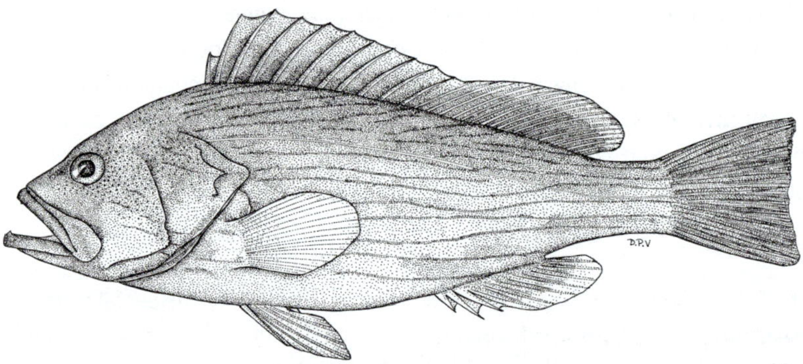

30 cm SL (India)

Body depth contained 2.7–3.1 times in standard length. Head length 2.5–2.7 times in standard length; interorbital area convex; preopercle angular, with a notch above the angle and the serrae at the angle enlarged; upper edge of operculum straight or slightly concave; posterior and anterior nostrils subequal or the posterior ones slightly larger; maxilla reaches to vertical at rear edge of eye; adults with a prominent knob or hook-like process on ventral edge of maxilla (covered by upper lip); midlateral part of lower jaw with 2 rows of teeth. Gill rakers longer and more numerous than other species of *Epinephelus*, 12–16 on upper limb, 20–23 on lower limb, 32–38 total; 1 or 2 rudimentary rakers in juveniles, but none in adults. Dorsal fin with 11 spines and 17–19 rays, the 3rd or 4th spine longest, its length contained 2.5–3.4 times in head length and longer than longest dorsal-fin ray; interspinous membranes of dorsal fin not incised;

anal fin with 3 spines and 8 rays; pectoral-fin rays 18 or 19; pectoral fins shorter than pelvic fins, pectoral-fin length 1.7–2.3 times in head length; pelvic fins not reaching anus (except for juveniles), pelvic-fin length 1.6–2.2 times in head length; caudal fin truncate to emarginate. Lateral-body scales distinctly ctenoid, with numerous auxiliary scales; lateral-line scales 63–76; lateral-scale series 124–150.

MAXIMUM RECORDED SIZE 120 cm TL

COLOUR IN LIFE Head, body and fins purplish grey to brownish grey, with brown to golden-brown dots on the head and wavy longitudinal lines of the same colour on the dorsal part of the body (lines faint or absent on large specimens); margin of the spinous dorsal fin narrowly blackish.

COLOUR IN MARKET Body colour a drab grey with no distinguishing marks; wavy lines often fade in death.

© Kwang-Tsao Shao

Adult *E. undulosus* (Henchun, Pintung County, Taiwan)

© Robert F Myers

Adult *E. undulosus* (Bali, Indonesia)

POPULATION **GENERAL** *E. undulosus* is a common species throughout its range, especially in Sri Lanka and southern India and is taken by a wide range of gears: handlines, vertical longlines, trawls, gillnets and fish traps. **FISHERY DEPENDENT** *E. undulosus* is heavily fished in southern India and Sri Lanka, but it is not known what impact this has on the

Epinephelus undulosus has a wide tropical Indo-West Pacific distribution, ranging from East Africa to southern India and Sri Lanka in the northern Indian Ocean, extending north to Taiwan and east to the Solomon Islands in the central western Pacific Ocean.

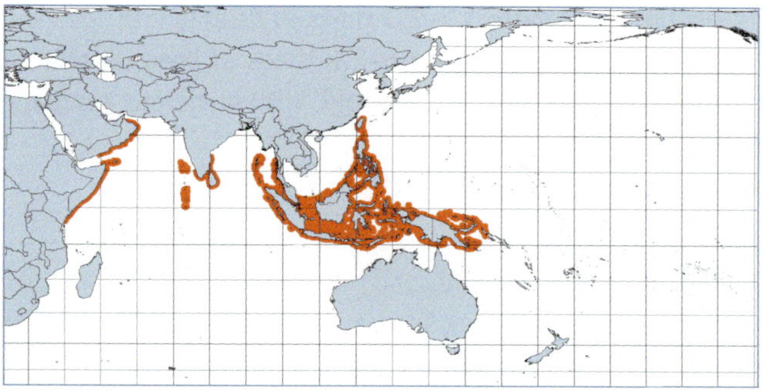

global population and the species is likely fished throughout all or most of its range. According to Deveraj and Murty (undated NACA library), in the Gulf of Mannar this species comprises ~6% of the grouper catch, which in turn comprises ~25% of the overall reef fish catch, in this primarily hook-and-line fishery; the species has been exported live to Hong Kong from India. Devaraj and Murty (no date) report on 'grow-out' aquaculture for fattening of this species after catching juveniles in the wild to be used in the live reef food fish trade. The species was common on the North Kenya Banks off Lamu and in the Wadge Bank trawl fishery off the west coast of Sri Lanka prior to 1993 but its current status is unknown (Munasinghe 1972, Heemstra and Randall 1993).

POPULATION TREND Unknown

MAJOR THREATS No major threats have been identified for this species although given its large size and increasing fishing pressure on groupers generally, fishing is a potential threat and there is no species-specific management.

HABITAT AND ECOLOGY **GENERAL** *E. undulosus* inhabits offshore banks and open silty sand and mud substrates where it seeks refuge in holes in the bottom or around solid structures, including shipwrecks or other bottom debris, in depths of 25–90 m. Juveniles occur as shallow as 5 m and reside near *Montipora* coral colonies. **FEEDING** Feeds on a variety of small fishes and small crustaceans, especially stomatopods and shrimps, and pelagic tunicates (*Thalia* and *Pyrosoma* spp.).

2008 IUCN RED LIST STATUS Data Deficient

Authors: *JH Choat, MT Craig, BP Ferreira, DA Pollard and LA Rocha*

Gonioplectrus hispanus (Cuvier 1828)

Spanish flag (E), pavillon espagnol (F), bandera española (S)

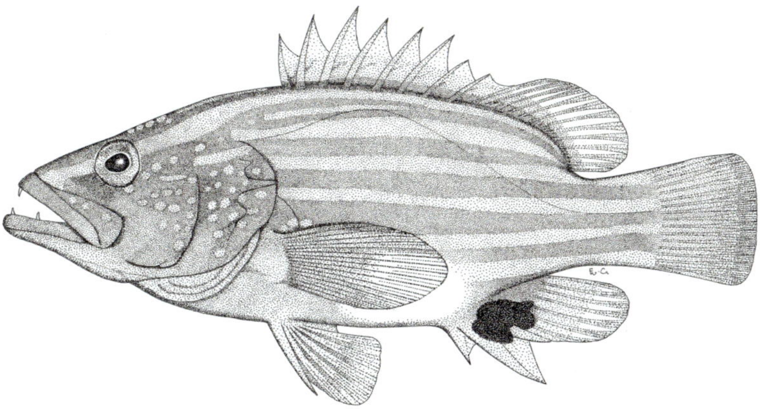

11 cm SL (Puerto Rico)
© SAIAB

Body depth subequal to head length, depth contained 2.3–2.5 times in standard length; body width 2.3–2.4 times in body depth. Head length 2.2–2.3 times in standard length; interorbital area convex, the dorsal head profile convex; preopercle rounded, with a large, curved (antrorse) spine at the angle, the lower edge fleshy; middle opercle spine very large, extending to or beyond rear edge of opercular membrane; posterior and anterior nostrils small, subequal; maxilla with a well-developed supramaxilla and a prominent bony knob (hidden by upper lip) on lower rear corner; a short, curved, fixed canine tooth on each side of symphysis of both jaws and 1–3 similar fixed canines at midside of lower jaw. Gill rakers 5–7 on upper limb, 16 on lower limb, the longest slightly longer than longest gill filaments. Dorsal fin with 8 spines and 13 rays, the 3rd or 4th spine longest; interspinous membranes deeply incised; base of spinous part of fin longer than soft-rayed part; anal fin with 3 spines and 7 rays; pectoral fins with 16 or 17 rays, the middle rays longest; pelvic fins reach anus. Lateral line distinctly curved upward below spinous dorsal fin. Scales on body large, distinctly ctenoid and without auxiliary scales; lateral-line scales 47–49; lateral-scales series 78–83; no scales on top of head, preorbital, snout, maxilla or lower jaw.

MAXIMUM RECORDED SIZE 30 cm TL

COLOUR IN LIFE Head, body, dorsal and caudal fins yellow; 6 or 7 salmon-coloured stripes from the head to the soft dorsal and caudal fins; blood-red blotch on the front half of the anal fin; white blotch on the side of the belly; pelvic fins pinkish purple.

COLOUR IN MARKET *G. hispanus* is one of the few groupers to maintain nearly identical colouration in death.

© Matthew T Craig

Adult *G. hispanus* (Puerto Rico)

Gonioplectrus hispanus is a western Atlantic species found from North Carolina (USA; pelagic post-larva, possibly drifted north from Florida or the Bahamas) throughout the Gulf of Mexico and Caribbean Sea, extending south to Vitoria and Trindade Island (Brazil).

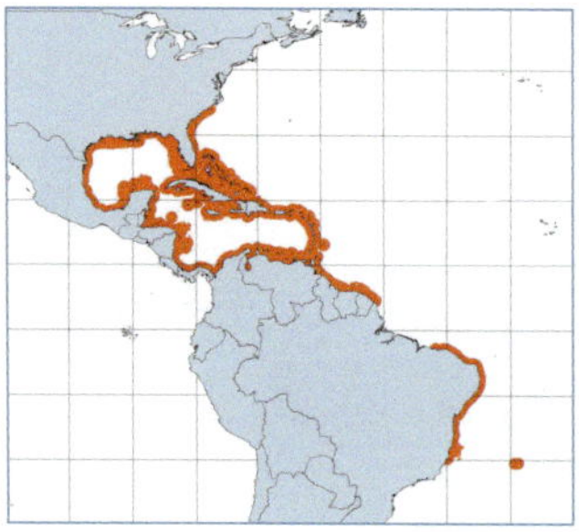

POPULATION **GENERAL** *G. hispanus* is considered rare throughout its range; however, its deep habitat precludes an accurate assessment of its population. **FISHERY DEPENDENT** *G. hispanus* is only occasionally taken by deep-water fisheries as bycatch; there is no dedicated fishery.

POPULATION TREND Unknown

MAJOR THREATS No major threats have been identified for this species.

HABITAT AND ECOLOGY **GENERAL** *G. hispanus* is a deep-water species inhabiting sandy bottoms and reefs in depths of 60–360 m.

2008 IUCN RED LIST STATUS Least Concern

Authors: *A Bertoncini, JH Choat, MT Craig, BP Ferreira, LA Rocha and W-L To*

Gracila albomarginata (Lowe 1834)

masked grouper (E), mérou bord rouge (F), mero paranjero (S)

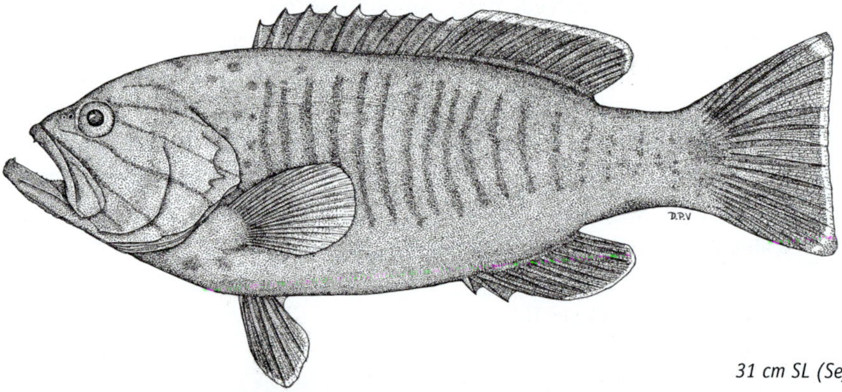

31 cm SL (Seychelles)
© SAIAB

Body oblong, somewhat compressed, depth contained 2.6–3.3 times in standard length; body width 1.8–2.3 times in the depth. Head smaller than most other groupers, head length 2.9–3.2 times in standard length; dorsal head profile evenly convex; interorbital area slightly convex, its width greater than eye diameter; preorbital narrow, its depth distinctly less than eye diameter, preorbital depth 10–15 times in head length; preopercle rounded, finely serrate, the lower edge smooth and fleshy; ventral edge of subopercle and interopercle smooth; middle opercular spine located one-third distance from lower to upper spine, the upper edge of operculum distinctly convex; posterior and anterior nostrils subequal in size; posterior part of maxilla in adults with a hook-shaped or step-like expansion on ventral edge; supramaxilla well developed; jaws with a pair of small canines at the front; palatine teeth present; midside of lower jaw with 2 rows of teeth, the inner teeth depressible and twice as long as outer teeth. Gill rakers 8–10 on upper limb, 14–16 on lower limb. Dorsal fin with 9 spines and 14–16 rays, the fin origin over rear end of operculum; dorsal-fin membranes not (adults) or slightly (juveniles) incised between the spines, dorsal-fin spines slender, the dorsal-fin spines not much thicker than the dorsal-fin rays; anal fin with 3 spines and 9 or 10 rays, anal-fin spines slender; pectoral fins rounded, with 18 or 19 rays, the middle rays longest; pectoral fins slightly longer than pelvic fins; pelvic fins not reaching anus; caudal fin truncate to slightly emarginate, with 8 branched rays and 9 or 10 procurrent rays in upper part and 7 branched rays and 9 or 10 procurrent rays in lower part. Midlateral-body scales ctenoid, with numerous auxiliary scales; lateral-line scales 66–76; lateral-scale series 101–114.

MAXIMUM RECORDED SIZE 50 cm TL

Gracila albomarginata is an Indo-Pacific species ranging from northern Mozambique to French Polynesia, extending north to Okinawa (Japan) and south to the northern Great Barrier Reef (Australia). It is unknown from the Red Sea and the Persian Gulf.

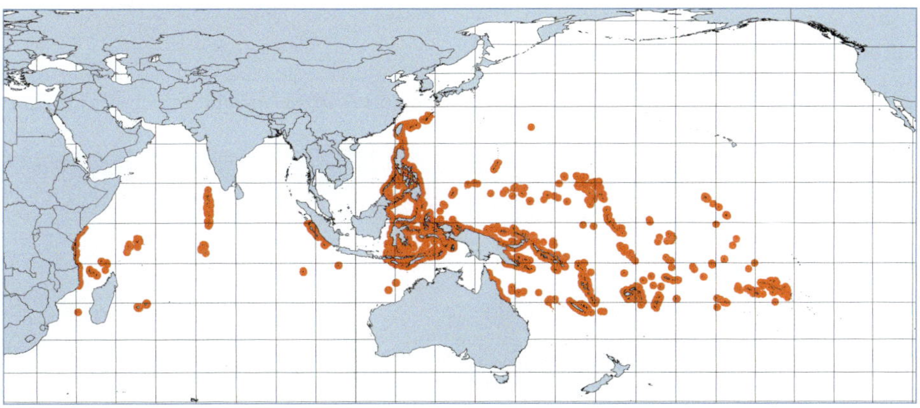

POPULATION **General** *G. albomarginata* is a patchily distributed species that is generally uncommon. **Fishery dependent** *G. albomarginata* is only occasionally taken by small-scale artisanal fisheries. Juveniles are occasionally taken for the aquarium trade.

POPULATION TREND Unknown

MAJOR THREATS No major threats have been identified for this species.

HABITAT AND ECOLOGY **General** *G. albomarginata* is a reef-associated species and an active swimmer usually found on outer reef slopes and channels adjacent to deep water. The species has been observed and collected between 15 and 120 m depth. It hovers above the bottom and is usually solitary. **Feeding** Feeds primarily on fishes.

COMMENTS Craig and Hastings (2007) demonstrate a close, nested relationship between this genus and the genus *Cephalopholis*, and suggest that *G. albomarginata* is best treated as a member of the genus *Cephalopholis*. We retain it here in *Gracila* pending further study.

2008 IUCN RED LIST STATUS Data Deficient

COLOUR IN LIFE Adults greenish or reddish brown to brownish grey, with several narrow curved dark bars on the midlateral part of the body and dark brown spot midlaterally on the caudal peduncle; 3–5 blue lines (often broken into spots) across the head (the upper 2 lines may serve as margins for a mask-like dark band from the snout to the end of the opercle, enclosing the eye); region under the maxilla and along the edge of the gill opening often orange; soft dorsal and anal fins with narrow blue margin; upper and lower edges of the caudal fin darker than the rest of the fin. A transient colour phase has the caudal peduncle white, with a prominent black spot, and a large white area flanked by black areas on the dorsal part of the body. Juveniles brown or violet, with a bright reddish orange stripe in the dorsal and anal fins and along the upper and lower edges of the caudal peduncle and fin.

COLOUR IN MARKET *G. albomarginata* may change drastically in death. All distinguishing characteristics may be lost. The species is best recognised by its rounded face, enlarged canine teeth, and relatively long and narrow caudal peduncle.

© Kwang-Tsao Shao

Adult *G. albomarginata* (Henchun, Pintung County, Taiwan)

© Matthew T Craig

Juvenile *G. albomarginata* (Los Angeles fish store, USA)

© John E Randall

Juvenile *G. albomarginata* (Kakabia, Indonesia)

© Robert F Myers

Adult *G. albomarginata* (Fiji)

Authors: *RF Myers, KL Rhodes, and BM Yeeting*

Hyporthodus acanthistius (Gilbert 1892)

rooster hind (E), mérou coq (F), baqueta (S)

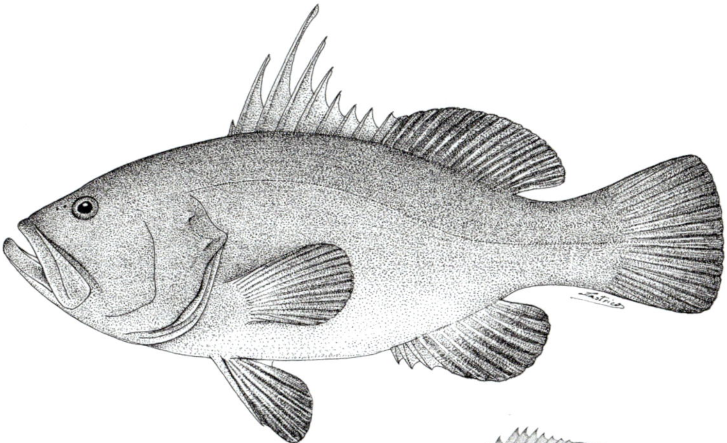

~90 cm TL
© FAO

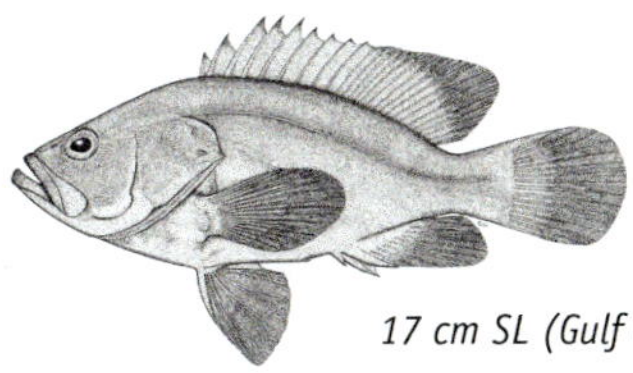

17 cm SL (Gulf of California, Mexico)
© SAIAB

Body depth contained 2.4–2.6 times in standard length (in fish 10–50 cm SL). Head length 2.3–2.4 times in standard length; interorbital area convex; preopercle rounded, finely serrate, the serrae at the angle enlarged but covered by skin; upper edge of operculum distinctly convex. Gill rakers 10 or 11 on upper limb, 14–17 on lower limb, 25–27 total. Dorsal fin with 9 spines and 17 rays; adults with 2nd to 4th dorsal-fin spines greatly elongated and the interspinous membranes deeply incised; anal fin with 3 spines (the 3rd longest) and 9 rays; pectoral-fin rays 18 or 19; pelvic-fin origin anterior to lower end of pectoral-fin base; pelvic fins shorter than pectoral fins; caudal fin rounded. Midlateral-body scales ctenoid; distal parts of soft dorsal, anal and caudal fins without scales; lateral-line scales 61–64; lateral-scale series 93–98.

MAXIMUM RECORDED SIZE At least 100 cm TL

COLOUR IN LIFE Head and body dark red or brown. Pectoral fins darker than the body. Tips of the interspinous dorsal-fin membranes darker than the rest of the membranes. Naked distal portion of the soft dorsal, anal and caudal fins darker than the proximal (scaly) parts of these fins. A prominent black moustache streak occurs above the maxilla.

COLOUR IN MARKET Head, body and fins uniform brown. Black moustache streak clearly visible. Body may become blotchy with light-coloured patches.

© Thierry Brulé

Adult *H. acanthistius* (Merida Fish Market, Mexico)

POPULATION **GENERAL** *H. acanthistius* is a common species in the Gulf of California and central Mexico. It is considered rare on the Pacific coast of Baja California (Mexico), and there is only a single record in southern California. **FISHERY DEPENDENT** *H. acanthistius* is a major component of fisheries in the northern Gulf of California.

POPULATION TREND Unknown

MAJOR THREATS The threat to *H. acanthistius* is overfishing and lack of management.

HABITAT AND ECOLOGY **GENERAL** *H. acanthistius* is found on reefs and sandy bottoms in depths of up to 90 m. It is generally uncommon in shallow waters.

COMMENTS *H. acanthistius* is unique among the deep-bodied groupers in having only 9 dorsal-fin spines. However, this species shares several morphological and genetic characters with other genera having 10 and 11 dorsal-fin spines, and is clearly unrelated to other grouper genera with 9 dorsal spines (e.g. *Cephalopholis*; Heemstra and Randall 1993, Craig and Hastings 2007). Craig and Hastings (2007) confirm the close relationship among the '*niveatus*' species-complex, which includes *H. acanthistius*, based on genetic data.

H. acanthistius is targeted by recreational fishing, but is only incidentally caught in commercial fisheries in the southern portion of its range. In the Gulf of California, it is targeted by commercial and artisanal fishermen at a level constituting a major threat.

This species was formerly placed in the genus *Epinephelus*. However, Craig and Hastings (2007) demonstrate the monophyly of this clade and associated diagnostic morphology.

2008 IUCN RED LIST STATUS Least Concern

Hyporthodus acanthistius is an eastern Pacific species ranging from southern California (USA) to central Peru.

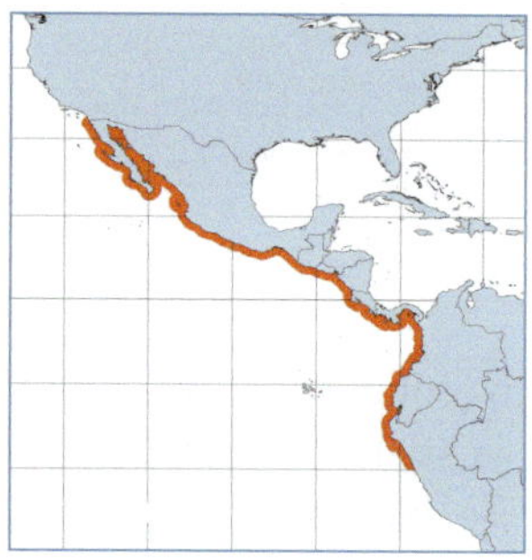

Authors: *AA Bertoncini, JH Choat, MT Craig, BP Ferreira, PC Heemstra and LA Rocha*

Hyporthodus darwinensis

(Randall and Heemstra 1991)

Darwin grouper (E), mérou Darwin (F), mero de Darwin (S)

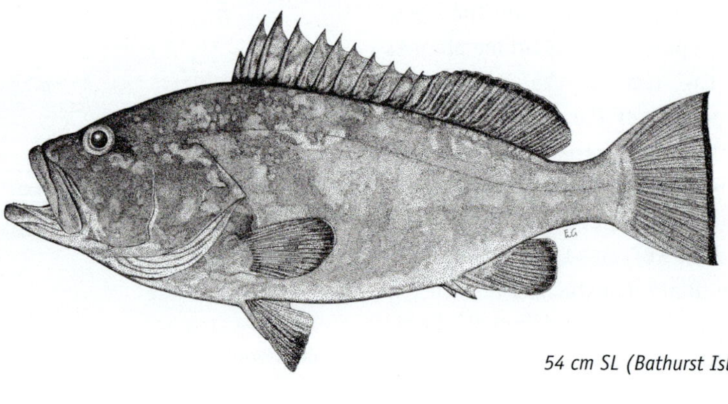

54 cm SL (Bathurst Island, Timor Sea)

Body depth contained 2.8 times in standard length (holotype 535 mm SL). Head length 2.6 times in standard length); interorbital area convex; dorsal head profile almost straight; preopercle angle rounded, with a shallow indentation above the angle and no enlarged serrae; upper edge of operculum straight; maxilla reaching almost to vertical at rear edge of eye; posterior nostrils 2.5 times larger than anterior nostrils; maxilla rounded posteriorly, with a prominent angle (hidden by upper lip) on lower edge; anterior canines in jaws small; midlateral part of lower jaw with 3 or 4 rows of small teeth. Gill rakers 9 on upper limb, 16 on lower limb; longest gill raker (1st on lower limb next to raker at angle) shorter than longest gill filaments; small bony platelets present on side of 1st gill arch. Dorsal fin with 11 spines and 16 rays, the 3rd spine longest, twice length of last spine, distinctly longer than longest rays and contained 2.9 times in head length; interspinous dorsal-fin membranes moderately incised; anal fin with 3 spines and 8 rays; pectoral-fin rays 18; pectoral fins equal to pelvic fins, their length contained 2.35 times in head length; caudal fin truncate. Lateral-body scales ctenoid, with numerous auxiliary scales; lateral-line scales 71; lateral-scale series 147.

MAXIMUM RECORDED SIZE 62 cm TL

COLOUR IN LIFE Known only from a single specimen.

COLOUR IN MARKET Yellowish brown, shading to yellowish ventrally; small faint irregular pale spots scattered over the head and body; dorsal and caudal fins yellowish brown, the caudal fin with a black posterior margin; soft dorsal and anal fins with traces of a white line along the distal margin; pectoral and pelvic fins blackish distally.

© Phillip C Heemstra

Adult *H. darwinensis* (Northern Territory, Australia)

Hyporthodus darwinensis is only known from a single specimen caught off the north-western coast of Australia and landed at Darwin.

POPULATION Unknown

POPULATION TREND Unknown

MAJOR THREATS Unknown

HABITAT AND ECOLOGY Unknown

COMMENTS This species was formerly placed in the genus *Epinephelus*. However, Craig and Hastings (2007) demonstrate the monophyly of this clade and associated diagnostic morphology.

2008 IUCN RED LIST STATUS Data Deficient

Author: *BP Ferreira*

Hyporthodus ergastularius (Whitley 1930)

sevenbar grouper (E), mérou sept raies (F), mero de siete bandas (S)

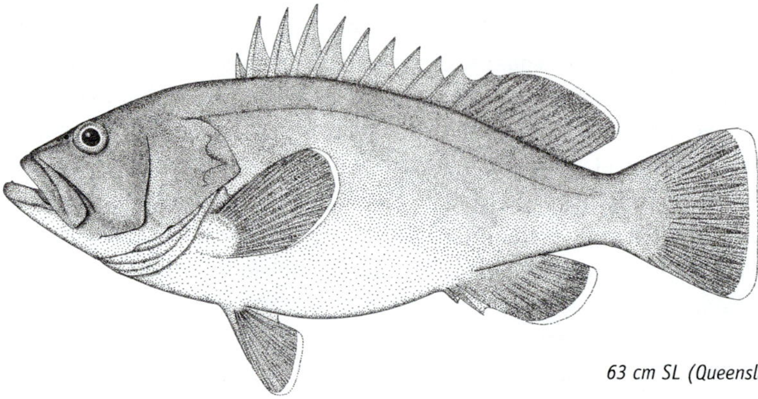

63 cm SL (Queensland, Australia)
© SAIAB

~20 cm TL (south-eastern Australia)
© SAIAB

Body depth contained 2.6–2.9 times in standard length (for fish 19–89 cm SL). Head length 2.4–2.6 times in standard length; interorbital area and dorsal head profile slightly convex; preopercle subangular, finely serrate, the serrae on the rounded corner not enlarged and the ventral edge with 1–4 small serrae; upper edge of operculum almost straight; posterior nostrils 2–4 times larger than anterior nostrils; maxilla reaches to or almost to vertical at rear edge of eye; midlateral part of lower jaw with 2 rows of teeth. Gill rakers 7–9 on upper limb, 14 or 15 on lower limb; sides of gill arch with small bony tooth plates. Dorsal fin with 11 spines and 14 or 15 rays, 3rd spine usually longest, its length contained 2.6–3.2 times in head length and subequal to length of longest ray, interspinous membranes deeply incised; fine dermal ridges present on body at base of dorsal fin; anal fin with 3 spines and 9 or 10 rays; pectoral-fin rays 18 or 19; pectoral-fin length 1.7–2.4 times in head length; pelvic fins distinctly shorter than pectoral fins; caudal fin slightly rounded in juveniles, truncate to slightly emarginate in adults. Lateral-body scales ctenoid, no auxiliary scales; lateral-line scales 63–70; lateral-scale series 103–116.

MAXIMUM RECORDED SIZE 157 cm TL

POPULATION GENERAL *H. ergastularius* is considered a relatively rare species within its limited range in Australian waters. **FISHERY DEPENDENT** *H. ergastularius* is an important component of dropline fisheries in eastern Australia despite low catch rates. Annual landings varied considerably in two major ports from 1990 to 2005, ranging from 0 to 30 t (Pogonoski et al. 2002, Sumpton and Ryan 2004). There is no definite indication that the population is decreasing, but catches are dominated by immature fish of small size (less than 60 cm FL); this may reflect either overfishing of larger fish, gear type or fishing

COLOUR IN LIFE Body of juveniles pale brown, with 7 broad dark bars, the last covering most of the caudal peduncle, its upper half black; prominent black maxillary streak; median and pelvic fins dark brown; pectoral fins translucent; 3 irregular dark bands extending posterior-ventrally from the eye. Adults pinkish grey, dark bars may or may not be present; fins darker than body and with white margins (except along the spinous part of the dorsal fin).

COLOUR IN MARKET Generally similar to colour in life; however, the dark bars may fade or disappear altogether.

© Stephen Newman

Adult *H. ergastularius* (Sydney Fish Market, Australia)

depth/location. According to Pogonoski et al. (2002), because of a consistently high market price (AU$16 per kg for whole fish, AU$23 per kg for cutlets at the Sydney fish markets, October 1999), this species has been heavily targeted since about 1990. In New South Wales, reported landings reached a maximum of about 29 t in 1990/1991, but declined to about 11 t in 1994/1995. Juveniles are taken as bycatch of traps and prawn trawls. There has been anecdotal evidence of a decrease in the average size of this species in catches (KR Rowling pers. comm.). There appears to be marked fluctuations in catches of this species from season to season which might relate to its aggregating and/or spawning behaviours (Pogonoski et al. 2002).

POPULATION TREND Unknown

Hyporthodus ergastularius is known only from the east coast of Australia between 18° and 36° S.

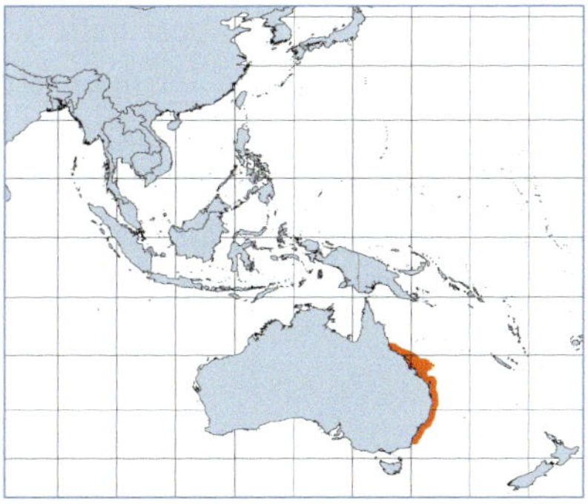

MAJOR THREATS No major threats have been identified for this species, although fishing may be a factor in population declines in at least some areas of its limited distribution, especially if aggregations are targeted.

HABITAT AND ECOLOGY **General** *H. ergastularius* is a deep-water species inhabiting rocky reefs on the continental shelf and upper slope in depths ranging from 15 m to over 370 m. Adults are mainly found deeper than 110 m, while juveniles and smaller fish are found shallower (~15–130 m). **Feeding** The diet consists of fishes, crustaceans and molluscs (Pogonoski et al. 2002). **Reproduction** *H. ergastularius* is a protogynous hermaphrodite, and females are mature between 70 and 80 cm FL in length (6–8 kg) (Rowling 1996).

COMMENTS In fish markets, this species is sometimes confused with *Epinephelus daemelii* and *H. octofasciatus* (Pogonoski et al. 2002), and is also easily confused with *H. septemfasciatus*. The latter species is known only from Japan, Korea and China, and *H. octofasciatus* is known from South Africa to New Zealand. These two species have a convex caudal fin in adults, and lack the fine dermal ridges present on the body of *H. ergastularius* at the base of the dorsal fin. Juveniles can be distinguished by the spacing of the dark bars on the body. In *H. ergastularius* the pale space separating the 2nd and 3rd dark bars and that between the 4th and 5th bars is distinctly narrower than the interspaces between the 3rd and 4th or 5th and 6th bars. In *H. octofasciatus* all the pale interspaces are about the same width, and in *H. septemfasciatus* the pale interspaces are about the same width, except for that between the 5th and 6th dark bars, which is distinctly narrower than the others.

This species was formerly placed in the genus *Epinephelus*. However, Craig and Hastings (2007) demonstrate the monophyly of this clade and associated diagnostic morphology.

2008 IUCN RED LIST STATUS Least Concern

Authors: *DA Pollard and KL Rhodes*

Hyporthodus exsul (Fowler 1944)

ten-spine grouper (E), mérou dix épines (F), mero diez espinas (S)

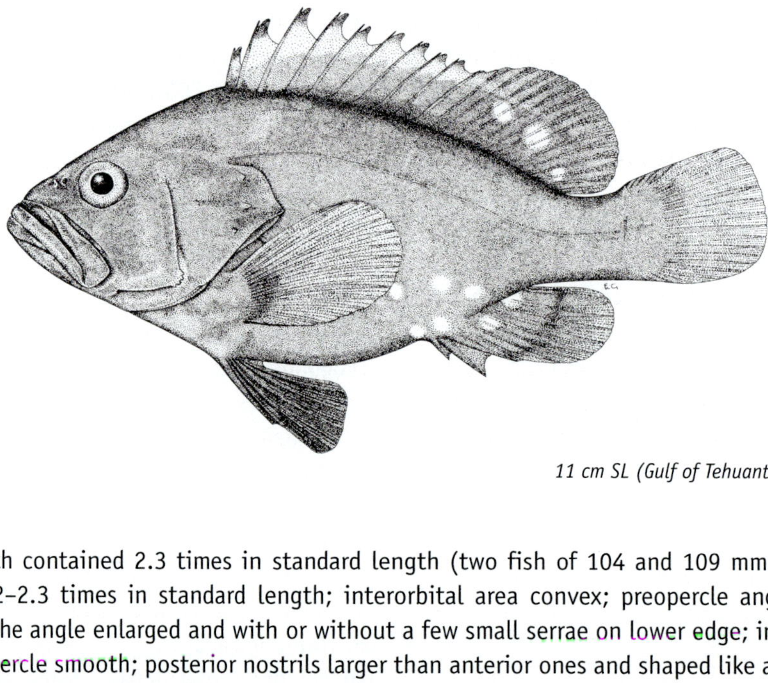

11 cm SL (Gulf of Tehuantepec, Mexico)
© SAIAB

Body depth contained 2.3 times in standard length (two fish of 104 and 109 mm SL). Head length 2.2–2.3 times in standard length; interorbital area convex; preopercle angular, with serrae at the angle enlarged and with or without a few small serrae on lower edge; interopercle and subopercle smooth; posterior nostrils larger than anterior ones and shaped like an inverted tear drop; maxilla reaches vertical at rear edge of eye. Gill rakers 8–10 on upper limb, 14–17 on lower limb, 24–27 total. Dorsal fin with 10 spines and 13 or 14 rays, the interspinous membranes distinctly indented; anal fin with 3 spines and 8 rays; pectoral-fin rays 18–20, the fin reaching vertical at anal-fin origin of juveniles; pelvic fins subequal to pectoral fins, pelvic-fin length contained 1.6–1.8 times in head length, the pelvic-fin origin anterior to pectoral-fin base; caudal fin rounded. Midlateral-body scales weakly ctenoid; lateral-line scales 64–68; lateral-scale series 87–92.

MAXIMUM RECORDED SIZE 125 cm TL

COLOUR IN LIFE Adults dark brown with no distinct markings. Juveniles dark brown with several white spots and occasionally a paler caudal fin.

COLOUR IN MARKET Drab brown with no distinct markings. Juveniles having spots may retain them.

© D Ross Robertson

Juvenile *H. exsul* (El Salvador)

© D Ross Robertson

Juvenile *H. exsul* (El Salvador)

© Matthew T Craig

Adult *H. exsul* (filleted; Panama City Fish Market, Central America)

POPULATION **General** *H. exsul* is a rare species throughout its range. **Fishery dependent** *H. exsul* only rarely appears in fish markets. A 125 cm specimen was purchased by MT Craig at the Panama City Fish Market (Panama) in 2001.

Hyporthodus exsul is an eastern central Pacific species occurring from the Gulf of California to Panama.

POPULATION TREND Unknown

MAJOR THREATS No major threats have been identified for this species.

HABITAT AND ECOLOGY **General** *H. exsul* is a reef-associated species inhabiting depths of at least 55 m.

COMMENTS *H. exsul* is one of two species of grouper in the eastern Pacific with 10 dorsal-fin spines, the other being *Epinephelus analogus*. These two species cannot be confused given the extreme differences in colour pattern and general body shape, *H. exsul* being much deeper-bodied. *H. exsul* is the sister species to *H. nigritus* of the western Atlantic and the two represent a transisthmian geminate group.

This species was formerly placed in the genus *Epinephelus*. However, Craig and Hastings (2007) demonstrate the monophyly of this clade and associated diagnostic morphology.

2008 IUCN RED LIST STATUS Data Deficient

Authors: *AA Bertoncini, JH Choat, MT Craig, BP Ferreira, PC Heemstra and LA Rocha*

Hyporthodus flavolimbatus (Poey 1865)

yellow-edge grouper (E), mérou aile jaune (F), cherna blanca (S)

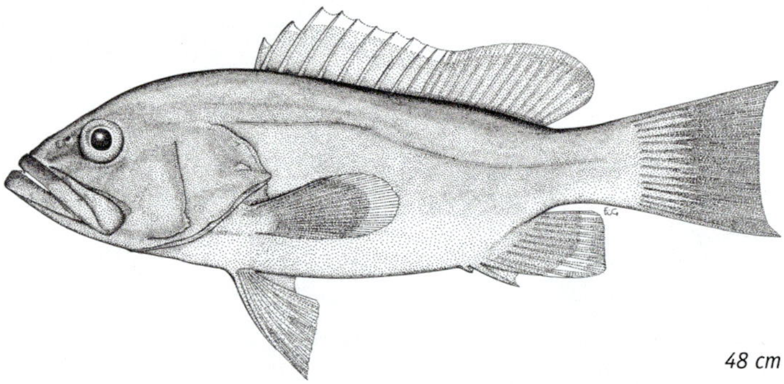

48 cm SL (Gulf of Mexico)
© SAIAB

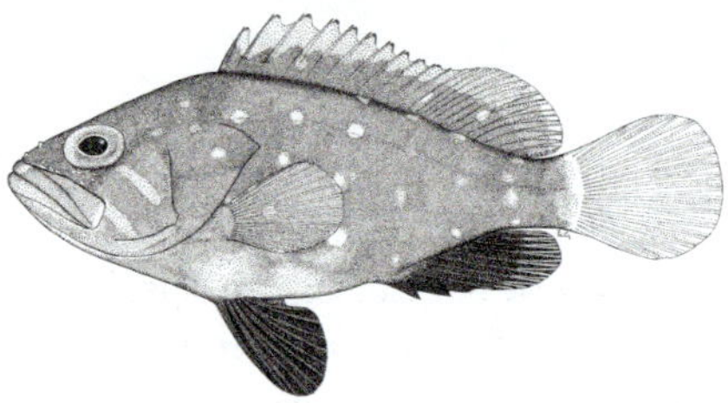

~5 cm SL (Gulf of Mexico)
© SAIAB

Body depth distinctly less than head length, the depth contained 2.7–2.9 times in standard length (for fish 13–64 cm SL). Head length 2.4–2.6 times in standard length; interorbital area convex; preopercle angular, the serrae at the angle distinctly enlarged and, in large fish, the angle is produced into a flat serrate lobe; eye diameter equal to or greater than interorbital width in fish less than 45 cm SL; posterior and anterior nostrils subequal in size. Gill rakers 8 or 9 on upper limb, 15–17 on lower limb, 23–25 total. Dorsal fin with 11 spines and 13–15 rays, the 3rd or 4th spine longest and the membranes only slightly incised between the anterior spines; anal fin with 3 spines and 9 rays; pectoral-fin rays 17 or 18; rear margin of caudal fin convex in fish less than 30 cm SL, truncate or slightly concave in larger fish. Lateral-body scales ctenoid; lateral-line scales about 65; lateral-scale series 82–99.

MAXIMUM RECORDED SIZE 115 cm TL

POPULATION **GENERAL** *H. flavolimbatus* has declined in abundance over the past several decades in many parts of its range, including the Gulf of Mexico, the Caribbean and Brazil. It is considered overexploited in the south-eastern USA (Ault et al. 1998). **FISHERY DEPENDENT** *H. flavolimbatus* has shown marked declines in catches throughout its range, with an overall inferred population decline throughout much of its range and catch data suggesting substantial declines in some areas. Its habitat has been fished by roller-rigged trawls, resulting in habitat destruction. Although abundance trends are lacking in many areas for this species and its catch is often pooled with other groupers, population trends for other more abundant species targeted by the same fisheries have shown drastic declines

COLOUR IN LIFE Head and body buff or greyish brown, whitish ventrally; margins of the dorsal and pectoral fins, and sometimes the anal and caudal fins, yellow (yellow margins reduced or absent in large adults); pale blue line from the eye to the angle of the preopercle (very faint in adults). Juveniles 5–10 cm SL with pearly spots arranged in 4 longitudinal rows and 7 vertical columns; dorsal fin with broad yellow margin; caudal fin white; anal and pelvic fins blackish; black saddle on the peduncle (if present) ends abruptly at the lateral line. Adults generally without distinguishing markings but sometimes they may (momentarily) display the white-spotted grid pattern of juveniles.

COLOUR IN MARKET Generally brown to light brown with white underside. Yellow edges of fins remain prominent, particularly pectoral fins.

© Mickey Amador

Adult *H. flavolimbatus* (Puerto Rico)

© NOAA

Adult *H. flavolimbatus* (south-eastern slope, USA)

(IUCN Red List Assessment 2008). In Brazil, catches in São Paulo declined from nearly 100 000 kg per year in 1998 to less than 20 000 kg in 2005, and similar trends were reported in Santa Catarina (nearly 250 000 kg in 2002 to less than 10 000 kg in 2006) and Rio Grande do Sul (over 400 000 kg in 1996 to less than 50 000 kg in 2006) (Instituto de Pesca 2006, GEP/UNIVALI 2006 and IBAMA-CEPERG 2006, respectively). In the Gulf of Mexico, CPUE dropped by 50% from 1992 to 1995 (Cass-Calay and Bahnick 2002), while elsewhere in the south-eastern USA landings declined five-fold over the past 20 years

(Williams and Dixon 2003). **FISHERY INDEPENDENT** Fishery-independent data for the deep-water *H. flavolimbatus* are scattered and do not show any trends. However, all reports indicate that this species is generally uncommon even in areas with previously healthy population sizes.

POPULATION TREND Decreasing

MAJOR THREATS *H. flavolimbatus* is threatened by fishing pressure.

HABITAT AND ECOLOGY **GENERAL** *H. flavolimbatus* is a deep-water species inhabiting a variety of habitats in depths ranging from 90 m to over 360 m. It is a demersal, solitary species occurring in rocky areas and on sand or mud bottoms. On soft substrates it is often seen in or near trenches or burrow-like excavations. **FEEDING** *H. flavolimbatus* feeds on invertebrates (mainly brachyuran crabs) and fishes (Bullock and Smith 1991). **AGE AND GROWTH** Manickchand-Heileman and Phillip (2000) report *H. flavolimbatus* as old as 35 years off Trinidad and Tobago; the von Bertalanffy growth equation for fish 3–35 years old is $L = 963(1 - e^{-0.057(t+4.6)})$, where L is the fork length in mm, and the length–weight relationship is $W = 5 \times 10^{-5}L^{2.80}$ (L is the fork length in mm and W is the weight in g). However, a recent investigation in the Gulf of Mexico using carbon-14 validation indicates that *H. flavolimbatus* may live as long as 85 years (Cook et al. 2009). **REPRODUCTION** *H. flavolimbatus* is a protogynous hermaphrodite (Bullock et al. 1996); 50% of fishes are mature at 58 cm, and 50% of females transform into males by the time they reach 81 cm TL in the Gulf of Mexico. Spawning occurs from April through October in the South Atlantic (Keener-Chavis 1984, Parker and Mays 1998). Mature females were found in the eastern Gulf of Mexico from May through September (Bullock et al. 1996). The species forms spawning aggregations in US waters (RG Gilmore pers. comm., cited in Musick et al. 2000).

COMMENTS This species was formerly placed in the genus *Epinephelus*. However, Craig and Hastings (2007) demonstrate the monophyly of this clade and associated diagnostic morphology.

2008 IUCN RED LIST STATUS Vulnerable (A2d+3d)

Hyporthodus flavolimbatus is distributed in the western Atlantic from North Carolina (USA) to French Guiana, including the Gulf of Mexico and the Caribbean, and from central Brazil (12° S) south to the border of Uruguay (Rio Grande do Sul; 32°32′ S).

Authors: *BP Ferreira and MB Peres*

Hyporthodus haifensis (Ben Tuvia 1953)

Haifa grouper (E), mérou d'Haifa (F), mero de Haifa (S)

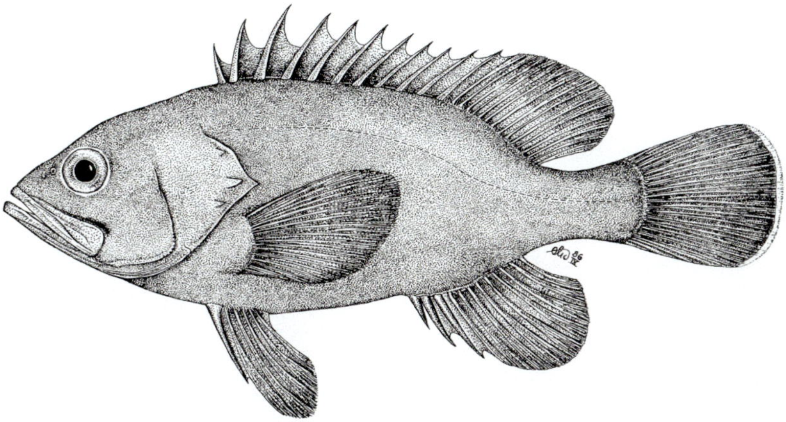

23 cm SL
© FAO

Body depth contained 2.4–2.8 times in standard length (for fish 10–39 cm SL). Head length 2.2–2.4 times in standard length; interorbital area convex; eye diameter greater than interorbital width in fish less than 30 cm SL, distinctly less than interorbital width in a fish of 39 cm SL; preopercle subangular, with the serrae at the angle enlarged, and 1–6 small serrae (usually covered by skin) on lower edge; posterior nostrils 2 or 3 times larger than anterior ones; maxilla naked, not reaching posterior to eye; midlateral part of lower jaw with 2 rows of teeth, the inner teeth distinctly larger than outer teeth. Gill rakers 7–10 on upper limb, 13–15 on lower limb, 20–25 total. Dorsal fin with 11 spines and 14 or 15 rays, the 3rd or 4th spine longest, the interspinous membranes deeply incised; anal fin with 3 spines and 9 rays; pectoral-fin rays 18–21, the fin length contained 1.4–1.9 times in head length; pelvic fins subequal to pectoral fins, reaching to or beyond anus; caudal fin rounded. Body scales distinctly ctenoid; lateral-line scales 64–75; lateral-scale series 104–112. Pyloric caeca very numerous, forming a large dendritic mass.

MAXIMUM RECORDED SIZE 110 cm TL

Hyporthodus haifensis is found from the central Mediterranean southwards along the coastline of North and West Africa to southern Angola.

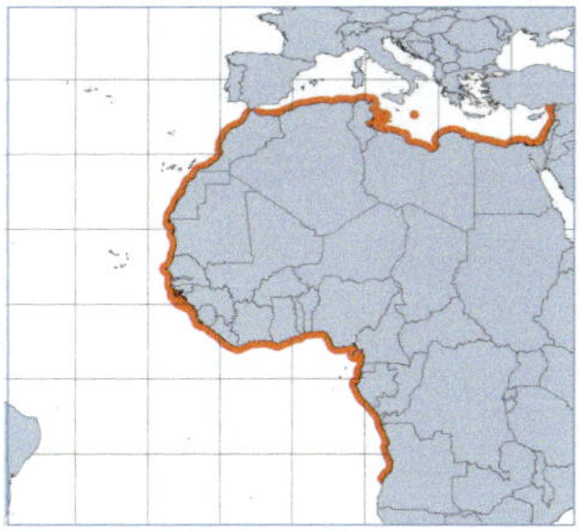

POPULATION General Little information is available on the population of *H. haifensis*.

POPULATION TREND Unknown

MAJOR THREATS No major threats have been identified for this species.

HABITAT AND ECOLOGY General *H. haifensis* is a deep-water species found over mud, sand or rock bottoms in depths of 90–220 m.

COMMENTS This species was formerly placed in the genus *Epinephelus*. However, Craig and Hastings (2007) demonstrate the monophyly of this clade and associated diagnostic morphology.

2008 IUCN RED LIST STATUS Data Deficient

COLOUR IN LIFE Head and body dark brown; soft dorsal, caudal and anal fins blackish distally (where there are no scales), the basal (scaly) part of these fins not so dark; caudal and pectoral fins with white edge; pelvic fins blackish; prominent black streak on the cheek at the upper edge of maxilla.

COLOUR IN MARKET Generally drab brown with few distinguishing markings. The black streak on the cheek remains prominent and the white edge of the caudal and pectoral fins may be dull but paler than the general colour of the fins.

Unknown

Adult *H. haifensis* (Turkey)

Authors: *JH Choat, MT Craig, DA Pollard and MA Samoilys*

Hyporthodus mystacinus (Poey 1852)

misty grouper (E), mérou brouillard (F), mero listado (S)

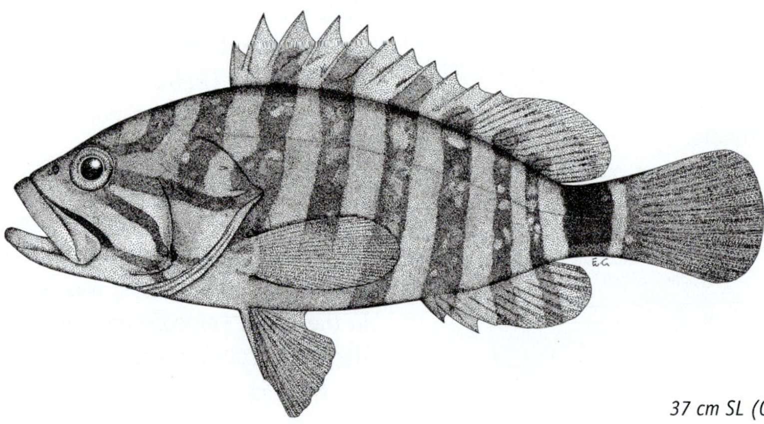

37 cm SL (Gulf of Mexico)

Body depth contained 2.4–2.9 times in standard length (for fish 14–40 cm SL). Head length 2.3–2.5 times in standard length; interorbital area convex; eye diameter greater than interorbital width for fish less than 30 cm SL, but less than interorbital in fish greater than 40 cm SL; preopercle angle rounded, with enlarged serrae, the lower edge with 1 or 2 small serrae (usually hidden by skin) just in front of the angle; interopercle and subopercle serrate; posterior nostrils greatly enlarged, their diameter 4 or more times that of anterior nostrils. Gill rakers 8–10 on upper limb, 14–16 on lower limb, 22–26 total. Dorsal fin with 11 spines and 14 or 15 rays, the 3rd spine longest and the membranes distinctly incised between the spines; anal fin with 3 spines and 9 (rarely 8) rays; pectoral-fin rays 18 or 19; pelvic fins shorter than pectoral fins and inserted slightly in front of lower end of pectoral-fin base; pelvic fins not reaching anus (except on fish less than 16 cm SL); caudal fin rounded. Body scales distinctly ctenoid, without auxiliary scales; lateral-line scales 58–69; lateral-scale series 99–112.

MAXIMUM RECORDED SIZE 160 cm TL

COLOUR IN LIFE Head and body buff, with 8 or 9 dark subvertical bars on body, the first on the nape, the last two (which are darker than the other bars) may be fused to form a wide dark band around the caudal peduncle; width of the dark body bars equal to or less than the width of the pale interspaces; dark bars faint on large adults. Pelvic fins blackish brown. Prominent blackish brown maxillary streak on the cheek along the upper edge of the maxilla. In the eastern Pacific, the margins of the median fins are orange to red.

COLOUR IN MARKET Dark brown to grey, vertical bars on the body often faded and only visible near the base of the dorsal fin. The maxillary streak remains prominent in some specimens.

Unknown

Adult *H. mystacinus* (Ecuador)

Hyporthodus mystacinus is known from the western Atlantic from North Carolina (USA) to southern Florida (USA), and throughout the Gulf of Mexico and Caribbean Sea, including the north-east coast of Brazil and Trindade Island. In the eastern Pacific, it is reported from Cocos Island and Paramount Seamount, as well as the Galápagos Islands and coastal Ecuador.

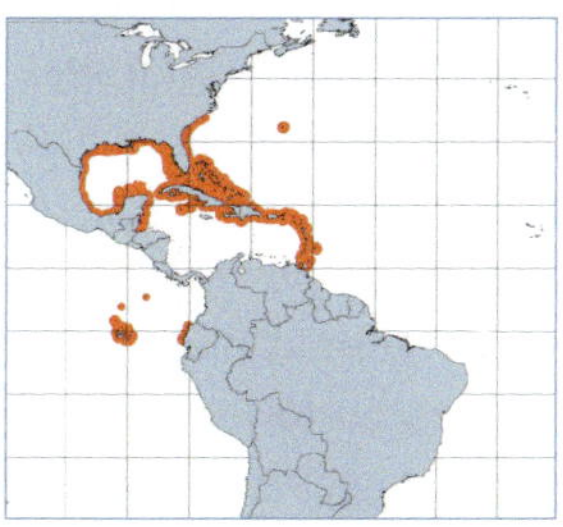

POPULATION **General** *H. mystacinus* is a deep-water species for which there is little population information. **Fishery dependent** From 1990 to 2003 the landings of *H. mystacinus* showed no trends in the south-eastern USA (NMFS database). Both commercial and recreational catches are low in the USA: annual landings of commercial catches of *H. mystacinus* amounted to 0.8–2.0 t for the six years between 1990 and 2005 for which there were data (NMFS database). The CPUE for *H. mystacinus* in Brazil is 3.2 fish per 1000 hooks between depths of 100 and 300 m (IUCN Red List Assessment 2008). **Fishery independent** Fishery-independent data indicate that *H. mystacinus* is relatively rare in SCUBA-diving depths.

POPULATION TREND Stable
MAJOR THREATS Fishing pressure is the main threat to this species and barotrauma associated with catch and release from its deep habitat means that released fish are unlikely to survive.
HABITAT AND ECOLOGY **GENERAL** *H. mystacinus* is a deep-water species that is typically solitary and occurs from 100 to 400 m. Juveniles are sometimes found in 30 m. **FEEDING** Feeds on fishes, crustaceans and squid. **AGE AND GROWTH** *H. mystacinus* attains a maximum body weight of 107 kg, with males attaining a maximum size of 160 cm TL and females at least 100 cm TL (Appeldoorn et al. 1987, Heemstra and Randall 1993). Natural mortality is estimated to be 0.14 (JC McGovern pers. comm.)
COMMENTS The only obvious morphological differences between the Atlantic and Pacific populations of this species is the orange to red margin of the median fins in Pacific populations which is absent in Atlantic populations. It is likely that future studies will show that these populations are distinct species.

This species was formerly placed in the genus *Epinephelus*. However, Craig and Hastings (2007) demonstrate the monophyly of this clade and associated diagnostic morphology.
2008 IUCN RED LIST STATUS Least Concern

Authors: *JH Choat, MT Craig, BP Ferreira and LA Rocha*

Hyporthodus nigritus (Holbrook 1855)

Warsaw grouper (E), mérou Varsovie (F), mero negro (S)

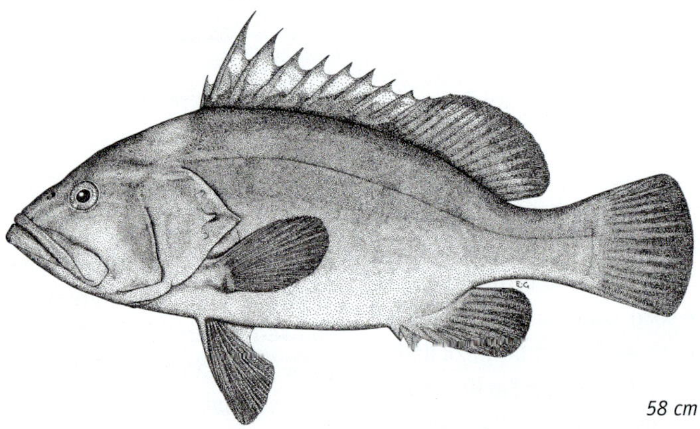

58 cm SL (Gulf of Mexico)
© SAIAB

Body depth less than head length, the depth contained 2.4–2.6 times in standard length (for fish 11–58 cm SL). Head length 2.1–2.5 times in standard length; interorbital area distinctly convex, wider than eye diameter for fish larger than 15 cm SL; preopercle angular, but the angle rounded, with slightly enlarged serrae and 1 or 2 small spines on lower edge just in front of angle; interopercle and subopercle smooth; nostrils subequal, or posterior nostril slightly larger; maxilla reaches well past eye; adults with 4 or 5 rows of teeth at midside of lower jaw, juveniles with 2 or 3 rows. Gill rakers 9–11 on upper limb, 14–16 on lower limb, 23–25 total. Dorsal fin with 10 spines and 13–15 rays, the 2nd spine distinctly longer than the others (in fish greater than 40 cm SL) and the interspinous membranes deeply incised; anal fin with 3 spines and 9 rays; pectoral-fin rays 18 or 19; pelvic fins longer than pectoral fins in fish 7–58 cm SL and inserted in front of vertical from lower end of pectoral-fin base; rear margin of caudal fin convex or truncate with rounded corners. Midlateral-body scales distinctly ctenoid, with auxiliary scales in adults; lateral-line scales 62–71; lateral-scale series 99–107. Pyloric caeca very numerous, in a large dendritic mass.

MAXIMUM RECORDED SIZE 230 cm TL

COLOUR IN LIFE Dark reddish brown or brownish grey to almost black dorsally, dull reddish grey below. Juveniles with a yellow caudal fin and a few randomly scattered whitish spots on the body; no dark saddle blotch on the peduncle.

COLOUR IN MARKET Generally brown to dark brown with no distinguishing marks.

© NOAA

Adult *H. nigritus* (Flower Garden Banks, Gulf of Mexico)

POPULATION **General** *H. nigritus* is a deep-water species and it is unclear whether populations are abundant at these depths. **Fishery dependent** Landings data for *H. nigritus* show no trends from 1950 to 2003. However, fishing effort during this time period has increased, and the species is considered overfished in USA waters by the National Marine Fisheries Service. *H. nigritus* is primarily caught by handlines and bottom longlines and incidentally in the deep-water snapper/grouper fishery. The major threat to this species is overfishing and mortality as a result of fishing or bycatch release mortality (due to barotrauma since it is deep-living). Despite the introduction of management, there continues to be killing of the species in bycatch which is unlikely to be reported. Musick et al. (2000) consider the species to be endangered.

There is also a trend for the recreational fishery to operate in deeper water as shallower stocks become depleted; there are so many recreational fishermen (over one million salt-water licences in Florida alone) that the potential impact on the already depleted *H. nigritus* population is serious.

The limited dataset used to determine the status of the Atlantic stock, in the absence of a full stock assessment (M Prager pers. comm.), indicates that the spawning potential ratios for *H. nigritus* in the Atlantic is between 6% and 14%, well below the threshold level indicating overfishing (NMFS 2003).

Hyporthodus nigritus occurs in the western Atlantic from Massachusetts to the Gulf of Mexico, Cuba, Trinidad, and as far south as Rio de Janeiro.

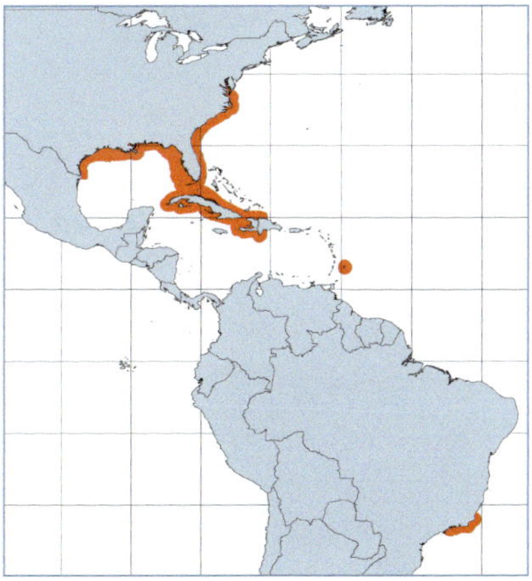

POPULATION TREND Decreasing
MAJOR THREATS Fishing pressure is a major threat to this species and appears to be increasing. Barotrauma for released animals compounds this threat.
HABITAT AND ECOLOGY **GENERAL** *H. nigritus* is a deep-water grouper that inhabits reefs (such as deep *Oculina* reef areas and other growth-encrusted hard bottoms) on the continental shelf break, in waters 76–219 m deep (Manooch and Mason 1987). Juveniles can be found in nearshore areas, occasionally seen on jetties and shallow-water reefs (Heemstra and Randall 1993). The species is rare in the West Indies, possibly due to lack of suitable habitat (IUCN Red List Assessment 2006). **FEEDING** Feeds on fishes and invertebrates. **AGE AND GROWTH** *H. nigritus* reaches at least 41 years (Manooch and Mason 1987, Musick et al. 2000). **REPRODUCTION** *H. nigritus* is a protogynous hermaphrodite; age of maturity is 9 years (Parker and Mays 1998, Musick et al. 2000). Spawning has been reported to occur around April and May in Cuba (García-Cagide et al. 1994).
COMMENTS This species was formerly placed in the genus *Epinephelus*. However, Craig and Hastings (2007) demonstrate the monophyly of this clade and associated diagnostic morphology.
2006 IUCN RED LIST STATUS Critically Endangered (A2d+3d)

Authors: *GR Huntsman and W-C Ng*

Hyporthodus niphobles

(Gilbert and Starks 1897)

starstudded grouper (E), mérou tacheté (F), mero machado (S)

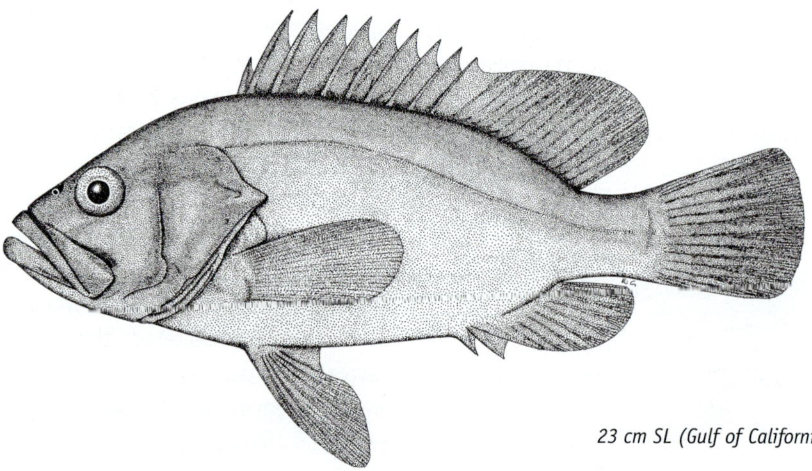

23 cm SL (Gulf of California, Mexico)
© SAIAB

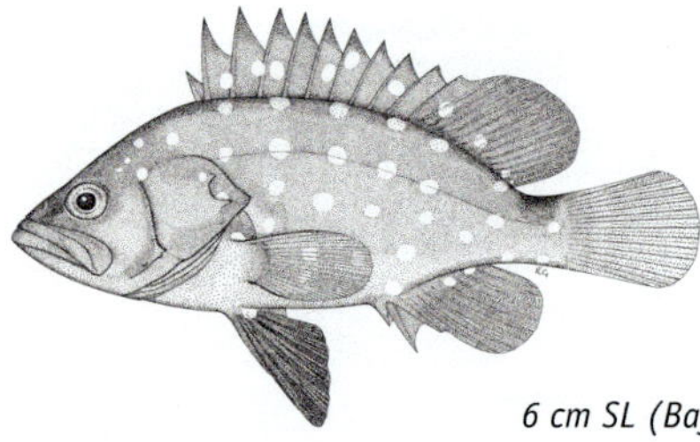

6 cm SL (Bay of Panama)
© SAIAB

Body depth contained 2.3–2.7 times in standard length (for fish 14–47 cm SL). Head length 2.2–2.5 times in standard length; interorbital convex, the width subequal to eye diameter for fish 5–22 cm SL and greater than eye diameter for fish larger than 30 cm SL; angle of preopercle slightly produced, with enlarged serrae; some specimens with a tiny spine (hidden by skin) on ventral edge of preopercle; upper edge of operculum distinctly convex; posterior nostrils vertically elongate, 2–5 times larger than anterior ones; maxilla reaches to or beyond vertical at rear edge of eye; midlateral part of lower jaw with 2 rows of teeth. Gill rakers 8–9 on upper limb, 15–17 on lower limb, 23–26 total. Dorsal fin with 11 spines and 13–15 rays, the 2nd spine longest in adults, and the membranes deeply incised between the spines; anal fin with 3 spines and 9 rays; pectoral-fin rays 17–21; pelvic-fin origin below or anterior to pectoral-fin base; pectoral fins subequal to pelvic fins, pectoral-fin length contained 1.6–2.1 times in head length; caudal fin convex in juveniles, truncate in adults. Midlateral-body scales rough, no auxiliary scales; lateral-line scales 63–71; lateral-scale series, 100–106.

MAXIMUM RECORDED SIZE 54 cm TL

COLOUR IN LIFE Juveniles (5–15 cm TL) dark reddish brown with white spots in a grid-like pattern of vertical series and horizontal rows on the body and dorsal fin; dark saddle blotch usually visible on the caudal peduncle; caudal and pectoral fins pale (possibly yellow to orange in life); anal and pelvic fins dark brown; maxillary groove black. Adults dark brown, without white spots.

COLOUR IN MARKET Generally dark brown with no distinguishing marks.

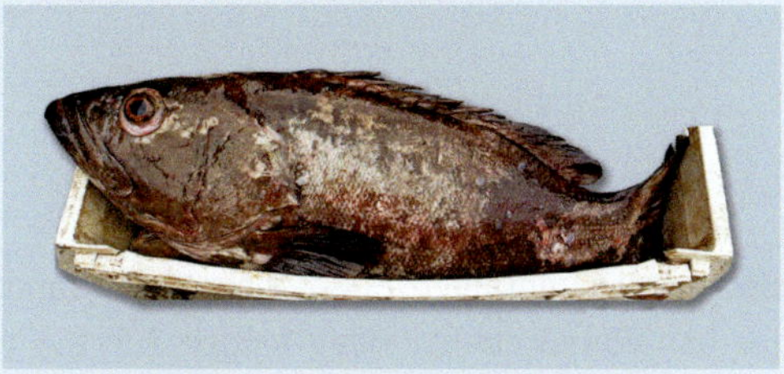

© Octavio Aburto

Adult *H. niphobles* (Gulf of California, Mexico)

POPULATION General *H. niphobles* is an uncommon species throughout its range, but may be locally abundant. **Fishery dependent** Incidentally taken as bycatch in trawl fisheries, and occasionally in recreational fisheries. There are no known directed fisheries for this species.

Hyporthodus niphobles occurs in the eastern Pacific, from northern California (USA) to Peru.

POPULATION TREND Unknown

MAJOR THREATS No major threats have been identified for this species.

HABITAT AND ECOLOGY General *H. niphobles* is a deep-water species found on rocky reefs and soft bottoms in depths to at least 100 m. Juveniles have been collected in relatively shallow water.

COMMENTS This species was formerly placed in the genus *Epinephelus*. However, Craig and Hastings (2007) demonstrate the monophyly of this clade and associated diagnostic morphology.

2008 IUCN RED LIST STATUS Data Deficient

Authors: *AA Bertoncini, JH Choat, MT Craig, BP Ferreira and LA Rocha*

Hyporthodus niveatus (Valenciennes 1828)

snowy grouper (E), mérou neige (F), cherna pintada (S)

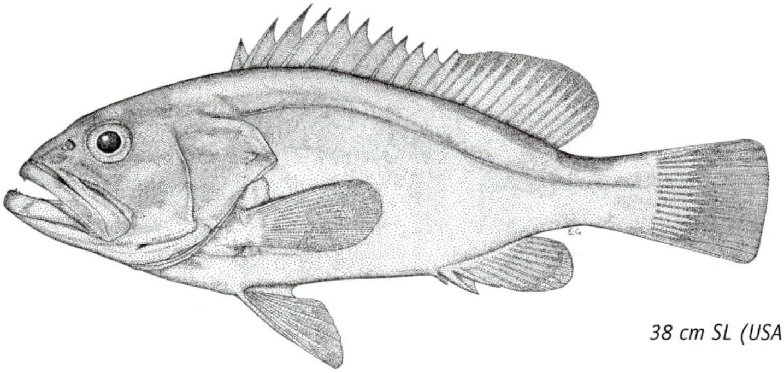

38 cm SL (USA off Georgia)
© SAIAB

9 cm SL
© FAO

Body depth contained 2.4–2.8 times in standard length (for fish 10–38 cm SL), 2.8–3.1 times for fish 75–93 cm SL. Head length 2.2–2.4 times in standard length (for fish 10–38 cm SL), 2.4–2.6 times for fish 75–93 cm SL; interorbital area convex, the width less than or subequal to eye diameter in fish 10–20 cm SL; in fish 30–93 cm SL, the interorbital width is 1.1–1.9 times the eye diameter; preopercle with enlarged serrae at the angle; often a tiny antrorse spine covered with skin on lower edge just in front of the angle; upper edge of operculum distinctly convex; posterior nostrils 2–5 times larger than anterior ones. Gill rakers 7–10 on upper limb, 15–17 on lower limb, 22–26 total. Dorsal fin with 11 spines and 13–15 rays, the 2nd or 3rd spine longest, the membranes deeply incised between the spines; anal fin with 3 spines and 9 rays; juveniles with pelvic fins longer than pectoral fins; in fish 40–60 cm SL, the pectoral and pelvic fins are about equal, and in large adults (over 60 cm SL), the pectoral fins are slightly longer than the pelvic fins; pelvic-fin origin anterior to pectoral-fin base; pectoral-fin rays 17–19); rear margin of caudal fin convex in juveniles (less than 30 cm SL), straight or concave in adults. Midlateral-body scales rough; lateral-line scales 64–73; lateral-scale series 102–128. Pyloric caeca numerous.

MAXIMUM RECORDED SIZE 120 cm TL

POPULATION **GENERAL** *H. niveatus* is a common species of deep-water grouper that is heavily fished throughout its range, particularly in the south-eastern United States. **FISHERY DEPENDENT** Fishery-dependent data are scattered throughout the literature and indicate substantial landings of this species. The degree of effort has not been quantified, but it is likely that this species is currently in decline and it is predicted to continue to decline under current conditions. Although commercial quotas and trip limits have been enacted in the USA since 1993, stricter conservation measures are needed because the existing population is still vulnerable to overfishing (IUCN Red List Assessment 2008). Epperly and Dodrill (1995) estimated that at one newly exploited site over 80% of the exploitable biomass was removed. Wyanski et al. (1999) suggested that the population of *H. niveatus* was overfished because of changes in size at age, high juvenile catches in the early 1990s, a marked decline in the percentage of mature males caught and decline in mean length between the early 1980s and early 1990s. In Brazil, *H. niveatus* comprised 12.1% of the total catch of benthic fishes by longliners from 1986 to 1995 (Paiva and Andrade-Tubino 1998). **FISHERY INDEPENDENT** Figures suggest that the very low abundance of *H. niveatus* in the natural environment needs to be better protected from further fishing activities, especially if there is fishing on spawning aggregations.

POPULATION TREND Decreasing

MAJOR THREATS Fishing pressure is the major threat to this species, which is further complicated by barotrauma for released individuals.

HABITAT AND ECOLOGY GENERAL Adult *H. niveatus* occur offshore on rocky bottoms at depths of 30–525 m, but are most common

COLOUR IN LIFE Adults dark brown, the spinous dorsal-fin margin black (except in large adults). Juveniles dark brown with conspicuous white spots in about 11 vertical series and 5 or 6 longitudinal rows, usually extending onto the head and dorsal fin (white spots visible on fish 6–40 cm SL); caudal and pectoral fins pale yellow; a black saddle blotch on the caudal peduncle reaching below the lateral line.

COLOUR IN MARKET Generally dark brown with no distinguishing marks.

© Athila Bertoncini Andrade

Juvenile *H. niveatus* (Bombinhas, Brazil)

© Matthew T Craig

Adult *H. niveatus* (Florida, USA)

© NOAA

Subadult *H. niveatus* (south-eastern slope, USA)

© NOAA *Ocean Explorer*

Adult *H. niveatus* (100 m depth, Gulf of Mexico)

between 100 and 200 m. Juveniles may be found inshore. Huntsman et al. (1999) state that [*H.*] *niveatus* is very rare and that only small individuals are observed in the US South Atlantic. **FEEDING** Adults feed mainly on fishes (47% by number), cephalopods (18%) and brachyuran crabs (Bielsa and Labisky 1987). **AGE AND GROWTH** The von Bertalanffy growth equation for *H. niveatus* is estimated as $L = 1255\ (1 - e^{-0.074(\mathrm{Age} + 1.92)})$, where L is the total length in mm (Matheson and Huntsman 1984). *H. niveatus* reaches ages of at least 29 years (Wyanski et al. 1999). **REPRODUCTION** *H. niveatus* is a protogynous hermaphrodite (Moore and Labisky 1984, Wyanski et al. 1999). In USA waters, female *H. niveatus* mature at 45–58 cm TL, and sexual transition occurs by about 70 cm TL. Spawning season in the USA is between April and July and may occur in spawning aggregations (Wyanski et al. 1999). **COMMENTS** This species was formerly placed in the genus *Epinephelus*. However, Craig and Hastings (2007) demonstrate the monophyly of this clade and associated diagnostic morphology. **2008 IUCN RED LIST STATUS** Vulnerable (A2d+3d)

Hyporthodus niveatus is found in the western Atlantic from Massachusetts (USA) to the Gulf of Mexico, Bermuda, the Caribbean Sea, southern Brazil, and Uruguay.

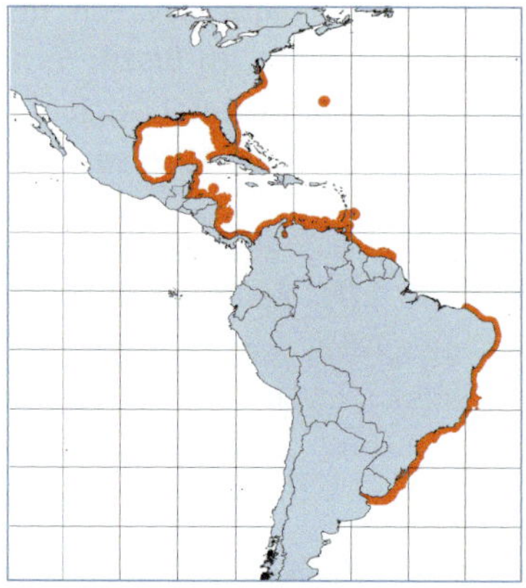

Authors: *AA Bertoncini, T-C Chan, JH Choat, MT Craig, BP Ferreira and LA Rocha*

Hyporthodus octofasciatus (Griffen 1926)

eightbar grouper (E), mérou huit raies (F), mero de ocho bandas (S)

23 cm SL (South Africa)
© SAIAB

Body depth contained 2.2–2.7 times in standard length (for fish 10–47 cm SL). Head length 2.4–2.5 times in standard length; interorbital distinctly convex, the dorsal head profile also convex; preopercle angular, the serrae at the angle enlarged; lower edge of preopercle with 0–3 small spines anterior to angle; lower edge of subopercle and rear edge of interopercle distinctly serrate; upper edge of operculum convex; diameter of posterior nostrils 2 or 3 times that of anterior nostrils but not more than distance from posterior nostril to edge of orbit; maxilla reaches to below rear half of eye; no distinct step on lower edge of maxilla; midlateral part of lower jaw with 2 rows of teeth. Gill rakers 7–9 on upper limb, 15–17 on lower limb. Dorsal fin with 11 spines and 14 or 15 rays, the interspinous membranes deeply incised, the 3rd or 4th spine longest, contained 2.5–2.9 times in head length and subequal to longest dorsal-fin rays; anal fin with 3 spines and 9 rays; pectoral-fin rays 18 or 19; pelvic fins subequal to or larger than pectoral fins, pectoral-fin length 1.8–2.0 times in head length; caudal-peduncle depth 0.9–1.3 times in peduncle length; caudal fin rounded. Lateral-body scales ctenoid, with a few auxiliaries; lateral-line scales 65–71; lateral-scale series 114–126. Pyloric caeca very numerous, in 3 or 4 dendritic bundles with more than 80 terminal branches.

MAXIMUM RECORDED SIZE 130 cm TL

Hyporthodus octofasciatus is an Indo-West Pacific species that is known from Somalia and South Africa to the Marquesas Islands, extending north to Japan and south to Australia and New Zealand.

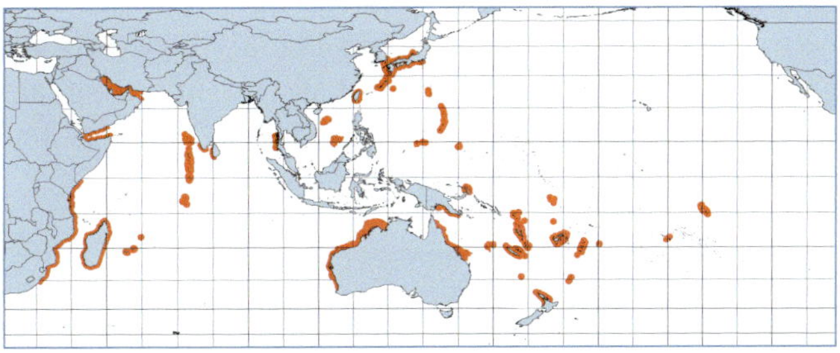

POPULATION **GENERAL** *H. octofasciatus* is a rare species throughout its range. **FISHERY DEPENDENT** Landings data are scattered for this species which is lumped with other groupers in many catch statistics. It may, however, be locally important for fisheries. For example, in Guam, landings of this species from 1998 to 2002 fluctuated biannually, but dramatically increased from 3 kg in 1992 to 1530 kg in 1994 (Sea Around Us Project Database), and in Western Australia, both CPUE and landings volume have increased (5800 kg in 2001 to 72 100 kg in 2005) (M Cliff pers. comm.) with similar effort. There is a reported increase of landings of deep-water groupers in Western Australia in recent years which includes this species (often misidentified as *H. septemfasciatus*).

POPULATION TREND Unknown

MAJOR THREATS *H. octofasciatus* is likely threatened by fishing pressure.

HABITAT AND ECOLOGY **GENERAL** *H. octofasciatus* is a poorly known deep-water species inhabiting rocky reefs in depths of 80–200 m. **AGE AND GROWTH** *H. octofasciatus* can reach at least 56 years according to bomb radiocarbon chronometer ageing techniques (Kalish et al. 2002).

COMMENTS *H. octofasciatus* is similar in appearance to *H. septemfasciatus*, which differs in having the body depth usually less (depth contained 2.5–3.1 times in standard length), caudal peduncle usually more slender (peduncle depth 1.2–1.5 times in its length), diameter of rear nostrils greater than the distance from the nostril to the orbit, and in the configuration of the dark body bars (6 dark bars wholly below the dorsal fin and 2 narrow dark bars below the middle rays of the soft dorsal fin; 3 pale interspaces below the soft dorsal fin). It is also similar to *H. mystacinus* which also has 8 dark bars on the body; but *H. mystacinus* has fewer lateral-scale series (99–112), the pelvic fins are shorter than the pectoral fins, and the dark bars on the body are narrower (their width equal to or less than the pale interspaces). *H. mystacinus* is known from the western Atlantic and the Galápagos Islands. *H. ergastularius*, which is known only from the east coast of Australia, is another similar species. Adults have a truncate or emarginate caudal fin and fine dermal ridges on the body at the base of the dorsal fin; also, the fins have white margins.

COLOUR IN LIFE Body buff with 8 broad dark brown bars, the 1st on the nape, the 2nd at the dorsal-fin origin and covering the first 2 dorsal spines, 7th bar broader than 6th and extending from the base of the middle 5–7 dorsal-fin rays to the rear half of the anal-fin base, 2 pale interspaces below the soft dorsal fin, last bar covering most of the caudal peduncle and the base of the last 2 or 3 dorsal-fin rays; pelvic fins and distal half of the soft dorsal and anal fins blackish brown; faint dark brown band from the eye to the middle opercle spine; blackish maxillary streak present; some specimens with a narrow white margin at the corners of the caudal fin, the lower edge of the anal fin and the anterior two-thirds of the edge of the soft dorsal fin. Gill rakers and gill arches covered with minute melanophores.

COLOUR IN MARKET Body a dark brown or grey. Bars may fade instantly upon death or, if present, are most prominent at the base of the dorsal fin.

© Phil Richmond

Adult *H. octofasciatus* (colour similar to life; Diego Garcia, Chagos Archipelago)

© Matthew T Craig

Adult *H. octofasciatus* (Okinawa Fish Market, Japan)

Unknown (supplied by Sean T Fennessy)

Juvenile *H. octofasciatus* (Durban, South Africa)

Juveniles can be distinguished by the spacing of the dark brown bars on the body: on *H. ergastularius* the pale space separating the 2nd and 3rd dark bars and that between the 4th and 5th bars is distinctly narrower than the interspaces between the 3rd and 4th or 5th and 6th bars; on *H. octofasciatus* all the pale interspaces are about the same width.

This species was formerly placed in the genus *Epinephelus*. However, Craig and Hastings (2007) demonstrate the monophyly of this clade and associated diagnostic morphology.

2008 IUCN RED LIST STATUS Data Deficient

Authors: *DA Pollard and W-L To*

Hyporthodus perplexus

(Randall, Hoese and Last 1991)

puzzling grouper (E), mérou curieux (F), mero acertijo (S)

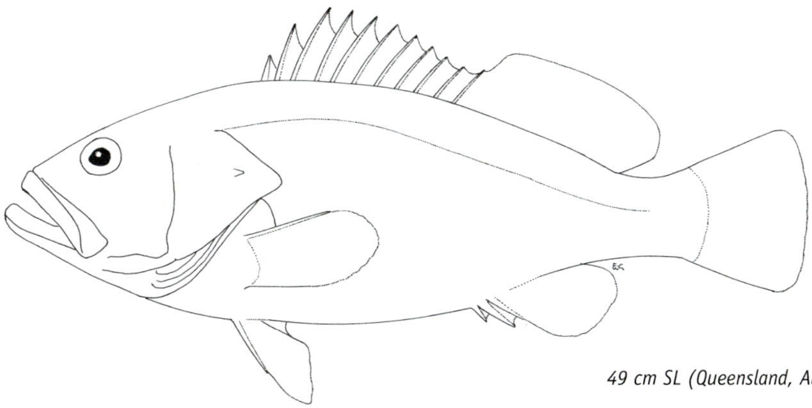

49 cm SL (Queensland, Australia)

(Data from a 465 mm SL fish that was fixed with its mouth wide open; hence ratios of standard length and head length are only approximate.) Body depth contained 2.9 times in standard length. Head length 2.3–2.4 times in standard length; snout length about 3.8 times, eye diameter 6.0 times, interorbital width 6.8 times and suborbital depth 9.5 times in head length; interorbital area slightly convex; preopercle finely serrate, the angle produced into a bulge bearing 7 moderately enlarged serrae; upper edge of operculum straight or nearly straight; posterior nostrils ovate, about twice size of anterior nostrils; midlateral part of lower jaw with 2 rows of teeth. Gill rakers 10 on upper limb, 17 on lower limb. Dorsal fin with 11 spines and 13 rays, the 3rd spine longest, its length contained 3.3–3.4 times in head length and slightly longer than longest dorsal-fin ray; interspinous dorsal-fin membranes deeply incised; anal fin with 3 spines and 8 rays; pectoral fins not fleshy, with 18 rays; pectoral-fin length 2.0–2.1 times and pelvic-fin length 2.5 times in head length; caudal fin slightly rounded, longest caudal-fin ray 2.2 times in head length. Lateral-body scales ctenoid, without auxiliary scales; lateral-line scales 52; lateral-scale series 104.

MAXIMUM RECORDED SIZE 56 cm TL

COLOUR IN LIFE Known only from a preserved specimen. No life colour known.

COLOUR IN MARKET '... after long preservation in formalin, light brown. Margins of unpaired fins dark brown. A dark spot on each body-scale. No dark moustache or other conspicuous markings.' (Whitley 1948)

Hyporthodus perplexus is known only from a single specimen caught off southern Queensland, Australia.

POPULATION GENERAL Nothing is known regarding this species.

POPULATION TREND Unknown

MAJOR THREATS Unknown

HABITAT AND ECOLOGY GENERAL The only known specimen was caught in about 130 m.

COMMENTS This species was formerly placed in the genus *Epinephelus*. However, Craig and Hastings (2007) demonstrate the monophyly of this clade and associated diagnostic morphology.

2008 IUCN RED LIST STATUS Data Deficient

Authors: *PC Heemstra, DA Pollard, KL Rhodes and BC Russell*

Hyporthodus quernus (Seale 1901)

Hawaiian grouper (E), mérou hawaiien (F), mero hawaiano (S)

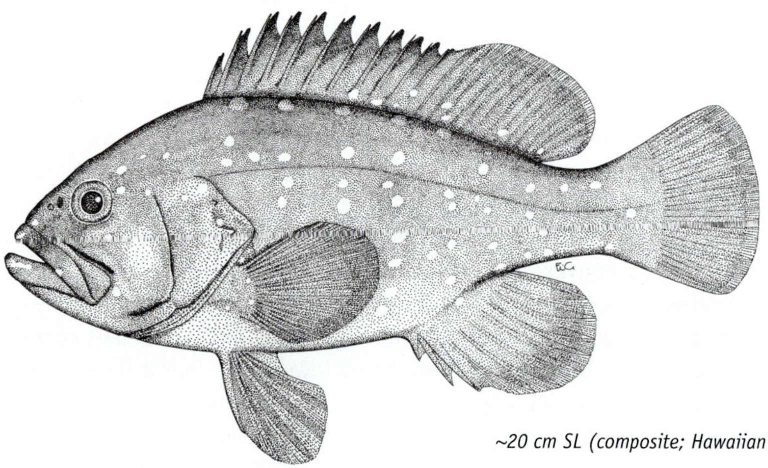

~20 cm SL (composite; Hawaiian Islands, USA)

Body depth contained 2.3–2.7 times in standard length (for fish 10–39 cm SL); body width 2.0–2.5 times in the depth. Head length 2.3–2.5 times in standard length; interorbital area convex, the width subequal to eye diameter for fish 10–20 cm SL; preopercle subangular, the angle with 3–5 enlarged serrae, the lowest directed ventrally; ventral edge of preopercle fleshy, without serrae; edge of interopercle and subopercle smooth or with a few serrae; upper edge of operculum convex; posterior nostrils of adults ovate and enlarged, its greatest diameter 2–4 larger than diameter of anterior nostrils. Gill rakers 8 or 9 on upper limb, 15 or 16 on lower limb, 23–24 total. Dorsal fin with 11 spines and 14 or 15 rays, the 3rd or 4th spine longest, its length contained 2.4–3.1 times in head length and slightly shorter than the longest dorsal-fin rays; interspinous dorsal-fin membranes deeply incised; anal fin with 3 spines and 9 rays; pectoral-fin rays 19 or 20; pelvic-fin origin below base of pectoral fin; pectoral fins subequal to pelvic fins, pectoral-fin length 1.7–2.1 times in head length; caudal fin rounded. Midlateral-body scales ctenoid, with auxiliary scales; lateral-line scales 66–73; lateral-scale series 122–142. Pyloric caeca numerous.

MAXIMUM RECORDED SIZE 80 cm TL

POPULATION **General** *H. quernus* is a common species in the Hawaiian Islands, but it is only rarely observed in the southern portion of the chain where it inhabits deeper waters. **Fishery dependent** *H. quernus* landings and CPUE have shown dramatic declines in the Hawaiian Islands; however, lack of information precludes determining the cause. In the two primary fishing zones in Hawaii, declines of 53% and 33% occurred from 1992 to 2001 (Western Pacific Regional Fishery Management Council 2003). **Fishery independent** *H. quernus* is rarely observed by divers in the southern Hawaiian Islands. It is, however, commonly observed in shallow waters in the northern portion of its range. **Genetics** Subtle population differences in *H. quernus* have been demonstrated; however, it is unclear whether this represents distinct stocks or relic evolutionary partitions (Rivera 2002).

Hyporthodus quernus is found in the central Pacific at the Hawaiian Islands and Johnston Atoll.

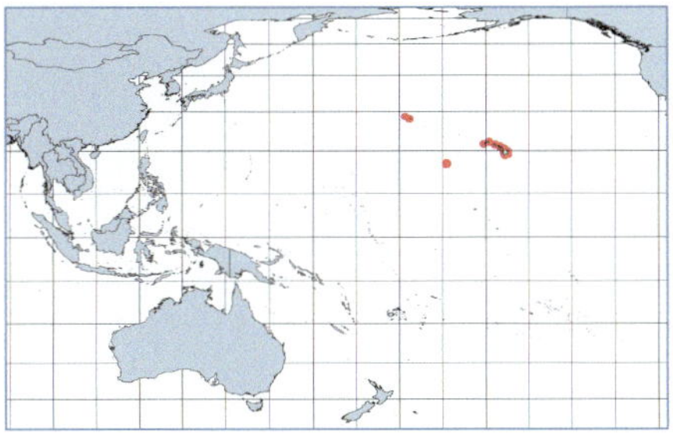

POPULATION TREND Decreasing

MAJOR THREATS Fishing pressure is the primary threat to this species.

HABITAT AND ECOLOGY **General** *H. quernus* inhabits deep coral and rocky reefs in depths of 20–380 m. **Feeding** Feeds on fishes and crustaceans. **Age and growth** The von Bertalanffy growth function fits (r^2 = 0.98) mean back-calculated lengths-at-age as follows: $L = 82.3\ (1 - e^{-0.157(t + 1.45)})$ for age *t*; back-calculated total lengths ranged from 27.5 cm at age 1 year to 87.0 cm at an observed maximum age of 34 years; the best-fit relationship is $W = 0.00002\ L^{2.997}$, where W = total round weight of fish in kg, and L = total length in cm (r^2 = 0.98) (Nichols and DeMartini 2008). **Reproduction** *H. quernus* is a protogynous hermaphrodite, changing sex from female to male at a large size. A study examining 133 individuals in 1978–1981 found that 50% of females had reached sexual maturity by 58–62 cm, while the smallest male was 75 cm and 100% of individuals above 95 cm were male (Everson 1992). *H. quernus* has a protracted spawning season that extends from January through June. Spawning reaches a peak in April, and again in June (Everson 1992).

COLOUR IN LIFE Juveniles greyish brown to dark brown, with 8 vertical series of white spots on the body; adults dark brown with the same vertical series of white spots, but they are less distinct and obscured by numerous additional pale spots and blotches of variable size; fins mostly unspotted and coloured like the body, except for a few pale spots on the basal part of the dorsal fin.

COLOUR IN MARKET Generally brown with only faded and irregular white spots. Often have pale blotches from lying in contact with the ground or the deck of a boat.

© Meagan Sundberg

Adult *H. quernus* (UFA fish auction, Honolulu, Hawaii, USA)

© Keoki Stender

Juvenile *H. quernus* (Waikiki Aquarium, Hawaii, USA)

Adult *H. quernus* (Fish Hole, Midway Atoll, Hawaii, USA)

© Keoki Stender

COMMENTS More than two-thirds of the range of *H. quernus* is now enclosed by the boundaries of the Papahānaoumokuākea Marine National Monument, a no-take marine reserve, and bottom-fishing regulations have become stricter in recent years. It is likely that this combination of protected areas and regulated harvest will reduce the impact of fishing on this species.

This species was formerly placed in the genus *Epinephelus*. However, Craig and Hastings (2007) demonstrate the monophyly of this clade and associated diagnostic morphology.

2004 IUCN RED LIST STATUS Near Threatened

Author: *AS Cornish*

Hyporthodus septemfasciatus

(Thunberg 1793)

convict grouper (E), mérou bagnard (F), mero carcelario (S)

60 cm SL (Japan)

Body depth contained 2.5–3.1 times in standard length (for fish 10–47 cm SL). Head length 2.4–2.6 times in standard length; interorbital area distinctly convex, the dorsal head profile also convex; diameter of posterior nostrils of adults 3 or 4 times that of anterior nostrils and greater than distance from posterior nostril to edge of orbit; preopercle angular, the serrae at angle enlarged; lower edge of preopercle with 1 or 2 small spines anterior to angle; lower edge of subopercle and rear edge of interopercle serrate or smooth, upper edge of operculum convex; maxilla reaches to below rear half of eye, a distinct step on lower edge of maxilla; midlateral part of lower jaw with 2 rows of teeth. Gill rakers 7–9 on upper limb, 15–17 on lower limb. Dorsal fin with 11 spines and 14 or 15 rays, the interspinous membranes deeply incised, the 3rd or 4th spine longest, its length contained 2.1–2.7 times in head length and subequal to longest dorsal-fin rays; anal fin with 3 spines and 9 rays; pectoral-fin rays 17 to 19; pelvic fins shorter than pectoral fins, pelvic-fin length 1.8–2.0 times in head length; caudal-peduncle depth 1.2–1.5 times in peduncle length; caudal fin rounded. Lateral-body scales ctenoid, with a few auxiliaries; lateral-line scales 63–70; lateral-scale series 105–118. Pyloric caeca numerous, in 3 or 4 dendritic bundles with more than 80 terminal branches.

MAXIMUM RECORDED SIZE 155 cm TL

COLOUR IN LIFE Body buff with 7 or 8 dark bars, the 1st on the nape, often merging with the dark colour of the head, the 2nd dark bar below the 3rd and 4th dorsal-fin spines, the 6th and 7th dark bars narrower than the other bars and closer together, extending from the middle rays of the soft dorsal fin to the anal-fin base; 3 pale interspaces below the soft dorsal fin; the last dark bar covers most of the caudal peduncle and extends to the base of the last 2 or 3 dorsal-fin rays; pelvic fins blackish brown; median fins blackish brown distally with narrow pale margin; dark maxillary streak usually hidden by maxilla. Gill rakers and gill arches pale, without minute melanophores.

COLOUR IN MARKET Body generally dark brown; bars on the body may disappear instantly upon death, or may remain but are indistinct.

Adult *H. septemfasciatus* (Tsukiji Fish Market, Tokyo)

POPULATION **General** The population status of *H. septemfasciatus* is unknown, but wild populations have been reputedly depleted to some extent by fishing in some parts of its range. **Fishery dependent** This species is of some commercial fisheries importance in Japan, and probably also in Korea and China (Heemstra and Randall 1993), although fisheries data are unavailable. Much interest has been shown in it more recently, in Japan and elsewhere, as an aquaculture species (e.g. Moon et al. 1999). Although *H. septemfasciatus* is now being bred and released back into the wild for stock enhancement purposes in Japan (Anon. 2003), there is no indication this increases wild populations.

Hyporthodus septemfasciatus is a north-western Pacific species known with certainty only from Japan, Korea and China.

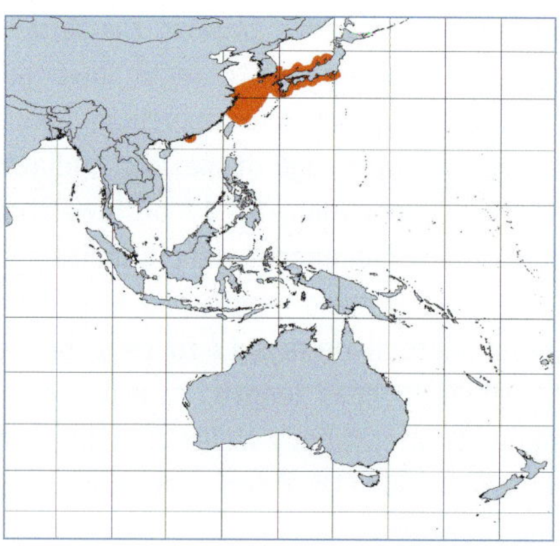

POPULATION TREND Unknown

MAJOR THREATS Fishing pressure is the primary threat to this species and it is likely heavily exploited throughout its range. It is nowhere specifically managed.

HABITAT AND ECOLOGY **General** *H. septemfasciatus* is a reef-associated species that occurs nearshore, including semi-enclosed sea areas on rocky reefs in depths of 5–30 m. **Feeding** Feeds on fishes and crustaceans. **Reproduction** *H. septemfasciatus* is a protogynous hermaphrodite and has been bred and raised in captivity (Kitajima et al. 1991, Anon. 2003).

COMMENTS *H. septemfasciatus* is very similar to *H. octofasciatus*, and to a lesser degree to *H. mystacinus* and *H. ergastularius*. However, *H. septemfasciatus* is more common in shallow water than the other species. *H. octofasciatus* differs in having a deeper body (depth contained 2.2–2.7 times in standard length), deeper caudal peduncle (peduncle depth 0.9–1.3 times in its length), pelvic fins subequal to or longer than pectoral fins, the diameter of the posterior nostrils is not greater than distance from nostril to orbit, and in the configuration of the dark bars on the body (only 5 bars wholly below the dorsal fin, the last one broader than the others and extending from the middle 5–7 dorsal-fin rays to the rear half of the anal fin; 2 pale interspaces below the soft dorsal fin). *H. mystacinus* (of the western Atlantic and eastern Pacific) has the dark body bars narrower than the pale interspaces, and the broad dark bar covering the peduncle does not extend over the bases of the last 2 or 3 dorsal-fin rays. *H. ergastularius*, known only from the east coast of Australia, has a truncate or emarginate caudal fin and fine dermal ridges on the body at the base of the dorsal fin; also, the fins have white margins.

This species was formerly placed in the genus *Epinephelus*. However, Craig and Hastings (2007) demonstrate the monophyly of this clade and associated diagnostic morphology.

2008 IUCN RED LIST STATUS Least Concern

Authors: *M Liu and DA Pollard*

Mycteroperca acutirostris (Valenciennes 1828)

comb grouper (E), badèche peigné (F), cuna negra (S)

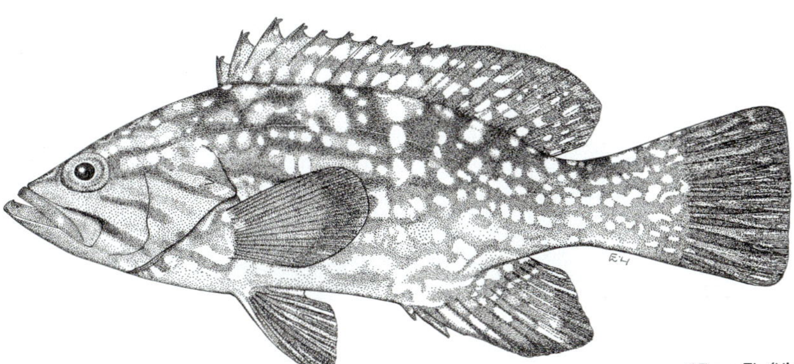

17 cm TL (Virgin Islands)
© SAIAB

Body oblong, compressed, depth contained 2.7–3.2 times in standard length. Head length 2.5–2.7 times in standard length; maxilla width 4.4–5.8% of standard length; interorbital area convex; preopercle angular, with serrae at the angle enlarged, forming a weak lobe; posterior and anterior nostrils subequal in size. Gill rakers 16–20 on upper limb, 32–36 on lower limb, 48–55 total (mean 51.9, n = 25). Dorsal fin with 11 spines and 15–17 rays, the interspinous membranes slightly indented, the margin of posterior part of fin rounded; anal fin with 3 spines and 10–12 rays, the fin margin distinctly pointed in adults; pectoral-fin rays 15–17; caudal fin rounded in fish less than 10 cm SL, truncate in fish 12–20 cm and concave in fish larger than 25 cm SL; no exserted caudal-fin rays, although the fin corners are pointed in large adults. Lateral-line scales 67–77; lateral-scale series 85–106.

MAXIMUM RECORDED SIZE 80 cm TL

COLOUR IN LIFE Head and body greyish brown, covered with irregular white spots and blotches; 3 or 4 dark brown stripes radiating posteriorly from the eye and continuing along the ventral half of the body as wavy dark stripes; another dark brown stripe continuing backwards from the maxillary streak to the edge of the preopercle; median fins darker than the body and also with white spots and streaks; juveniles less than 15 cm TL with a small black saddle on the peduncle; large adults mostly uniform greyish.

COLOUR IN MARKET Body generally brown to pale brown with a subtle mottling pattern comprising irregular pale spots and blotches; the bands radiating from the eye are faded but remain visible; the maxillary streak remains prominent.

© Athila Bertoncini Andrade

Juvenile *M. acutirostris* (Arvoredo Island, Brazil)

© Trevor Meyer

Adult *M. acutirostris* (Bocas del Toro, Panama)

© Luiz A Rocha

Adult *M. acutirostris* (Venezuela)

© Athila Bertoncini Andrade

Adult *M. acutirostris* (Arvoredo Island, Brazil)

POPULATION **General** *M. acutirostris* is generally uncommon in the northern portion of its range, but common in the southern portion (e.g. Brazil). **Fishery dependent** *M. acutirostris* is taken opportunistically throughout its range, and is heavily fished in Brazilian waters. Catch statistics are not available for this fishery, thus the impacts of fishing are difficult to interpret. **Fishery independent** In Brazil, underwater visual censuses recorded densities of up to 0.35 individuals per 100 m^2 (AA Bertoncini unpublished data).

POPULATION TREND Stable

MAJOR THREATS Fishing pressure is the primary threat to this species.

HABITAT AND ECOLOGY **General** Juveniles occur in turtle grass beds, mangrove areas, and in shallow water among soft corals and coral reefs, as well as in tide pools in southern Brazil (Barreiros et al. 2004); adults are found on rocky bottoms with high relief to depths of at least 30 m. The species can be found in groups of dozens of individuals. Small juveniles (2–5 cm) are usually found among floating objects. Adults are often sighted in the water column predating on sardine schools. **Feeding** *M. acutirostris* feeds on macrozooplankton, sardine and squid. **Age and growth** The species reaches approximately 4 kg, although reports of specimens up to 10 kg exist (Carvalho-Filho 1999, Gibran 2004). **Reproduction** Preliminary results of reproductive biology (LC Gerhardinger unpublished data) indicate that 2 kg fish are still juveniles. It is suggested that they spawn during springtime in Brazil (September–December) (Silvano et al. 2006).

COMMENTS *M. acutirostris* is important for small-scale artisanal fisheries as well as amateur fishing in Brazil and is regarded as a target of a 'heavy' fishing pressure by Floeter et al. (2006). The species is also a popular game fish in southern Brazil for both spearfishermen and rod-and-reel anglers. Fish of 4–7 kg are common, with the spearfishing record at 10.1 kg (Capt. Eduardo Baumeier, pers. comm., 2001).

2008 IUCN RED LIST STATUS Least Concern

Mycteroperca acutirostris has a disjunct distribution in the western Atlantic. It is known from Bermuda and the north-western Gulf of Mexico (where it is rare), Cuba, Jamaica, Virgin Islands, Leeward Islands, Panama, Colombia, Venezuela, Curaçao, and southern Brazil. It is absent from north-eastern Brazil and much of the Caribbean Sea.

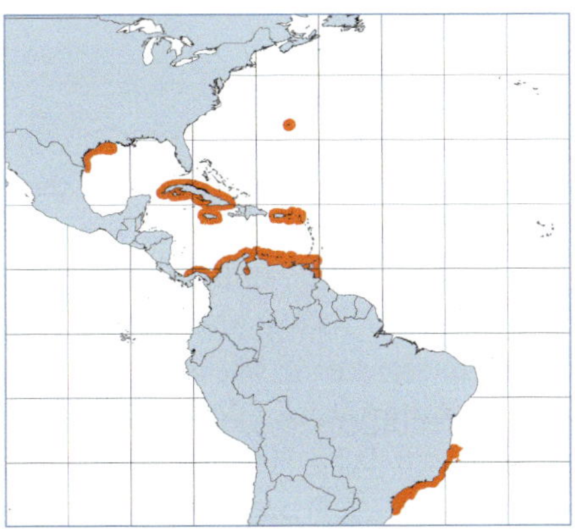

Authors: *AA Bertoncini, JH Choat, MT Craig, BP Ferreira, LC Gerhardinger and LA Rocha*

Mycteroperca bonaci (Poey 1860)

black grouper (E), badèche bonaci (F), cuna bonaci (S)

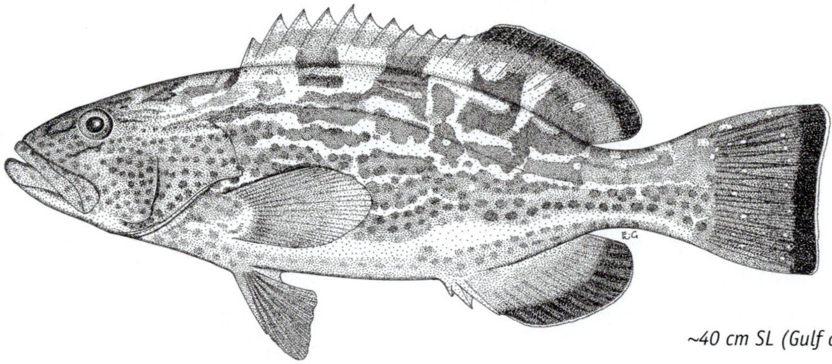

~40 cm SL (Gulf of Mexico)
© SAIAB

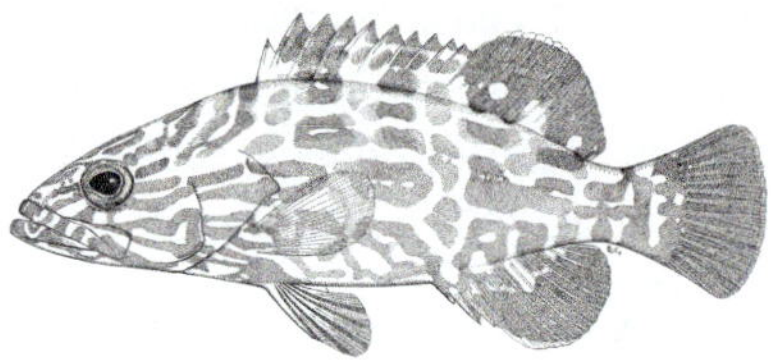

~9 cm SL (Gulf of Mexico)
© SAIAB

Body depth distinctly less than head length, the depth contained 3.3–3.5 times in standard length. Head length 2.5–2.8 times in standard length; preopercle evenly rounded, without a distinct notch or lobe at the angle; posterior and anterior nostrils subequal in size. Developed gill rakers 2–5 on upper limb, 8–12 on lower limb. Dorsal fin with 11 spines and 15–17 rays, the interspinous membranes distinctly incised; anal fin with 3 spines and 11–13 rays; pectoral-fin rays 16 or 17; no median fin rays exserted; margins of dorsal and anal fins rounded; caudal fin truncate (convex if widely spread) to slightly emarginate. Lateral-line scales 78–83; lateral-scale series 119–126.

MAXIMUM RECORDED SIZE 152 cm TL

POPULATION **GENERAL** *M. bonaci* was probably the most abundant large grouper of the shallow-water serranid-lutjanid complex throughout most of its area of occurrence. **FISHERY DEPENDENT** This species was once considered very common in fisheries of the Gulf of Mexico and South Atlantic, among other locations, both for recreational and commercial fisheries. In the 1980s, population(s) apparently underwent significant reductions, and in the early- to mid-1980s many fisheries for *M. bonaci* crashed. For example, in Bermuda, landings showed a sixfold decrease over a seven-year period (Bannerot et al. 1987). In Brazil, the collapse of the *M. bonaci* fishery precipitated a shift in targeted resources to smaller fishes such as snappers (Martins et al. 2005). Declines were also reported in Cuba, Mexico and parts of the USA, as well as from spawning aggregations (NMFS 2006, IUCN Red List Assessment 2008).

POPULATION TREND Decreasing

COLOUR IN LIFE Head and body greyish or dark brown, with close-set irregular spots separated by a bluish border (some spots join to form a longitudinal 'chain' pattern); dorsolateral part of the body sometimes with 7 or 8 columns of rectangular dark blotches, the first above the opercle and the last on the caudal peduncle. Pectoral fins dusky brown, gradually becoming orange at the margin; soft dorsal and anal fins and leading edge of pelvic fins with a dark margin.

COLOUR IN MARKET Body colour fades to a dull blackish brown. The blotches on the dorsolateral portion of the body are often visible, but faded.

Juvenile *M. bonaci*
(Galé Island, Brazil)

© Athila Bertoncini Andrade

© Robert F Myers

Adult *M. bonaci* (Roatan, Honduras)

© Robert F Myers

Adult *M. bonaci* (Roatan, Honduras)

MAJOR THREATS The major threats to this species are overfishing, particularly on its spawning aggregations, and the absence of effective management in many locations.

HABITAT AND ECOLOGY **GENERAL** *M. bonaci* is a reef-associated, solitary species inhabiting rocky and coral reefs to depths of 250 m. Juveniles sometimes occur in estuarine seagrass and oyster rubble habitat in North and South Carolina (USA) (Keener et al. 1988, Ross and Moser 1995), while in the Florida Keys and Brazil, juveniles settle on patch reefs (Sluka et al. 1994, Schwamborn 2005). **FEEDING** Adults feed primarily on fishes, including grunts, snappers and herring. Juveniles feed solely on crustaceans. **AGE AND GROWTH** In south Florida, Crabtree and Bullock (1998) report that this species reaches a maximum age of at least 33 years; the von Bertalanffy growth equation is $L = 1.306(1 - e^{-0.169(t + 0.768)})$ where L is the total length in mm at age t. In the Florida Keys, Manooch and Mason (1987) obtained the following von Bertalanffy growth equation for *M. bonaci*: $L = 1.352 (1 - e^{-0.1156(t + 0.927)})$ where L is the total length in mm at age t. The maximum age estimated was 14 years. **REPRODUCTION** The species is a monandric, protogynous hermaphrodite that forms spawning aggregations (García-Cagide and García 1996, Crabtree and Bullock 1998, Teixeira et al. 2004). At Campeche Bank (Gulf of Mexico), females changed sex between 85.5 and 125.0 cm FL, with median length of 103.3 cm FL. At 114.5 cm FL, 50% of the females in the sample had transformed into males (Brulé et al. 2003). The age at sex change was 15.5 years. Peak spawning varies by locality: in Brazil, peak spawning period occurs between April and September (Teixeira et al. 2004), in Cuba between November and February (García-Cagide and García 1996), in Florida from December to March (Crabtree and Bullock 1998), and at Campeche Bank (Mexico) the spawning season is from December to March (Brulé et al. 2003).

2008 IUCN RED LIST STATUS Near Threatened

Mycteroperca bonaci is a western Atlantic species distributed from Bermuda and Massachusetts (USA) to Brazil (to about 28° S), including the Brazilian offshore islands, the southern Gulf of Mexico, the Florida Keys, the Bahamas, Cuba, and throughout the Caribbean.

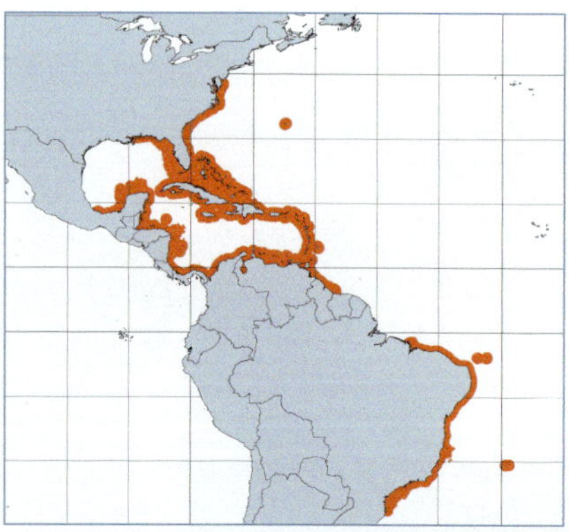

Authors: *MT Craig, BP Ferreira, ALB Gaspar and LA Rocha*

Mycteroperca cidi (Cervigón 1966)

Venezuelan grouper (E), badèche blanche (F), cuna blanca (S)

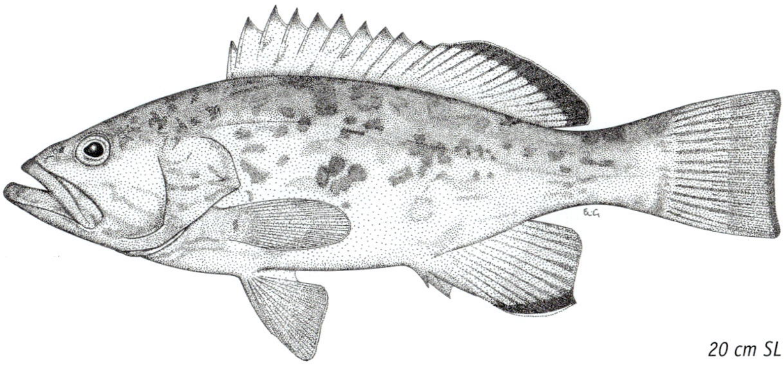

20 cm SL (Venezuela)
© SAIAB

Body depth less than head length, depth contained 3.1–3.3 times in standard length. Interorbital area distinctly convex; preopercle angular, with a distinct lobe bearing enlarged serrae at the angle; nostrils subequal in fish less than 25 cm SL, but in fish larger than 40 cm SL the posterior nostrils are much larger than the anterior nostrils. Gill rakers well developed, the longest is longer than eye diameter; 9–13 on upper limb, 18–23 on lower limb (not counting 2–4 rudiments on each limb). Dorsal fin with 11 spines and 15–17 rays, the 9th to 11th rays distinctly longer than the others in adults, giving the fin an angular aspect; anal fin with 3 spines and 10–12 rays, the fin margin pointed in adults with the 4th to 6th rays elongated; pectoral-fin rays 15–17; caudal fin truncate to slightly emarginate, with greatly elongated exserted rays in adults (greater than 40 cm TL). Lateral-body scales smooth; lateral-line scales about 75; lateral-scale series 108–126.

MAXIMUM RECORDED SIZE 114 cm TL

Mycteroperca cidi is found only on the Caribbean coast of Venezuela and was recently reported from the Port Royal reefs off Jamaica, indicating that it may be found in other Caribbean localities.

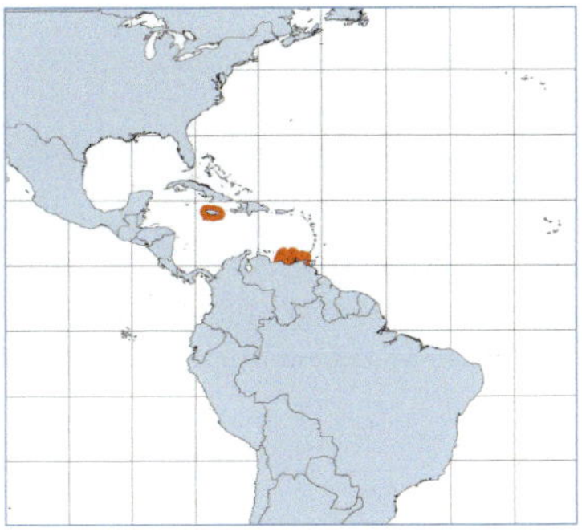

POPULATION GENERAL *M. cidi* is locally common in Venezuela and known only from a single specimen in Jamaica. There are no data on its population numbers.

POPULATION TREND Unknown

MAJOR THREATS No major threats have been identified for this species, although fishing is a possible threat and needs to be evaluated.

HABITAT AND ECOLOGY GENERAL *M. cidi* is common within its restricted range in Venezuela on coral reefs in depths of 5–40 m. Juveniles are common in shallow water over sandy bottoms near and on coral reefs and seagrass beds.

2008 IUCN RED LIST STATUS Data Deficient

COLOUR IN LIFE Adults pale greyish brown; juveniles greenish brown with irregular brown spots on the body; soft dorsal and anal fins with a white edge and dark submarginal band.

COLOUR IN MARKET Pale brown to grey with scattered irregular spots that may be faded and interrupted with pale patches.

© JAMARC

Adult *M. cidi* (Suriname)

© Luiz A Rocha

Juvenile *M. cidi* (Venezuela)

© Luiz A Rocha

Adult? *M. cidi* (Venezuela)

Authors: *JH Choat, MT Craig, BP Ferreira and LA Rocha*

Mycteroperca fusca (Lowe 1836)

island grouper (E), mérou d'île (F), abadejo (S)

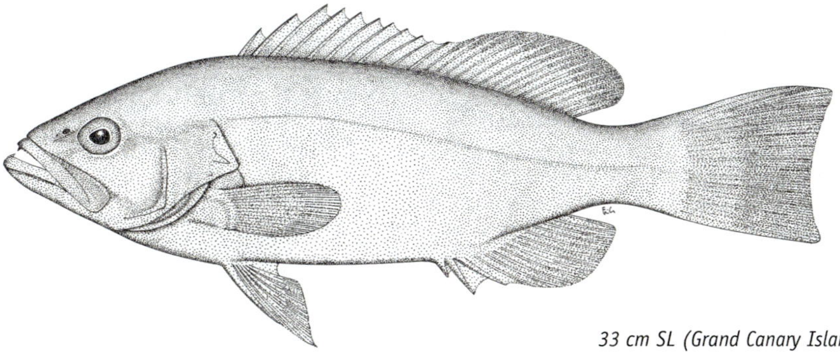

33 cm SL (Grand Canary Island, Spain)

Body oblong, compressed, the depth less than head length, depth contained 3.0–3.3 times in standard length. Head length 2.6–2.9 times in standard length; interorbital area convex; preopercle serrae enlarged at the angle, forming a rounded lobe below a shallow indentation on vertical limb; subopercle and interopercle with a few small serrae; nostrils subequal in juveniles, the diameter of posterior ones about 3 times larger than anterior nostrils in fish larger than 45 cm SL; maxilla width 3.8–5.0% of standard length; lower jaw extends well in front of upper jaw. Gill rakers 11–14 on upper limb, 20–24 on lower limb, 32–36 total. Dorsal fin with 11 spines and 14–16 rays, the interspinous membranes slightly indented, the margin of soft-rayed part rounded; anal fin with 3 spines and 10–12 rays, the fin margin rounded; pectoral-fin rays 15–17; caudal-fin rear margin truncate (juveniles) to concave (adults). Lateral-line scales 72–78; lateral-scale series. 96–106.

MAXIMUM RECORDED SIZE 80 cm TL

Mycteroperca fusca is known only from the eastern Atlantic. The species is known with certainty only from the Azores and Madeira (Portugal), Cape Verde and the Canary Islands (Spain).

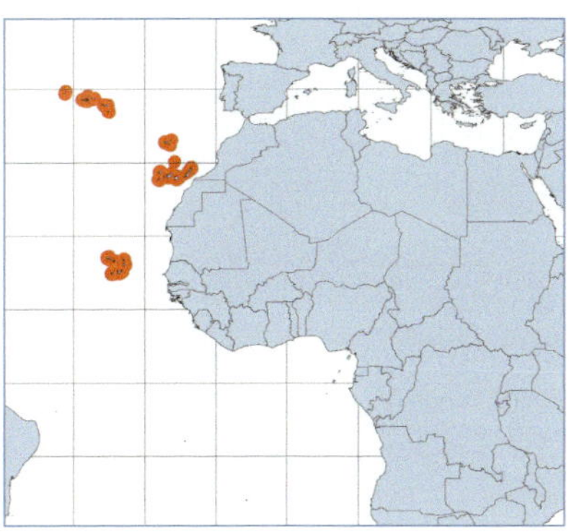

POPULATION GENERAL *M. fusca* is an uncommon species throughout its limited range **FISHERY DEPENDENT** While no detailed fishery dependent data are currently available, the species was previously abundant, but due to fishing pressure it is now locally rare. There has been local extirpation from the most intensively fished areas in the Canary Islands and a significant part of its range and there are significant declines in mature-sized fish. Both *M. fusca* (along with *Epinephelus marginatus*) showed the strongest negative responses to variations in fishing intensity and human population among the Canary Islands, thus supporting the hypothesis that major human intervention has affected the abundance and biomass of both species across the archipelago

COLOUR IN LIFE Adults are brownish or dark grey, with irregular pale blotches and spots and a prominent maxillary streak; a live fish under stress may reverse this pattern so that the head and body are pale, with irregular dark markings. Juveniles mottled greenish brown, with prominent white spots on the head and body, white streaks on the median fins and hyaline golden pectoral fins. *M. fusca* may also display a yellow, xanthic, colour phase.

COLOUR IN MARKET Uniform grey or brown; a very nondescript species in death.

© Rui Freitas

Adult *M. fusca* (Cape Verde)

© Carlos Rangel

Adult *M. fusca* (Cape Verde)

© Carlos Rangel

Adult *M. fusca* (colour variant; Cape Verde)

(Tuya et al. 2006). **FISHERY INDEPENDENT** In the Canary archipelago, among all groupers surveyed by underwater visual census, *M. fusca* had the greatest mean biomass by area (464.63 g per 100 m², SD 1464.25; *n* = 768 transects), while the mean abundance was 0.32 individuals per 100 m² (SD 0.76; *n* = 768). Significant differences in abundances and biomasses were observed among the islands surveyed (Tuya et al. 2006).

POPULATION TREND Decreasing

MAJOR THREATS Fishing pressure is the major threat to this species, combined with lack of management, especially during spawning periods.

HABITAT AND ECOLOGY **GENERAL** *M. fusca* occurs in rocky areas at depths from 1 to 200 m. Juveniles may be found in tide pools. **AGE AND GROWTH** Age and growth data on 214 fish collected between January 2004 and December 2005 show *M. fusca* to be slow-growing and long-lived, with ages up to 20 years recorded. The von Bertalanffy growth parameters are as follows: L_∞ = 898 mm, k = 0.062 y^{-1} and t_0 = –3.83 y; males and females show differences in growth (Bustos et al. 2009). **REPRODUCTION** The species forms spawning aggregations which are targeted by fisheries and some no longer form (IUCN Red List Assessment 2008).

2008 IUCN RED LIST STATUS Endangered (B1ad(v))

Authors: *AA Bertoncini, JH Choat, MT Craig, BP Ferreira, LA Rocha and CLS Sampaio*

Mycteroperca interstitialis (Poey 1860)

yellowmouth grouper (E), badèche gueule jaune (F), cuna amarilla (S)

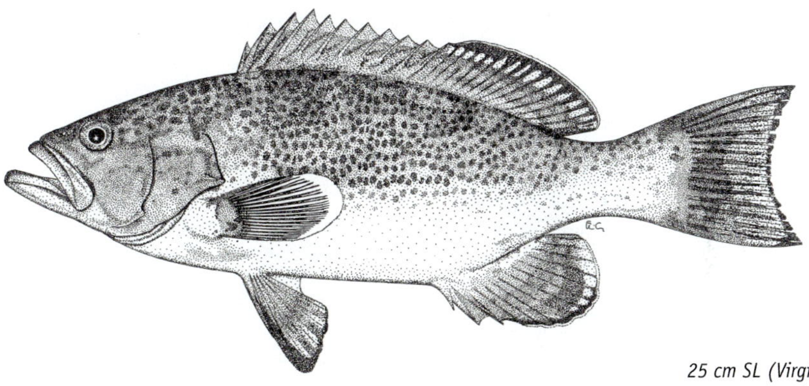

25 cm SL (Virgin Islands)
© SAIAB

Body depth less than head length, depth contained 3.0–3.4 times in standard length. Head length 2.8–3.0 times in standard length; preopercle angular, with a prominent lobe bearing enlarged serrae at the angle and a distinct notch above the serrate lobe; posterior nostrils greatly enlarged in adults. Gill rakers 4–6 on upper limb, 11–15 on lower limb, plus 3–5 rudiments on each limb, 23–27 total. Dorsal fin with 11 spines and 16–18 rays, the 12th and 13th rays elongated in adults, giving the fin a pointed rear margin; anal fin with 3 spines and 10–12 rays, the fin margin pointed in adults, with 6th to 8th rays distinctly longer than others; pectoral-fin rays 16 or 17; caudal fin emarginate, with evenly exserted rays in adults. Lateral-body scales ctenoid (weakly so in large adults); lateral-line scales 70–74; lateral-scale series 112–128.

MAXIMUM RECORDED SIZE 84 cm TL

COLOUR IN LIFE Pale brownish grey, with close-set small brown spots on the dorsal part of the body; margin of the spinous dorsal fin and mouth yellowish; pectoral-fin membranes clear, the rays dark and the fin margin with a white edge; dark moustache streak present above the maxilla. Some fish uniform brown dorsally, and others may have faint irregular vertical bars. Juveniles bicoloured; head and body dark brown dorsally (the dark dorsal colour may be broken into broad dark bars or saddle blotches) and abruptly white below; a white mid-dorsal stripe from the tip of the lower jaw along the top of the snout, head and base of the dorsal fin.

COLOUR IN MARKET Generally pale brown to grey with no distinguishing marks on the body. Mouth retains yellow colour which is usually faded on the dorsal fin in dead specimens. Bicoloured juveniles will retain this colour pattern in death.

© Thierry Brulé

Adult *M. interstitialis* (Merida Fish Market, Mexico)

© Athila Bertoncini Andrade

Subadult *M. interstitialis* (Arvoredo Island, Brazil)

© Robert F Myers

Adult *M. interstitialis* (Bonaire)

© Robert F Myers

Juvenile *M. interstitialis* (Bonaire)

POPULATION **General** *M. interstitialis* is a naturally uncommon species throughout its range. **Fishery dependent** Despite its rarity, *M. interstitialis* is an important component of fisheries throughout its range (IUCN Red List Assessment 2008). Catch data indicate significant declines in landings. For example, in the 1980s, a rapid development of trap fishing on shallow-water grouper populations in Bermuda led to a seven-year decline in grouper catches from about 18 000 kg to about 4500 kg per year (Bannerot et al. 1987). Similar declines in both landings and catch rates have been noted throughout the species' range (Mendoza and Larez 1996, Manickchand-Heileman and Phillip 2000, Costa et al. 2003). Extirpation is possible for this uncommon species, since it is taken along with similar yet more abundant and more persistent species, such as *M. phenax* (Musick et al. 2000) and *M. bonaci* (Ferreira et al. 1998). In Brazil, a sharp decline was observed in the grouper fishery in the Abrolhos Bank (*Mycteroperca* spp.). The present CPUE (1.4 kg per fisher per day) is considerably lower than in the late 1970s (12 kg per fisher per day) (Costa et al. 2003). **Fishery independent** Fishery-independent data (SCUBA observations and experimental fishing) confirm the natural rarity of this species.

POPULATION TREND Decreasing

MAJOR THREATS Fishing pressure is the major threat to this species, and management is lacking in most areas where it occurs.

HABITAT AND ECOLOGY **General** *M. interstitialis* is an insular, reef-associated species found mainly on rocky or coral bottoms from the shoreline to depths of at least 55 m. Small and medium-sized individuals commonly occur in mangrove-lined lagoons. **Feeding** Feeds mainly on fishes. **Age and growth** *M. interstitialis* may reach at least 41 years (Manickchand-Heileman and Phillip 2000). The weight–length relationship for *M. interstitialis* from the Gulf of Mexico was given by Bullock and Smith (1991) as follows: $W = 4.7 \times 10^{-8} L^{2.893}$ where *W* is the weight in kg and *L* is the standard length in mm.

COMMENTS *M. interstitialis* is similar in appearance to *M. phenax*. However, *M. interstitialis* of 20–57 cm SL usually has shorter pelvic fins and a shorter head (15–17% and 30–36% of standard length, vs 18–20% and 36–38% of standard length, respectively, in *M. phenax*). *M. interstitialis* also lacks dark spots on the dorsal and anal fins, and the exserted caudal-fin rays are of similar length, whereas *M. phenax* has small brown spots on these fins and the exserted caudal-fin rays are more uneven.

2008 IUCN RED LIST STATUS Vulnerable (A2d+3d)

Mycteroperca interstitialis is distributed in the western Atlantic from Florida (USA) and Bermuda to southern Brazil, including the Brazilian offshore islands, the southern Gulf of Mexico, Florida Keys, Bahamas, Cuba and the Caribbean.

Authors: *JH Choat, MT Craig, BP Ferreira and LA Rocha*

Mycteroperca jordani

(Jenkins and Evermann 1889)

gulf grouper (E), mérou golfe (F), mero baya (S)

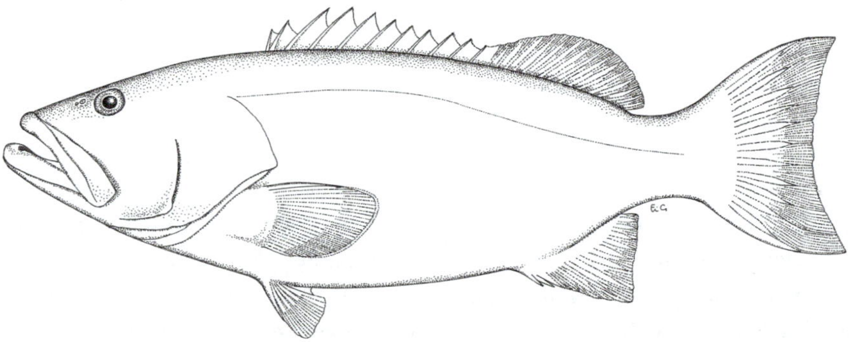

150 cm SL (Baja California, Mexico)
© SAIAB

Body depth much less than head length, the depth contained 3.1–3.4 times in standard length. Head length 2.5–2.7 times in standard length; preopercle rounded, finely serrate, without a distinct lobe at the angle; posterior nostrils not notably enlarged, their diameter less than twice that of anterior nostrils. Total gill rakers 21–26, not counting rudiments. Dorsal fin with 11 spines (4th or 5th longest) and 16 or 17 rays, the interspinous membranes deeply incised in adults; posterior margin of dorsal fin rounded; anal fin with 3 spines and 10 or 11 rays, the fin margin pointed in large adults; pectoral-fin rays 16–18; no fin rays exserted; caudal fin rounded in juveniles, truncate to emarginate in adult. Midlateral-body scales ctenoid on small juveniles, smooth on adults; lateral-line scales 75–85; lateral-scale series 125–128.

MAXIMUM RECORDED SIZE 198 cm TL

Mycteroperca jordani occurs in the eastern Pacific from La Jolla, California (USA), to Mazatlán (Mexico).

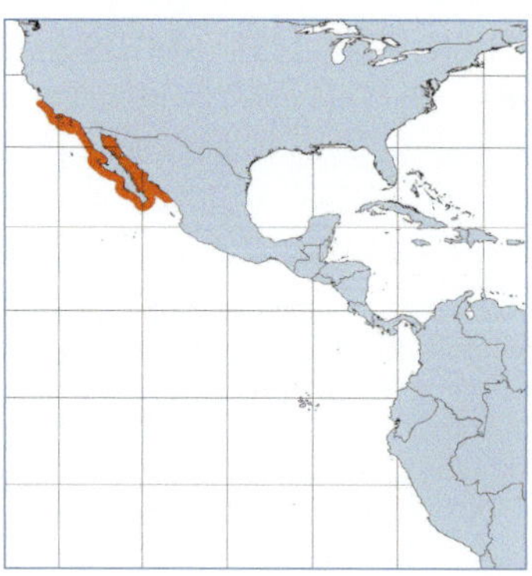

POPULATION General Naturally rare in the northern portion of its range. Sáenz-Arroyo et al. (2005) reveal that *M. jordani* in central Baja California was abundant in the past and probably dominated the rocky-reef fish community in terms of biomass. **Fishery dependent** *M. jordani* has been an important component of fisheries in Mexico. Sáenz-Arroyo et al. (2005) report that *M. jordani* comprised 45% of the total state finfish production between 1959 and 1960. By early 1972, the percentage had dropped to 6% (Sáenz-Arroyo et al. 2005). More recent estimates suggest that *M. jordani* comprises less than 1% of the total finfish

COLOUR IN LIFE Adults usually uniform dark brown or grey, but they can rapidly assume the juvenile pattern if disturbed or excited. In large adults, the pectoral-fin margin is white and the median fins have a narrow white edge. Juveniles greyish brown, with large dark spots along the upper half of the body.
COLOUR IN MARKET Most specimens are uniform brown to grey, especially larger specimens. Smaller fish may retain spotting pattern but this may be faded and interrupted by patches of pale skin.

© Matthew T Craig

Adult *M. jordani* (Gulf of California, Mexico)

© Octavio Aburto

Adult *M. jordani* (Gulf of California, Mexico)

catch (Rámirez-Rodriguez 1996). Based on all available evidence, there has been a 10-fold reduction in daily landings and CPUE between 1950s and 2000s, and the fishery is now comprised of juvenile fish. Interviews with fishermen indicated a 50–70% decline since 1950 in the Gulf of California. Based on changes in the numbers of individuals within spawning aggregations, the population decline from 1940s to the present could be greater than 99% (IUCN Red List Assessment 2008). **FISHERY INDEPENDENT** Fishery-independent data confirm the rarity of this species.
POPULATION TREND Decreasing
MAJOR THREATS Fishing pressure is the major threat to this species in the absence of effective management, because it is anticipated to increase and because spawning aggregations are targeted.
HABITAT AND ECOLOGY **GENERAL** *M. jordani* is found on rocky reefs and in kelp beds in depths to at least 30 m. **FEEDING** Feeds on fishes and crustaceans. **REPRODUCTION** *M. jordani* is an aggregating spawner (Sáenz-Arroyo et al. 2005), and there is a skewed sex ratio (males significantly fewer than females) in the Gulf of California. In the Gulf of California, 62% of captured *M. jordani* are near 60 cm TL in length and are smaller than the size of sexual maturity, which occurs around 97.9 cm TL in length. Immature individuals comprise approximately 99% of

the commercial catch. The average lifespan of this species is estimated to be 48.3 years (Aburto-Oropeza et al. 2008).

COMMENTS *M. jordani* is heavily targeted by recreational and subnational fisheries throughout its range. It is also incidentally caught by shrimp trawlers in the Gulf of California. Recreational fishers from the USA are a major contributor to the overfishing of remaining stocks.

Few large adults, and no juveniles, of *M. jordani* have been taken in California, and these likely represent expatriates from a more southerly breeding population (Rosenblatt and Zahuranec 1967). Despite these sporadic records, there is no evidence of historically abundant populations in this northern portion of the species range. Some confusion exists about this as there are historical landings data for *M. jordani* in California; however, these data represent landing of fish caught further south (Fitch 1949, Craig et al. 2006c).

2008 IUCN RED LIST STATUS Endangered (A2d+3d)

Authors: *AA Bertoncini, JH Choat, MT Craig, BP Ferreira, PC Heemstra and LA Rocha*

Mycteroperca microlepis

(Goode and Bean 1880)

gag (E), badèche baillou (F), cuna aguají (S)

40 cm SL (Gulf of Mexico)

Body depth much less than head length, the depth contained 3.0–3.5 times in standard length. Head length 2.5–2.7 times in standard length; interorbital area convex; angle of preopercle (on fish larger than about 40 cm SL) produced into a rounded lobe bearing enlarged serrae; adults (larger than 60 cm) with posterior nostrils much larger than anterior ones. Gill rakers 8 or 9 on upper limb, 16 on lower limb, including 5 or 6 rudiments on each limb. Dorsal fin with 11 spines and 16–18 rays, the 3rd or 4th spine longest, but shorter than longest dorsal rays; and the membranes distinctly incised between the spines; anal fin with 3 spines and 10–12 rays; dorsal and anal fins rounded at all sizes; no exserted dorsal- and anal-fin rays; pectoral-fin rays 16–18; caudal fin emarginate. Lateral-body scales smooth, except those covered by pectoral fin; lateral-line scales 88–96; lateral-scale series 128–146.

MAXIMUM RECORDED SIZE 145 cm TL

COLOUR IN LIFE Adult females and juveniles are generally brownish grey with darker vermiculations. A resting or 'camouflage phase' has 5 dark brown saddles separated by short white bars below the dorsal fin. Large adults (males) may display a 'black-belly' and 'black-back' phase. The black-belly phase is mostly pale grey, with faint dark reticulations below the soft dorsal fin, the belly and ventral part of the body above the anal fin are black, as are the margin of the soft dorsal fin, central rear part of the caudal fin and rear margins of the pectoral and pelvic fins. The black-back phase is similar to the black-belly phase, but with more black pigment on the rear part of body, the dorsal half of the peduncle, all of the soft dorsal and anal fins, and also over the snout and front of the jaws; the caudal fin is white with a black margin posteriorly.

COLOUR IN MARKET Uniform brown to grey; may retain alternate colour phases if displaying these at death.

© Thierry Brulé

Adult *M. microlepis* (Merida Fish Market, Mexico)

© Leonardo Machado

Adult *M. microlepis* (Santa Catarina, Brazil)

POPULATION **GENERAL** *M. microlepis* is a common species throughout its range where it is extremely important economically for commercial and recreational use. It is the second most common commercial grouper species in commercial catches of the USA. **FISHERY DEPENDENT** Generally, fishery-dependent data do not show any clear patterns although the species is taken in considerable numbers. There has been local decline due to overfishing in the Gulf of Mexico; according to interviews with grouper fishers in the early 1990s, those in north-western Florida reported that there was a steady decline in the landings over the past decades (Koenig et al. 1996). However, data on landings of *M. microlepis* quoted from different studies and sources show that the quantity of commercial and recreational catches of *M. micolepis* increased steadily from 1980 to 2004, with recreational catches averaging 55% from 1995 to 2004. Integrated populations analysis of *M. microlepis* stock suggests that the recruitment of *M. microlepis* is highly variable and that very young fish suffer little fishing mortality, while older fish face increasing fishing pressure with age such that very few fish survive longer than 10 years. When compared with sex-size data collected before the 1980s, sex-size frequency data from *M. microlepis* collection in 1991–1993 show reductions in mean female and male sizes. The Gulf of Mexico stock has been classified as overfished, but the south Atlantic stock appears to be relatively stable (NMFS/NOAA Fish Stock Sustainability Index 2006, IUCN Red List Assessment 2008). **FISHERY INDEPENDENT** Fishery-independent data indicate a steady decline in *M. microlepis* populations since the 1980s.

POPULATION TREND Decreasing

MAJOR THREATS Fishing pressure is the main threat to this species.

HABITAT AND ECOLOGY **GENERAL** *M. microlepis* is a common species found on soft bottoms and rocky ledges in depths of 30 to at least 160 m. Juveniles occasionally occur in estuaries and seagrass beds. Adults are either solitary or occur in groups of 5–50 individuals. *M. microlepis* appears to prefer habitats characterised by maximum structural complexity, such as living *Oculina* coral reefs, at depths between 70 and 100 m (McGovern et al. 1998). In the southern Gulf of Mexico, juvenile females have been taken inshore at depths of 1–10 m, while adult females, males and transitioning intersex individuals have been taken offshore at depths of 33–167 m (Brulé et al. 2003). **FEEDING** Feeds on a variety of fishes and invertebrates. **AGE AND GROWTH** *M. microlepis* can exceed 20 years; estimates of the von Bertalanffy growth equation parameters for this species are as follows: L_∞ = 1190 mm TL, k = 0.166 y^{-1} and t_0 = –0.62 y (Hood and Schlieder 1992). **REPRODUCTION** *M. microlepis* is a protogynous hermaphrodite. Sex change occurs between 3 and 5 years (about 72 cm TL). Spawning occurs in aggregations from December to May, with peak activity occurring during February and March in the eastern Gulf of Mexico (Hood and Schlieder 1992). In the southern Gulf of Mexico, females reach first maturity at 72.1 cm FL (Brulé et al. 2003) and spawn at depths of 50–53 m from January to March. There has been a dramatic decrease in the percentage of males in spawning aggregations (17% to 1% in 20 years) (Chapman et al. 1999, Koenig et al. 2000). Outside of the spawning season, males remain on spawning sites while females move into relatively more shallow water.

Mycteroperca microlepis is a western Atlantic species with a wide, disjunct distribution. It is known from North Carolina (USA) (with juveniles occurring as far north as Massachusetts) to the Yucatan Peninsula, Mexico, and from southern Brazil.

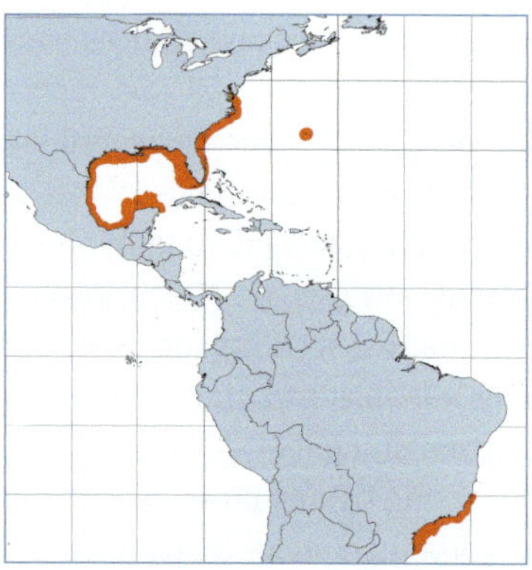

COMMENTS Juveniles of *M. microlepis* that have not yet developed the distinctive notch and rounded lobe at the angle of the preopercle are similar in appearance to *M. bonaci*, but *M. microlepis* has more gill rakers (8–12 on lower limb of *M. bonaci*) and higher scale counts (lateral-line scales 78–83 and lateral-scale series 119–126 for *M. bonaci*). Chapman et al. (1999) comment that the size-limited population of *M. microlepis* suffers from restricted gene flow among genetically differentiated populations.

2008 IUCN RED LIST STATUS Least Concern

Authors: *AA Bertoncini, JH Choat, MT Craig, BP Ferreira and LA Rocha*

Mycteroperca olfax (Jenyns 1843)

sailfin grouper (E), mérou voile (F), garropa parda (S)

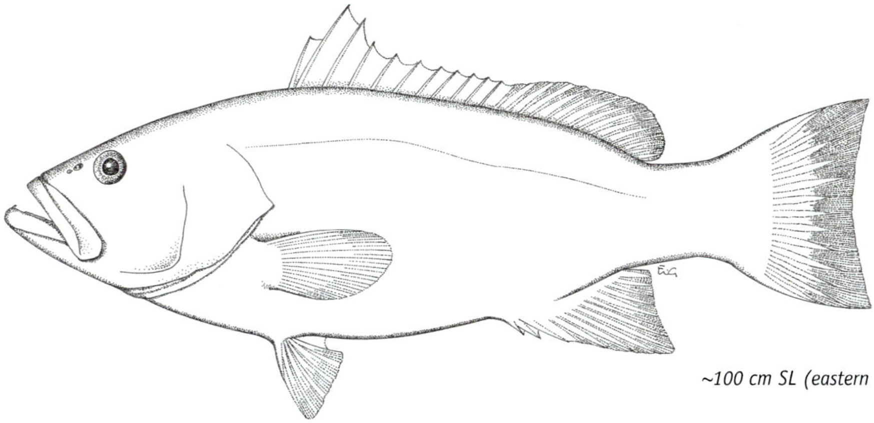

~100 cm SL (eastern Pacific)

Body depth less than head length, the depth contained 2.9–3.2 times in standard length. Head length 2.6–2.9 times in standard length; interorbital slightly convex; preopercle angular, with a serrate lobe at the angle; posterior nostrils of adults (more than 40 cm SL) greatly enlarged, more than twice size of anterior nostrils. Total gill rakers 24–29 (not counting rudiments). Dorsal fin with 11 spines and 16 or 17 rays, the 2nd and 3rd dorsal-fin spines elongated, forming an elevated lobe, the interspinous membranes not indented; soft-rayed part of dorsal fin rounded; anal fin with 3 spines and 11 rays, the fin pointed in fish longer than 25 cm SL; pectoral-fin rays 16 or 17; caudal fin truncate in juveniles, slightly emarginate in adults, without exserted rays. Lateral-scale series 92–110.

MAXIMUM RECORDED SIZE 120 cm TL

Mycteroperca olfax is mainly distributed at insular localities in the eastern Pacific. It is known with certainty from Cocos Island (Costa Rica), Malpelo Island (Colombia) and the Galápagos Islands (Ecuador). There is a single confirmed record from mainland Ecuador (Béarez and Jiménez Prado 2003).

POPULATION **GENERAL** *M. olfax* is common in the Galápagos Islands, but rare at other island sites within its range. **FISHERY DEPENDENT** *M. olfax* was once the most valuable and exploited fish in the Galápagos Islands artisanal fishery (Kasteleijin 1987), comprising over 40% of the catch (Reck 1983). Now, this species comprises <20% of the catch and fishers indicate that the CPUE and average individual size have declined (Bustamente 1998). **FISHERY INDEPENDENT** In the Galápagos, the estimated density of individuals is 2.97 per 500 m^2 within a marine reserve area, 2.86 per 500 m^2 in an area reserved for tourism activities, and 2.6 per 500 m^2 in an area open to fishing (Edgar et al. 2004a).

POPULATION TREND Unknown

MAJOR THREATS Fishing pressure and lack of effective management are the major threats to this species.

HABITAT AND ECOLOGY **GENERAL** *M. olfax* is found among rock walls, rocky reef drop-offs, underwater lava ridges, and around well-structured vertical rock formations (Reck 1983), in depths to at least 30 m. Juveniles are found in shallow sandy lagoons, mangrove lagoons, shallow lava reefs and inland lava ponds. **FEEDING** Feeds on fishes and crustaceans (Reck 1983). **REPRODUCTION** *M. olfax* is a protogynous hermaphrodite. Sexual maturity occurs at about 47 cm SL. Spawning peaks in October and December (Coello and Grimm 1993). First maturation occurs at 47.5 cm SL and at 4 years, with sex inversion at 83.1 cm SL, possibly at 12 years of age. The species has spawning peaks in October and December (Coello and Grimm 1993).

COMMENTS Although *M. olfax* has shown declines in landings over the past two decades, it is unclear whether this reflects a decrease in abundance since fishing effort information

COLOUR IN LIFE Typically dark to olive-brown body spotted with purplish and lighter brown, or plain dark brown. A xanthic morph is bright yellow all over.

COLOUR IN MARKET Generally a shade of brown with no distinguishing marks.

© Matthew T Craig

Adult *M. olfax* (Galápagos Islands, Ecuador)

© Chris Grossman

Adult *M. olfax* (Cocos Island, Costa Rica)

is not available. Nicolaides et al. (2002), for example, report that decreases in landings around the year 2000 reflected a change in the targeted resources, with a shift in focus to sea cucumbers and lobsters.

2008 IUCN RED LIST STATUS Vulnerable (D2)

Authors: *AA Bertoncini, JH Choat, MT Craig, BP Ferreira, LC Gerhardinger and LA Rocha*

Mycteroperca phenax (Jordan and Swain 1885)

scamp (E), badèche galopin (F), cuna garopa (S)

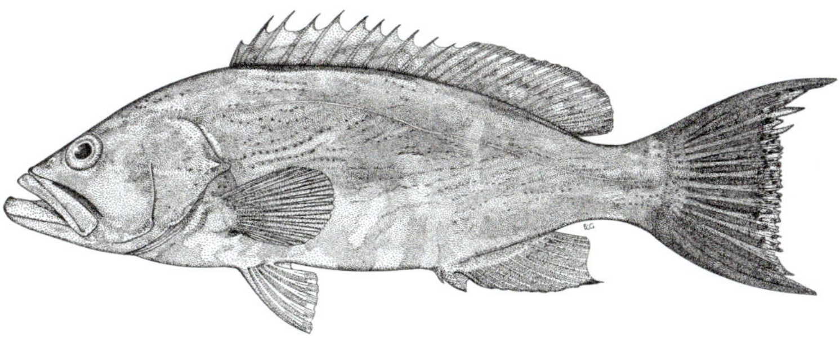

35 cm SL (Gulf of Mexico)
© SAIAB

Body depth less than head length, the depth contained 3.0–3.4 times in standard length. Head length 2.6–3.0 times in standard length; interorbital area convex; preopercle angular, with a distinct bony lobe at the angle (in fish larger than 40 cm SL); subopercle and interopercle serrate; posterior nostrils of adults 2–4 times larger than anterior nostrils. Gill rakers 8–10 on upper limb, 17–21 on lower limb, 26–31 (total, including 3 or 4 rudiments on each limb), the longest raker longer than the longest gill filament. Dorsal fin with 11 spines and 16–18 rays; anal fin with 3 spines and 10–12 rays, fin margin pointed in adult; pectoral-fin rays 15–17, fin length about half head length; caudal fin concave; adults with unevenly exserted rays in dorsal, anal and caudal fins. Lateral-body scales ctenoid; lateral-line scales about 76–82; lateral-scale series about 124.

MAXIMUM RECORDED SIZE 107 cm TL (not including exserted caudal-fin rays)

POPULATION **GENERAL** *M. phenax* is a common species throughout its range and is an important component of recreational and commercial fisheries. **FISHERY DEPENDENT** Landings data of *M. phenax* in the USA indicate a decrease in the average size of landing fish throughout the 1990s, but do not show marked signs of reduced number of fish taken. The US South Atlantic stock was assessed in 1997 and was not overfished at the time (Potts and Brennan 2001). The Gulf of Mexico stock has not been assessed.

POPULATION TREND Stable

MAJOR THREATS Fishing pressure is the major threat to this species.

HABITAT AND ECOLOGY **GENERAL** *M. phenax* is a reef-associated species found in a variety of habitats at depths of 30–100 m. **FEEDING** Feeds on fishes, cephalopods and crustaceans (Matheson et al. 1986). **AGE AND GROWTH** *M. phenax* reaches at least 27 years. While there were no differences in the mean lengths or ages of immature *M. phenax* between 1979–1989 and 1990–1997, mature *M. phenax* were significantly smaller (594 vs 557 mm TL) and younger (5.6 vs 4.9 years) during the later period (Harris et al. 2002). The percentage of *M. phenax* age 10 and older declined from 17% in 1979–1989 to 7% in 1990–1997 (Harris et al. 2002). **REPRODUCTION** *M. phenax* is a protogynous hermaphrodite. Spawning occurs in aggregations from February through July in the South Atlantic Bight and in the Gulf of Mexico, with a peak in March to mid-May.

COLOUR IN LIFE Four colour patterns were described and illustrated by Gilmore and Jones (1992); the usual pattern is the brown phase, with the head and body pale brown, covered (except ventrally) with small reddish brown spots (1 on each scale) that extend onto the median fins. The "cat's paw" phase is pale brown, the dorsolateral parts of body with several clusters of dark brown spots resembling the paw print of a cat. Larger adults displayed a grey-head phase, with the rear two-thirds of the body dark; the head and the body anterior to the 6th dorsal-fin spine is silvery grey with dark reticulations, the light areas being the same areas that were darkly pigmented in the "cat's paw" phase; belly and ventral part of the body above the anal fin with several white spots; fins white except for the black margin of the pectoral fins. The bicoloured phase, seen only once, was pale brown anteriorly and abruptly dark chocolate brown posteriorly, with the transition at the origin of the soft dorsal fin. All colour phases are ephemeral, but only large adults (over 50 cm SL) displayed the grey-head phase.

COLOUR IN MARKET Head and body uniform brown to grey with no distinguishing marks.

© Phillip C Heemstra

Adult *M. phenax* (Gulf of Mexico, Florida, USA)

© Robert F Myers

Adult *M. phenax* (Fort Lauderdale, Florida, USA)

Mycteroperca phenax is an occasional trans-Atlantic species occurring on the continental coast of North and Central America, from North Carolina (USA) to Venezuela, and at the Azores Islands (Portugal).

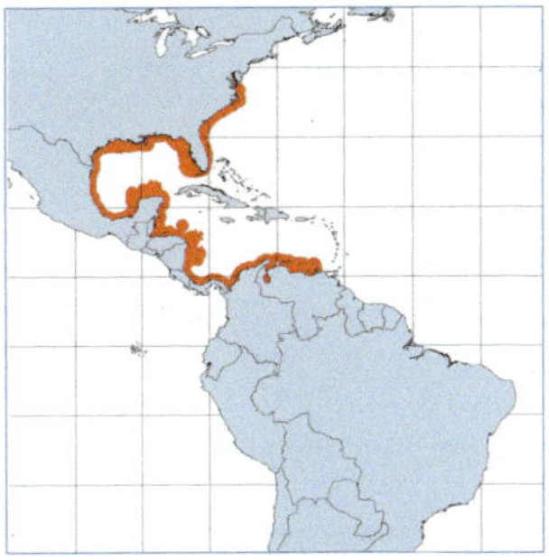

Length and age at first spawning of females off North Carolina to south-eastern Florida (USA) is 30–35 cm TL and age 1–2. Nearly all females were mature by 451–500 mm TL and age 4. The percentage of males declined from 34% in 1979–1989 to 21% 1990–1997 for specimens >500 mm TL (Harris et al. 2002).

2008 IUCN RED LIST STATUS Least Concern

Authors: *AA Bertoncini, JH Choat, MT Craig, BP Ferreira, JC McGovern and LA Rocha*

Mycteroperca prionura

(Rosenblatt and Zahuranec 1967)

sawtail grouper (E), mérou scie-queue (F), garropa aserrada (S)

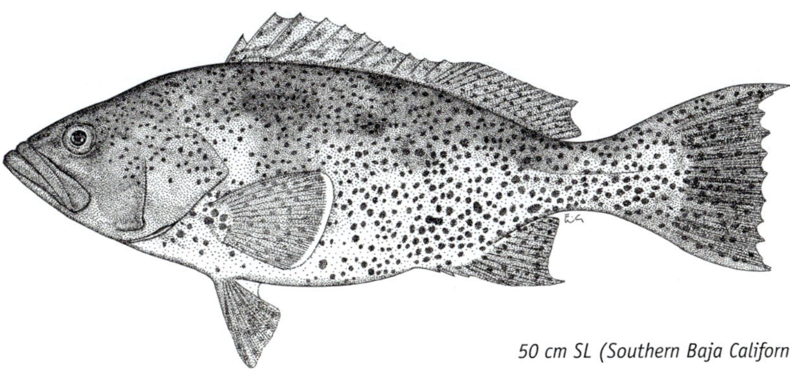

50 cm SL (Southern Baja California, Mexico)

Body depth less than head length. Preopercle angular, with a serrate lobe at the angle; posterior nostrils not greatly enlarged in adults. Total gill rakers 34–38, not counting rudiments. Dorsal fin with 11 spines and 16–18 rays, the 3rd spine longest, but not much longer than 4th spine, the following spines gradually shorter to the 10th, which is slightly shorter than the last spine, interspinous membranes not indented; anal fin with 3 spines, 10–12 rays; margin of soft dorsal and anal fins pointed in fish larger than 35 cm SL; pectoral-fin rays 15 or 16; caudal fin of juveniles (less than 20 cm SL) rounded or truncate; in adults (larger than 25 cm SL), rear margin of caudal fin scalloped, with exserted rays. Lateral-scale series 85–109.

MAXIMUM RECORDED SIZE 100 cm TL

COLOUR IN LIFE Adults pale grey to brown with a purplish tint underwater; the body, median fins and rear part of the head (except ventrally) covered with small dark reddish brown spots; pectoral fins and pelvic fins dusky, with faint irregular dark spots; dusky blotches on the dorsolateral part of the body. Juveniles less than 10 cm SL yellowish tan, with round, well-spaced, reddish brown spots on the body and similar but smaller spots on the head and median fins; dark stripe from the eye to the tip of the lower jaw. With growth, the dark spots on the body become smaller and more numerous.

COLOUR IN MARKET Generally pale grey to brown with a faint purplish tint; dark reddish spots remain, but less prominent than in life.

© Matthew T Craig

Adult *M. prionura* (Gulf of California, Mexico)

© Octavio Aburto

Adult *M. prionura* (Gulf of California, Mexico)

Mycteroperca prionura is an eastern central Pacific species occurring from the Gulf of California south to Jalisco (Mexico).

POPULATION **General** *M. prionura* is naturally rare and low in abundance within a relatively small distribution range (Heemstra and Randall 1993, Musick et al. 2000). It is more abundant in the central and northern Gulf of California than the southern gulf. **Fishery independent** Mean density in rocky habitats in 1999–2001 was 0.02 ± 0.01 (mean ± SE) individuals per 250 m^2 (Sala et al. 2003).

POPULATION TREND Decreasing

MAJOR THREATS Although not specifically targeted, fishing pressure poses a threat to this species and there is no species-specific management. Sportfishing, and illegal spearfishing using lights and hookah breathing apparatus at night from La Paz to the Midriff Islands in the Gulf of California, are particular threats.

HABITAT AND ECOLOGY **GENERAL** *M. prionura* is a reef-associated species found on rocky reefs to depths of 50 m. It is more common on fields of large boulders with gorgonians and black corals. **FEEDING** Feeds on fishes and crustaceans. **REPRODUCTION** *M. prionura* is known to occur in mixed species spawning aggregations with *M. rosacea* and *M. jordani*. Nothing is known of its reproductive biology.

COMMENTS Despite being uncommon, *M. prionura* is opportunistically targeted by fishers, both commercially and recreationally. Because it occurs in mixed-species aggregations with *M. rosacea*, which is heavily targeted during spawning, this species is indirectly targeted by the sportfishing industry.

2008 IUCN RED LIST STATUS Near Threatened

Authors: *MT Craig and DA Pollard*

Mycteroperca rosacea (Streets 1877)

leopard grouper (E), mérou léopard (F), cabrilla sardinera (S)

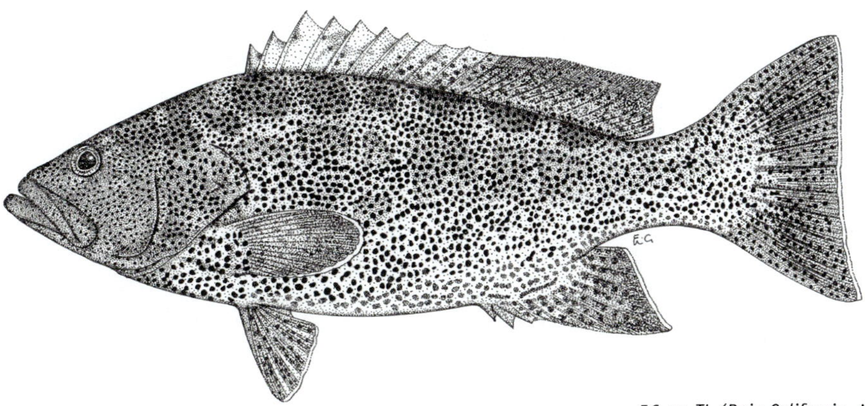

~56 cm TL (Baja California, Mexico)

Body depth less than or subequal to head length, the depth contained 2.7–3.1 times in standard length. Head length 2.6–2.8 times in standard length; preopercle angle of adults with a weakly-developed serrate lobe; posterior nostrils of large adults 2 or 3 times larger than anterior nostrils. Total gill rakers 38–43, more than 21 rakers on lower limb. Dorsal fin with 11 spines and 16–18 rays, the interspinous membranes slightly indented, the margin of soft-rayed part of fin in large adults angular posteriorly; anal fin with 3 spines and 10 or 11 rays, the fin margin pointed (except in small juveniles) with 5th or 6th ray elongated; pectoral-fin rays 15–17; caudal fin truncate or concave posteriorly; no exserted fin rays Midlateral-body scales ctenoid; lateral-line scales 72–77; lateral-scale series 106–120.

MAXIMUM RECORDED SIZE 98 cm TL

COLOUR IN LIFE Greenish to greyish brown, covered with small reddish brown spots and irregular pale spots and lines, the fins with a white margin. A second xanthic pattern is exhibited by a small portion of the population in which the fish is entirely bright yellow-orange, sometimes with a few irregular black spots.

COLOUR IN MARKET *M. rosacea* displays nearly identical colouration in death. Pale blotches may be present if the fish was left on the deck of a boat or floor of fish market.

© Tomas Plomozo

Adult *M. rosacea* (normal colour morph [top] and xanthic morph [bottom]; La Paz Fish Market, Baja California Sur, Mexico)

© Octavio Aburto

Adult *M. rosacea* (Gulf of California, Mexico)

Juvenile *M. rosacea* (Gulf of California, Mexico)

© Octavio Aburto

POPULATION **General** *M. rosacea* is very common throughout the Gulf of California. **Fishery dependent** *M. rosacea* is the most commonly taken grouper in the Gulf of California. While no explicit landings data are currently available, observations indicate a reduction in population abundance over the past 10 years of approximately 50% of the standing biomass due to intense subsistence, artisanal and recreational fishing throughout the range of the species (MO Aburto-Oropeza, B Erisman and M Buckhorn unpublished data). **Fishery independent** *M. rosacea* is commonly observed by divers in the Gulf of California (Mexico).

POPULATION TREND Decreasing

MAJOR THREATS Fishing pressure is the major threat to this species in the absence of management.

HABITAT AND ECOLOGY **General** *M. rosacea* is a reef-associated species that prefers rocky areas in shallow water to depths of about 50 m. **Feeding** Adults feed on schools of the flatiron herring *Harengula thrissina* and anchoveta *Cetengraulis mysticetus*. **Age and growth** *M. rosacea* reaches at least 21 years in the southern Gulf of California (Díaz-Uribe et al. 2001). That study described length–weight relationships, and listed the von Bertalanffy growth parameters as: L_∞ = 1226 mm (total length), k = 0.0659 y^{-1} and t_0 = –1.393 y. **Reproduction** *M. rosacea* is a gonochoric species (Erisman et al. 2008). Females and males mature at about 22 cm SL (Erisman et al. 2008). Spawning occurs in aggregations at dusk from April to June (Erisman et al. 2007).

2008 IUCN RED LIST STATUS Vulnerable (A2ad+4ad)

Mycteroperca rosacea is distributed in the eastern central Pacific, ranging throughout the Gulf of California to Jalisco (Mexico).

Authors: *MT Craig and YJ Sadovy de Mitcheson*

Mycteroperca rubra (Bloch 1793)

mottled grouper (E), badèche rouge (F), gitano (S)

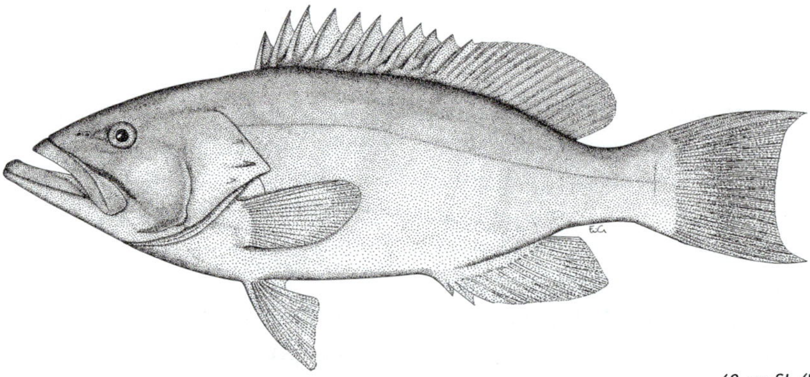

49 cm SL (West Africa)
© SAIAB

Body oblong, compressed, the depth less than head length, depth contained 2.8–3.2 times in standard length. Head length 2.5–2.7 times in standard length; interorbital area convex; preopercle serrae enlarged at the angle, forming a rounded lobe set off by the indentation immediately above; nostrils subequal in juveniles, the diameter of posterior nostrils about twice that of anterior ones in a fish of 59 cm SL; maxilla width 3.8–4.5% of standard length for fish 17–37 cm SL, 4.8% of standard length for a fish of 59 cm SL. Gill rakers 16–18 on upper limb and 28–31 on lower limb (including 1–3 rudiments on each limb), 44–49 total. Dorsal fin with 11 spines and 15–17 rays, the interspinous membranes distinctly indented, the posterior margin of fin rounded; anal fin with 3 spines and 11 or 12 rays, the fin margin angular in adults; pectoral-fin rays 16 or 17; caudal-fin margin convex in juveniles less than about 20 cm SL, truncate in fish of 20–50 cm SL, and concave in adults over 50 cm SL. Lateral-line scales 69–76; lateral-scale series 94–108.

MAXIMUM RECORDED SIZE 80 cm TL

Mycteroperca rubra is an eastern Atlantic species that occurs throughout the Mediterranean and from Portugal to southern Angola.

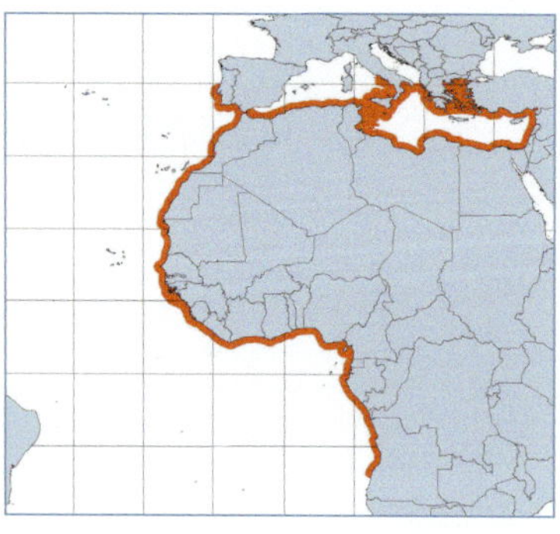

POPULATION **General** *M. rubra* is a widespread, but generally uncommon, species. **Fishery dependent** *M. rubra* is only occasionally taken in fisheries, likely due to its rarity.

POPULATION TREND Unknown

MAJOR THREATS No major threats have been identified for this species.

HABITAT AND ECOLOGY **General** *M. rubra* occurs over rocky and sandy bottoms from 15 to 200 m. **Feeding** *M. rubra* feeds on molluscs and small fishes (Freitas 1982, Diatta et al. 2003). **Reproduction** *M. rubra* is a protogynous hermaphrodite (Siau and Bouain 1994). Sex change occurs at 9 years and 53 cm TL. Siau and Bouain (1994) report that *M. rubra* spawns

COLOUR IN LIFE Generally uniform reddish brown, sometimes mottled with blackish or pale grey spots with a black streak above the maxilla. Juveniles with a black saddle blotch on the peduncle.

COLOUR IN MARKET Uniform brown to grey; maxillary streak remains prominent.

© Patrick Louisy

Adult *M. rubra* (Kaş, Turkey)

© Matthew T Craig

Adult *M. rubra* (Monaco Aquarium)

at the end of spring in the Mediterranean, with sexual maturation occurring at 27–32 cm TL and 4–5 years.

COMMENTS *M. rubra* is very similar in appearance to *M. fusca* of the eastern Atlantic but the latter has only 20–24 lower gill rakers.

2008 IUCN RED LIST STATUS Least Concern

Authors: *AA Bertoncini, JH Choat, MT Craig, BP Ferreira and LA Rocha*

Mycteroperca tigris (Valenciennes 1833)

tiger grouper (E), badèche tigre (F), cuna gata (S)

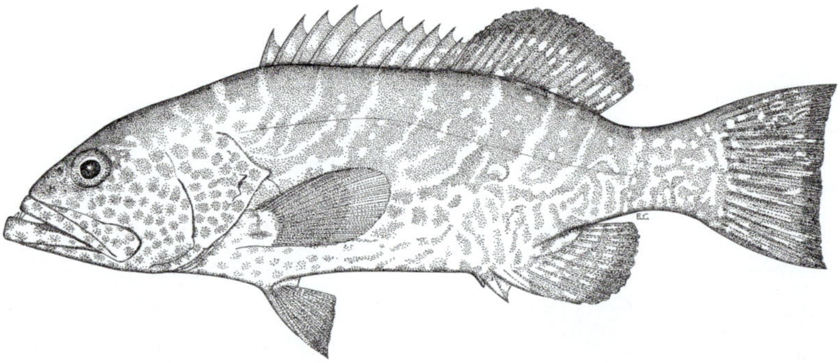

37 cm SL
© SAIAB

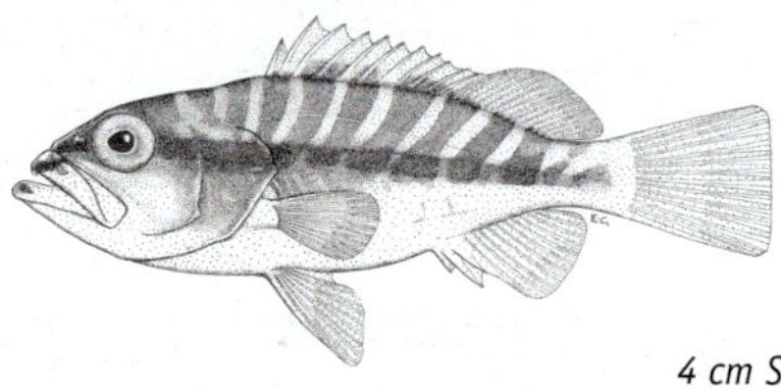

4 cm SL (Bermuda)
© SAIAB

Body depth less than head length, the depth contained 3.1–3.6 times in standard length (for fish 19–43 cm SL). Head length 2.5–2.8 times in standard length; interorbital area flat; preopercle rounded, without a lobe at the angle; diameter of posterior nostrils 3–5 times larger than anterior nostrils in fish larger than 40 cm SL; teeth large, canines well developed. Gill rakers short, 8 on upper limb, 15–17 on lower limb (including 5 or 6 rudiments on upper limb and 8 or 9 on lower limb), 23–25 total. Dorsal fin with 11 spines and 15–17 rays, the interspinous membranes distinctly indented; anal fin with 3 spines and 11 rays; in large adults the soft dorsal and anal fins are pointed, with the middle rays elongate; pectoral-fin rays 17; caudal fin truncate in juveniles, truncate to emarginate in adults; in fish larger than 60 cm SL, some caudal-fin rays are exserted. Midlateral-body scales ctenoid in juveniles, smooth in adults; lateral-line scales 82–83; lateral-scale series about 120.

MAXIMUM RECORDED SIZE 100 cm TL

POPULATION **GENERAL** *M. tigris* is a generally uncommon species throughout its range. **FISHERY DEPENDENT** *M. tigris* is taken opportunistically by fishermen and is targeted at some of its spawning aggregations. There are no indications of overall declines in landings although the current taking of the species from unmanaged spawning aggregations is likely to affect the species. At one monitored site landings declined from 5 t in 1995 to 2 t in 1998 prompting suggestion of need for possible management (Matos-Caraballo et al. 2006). **FISHERY INDEPENDENT** Data from underwater visual census suggest possible declines of this species in the Greater and Lesser Antilles.

POPULATION TREND Unknown

MAJOR THREATS No major threats have

COLOUR IN LIFE Adults greenish brown, grey or red with pale rivulations; head and body darker dorsally, with 9–11 wavy oblique pale stripes; median fins with irregular pale spots and stripes; pectoral fins pale yellow distally. Live colouration of small juveniles (25–100 mm SL) yellow, with a blackish brown midlateral stripe from the tip of the lower jaw through the eye and along body almost to the caudal fin (P Colin pers. comm.). The longitudinal stripes do not meet on the lower jaw, and they become fainter with growth, being mostly obscured by the dark oblique dorsal bars of the adult colour pattern on fish of about 20 cm SL. The temporary colour of the ripe male in the spawning aggregation is distinctive, with a pale head, yellow-tipped black pectoral fins and a large white ventro-posterior patch.

COLOUR IN MARKET *M. tigris* maintains nearly identical colouration in death.

© Thierry Brulé

Adult *M. tigris* (Merida Fish Market, Mexico)

© Robert F Myers

Adult *M. tigris* (Bonaire)

been identified for this species other than uncontrolled fishing on some of its spawning aggregations.

HABITAT AND ECOLOGY **GENERAL** *M. tigris* occurs over hard substrates in depths of at least 30 m. **FEEDING** Feeds primarily on fishes. **AGE AND GROWTH** Log W = –4.985 + 3.161 log L where W is the weight in g and L is the standard length in mm (n = 11; Sadovy et al. 1994). **REPRODUCTION** Spawning aggregations of several hundreds of fish take place approximately one week following each

Mycteroperca tigris is a western Atlantic species, occurring from south Florida (USA), the Gulf of Mexico, the Bahamas, Cuba, Haiti, Jamaica, Venezuela, and throughout the Caribbean to Brazil.

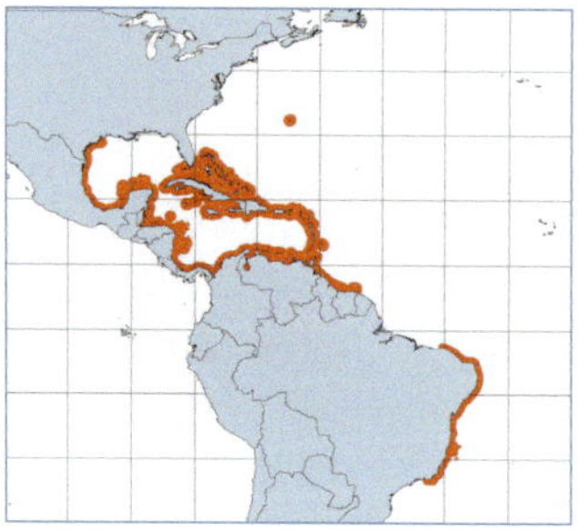

full moon between the months of January to April in Puerto Rico (Sadovy et al. 1994). Females reach sexual maturity at between 25 and 35 cm TL and the smallest mature male sampled at one aggregation was 357 mm SL (Sadovy et al. 1994). The species is a protogynous hermaphrodite.

2004 IUCN RED LIST STATUS Least Concern

Author: *G García-Moliner*

Mycteroperca venenosa (Linnaeus 1758)

yellowfin grouper (E), badèche de roche (F), cuna de piedra (S)

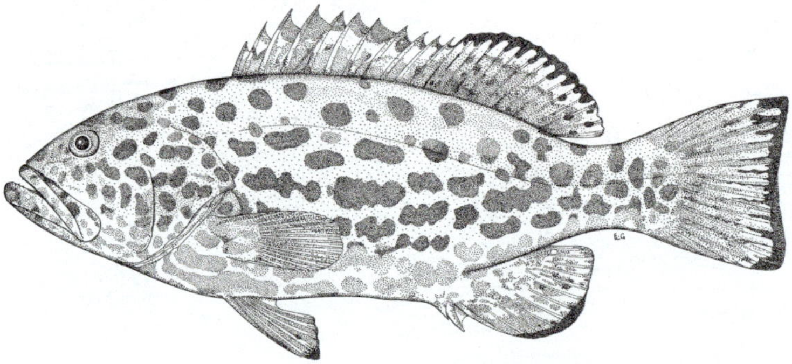

24 cm SL (Puerto Rico)
© SAIAB

Body depth less than head length, depth contained 2.9–3.2 times in standard length (for fish 15–54 cm SL). Head length 2.6–2.9 times in standard length; interorbital area convex; preopercle evenly rounded or with a slight notch, but no projecting bony lobe at the angle; nostrils subequal or posterior nostrils about twice diameter of anterior nostrils. Gill rakers 8–10 on upper limb and 17 or 18 on lower limb, including 4–7 rudiments on each limb, 24–27 total. Dorsal fin with 11 spines and 15 or 16 rays, the interspinous membranes distinctly indented between the spines; anal fin with 3 spines and 10–12 rays; soft dorsal- and anal-fin margins rounded; no exserted rays in the median fins; pectoral-fin rays 16–18; caudal fin truncate in juveniles, concave in adults. Midlateral-body scales ctenoid in juveniles, smooth in adults, with numerous auxiliary scales; lateral-line scales 72–81; lateral scale series 111–125.

MAXIMUM RECORDED SIZE 100 cm TL

COLOUR IN LIFE Two colour morphs: a deep-water reddish form and a shallow-water greenish form. Both colour morphs show the following markings: the head and body with oblong dark blotches overlying groups of small black spots; ventral part of the head and body with rosettes of small, dark red spots; soft dorsal, anal and caudal fins with a dark margin and white edge; the distal third of the pectoral fins abruptly yellow, the basal part of fin with small, irregular dark spots. The temporary colour of the ripe male in the spawning aggregation is distinctive, with a bright yellow blotch on both sides of the lower jaw.

COLOUR IN MARKET *M. venenosa* maintains similar colouration in death.

© Matthew T Craig

Adult *M. venenosa* (Dry Tortugas, Florida, USA)

© Ewald Lieske

Adult *M. venenosa* (locality unknown)

POPULATION **GENERAL** *M. venenosa* was once considered common but has undergone population reductions throughout its range. However, in some areas this species is ciguatoxic and not heavily taken for food. **FISHERY DEPENDENT** Landings data show steady declines in catches of between 80% and 90% from 1980 through 2000 in many countries throughout its range, with no indication of a reduction in effort or any other factor that could explain the landings other than overfishing (IUCN Red List Assessment 2004). Populations in the Caribbean area are suggested to be large historically (Heemstra and Randall 1993), although current population size may be considerably reduced as inferred from the landings of this species in the USA and its territories and the status of its spawning aggregations. Spawning aggregations are often targeted and many have undergone decline.

POPULATION TREND Decreasing

MAJOR THREATS Fishing pressure is the major threat to this species with little effective management.

Mycteroperca venenosa is distributed along the east coast of Central America within the Caribbean, extending south to São Paulo (Brazil).

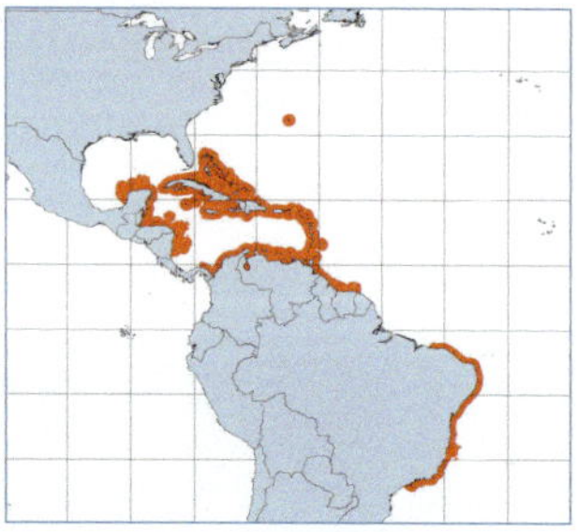

HABITAT AND ECOLOGY **General** Juveniles occur in shallow turtle grass beds; adults are usually found on rocky and coral reefs in depths of 2–137 m. **Feeding** Feeds on fishes and invertebrates. **Reproduction** Spawning occurs at different times of the year depending on location. For example, in Bermuda spawning occurs in July, whereas in the Florida Keys it occurs in March. In Cuba, this species spawns during winter and spring with peaks in January, April and May, and in February through April in the US Virgin Islands (IUCN Red List Assessment 2004).

2004 IUCN RED LIST STATUS Near Threatened

Authors: *T Brulé and G García-Moliner*

Mycteroperca xenarcha (Jordan 1888)

broomtail grouper (E), mérou genêt-queue (F), mero brujo (S)

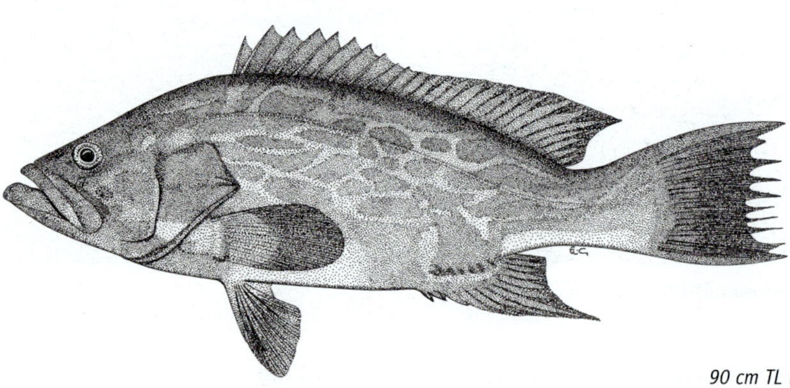

90 cm TL (Costa Rica)
© SAIAB

Body depth less than head length, depth contained 2.9–3.1 times in standard length. Head length 2.6–2.8 times in standard length; preopercle angular, with a distinct serrate lobe at the angle; posterior nostrils not notably larger than anterior nostrils. Gill rakers 29–33 (total, not counting rudiments). Dorsal fin with 11 spines and 15 or 16 rays, the interspinous membranes not indented, the margin of the soft-rayed part becomes pointed (with rays 9–11 elongated) in fish larger than 20 cm SL; anal fin with 3 spines and 10 or 11 rays, the fin margin also pointed (rays 4–6 elongated) in fish larger than 10 cm SL; pectoral-fin rays 16–18; caudal fin truncate, with exserted rays in fish larger than 20 cm SL.

MAXIMUM RECORDED SIZE 150 cm TL

COLOUR IN LIFE There are two colour patterns. The blotched pattern is brown, grey or greyish green, with oblong dark blotches (most enclosing a pale line) dorsally, close-set in juveniles, forming an irregular maze-like, pale reticulum; in adults the dark blotches are more widely spaced, with the pale reticulum expanded to form a general pale ground colour; the ventral part of the body with irregular dark elongate markings; juveniles with a conspicuous black saddle on the caudal peduncle. The plain colour pattern is uniform greyish or brown, the fins darker and edged with white. Large adults are mostly plain, but may assume the blotched pattern briefly.

COLOUR IN MARKET Large individuals are uniform brown with no distinguishing marks. Small adults may retain blotches in death. Pale blotches often form in long-dead specimens.

© D Ross Robertson

Adult *M. xenarcha* (eastern Pacific)

© Daniel W Gotshall

Adult *M. xenarcha* (Gulf of California, Mexico)

POPULATION General *M. xenarcha* was once very common, particularly in the Gulf of California (Mexico), but has undergone population declines.

POPULATION TREND Unknown

MAJOR THREATS No major threats have been identified for this species.

HABITAT AND ECOLOGY General *M. xenarcha* inhabits mangrove areas, including estuaries (Thomson et al. 2000) and reefs from shallow waters to depths of up to 60 m.

COMMENTS While small, isolated groups of *M. xenarcha* are persistent in California (e.g. King Harbor in Redondo Beach and the La Jolla kelp bed, California), there is no evidence of historically abundant populations in this northern portion of its range. Some confusion exists about this as there are historical

Mycteroperca xenarcha is distributed in the eastern central Pacific from San Francisco Bay, California (USA), to southern Peru.

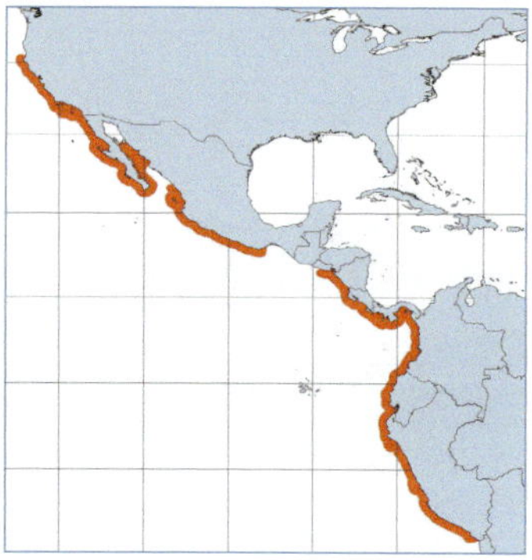

landings data for *M. xenarcha* in California; however, these data represent landing of fish caught further south (Fitch 1949, Craig et al. 2006c).

2008 IUCN RED LIST STATUS Least Concern

Authors: *AA Bertoncini, JH Choat, MT Craig, BP Ferreira, PC Heemstra and LA Rocha*

Paranthias colonus (Valenciennes 1855)

Pacific creolefish (E), badèche du Pacifique (F), indio (S)

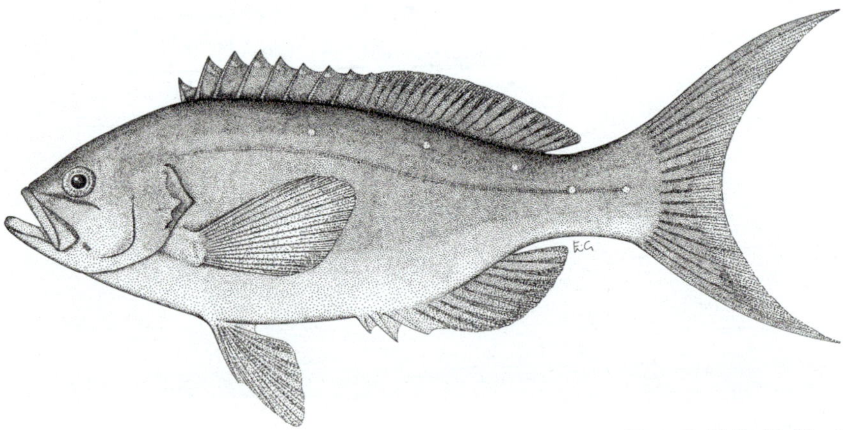

12 cm SL (Gulf of California, Mexico)

Body depth contained 2.9–3.4 times in standard length. Head length 3.2–4.0 times in standard length; interorbital area convex; preopercle subangular, with a shallow notch, the vertical edge and rear half of lower edge finely serrate; posterior and anterior nostrils subequal in size. Gill rakers 12–15 on upper limb, 24–29 on lower limb, 37–44 total. Dorsal fin with 9 spines and 18–21 rays, the interspinous membranes indented; anal fin with 3 spines and 9–11 rays; pectoral-fin rays 19–23; caudal fin deeply forked. Midlateral-body scales ctenoid; lateral-line scales 68–86; lateral-scale series 95–146.

MAXIMUM RECORDED SIZE 36 cm TL

COLOUR IN LIFE Reddish or reddish grey, with 2 or 3 bright blue or violet spots on the dorsal part of the body and another 2 on the midlateral part of the caudal peduncle; pectoral-fin axil with a bright blue spot; dorsal-fin base dark, the fin margin reddish green. Some fish with 2 blue lines on the cheek, one approximately horizontal and tangential to the lower edge of the eye, the other along upper edge of maxilla and continued onto the lower part of the cheek. Juveniles pinkish yellow with bright blue dorsal spots.

COLOUR IN MARKET Maintains nearly identical colouration in death. The reddish hue may fade.

© Octavio Aburto

Adult *P. colonus* (Gulf of California, Mexico)

© D Ross Robertson

Juvenile *P. colonus* (eastern Pacific)

© D Ross Robertson

Adult *P. colonus* (eastern Pacific)

POPULATION **GENERAL** *P. colonus* is very abundant throughout its range. **FISHERY DEPENDENT** *P. colonus* is not targeted by fishers but may be taken opportunistically in small-scale, mixed-species fisheries. No explicit fisheries data are available but there is no evidence of decline. **FISHERY INDEPENDENT** *P. colonus* is frequently seen by SCUBA divers.
POPULATION TREND Stable

Paranthias colonus is restricted to the eastern Pacific and is distributed from the Gulf of California to Peru, including the Revillagigedo, Galápagos, Clipperton, Cocos and Malpelo islands.

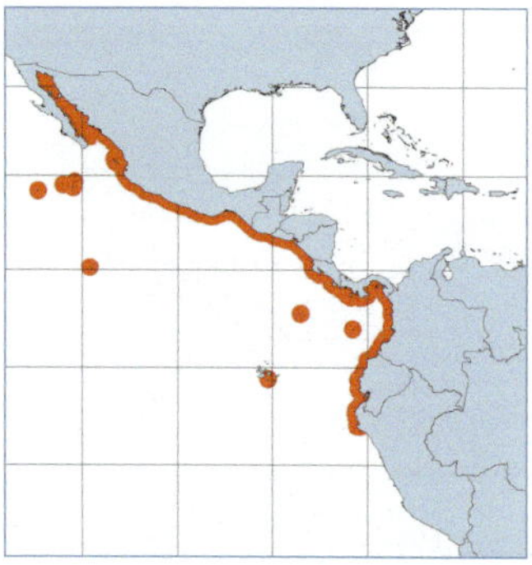

MAJOR THREATS No major threats have been identified for this species.
HABITAT AND ECOLOGY **GENERAL** *P. colonus* is a semi-pelagic, reef-associated species inhabiting depths of up to 30 m. **FEEDING** *P. colonus* and its congener *P. furcifer* are among the few species of grouper that feed on plankton.
COMMENTS Members of the genus *Paranthias* are unlike most groupers in their habitat and ecology. These species are semi-pelagic planktivores that typically shoal a few metres above the bottom. This behaviour is facilitated by several morphological features that distinguish these species from other groupers. These include a forked tail, small mouth, short snout and long numerous gill rakers.

Craig and Hastings (2007) demonstrate a close, nested relationship between this morphologically distinct genus and the genus *Cephalopholis*, and suggest that *Paranthias* spp. are best treated as a member of the genus *Cephalopholis*. We retain them here in *Paranthias* pending further study.
2008 IUCN RED LIST STATUS Least Concern

Authors: *JH Choat, MT Craig, BP Ferreira and LA Rocha*

Paranthias furcifer

(Valenciennes in Cuv. and Val. 1828)

Atlantic creolefish (E), badèche creole (F), cuna lucero (S)

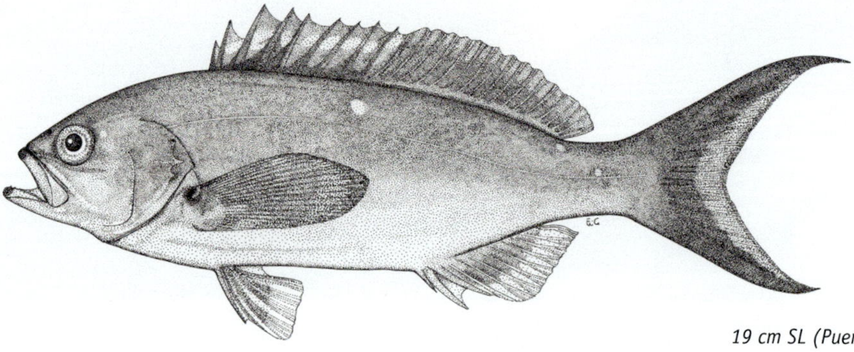

19 cm SL (Puerto Rico)
© SAIAB

Body depth greater than head length, depth contained 2.9–3.4 times in standard length. Head length 3.5–3.8 times in standard length; preopercle subangular, the vertical edge and rear half of lower edge finely serrate; posterior and anterior nostrils subequal in size. Gill rakers 12–14 on upper limb, 24–26 on lower limb, 38 total. Dorsal fin with 9 spines and 17–19 rays, the interspinous membranes slightly indented; length of dorsal-fin base 56–58% of standard length; anal fin with 3 spines and 8–10 rays; anal-fin base 16–19% of standard length; pectoral-fin rays 19 or 20; caudal fin deeply forked. Midlateral-body scales ctenoid; lateral-line scales 69–77; lateral-scale series 124–129.

MAXIMUM RECORDED SIZE 35 cm TL

COLOUR IN LIFE Head and body reddish brown, paler ventrally; bright orange-red spot at the upper end of the pectoral-fin base; 3 widely spaced white spots between the lateral line and dorsal-fin base; 2 blue lines on the cheek, one approximately horizontal and tangential to the lower edge of the eye, the other along the upper edge of the maxilla and continued onto the lower part of the cheek; yellow-green spot on each interspinous dorsal-fin membrane and continued on the soft-rayed part of the fin as a dark green submarginal line.

COLOUR IN MARKET *P. furcifer* maintains nearly identical colouration in death.

© Áthila Bertoncini Andrade

Adult *P. furcifer* (Campeche Bank, Gulf of Mexico)

© Áthila Bertoncini Andrade

Adult *C. fulvus* × *P. furcifer* hybrid (Fernando de Noronha Archipelago, Brazil)

POPULATION **GENERAL** *P. furcifer* is a very common species throughout its range. **FISHERY DEPENDENT** *P. furcifer* is not directly targeted by any fisheries although specimens may be taken opportunistically in small-scale, mixed-species fisheries. **FISHERY INDEPENDENT** *P. furcifer* is often seen by SCUBA divers.

POPULATION TREND Stable

MAJOR THREATS No major threats have been identified for this species.

HABITAT AND ECOLOGY **GENERAL** *P. furcifer* is a semi-pelagic, reef-associated species inhabiting depths of up to 30 m. **FEEDING** *P. furcifer* and its congener *P. colonus* are unusual groupers in that they feed on plankton. **AGE AND GROWTH** The species can grow to at least 8 years, with von Bertalanffy growth parameters of L_∞ = 223.2 mm FL, k = 0.306 y^{-1} and t_0 = 0.488 y, and a length-weight relationship of $W = 6.25 \times 10^{-5} L^{2.76}$ where W is the weight in g and L is the fork length in mm (Posada 1996). **REPRODUCTION** The species is functionally gonochoristic and spawns in aggregations at full moon in November and December in Puerto Rico (Posada 1996); the smallest mature female was reported as 144 mm FL and the smallest male was 138 mm FL. The size and age of 50% maturation was 165 and 175 mm FL, and 2–3 years, respectively.

COMMENTS Members of the genus *Paranthias* are unusual groupers in their habitat and ecology. These species are semi-pelagic, schooling planktivores that typically shoal a few metres above the bottom. This behaviour is facilitated by several morphological features that distinguish these species from other groupers. These include a forked tail, small mouth, short snout and long numerous gill rakers.

Craig and Hastings (2007) demonstrate a close, nested relationship between this morphologically distinct genus and the genus *Cephalopholis*, and suggest that *Paranthias* spp. are best treated as a member of the genus *Cephalopholis*. We retain them here in *Paranthias* pending further study.

2008 IUCN RED LIST STATUS Least Concern

Paranthias furcifer is known from the western Atlantic from Florida (USA) to Brazil, including the Gulf of Mexico, and from Ascension Island in the central Atlantic and the eastern Atlantic islands of São Tomé and Príncipe.

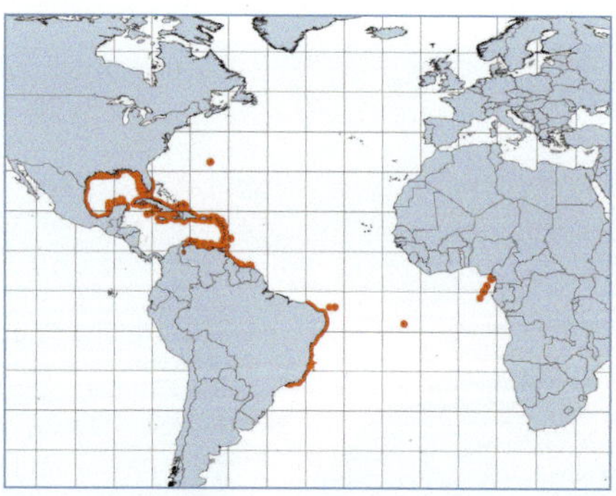

Authors: *JH Choat, MT Craig, BP Ferreira and LA Rocha*

Plectropomus areolatus (Rüppell 1830)

squaretail coralgrouper (E), mérou queue carrée (F), mero troncón (S)

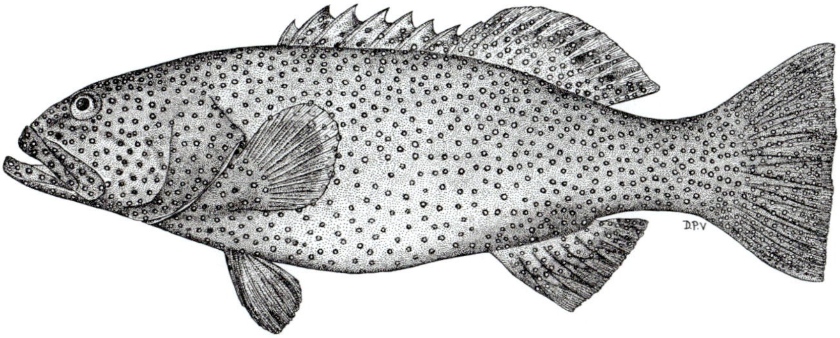

37 cm SL (Red Sea)
© SAIAB

Body elongate, robust, the depth contained 2.9–3.9 times in standard length (for fish 15–48 cm SL). Head length 2.7–3.1 times in standard length; snout length 2.8–3.6 times in head length; suborbital depth 5.6–10 times in head length; interorbital area flat (rounded at edges of orbits), with small, embedded scales; preopercle broadly rounded, with 3 large, ventrally directed spines along lower half; interopercle and subopercle smooth; opercle with 3 flat spines, the upper and lower spines covered by skin; posterior and anterior nostrils subequal in size, set in a shallow groove running forward from eye; midlateral part of upper jaw with 1–4 enlarged fixed canines. Lower limb developed gill rakers 2–7, gill raker at angle shorter than gill filaments at angle. Dorsal fin with 7 or 8 slender spines and 10–12 rays, the 3rd or 4th spine longest, its length contained 3.2–4.2 times in head length, the longest ray 2.3–2.6 times in head length; base of spinous part of dorsal fin subequal to that of soft-rayed part; anal fin with 3 slender spines and 8 rays, the first 1 or 2 spines embedded and difficult to see in large fish; pectoral-fin rays 15 or 16; pectoral fins subequal to pelvic fins, pectoral-fin length 2.0–2.4 times in head length; caudal fin truncate to slightly emarginate, the caudal concavity more than 13 times in head length. Lateral-line scales 83–97.

MAXIMUM RECORDED SIZE 73 cm TL

Plectropomus areolatus is an Indo-Pacific species primarily inhabiting insular localities (with the exception of the Red Sea and Australia).

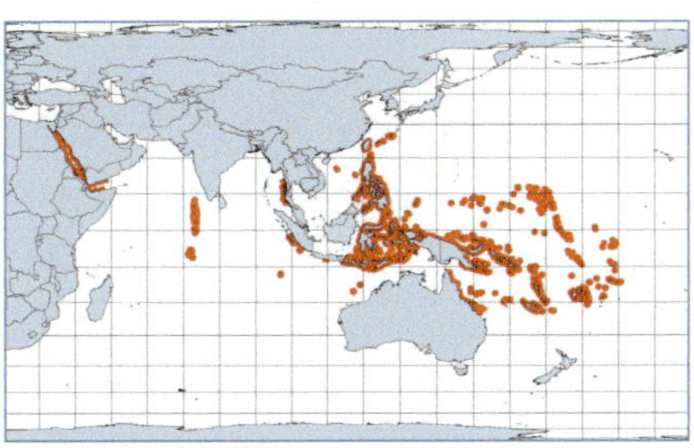

POPULATION **GENERAL** *P. areolatus* was once one of the most common species of *Plectropomus* in the central western Pacific. **FISHERY DEPENDENT** *P. areolatus* is intensively fished throughout its range and has experienced localised population declines in Palau, Pohnpei, Kiribati, Indonesia, Papua New Guinea, Maldives and the Solomon Islands, among other locations (Hamilton and Matawai 2006, IUCN Red List Assessment 2008, Rhodes and Tupper 2008, BM Yeeting pers. obs.). The species is valuable in the live reef food fish trade of Asia.

Fishery independent Fishery-independent surveys indicate that the density of *P. areolatus* has steadily declined in many areas as fishing pressure has increased (IUCN Red List Assessment 2008).

POPULATION TREND Decreasing

MAJOR THREATS Fishing pressure is the main threat to this species particularly when concentrated on spawning aggregations.

HABITAT AND ECOLOGY **General** *P. areolatus* is a reef-associated species inhabiting outer reef channels and slopes in depths

COLOUR IN LIFE Head, body and median fins greenish grey to brown or brownish red, with numerous round to oval dark-edged blue spots (the largest about equal to the pupil); most spots within a spot diameter of the adjacent spots; the pelvic fins with dark brown to blackish membranes; the rear margin of the caudal fin with a white edge and often with a blackish submarginal band.

COLOUR IN MARKET Generally a pale brown to red; dark spots on the body tend to fade but remain visible; the white edge of the caudal fin fades but remains visible as a pale or translucent margin.

© Stephen Newman

Adult *P. areolatus* (Scott Reef, Western Australia)

Juvenile *P. areolatus* (Chuuk, Federated States of Micronesia)

© Robert F Myers

Adult *P. areolatus* (Palau)

© Robert F Myers

of at least 30 m. **FEEDING** Feeds exclusively on fishes. **REPRODUCTION** *P. areolatus* spawns in aggregations numbering between a few hundred to several thousand individuals. The species is commonly found at the same aggregation sites as *Epinephelus fuscoguttatus* and *E. polyphekadion*, often associated with outer reef slopes and particularly reef channel habitat. Spawning season lasts from a few to many months of the year and varies in timing across different areas (e.g. Johannes et al. 1999, Pet et al. 2005). Fish are known to move up to 23 km between residential and spawning areas (Hutchison and Rhodes 2010). The onset of sexual maturation varies by locality, and females generally reach sexual maturity before males. Female maturation may take place in individuals as small as 32 cm SL, and for males at 39 cm SL (IUCN Red List Assessment 2008).

2008 IUCN RED LIST STATUS Vulnerable (A4d)

Authors: *T-C Chan, YJ Sadovy de Mitcheson and BM Yeeting*

Plectropomus laevis (Lacepède 1801)

blacksaddled coralgrouper (E), mérou sellé (F), mero ensillado (S)

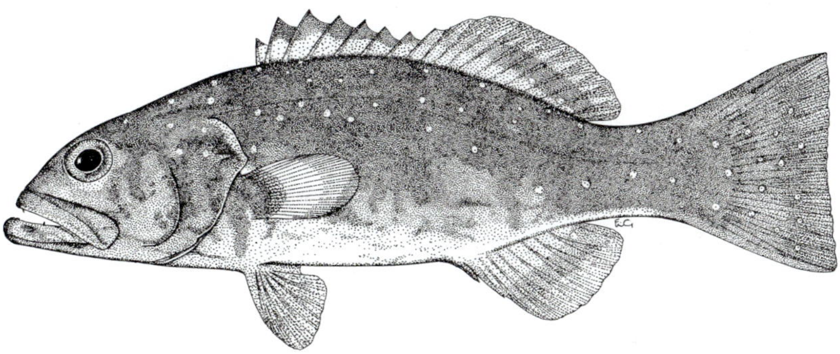

Dark form, 11 cm SL (Tuamotu Archipelago, French Polynesia)

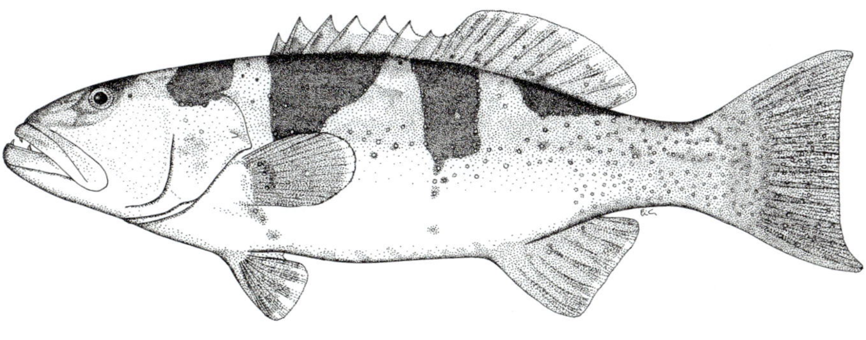

Pale black-saddle form, 60 cm SL (Marshall Islands)

Body elongate, robust, the depth contained 2.9–3.9 times in standard length (for fish 11–62 cm SL). Head length 2.7–3.1 times in standard length; snout length 2.8–3.6 times in head length; suborbital depth 5.6–10 times in head length; interorbital area flat (rounded at edges of orbits), without scales; preopercle broadly rounded, with 3 large, ventrally directed spines along lower half; interopercle and subopercle smooth; opercle with 3 flat spines, the upper and lower spines covered by skin; posterior and anterior nostrils subequal in size, set in a shallow groove running forward from eye; midlateral part of upper jaw with 1–4 enlarged fixed canines. Developed gill rakers on lower limb 4–10, gill raker at angle shorter than gill filaments at angle. Dorsal fin with 7 or 8 slender spines and 10–12 rays, the 3rd or 4th spine longest, its length contained 3.2–4.2 times in head length, the longest dorsal-fin ray 2.3–3.3 times in head length; base of spinous part of dorsal fin subequal to that of soft-rayed part; anal fin with 3 slender spines and 8 rays, the first 1 or 2 spines embedded and difficult to see in large fish; pectoral-fin rays 16–18; pectoral fins subequal to pelvic fins, pectoral-fin length 2.1–2.4 times in head length; caudal fin emarginate, the caudal concavity 5–10 times in head length. Lateral-line scales 83–97.

MAXIMUM RECORDED SIZE 130 cm TL

Plectropomus laevis is widespread in the Indo-Pacific from East Africa to the central and southern Pacific.

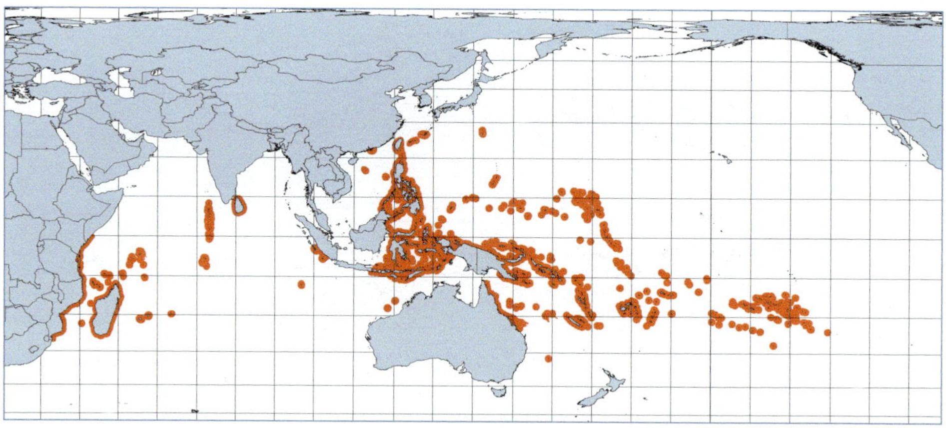

POPULATION **GENERAL** Except in the Coral Sea (Australia), *P. laevis* is a naturally rare species achieving abundances of less than one individual per 1000 m^2 (IUCN Red List Assessment 2008). **FISHERY DEPENDENT** *P. laevis* is exploited over much of its range through line fisheries and is occasionally taken as part of the live reef food fish trade. Regional catch data indicate substantial declines (IUCN Red List Assessment 2008). No explicit data are currently available as this species is typically 'lumped' with other groupers in landings data. **FISHERY INDEPENDENT** Fishery-independent data confirm the natural rarity of this species.

POPULATION TREND Decreasing

MAJOR THREATS Fishing pressure is the major threat to this species.

HABITAT AND ECOLOGY **GENERAL** *P. laevis* occurs on outer coral reef slopes in depths of 4–90 m. **FEEDING** Feeds on fishes and crustaceans. **AGE AND GROWTH** *P. laevis* is a relatively fast-growing species and reaches 50 cm FL in less than 4 years. The von Bertalanffy growth parameters are L_∞ = 948 mm FL, k = 0.14 y^{-1} for the Great

COLOUR IN LIFE There are two colour forms. The pale black-saddled form is whitish or pale yellowish with 5 dark brown to black, irregular, saddle-like or wedge-shaped bars or blotches on the dorsal part of the head and body, the 1st just behind the eyes, the 2nd on the nape, the 3rd to 5th under the dorsal fin; small dark-edged blue spots may be visible on the body (mainly posteriorly); fins, caudal peduncle, snout and jaws yellow; black blotch centrally at the base of the paired fins; small juveniles (7–15 cm SL) with membranes between 1st and 5th dorsal-fin spines mostly blackish.

The dark form is brown, olivaceous, red, or nearly black, the dark bars (as on the pale form) are usually faint or absent; numerous round, dark-edged blue spots (about half the pupil diameter) on the head and body (except ventrally), soft dorsal, caudal and anal fins, and base of pectoral fins; pectoral fins dark brown distally, the rays darker than membranes, the rear edge white. Large courting males in spawning aggregations have a distinctive colouration with white lips and a large light-coloured area on the midventral area of the body.

COLOUR IN MARKET *P. laevis* will retain similar colour patterns to those displayed at time of death.

© Stephen Newman

Adult *P. laevis* (dark, deep-water form; Rowley Shoals, Western Australia)

© Stephen Newman

Adult *P. laevis* (dark form; Scott Reef, Western Australia)

© Stephen Newman

Adult *P. laevis* (light form; Scott Reef, Western Australia)

© Robert F Myers

Adult *P. laevis* (light form; Bali, Indonesia)

© Robert F Myers

Juvenile *P. laevis* (light form; Pemba Island, Tanzania)

© Robert F Myers

Adult *P. laevis* (dark form; Maldives)

Barrier Reef (n = 346; Davies et al. 2006); and L_∞ = 1015 mm FL, k = 0.19 y^{-1} for Aldabra (n = 22; Grandcourt 2005). **REPRODUCTION** *P. laevis* is a monandric protogynous hermaphrodite (Adams 2002, 2003). Females achieve sexual maturity in less than 3 years and 40 cm SL (Davies et al. 2006). Sex change occurs at about 8 years. Spawning occurs in small groups of at least 20 fish in Fiji (YJ Sadovy de Mitcheson unpublished data).

COMMENTS Although widespread, *P. laevis* is rare over most of its range with specific habitat requirements. As with all other species of *Plectropomus*, *P. laevis* is subject to intense fishing pressure over most of the range.
2008 IUCN RED LIST STATUS Vulnerable (A2d+4d)

Authors: *JH Choat, MT Craig, M Liu, J Robinson and BM Yeeting*

Plectropomus leopardus (Lacepède 1802)

leopard coralgrouper (E), saumonée léopard (F), mero celestial (S)

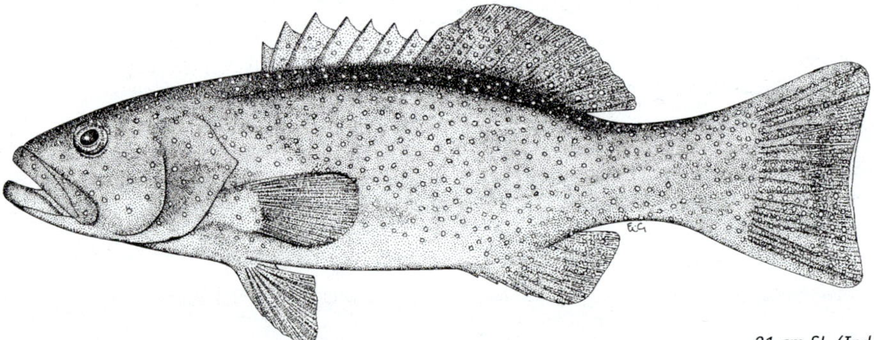

21 cm SL (Indonesia)
© SAIAB

Body elongate, robust, the depth contained 2.9–3.9 times in standard length (for fish 12–50 cm SL). Head length 2.7–3.1 times in standard length; snout length 2.8–3.6 times in head length; suborbital depth 5.6–10 times in head length; interorbital area flat (rounded at edges of orbits), without scales; preopercle broadly rounded, with 3 large, ventrally directed spines along lower half; interopercle and subopercle smooth; opercle with 3 flat spines, the upper and lower spines covered by skin; posterior and anterior nostrils subequal for fish less than 40 cm SL (at larger sizes, the posterior nostrils may be enlarged); midlateral part of upper jaw with 1–4 enlarged fixed canines. Developed gill rakers on lower limb 6–10, gill raker at angle longer than gill filaments at angle. Dorsal fin with 7 or 8 slender spines and 10–12 rays, the 3rd or 4th spine longest, contained 3.2–4.2 times in head length, the longest dorsal-fin ray 2.2–2.7 times in head length; base of spinous part of dorsal fin subequal to that of soft-rayed part; anal fin with 3 slender spines and 8 rays, the first 1 or 2 spines embedded and difficult to see in large fish; pectoral-fin rays 15–17; pectoral fins subequal to pelvic fins, pectoral-fin length 1.9–2.2 times in head length; caudal fin emarginate, the caudal concavity 5–12 times in head length. Lateral-line scales 89–99.
MAXIMUM RECORDED SIZE 120 cm TL

COLOUR IN LIFE Olive to reddish brown, orange-red, pink or red, with numerous small (nostril-sized and usually dark-edged) blue spots on the head and body (except ventrally) and median fins; more than 10 spots on the cheek (below and behind the eye to the rear edge of the preopercle; a blue ring on the edge of the orbit (some broken into segments); pectoral fins reddish or hyaline with darker rays; an indistinct dark band at the rear margin of the caudal fin, with a white line usually visible along the middle of the rear edge of the fin.

COLOUR IN MARKET Generally maintains similar colouration in life; ground colour may fade to pale red.

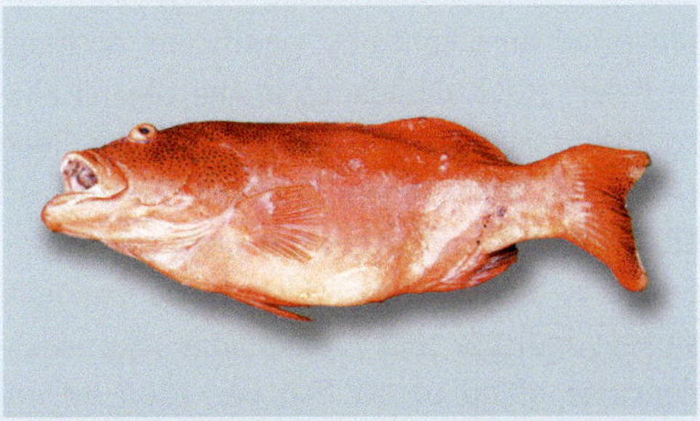
© Allen W-L To

Adult *P. leopardus* (Haikou, Hainan Province, China)

© Robert F Myers

Adult *P. leopardus* (New Caledonia)

POPULATION GENERAL *P. leopardus* is a common species throughout its range and is an important component of the live reef food fish trade. **FISHERY DEPENDENT** *P. leopardus* is heavily fished throughout its range mainly by hook and line, and speargun, but also occasionally by fish trap and in some areas of South-East Asia with cyanide. In the Philippines, declines in landings of over 40% were noted over a four-year period when the species was targeted for a live fish fishery, during which time average fish size decreased (Padilla et al. 2003); immature fish are now directly targeted for grow-out mariculture in which they are fed in captivity until they reach marketable size of between 0.5 and 1 kg. The species is mainly targeted for the live fish trade, with Australia and the Philippines, as well as Island nations of the western Pacific, as major exporting countries to the trade centres of Hong Kong and mainland China (IUCN Red List Assessment 2004). Despite extensive fisheries and protected area controls, significant declines in *P. leopardus* abundance in the central and southern sections of the Great Barrier Reef (Australia) have been reported (Ayling et al. 2000) but protection has been improved. **Fishery independent** Fishery-independent surveys indicate that *P. leopardus* may reach densities of up to 10 fish per 1000 m^2 on the Great Barrier Reef (Ayling et al. 2000). Population estimates by mark-capture data range from 1890 to 2134 fish for the area around Lizard Island (750 966 m^2), equivalent to 2.6 fish per 1000 m^2, whereas it is about 3988 (5.3 fish 1000 per m^2) when estimated by underwater visual census (Samoilys and Carlos 2000, Zeller and Russ 2000). On the Great Barrier Reef, *P. leopardus* is considered naturally common with densities ranging from 3.5 to 9.0 fish per 1000 m^2 (Ayling et al. 2000). In contrast, the species is relatively uncommon in Solomon Islands, Fiji and Papua New Guinea (Samoilys et al. 1995). **Genetics** Genetic data indicate population subdivision throughout the species' range in Australia (van Herwerden et al. 2006).

POPULATION TREND Decreasing

Plectropomus leopardus occurs in the western Pacific from southern Japan to northern Australia and eastwards to the Caroline Islands and Fiji.

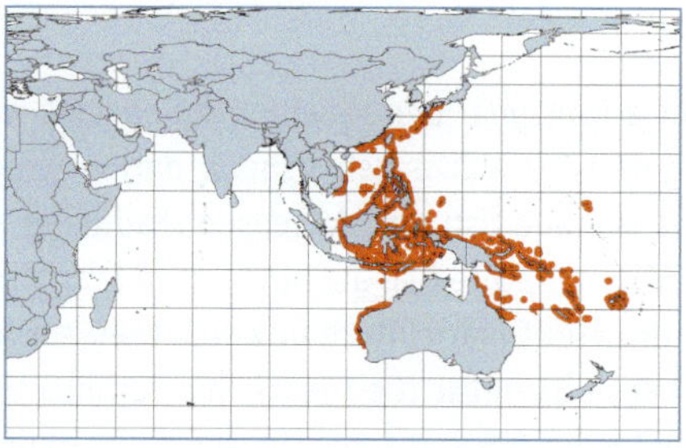

MAJOR THREATS Fishing pressure is the major threat to this species.

HABITAT AND ECOLOGY **General** *P. leopardus* inhabits coral reefs in depths of up to 100 m with adults making only local movements around resident reefs, even in relation to spawning aggregation sites (Zeller and Russ 1998). **Feeding** Feeds on fishes and crustaceans. **Age and growth** The species can live 14–19 years, grows rapidly in its first 2–3 years, matures relatively early as a female, changing to male at around 7 years; von Bertalanffy growth parameters from equations derived from two locations in Australia are: $L_\infty = 52.2$ cm FL, $k = 0.35$ y^{-1} and $t_0 = -0.766$ (Ferreira and Russ 1994). $W = 2.36 \times L^{3.30}$ where W is the weight in g and L is the fork length in mm (Samoilys 2000). **Reproduction** *P. leopardus* is a protogynous hermaphrodite (Ferreira 1995, Samoilys 1997, Adams et al. 2000). Spawning occurs in relatively small aggregations, tens to several hundreds of fish, from September to November. Size at first reproduction varies from 20–25 cm FL (Samoilys 2000) to 32–36 cm FL according to histological criteria used (Ferreira 1995). In the Cairns area of the Great Barrier Reef, the best studied aggregation sites for this species occur from September to November (Samoilys 1997).

COMMENTS *P. leopardus* superficially resembles other species of *Plectropomus*. This species has no spots ventrally, and those on the head are never elongated. The blue ring around the eye is a distinctive characteristic for this species.

2004 IUCN RED LIST STATUS Near Threatened

Authors: *AS Cornish and DA Pollard*

Plectropomus maculatus (Bloch 1790)

spotted coralgrouper (E), vielle Saint-Silac (F), mero con pintas (S)

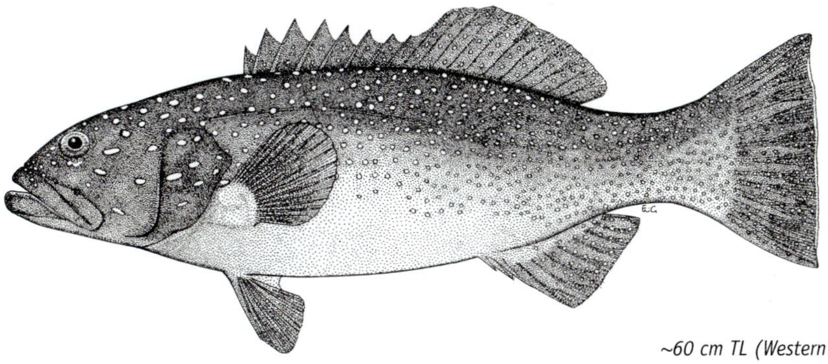

~60 cm TL (Western Australia)
© SAIAB

Body elongate, robust, depth contained 2.9–3.9 times in standard. Head length 2.7–3.1 times in standard length; snout length 2.8–3.6 times in head length; suborbital depth 5.6–10 times in head length; interorbital area flat (rounded at edges of orbits), without scales; preopercle broadly rounded, with 3 large, ventrally directed spines along lower half; interopercle and subopercle smooth; opercle with 3 flat spines, the upper and lower spines covered by skin; posterior and anterior nostrils subequal in size, set in a shallow groove running forward from eye; midlateral part of upper jaw with 1–4 enlarged fixed canines. Developed gill rakers on lower limb 6–9, gill raker at angle distinctly longer than gill filaments at angle. Dorsal fin with 7 or 8 slender spines and 10–12 rays, the 3rd or 4th spine longest, its length contained 3.2–4.2 times in head length, the longest dorsal-fin ray 2.2–2.5 times in head length; base of spinous part of dorsal fin subequal to that of soft-rayed part; anal fin with 3 slender spines and 8 rays, the first 1 or 2 spines embedded and difficult to see in large fish; pectoral-fin rays 15–17; pectoral fins subequal to pelvic fins, pectoral-fin length 1.7–2.2 times in head length; caudal fin emarginate, the caudal concavity 5–10 times in head length. Lateral-line scales 83–97.

MAXIMUM RECORDED SIZE 120 cm TL

POPULATION **General** *P. maculatus* is a common species on the western Australian plate and occurs less frequently in other portions of its range. *P. maculatus* is subject to fishing over most of its range including for the live reef food fish trade and was the fourth most abundant epinephelid recorded in a survey of the north-western Australia coastal trap- and linefishery (Moran et al. 1988, Sadovy et al. 2003). Records of capture may be difficult to interpret as this species is usually lumped with *P. leopardus* in the eastern Australian fishery. **Fishery dependent** *P. maculatus* is fished intensively across its range, particularly in Australia, and for the live reef food fish trade. **Fishery independent** Fishery-independent data show that this species is more common in inshore localities and may be considered a continental inhabitant.

POPULATION TREND Decreasing

MAJOR THREATS Fishing pressure is the primary threat to this species.

HABITAT AND ECOLOGY **General** *P. maculatus* inhabits inshore reefs in depths of 5–50 m. *P. maculatus* inhabits largely coastal and inshore reefs contrasting with other species of *Plectropomus* that are mainly associated with offshore and oceanic reefs. **Feeding** Feeds primarily on fishes. **Age and growth** *P. maculatus* reaches ages of at least 12 years (Ferreira and

COLOUR IN LIFE Head, body and median fins greenish grey, brown, red, or orange-red, with dark-edged blue spots; the spots on the head and anterior part of the body about half the pupil diameter or larger (and many are oval or horizontally elongated), becoming much smaller, more uniformly round, and more numerous on the rear part of the body and on the fins; a few spots on the head, 3–7 on the cheeks (below and behind the eye to the rear edge of the preopercle); no blue spots on the ventral parts of the head and body; usually a single blue spot on the pectoral-fin base and another on the axil; no blue spots on the pelvic fins; the rear margin of the caudal fin with a white edge.

COLOUR IN MARKET Head and body pale grey to orange, red or brown; blue spots fade to a common dark hue with only a faint appearance of blue on the edges; pale blotches lacking any pigment often form on the body.

© Matthew T Craig

Adult *P. maculatus* (Sulawesi, Indonesia)

© Robert F Myers

Subadult *P. maculatus* (Dampier Straits, West Papua New Guinea)

Russ 1992). Maximum age of females is 6 years, and of males 10 years, suggesting all females change sex. Ferreira and Russ (1992) give von Bertalanffy growth parameters of L_{∞} = 732 mm FL, k = 2.1 y^{-1}; the median size of the sample was 370 mm FL and median age was 4 years. *P. maculatus* is relatively fast-growing with males achieving a greater growth rate than females (Adams and Williams 2001). Adams (2002) estimates age-based reproductive parameters with 50% female maturity at 18 months and 300 mm FL. **Reproduction** Adams (2002) reports that *P. maculatus* is a diandric protogynous hermaphrodite. Secondary sexual transition occurs between 5 and 6 years of age; 50% female maturity occurs at 18 months and 30 cm FL. *P. maculatus* is known to hybridise with *P. leopardus*. Adams (2002) concludes that this species is a diandric protogynous hermaphrodite on the Great Barrier Reef, with immature males identifiable at 1 year of age, and transitional gonads in females 5–6 years of age; the sex ratio (F:M) is 3.1:1.0.

Plectropomus maculatus occurs in the western tropical Pacific from Thailand, Singapore, Philippines, Indonesia, Papua New Guinea, Solomon Islands, and along the northern Australia coast (north-western Australia to southern Queensland).

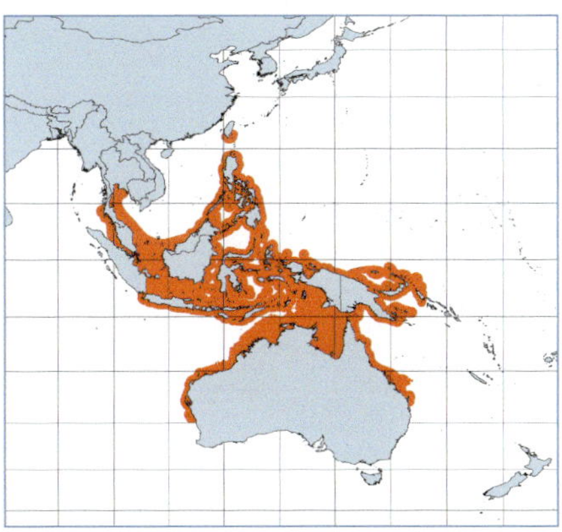

COMMENTS There is evidence for hybridisation between *P. leopardus* and *P. maculatus* on the east coast of Australia (Frisch and van Herwerden 2006, van Herwerden et al. 2006); analyses suggest that *P. leopardus* females have hybridised historically with *P. maculatus* males and that *P. maculatus* mitochondria were displaced through introgressive hybridisation and fixation in the *P. maculatus* founder population on the Great Barrier Reef. Both species co-occur on midshelf reefs of the east coast but not on the west coast where hybridisation has not been recorded. It is unknown whether hybridisation occurs at any other localities where these two species co-occur.

2008 IUCN RED LIST STATUS Least Concern

Authors: *JH Choat, MT Craig, BP Ferreira and LA Rocha*

Plectropomus oligacanthus (Bleeker 1854)

highfin coralgrouper (E), mérou-loche cacatois (F), mero vela (S)

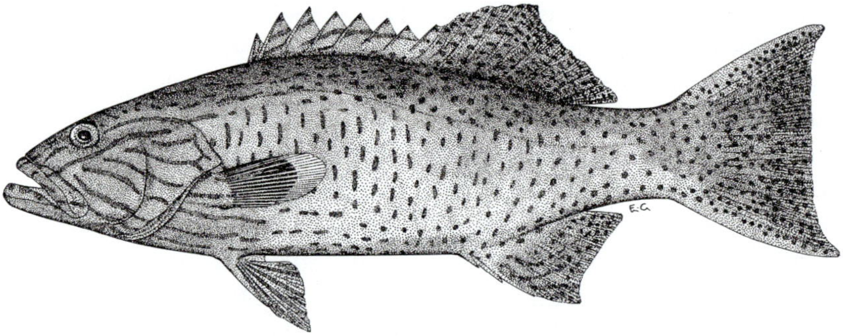

24 cm SL (Solomon Islands)
© SAIAB

Body elongate, robust, the depth contained 2.9–3.9 times in standard length. Head length 2.7–3.1 times in standard length; snout length 2.8–3.6 times in head length; suborbital depth 3.3–7.2 times in head length; interorbital area flat (rounded at edges of orbits), without scales; preopercle broadly rounded, with 3 large, ventrally directed spines along lower half; interopercle and subopercle smooth; opercle with 3 flat spines, the upper and lower spines covered by skin; midlateral part of upper jaw with 1–4 enlarged fixed canines. Developed gill rakers on lower limb 7–9, gill raker at angle subequal to gill filaments at angle. Dorsal fin with 7 or 8 slender spines and 10–12 rays, the 3rd or 4th spine longest, its length contained 3.2–4.2 times in head length; base of spinous part of dorsal fin subequal to that of soft-rayed part; anal fin with 3 slender spines and 8 rays, the first 1 or 2 spines embedded and difficult to see in large fish; soft dorsal and anal fins pointed anteriorly, the 2nd to 4th dorsal- and anal-fin rays elongated, contained 1.5–2.1 times in head length; pectoral-fin rays 14–16; pectoral fins shorter than pelvic fins, contained 2.1–2.4 times in head length; caudal fin emarginate, the caudal concavity 3–7 times in head length. Lateral-line scales 86–96.

MAXIMUM RECORDED SIZE 75 cm TL

Plectropomus oligacanthus is a western Pacific species found in the Philippines, Indonesia, Papua New Guinea, north-eastern Australia (Cape York to the northern Great Barrier Reef), Palau, Chuuk, Caroline Islands, Marshall Islands, and the Solomon Islands, Malaysia, Western Australia (Scott Reef), and the Timor Sea.

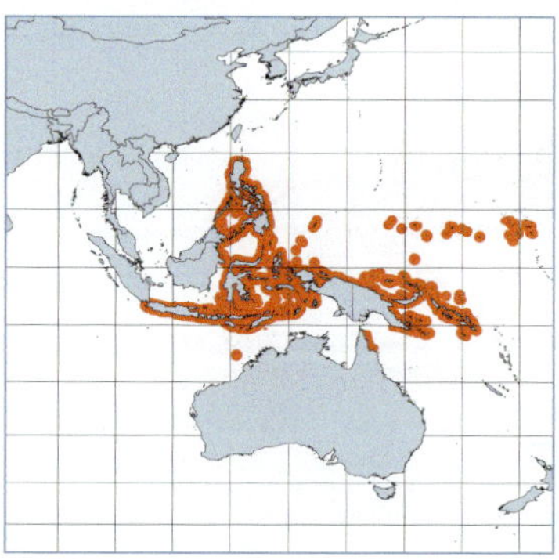

COLOUR IN LIFE Head, body and fins reddish brown to red, with horizontal to oblique blue lines on the rear of the head and anterodorsally on the body; vertical blue lines (or dashes) anteriorly on the side of the body (at least in adults), and numerous blue spots elsewhere on the body and on the caudal fin; dorsal and anal fins with blue lines and spots; pectoral fins pale yellowish, with the basal two-thirds of the rays dark brown and a few blue lines at the base of the fin; pelvic fins with brownish red rays and blue membranes.

COLOUR IN MARKET Head and body pale brown to red; oblique bars on the head and body fade but remain prominent, particularly on the anterior portion of the fish; blue spots on the body may lose colour but generally remain visible.

© Michael Gawel

Adult *P. oligacanthus* (Pohnpei, Federated States of Micronesia)

© Robert F Myers

Adult *P. oligacanthus* (Dampier Straits, West Papua New Guinea)

Adult *P. oligacanthus* (New Britain, Papua New Guinea)

© Robert F Myers

POPULATION **GENERAL** *P. oligacanthus* is a naturally rare species throughout its range. **FISHERY DEPENDENT** While there are no explicit landings data for *P. oligacanthus*, it is known to be taken opportunistically in catches throughout its range, and is valued in the live reef food fish trade which is unmanaged. **FISHERY INDEPENDENT** Fishery-independent surveys confirm the natural rarity of this species.
POPULATION TREND Decreasing
MAJOR THREATS Fishing pressure is the primary threat to this species as is the absence of fishery management throughout most of its range.
HABITAT AND ECOLOGY **GENERAL** *P. oligacanthus* inhabits coral reefs to depths of at least 30 m. **FEEDING** Schroeder (1980) reports that groups of two or three may forage together for rock- and sand-living crustaceans and fishes.
2008 IUCN RED LIST STATUS Near Threatened

Authors: *AS Cabanban, YJ Sadovy de Mitcheson and MA Samoilys*

Plectropomus pessuliferus (Fowler 1904)

roving coralgrouper (E), mérou-loche vagabonde (F), mero errante (S)

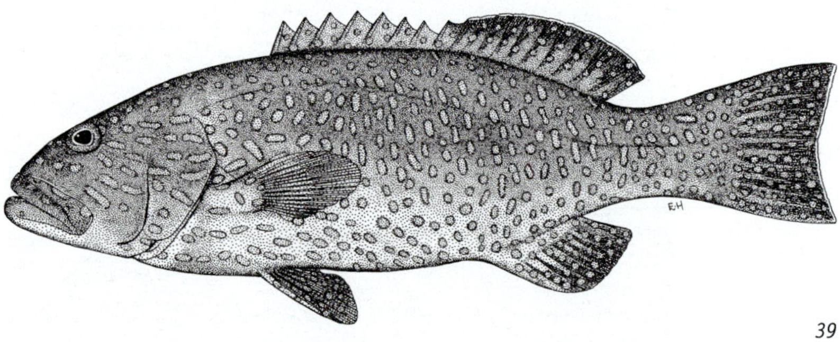

39 cm SL (Fiji)
© SAIAB

Body elongate, robust, the depth contained 2.9–3.9 times in standard. Head length 2.7–3.1 times in standard length; snout length 2.8–3.6 times in head length; suborbital depth 5.6–10 times in head length; interorbital area flat (rounded at edges of orbits), without scales; preopercle broadly rounded, with 3 large, ventrally directed spines along lower half; interopercle and subopercle smooth; opercle with 3 flat spines, the upper and lower spines covered by skin; nostrils set in a shallow groove running forward from eye; nostrils subequal for fish less than about 30 cm SL, the posterior nostrils often enlarged in larger fish; midlateral part of lower jaw with 1-4 enlarged fixed canines. Developed gill rakers on lower limb 7–10, the gill raker at angle shorter than gill filaments at angle. Dorsal fin with 7 or 8 slender spines and 10–12 rays, the 3rd or 4th spine longest, its length contained 3.2–4.2 times in head length, the longest ray 2.3–3.0 times in head length; base of spinous part of dorsal fin subequal to that of soft-rayed part; anal fin with 3 slender spines and 8 rays, the first 1 or 2 spines embedded and difficult to see in large fish; pectoral-fin rays 15 or 16; pectoral fins subequal to pelvic fins, pectoral-fin length contained 1.9–2.3 times in head length; caudal fin emarginate (truncate in juveniles), the caudal concavity 4.6–6.2 times in head length. Lateral-line scales 85–104. Pyloric caeca 3, large and finger-like.
MAXIMUM RECORDED SIZE 120 cm TL

COLOUR IN LIFE Head, body and fins brown to orange-red, with numerous small dark-edged blue spots; some spots on the head and sides of the body of adults are elongated (those on the body usually vertically elongate); spots few or absent on the ventral part of the body; the edge of the orbit is often blue (may be broken into segments).

COLOUR IN MARKET Head and body pale orange to red; spots may lose their blue edge but remain visible; elongate spots on the body retain their shape and are distinctive; may have several pale blotches lacking reddish colour but the spots remain.

© Allen W-L To

Adult *P. pessuliferus* (displaying near-life colours; Sai Kung Market, Hong Kong)

© Matthew T Craig

Adult *P. pessuliferus* (Sai Kung Market, Hong Kong)

© Klaus Fiedler

Adult *P. pessuliferus* (Red Sea)

© Robert F Myers

Adult *P. pessuliferus* (Maldives)

© Patrice Marker

Adult *P. pessuliferus* (Red Sea, at night)

Plectropomus pessuliferus has a widespread but disjunct distribution from the Red Sea to Fiji. It is known with certainty from the Red Sea (as subspecies *P. pessuliferus marisrubri*), Zanzibar, the Maldives, St Brandon's Shoals, Sri Lanka, Chagos, Nazareth Bank, Sumatra and Fiji.

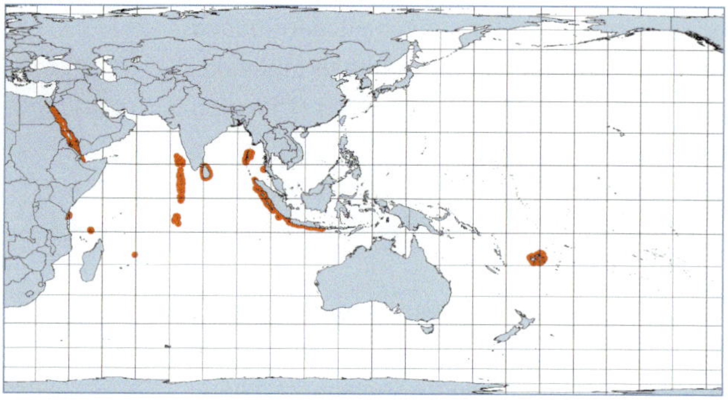

POPULATION **GENERAL** *P. pessuliferus* is rare to uncommon throughout most of its range, with higher abundances in the Maldives and the Red Sea. **FISHERY DEPENDENT** Few explicit landings data are available for this species. It is heavily fished in the Red Sea and annual landings in the Kingdom of Saudi Arabia averaged about 380 t from 1995 to 2002 with no signs of decline, but no estimate of CPUE was reported (Fallatah 2005). Interviews with fishermen in the Red Sea have also indicated qualitative declines in landings. **FISHERY INDEPENDENT** Fishery-independent surveys confirm the general rarity of this species.

POPULATION TREND Decreasing

MAJOR THREATS Fishing pressure is the primary threat to this species.

HABITAT AND ECOLOGY **GENERAL** *P. pessuliferus* occurs on coral reefs and to a lesser degree in shallow lagoons in depths of 3–147 m. **FEEDING** Feeds primarily on fishes.

COMMENTS Randall and Hoese (1986) recognise two subspecies: *P. pessuliferus marisrubri* Randall and Hoese (1986) in the Red Sea, and *P. pessuliferus pessuliferus* from the rest of the Indo-Pacific region. In the Red Sea, individuals have more numerous blue spots on the cheek (30–50 behind and below the centre of the eye to the edge of the preopercle, vs 5–12 blue spots on the cheek of the nominate subspecies).

P. pessuliferus is easily confused with *P. maculatus*. The former has the longest gill rakers shorter than the longest filaments (longest gill raker distinctly longer than longest gill filament in *P. maculatus*), vertically elongated spots on the body (spots on body exclusively round in *P. maculatus*), and blue spots on the pelvic fins (no spots on the pelvic fins in *P. maculatus*).

2008 IUCN RED LIST STATUS Near Threatened

Authors: *JH Choat, BP Ferreira, ALB Gaspar, RF Myers and MA Samoilys*

Plectropomus punctatus

(Quoy and Gaimard 1824)

marbled coralgrouper (E), mérou pointillé (F), mero pecoso (S)

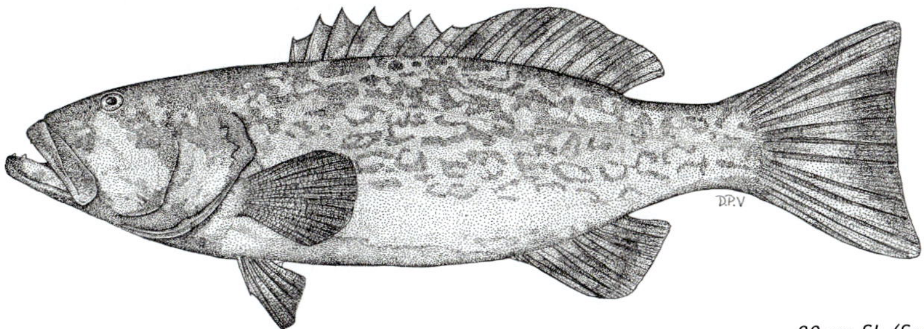

90 cm SL (South Africa)
© SAIAB

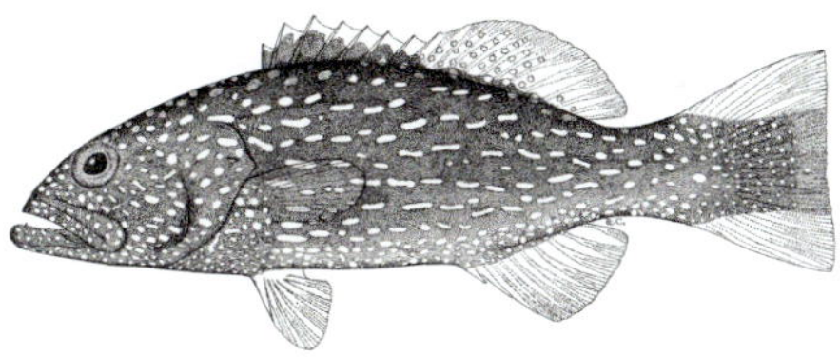

~8 cm SL (South Africa)
© SAIAB

Body elongate, robust, depth contained 2.9–3.9 times in standard length (for fish 15–61 cm SL). Head length 2.7–3.1 times in standard length; snout length 2.8–3.6 times in head length; suborbital depth 5.6–10 times in head length; interorbital area flat (rounded at edges of orbits), without scales; preopercle broadly rounded, with 3 large, ventrally directed spines along lower half; interopercle and subopercle smooth; opercle with 3 flat spines, the upper and lower spines covered by skin; nostrils set in a shallow groove in front of eye; posterior nostrils larger than anterior ones on fish more than 60 cm SL; midlateral part of upper jaw with 1–4 enlarged fixed canines. Developed gill rakers on lower limb 2–9, gill raker at angle slightly shorter than gill filaments at angle. Dorsal fin with 7 or 8 slender spines and 10–12 rays, the 3rd or 4th spine longest, its length contained 3.2–4.2 times in head length, the longest ray 2.3–3.2 times in head length; base of spinous part of dorsal fin subequal to that of soft-rayed part; anal fin with 3 slender spines and 8 rays, the first 1 or 2 spines embedded and difficult to see in large fish; pectoral-fin rays 16–18; pectoral fins subequal to pelvic fins, contained 2.0–2.4 times in head length; caudal fin truncate to slightly emarginate, the caudal concavity contained 7–16 times in head length. Lateral-line scales 88–95. Pyloric caeca 3, large and thick.

MAXIMUM RECORDED SIZE 96 cm TL

COLOUR IN LIFE Adults brown to reddish or purplish brown, often irregularly mottled with pale olive-green, orange or white; pectoral fins dark brown; pelvic and anal fins sometimes with a bluish band or row of blue spots distally. Body of juveniles (7–12 cm SL) brownish dorsally, with numerous round or oval pale spots (about half the eye diameter in size); interspinous membranes between the first 5 dorsal-fin spines black; subadults (14–20 cm SL) with numerous horizontal pale streaks and elongate pale spots (about an eye diameter in length) on the head and body.

COLOUR IN MARKET Pale red to brown with patches of pale colours; no distinctive markings in death.

© Matthew T Craig

Adult *P. punctatus* (Sodwana Bay, South Africa)

POPULATION GENERAL *P. punctatus* is a relatively uncommon species throughout its range. **FISHERY DEPENDENT** The species is taken by both commercial and recreational linefishing; it is included in the live reef fish trade, and as juveniles in the aquarium fish trade. According to Morgans (1982), *P. punctatus* was common in markets of Zanzibar (Tanzania) in the 1950s. It probably is (or was) of commercial importance in the Seychelles and Mauritius. Caught with hook and line, trolling, spear and on benthic longlines (Heemstra and Randall 1993).

Plectropomus punctatus occurs in the western Indian Ocean from Kenya to Mozambique, South Africa, Comoros Islands, Seychelles, Madagascar, the Mascarene Islands and the Chagos Archipelago, as well as from southern Oman, Socotra (Yemen) and Somalia. It is unknown from the Red Sea and Persian Gulf.

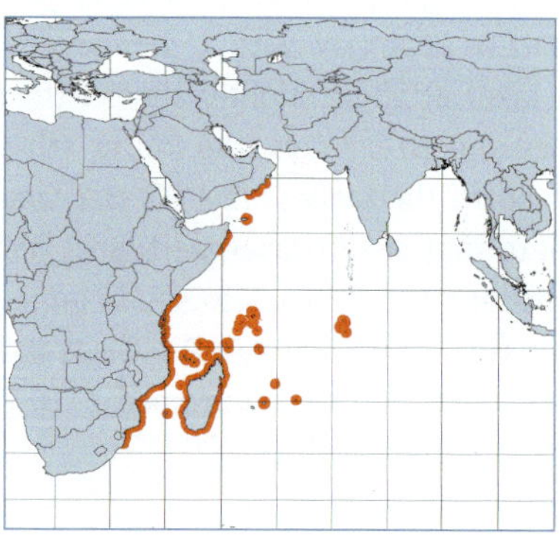

POPULATION TREND Unknown

MAJOR THREATS Fishing pressure is the primary threat to this species.

HABITAT AND ECOLOGY GENERAL *P. punctatus* is a reef-associated species inhabiting depths of at least 30 m. The species occurs solitarily or in small groups. It often drifts well above the bottom (Lieske and Myers 1994). **FEEDING** Feeds primarily on fishes. **REPRODUCTION** *P. punctatus* is known to spawn in aggregations but probably only gathers in small numbers rather than large aggregations (Robinson 2004).

2008 IUCN RED LIST STATUS Data Deficient

Authors: *JH Choat, ST Fennessy, BC Russell and MA Samoilys*

Saloptia powelli (Smith 1964)

golden grouper (E), mérou d'or (F), mero dorado (S)

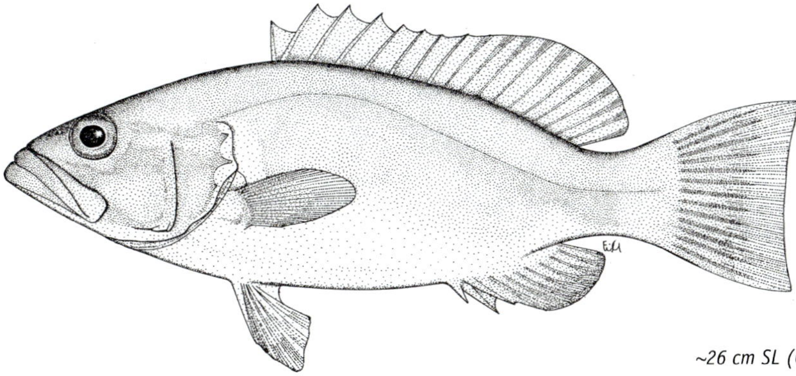

~26 cm SL (Cook Islands)

Body oblong, robust, depth less than head length and contained 2.6–3.0 times in standard length. Head length 2.5–2.6 times in standard length; snout distinctly longer than eye diameter; nostrils set in a shallow groove running forward from eye, the posterior nostrils about twice size of anterior nostrils; supramaxilla well developed; a pair of canines at front of both jaws; lower jaw with 2 rows of teeth, but no enlarged canines at midside of jaw. Gill rakers 8 or 9 on upper limb, 16 or 17 on lower limb. Dorsal fin with 8 spines and 11 rays, the 3rd to 8th spines subequal; anal fin with 3 spines and 8 rays; dorsal- and anal-fin spines strong, the anal spines quite distinct; pectoral fins short and rounded, with 14 or 15 rays, the middle rays longest, subequal to pelvic fins, pectoral-fin length contained 2.2–2.4 times in head length. Body scales small, distinctly ctenoid, even on belly; no auxiliary scales; lateral-line scales 70–78; lateral-scale series 115–133.

MAXIMUM RECORDED SIZE 47 cm TL

COLOUR IN LIFE Head, body and fins yellow to orange-yellow, shading ventrally to white or pink, the snout, lips and dorsal part of head suffused with red; dorsal-fin spines sometimes streaked with red.

COLOUR IN MARKET Generally pale yellow to orange with no distinguishing marks.

Adult *S. powelli* (Okinawa Fish Market, Japan)

POPULATION **General** *S. powelli* is a poorly known, deep-water species. **Fishery dependent** *S. powelli* is taken opportunistically in deep-water fisheries. It may be an important species in many South Pacific fisheries but is poorly documented there.

Saloptia powelli occurs in the South China Sea, extending east to French Polynesia.

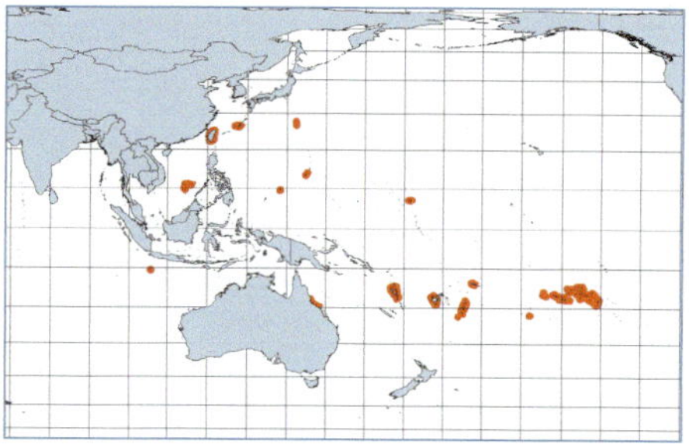

POPULATION TREND Unknown

MAJOR THREATS Fishing pressure may threaten this species.

HABITAT AND ECOLOGY GENERAL *S. powelli* inhabits hard substrate in depths of 100–360 m.

2008 IUCN RED LIST STATUS Data Deficient

Authors: *JH Choat, LA Rocha and MA Samoilys*

Triso dermopterus (Temminck and Schlegel 1842)

oval grouper (E), mérou ovale (F), mero ovalado (S)

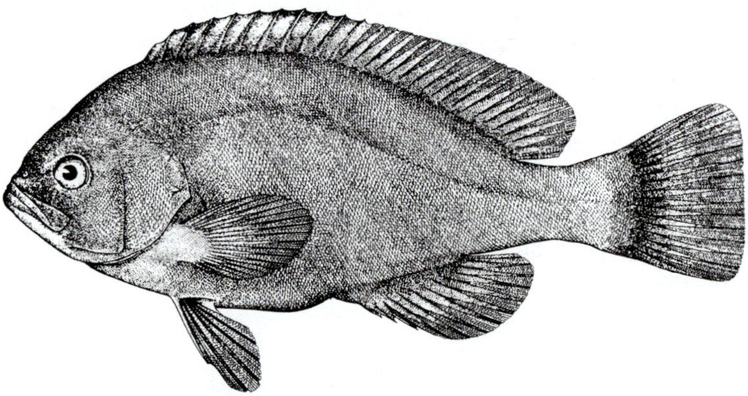

~50 cm TL
© FAO

Body oval (in lateral view), compressed, depth contained 2.4–2.7 times in the standard length. Head length 3.0–3.4 times in standard length; dorsal head profile distinctly convex; snout length distinctly longer than eye diameter; nostrils subequal in juveniles, the diameter of posterior nostrils twice that of anterior nostrils in a 56 cm SL adult; mouth small, oblique, a pair of short canines at front of each premaxilla. Gill rakers 8 or 9 on upper limb, 16–18 on lower limb, no rudiments, the longest raker slightly shorter than the longest filaments on 1st arch. Dorsal fin with 11 spines and 18–21 rays, the fin origin above opercle; dorsal-fin membranes not (or only slightly) incised between the spines, the 4th to 11th spines subequal; anal fin with 3 spines and 9–12 rays, the 3rd spine longest; pectoral fins asymmetrical, with 18–20 rays, the upper rays longest, their length contained 1.4–1.6 times in head length; a well-developed scaly flap of skin joining upper pectoral-fin rays to body; pelvic-fin length 1.4–1.7 times in head length; caudal fin truncate to emarginate, the corners rounded. Small ctenoid scales on head and body, no auxiliary scales; lateral-line scales 67–76; lateral-scale series 131–145.

MAXIMUM RECORDED SIZE 68 cm TL

COLOUR IN LIFE Dark brown or violet-black, the fins darker.
COLOUR IN MARKET Generally pale brown to grey with no distinguishing features.

© Andrew S Cornish

Adult *T. dermopterus* (Lamma Island, Hong Kong)

© Allen W-L To

Adult *T. dermopterus* (Lei Yue Mun Market, Hong Kong)

Triso dermopterus has an anti-tropical distribution in the eastern Indian Ocean and western Pacific. It is known in the Northern Hemisphere from Korea, Japan, Taiwan, Hong Kong and the Fujian Province of China; in the Southern Hemisphere it occurs off the coast of Western Australia (south of 19°30′ S) and off eastern Australia between 25° and 32° S.

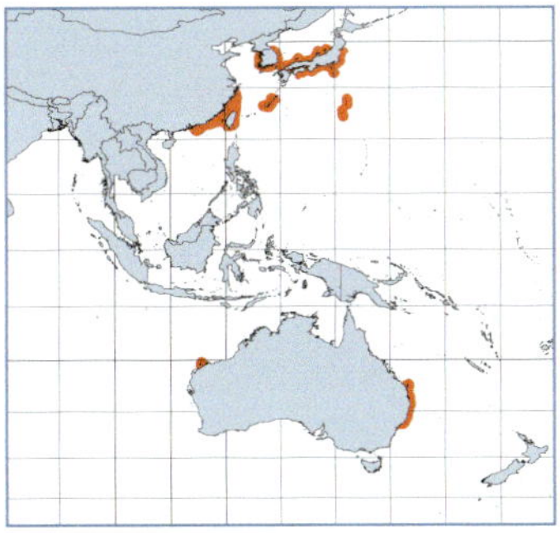

POPULATION **General** *T. dermopterus* is a naturally rare species throughout its range. **Fishery dependent** No fishery-dependent data are available. *T. dermopterus* is taken opportunistically.
POPULATION TREND Unknown
MAJOR THREATS No major threats have been identified for this species.
HABITAT AND ECOLOGY **General** *T. dermopterus* occurs on rocky or soft bottoms at depths of 22–103 m. **Feeding** Juveniles feed on zooplankton in the water column; feeding in adults has not been observed or studied.
2008 IUCN RED LIST STATUS Least Concern

Authors: *AS Cabanban, JH Choat and MT Craig*

Variola albimarginata (Baissac 1952)

white-edged lyretail (E), croissant queue blanche (F), mero rabiblanco (S)

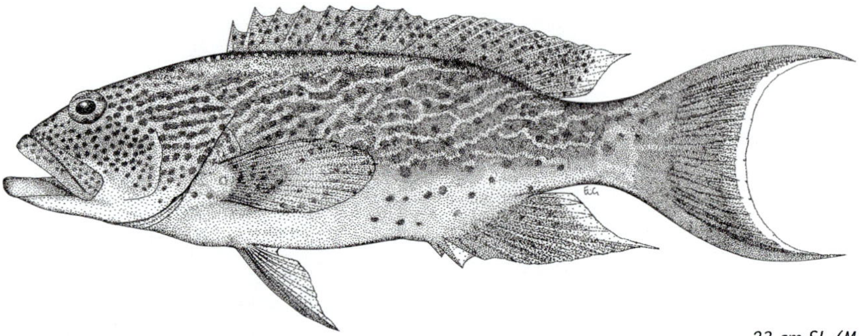

23 cm SL (Mauritius)
© SAIAB

Body oblong, depth less than head length and contained 2.8–3.5 times in standard length. Head length 2.6–2.8 times in standard length. Gill rakers 7–9 on upper limb, 13–16 on lower limb (including 4–7 rudiments on each limb). Dorsal fin with 9 spines and 14 rays; anal fin with 3 spines and 8 rays; pectoral-fin rays 17–19, the middle rays longest, contained 1.6–1.8 times in head length; pelvic fins usually not reaching anus, pelvic-fin length 1.4–1.8 times in head length; caudal fin lunate, upper and lower lobes produced, about twice the length of middle rays; branched rays 15. Midlateral-body scales ctenoid, without auxiliary scales; lateral-line scales 66–76; lateral-scale series 109–127.

MAXIMUM RECORDED SIZE 65 cm FL

Variola albimarginata is an insular Indo-Pacific species ranging from the east coast of Africa (Zanzibar and Mafia Island, Tanzania) to the Comoros Islands, Seychelles, Mascarene Islands, Chagos, Maldives, Sri Lanka to Indonesia, Philippines, Taiwan and Samoa, extending north to the Ryukyu Islands (Japan) and south to Queensland (Australia), New Guinea, New Caledonia and the East Andaman Sea (Thailand).

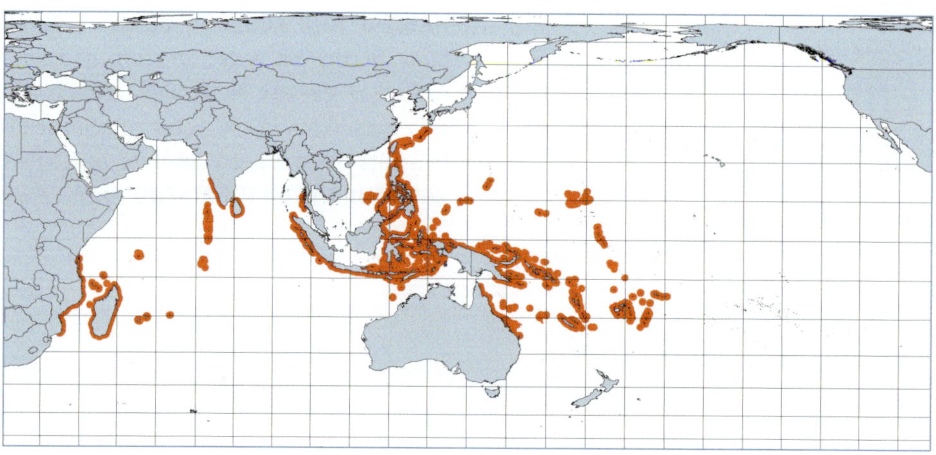

COLOUR IN LIFE Body brownish orange or reddish purple, with irregular red bands alternating with yellow lines, the red bands containing small irregular pale blue to pink spots; head orange-yellow, densely spotted with red; median fins with small pink and red spots; rear margin of the caudal fin usually dusky, with a narrow white edge; rear margin of the dorsal and anal fins hyaline, without a trace of yellow; pectoral fins yellow, the basal half of the rays often reddish or dark brown.

COLOUR IN MARKET This species maintains nearly identical colouration in death.

© Matthew T Craig

Subadult *V. albimarginata* (Sulawesi, Indonesia)

© Stephen Newman

Adult *V. albimarginata* (Scott Reef, Western Australia)

© Robert F Myers

Juvenile *V. albimarginata*
(Helen Atoll, Palau)

© Robert F Myers

Adult *V. albimarginata* (Bali, Indonesia)

POPULATION **GENERAL** *V. albimarginata* is patchily distributed throughout its range with locally abundant populations. **FISHERY DEPENDENT** *V. albimarginata* is taken opportunistically by fishers throughout its range.

POPULATION TREND Decreasing

MAJOR THREATS No major threats have been identified for this species.

HABITAT AND ECOLOGY **GENERAL** *V. albimarginata* is a reef-associated species found in depths ranging from 4 to 200 m. Juveniles are typically found inshore on algae and soft coral reefs, usually swimming well above the substrate and looking much like a basslet. **FEEDING** Feeds mainly on fishes. **AGE AND GROWTH** *V. albimarginata* reaches ages of at least 5 years (Fry et al. 2006).

COMMENTS *V. albimarginata* is easily confused with its congener *V. louti*. The two are distinguished primarily by the yellow (*V. louti*) or white (*V. albimarginata*) margin of the caudal fin.

2008 IUCN RED LIST STATUS Least Concern

Authors: *AS Cabanban, ST Fennessy, RF Myers, DA Pollard and KL Rhodes*

Variola louti (Forsskål 1775)

yellow-edged lyretail (E), croissant queue jaune (F), mero luna creciente (S)

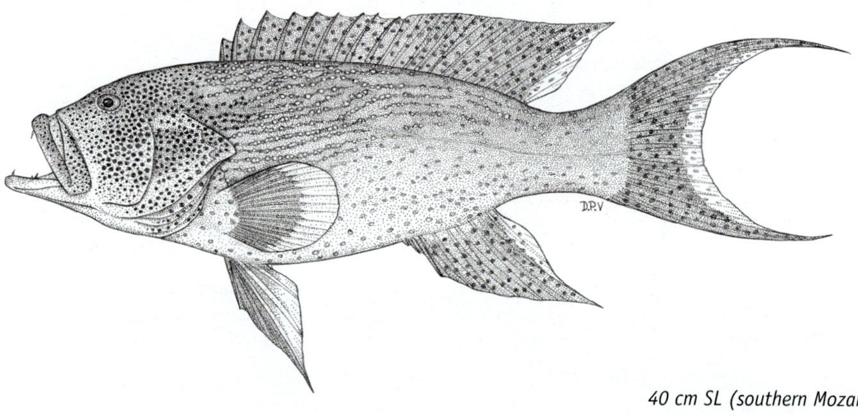

40 cm SL (southern Mozambique)

Body oblong, depth less than head length and contained 2.8–3.3 times in standard length. Head length 2.5–2.8 times in standard length. Gill rakers 7–10 on upper limb, 15–18 on lower limb (including 6–8 rudiments on each limb). Dorsal fin with 9 spines and 13 or 14 rays; anal fin with 3 spines and 8 rays; pectoral-fin rays 16–19, the middle rays longest, their length contained 1.6–1.8 times in head length; pelvic fins of adults reach past anus, pelvic-fin length 1.0–1.7 times in head length; caudal fin lunate, the upper and lower lobes produced, about twice length of middle rays. Midlateral-body scales ctenoid, without auxiliary scales; lateral-line scales 66–77; lateral-scale series 113–135.

MAXIMUM RECORDED SIZE 81 cm FL

COLOUR IN LIFE Head, body and median fins yellowish brown to orange with numerous small round or elongate spots of blue to lavender or pink; rear margin of the median fins broadly yellow; pectoral-fin rays red to brown, the distal third abruptly yellow. Large juveniles (8–18 cm SL) with an irregular black band along the dorsal part of the body ending below the rear of the dorsal fin and continuing on the head (up to the eye) as 3 irregular black spots; an irregular black blotch at the base of the upper caudal-fin rays; head and body (including black band) with small pale blue to pink spots; a pale yellow or white stripe mid-dorsally on the head from the tip of the lower jaw to the dorsal-fin origin. Small juveniles (less than 7 cm SL) lack the black band and black spots on the dorsal part of the head and body.

COLOUR IN MARKET *V. louti* maintains nearly identical colouration after death.

© Matthew T Craig

Adult *V. louti* (Sodwana Bay, South Africa)

© Patrick Louisy

Juvenile *V. louti* (Dahab, Gulf of Aqaba, Red Sea)

© Robert F Myers

Adult *V. louti* (Palau)

Variola louti occurs throughout the tropical Indo-Pacific region, extending from the east coast of Africa to the Red Sea and east to the Pitcairn Islands in the Pacific.

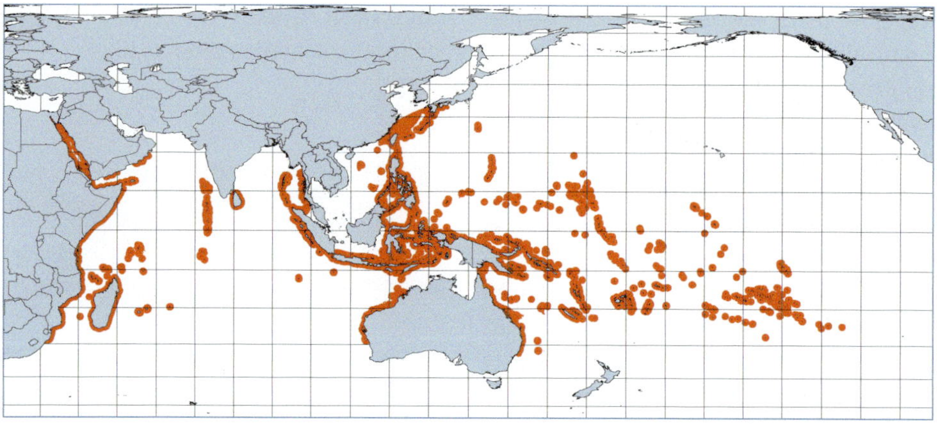

POPULATION **GENERAL** *V. louti* is a widespread grouper species that is abundant in the western Indian Ocean, gradually reaching low density in the western Pacific. **FISHERY DEPENDENT** Landings data for *V. louti* are sparse but indicate a decline in landings over the past 20 years. **FISHERY INDEPENDENT** Observations based on visual censuses indicate that *V. louti* has a patchy distribution.

POPULATION TREND Stable

MAJOR THREATS Fishing pressure is the primary threat to this species.

HABITAT AND ECOLOGY **GENERAL** *V. louti* is a coral reef-associated species found at depths of 3–240 m. **FEEDING** *V. louti* feeds primarily on fishes. **AGE AND GROWTH** *V. louti* reaches ages of at least 15 years (Grandcourt 2005). **REPRODUCTION** In the Indian Ocean, spawning peaks in March, April and October; the sexual pattern is not known.

COMMENTS *V. louti* is easily confused with its congener *V. albimarginata*. The two are distinguished primarily by the yellow (*V. louti*) or white (*V. albimarginata*) margin of the caudal fin.

2008 IUCN RED LIST STATUS Least Concern

Authors: *JH Choat, MT Craig and BP Ferreira*

Appendix

Bibliography

This bibliography lists the published research dealing with groupers and includes all the literature cited in the book as well as the many additional references that were consulted in producing the species assessments.

Abdula RG, van der Elst RP, Lichucha IDLT, Govender A, Cuco C. 2000. The industrial trapfishery of southern Mozambique. Results of experimental phase: 1997–1999. Unpublished report. Instituto de Investigaçao Pesqueira, Maputo, Mozambique.

Abo T, Bender B, Capelle A, DeBrum T. 1976. *Marshallese English dictionary*. Micronesia: Pail Language Texts.

Abu-Hakima R. 1987. Aspects of the reproductive biology of the grouper, *Epinephelus tauvina* (Forsskål), in Kuwait waters. *Journal of Fish Biology* 30: 213–222.

Aburto-Oropeza O, Erisman B, Valdez-Ornelas V, Danemann G. 2008. Commercially important serranid fishes from the Gulf of California: ecology, fisheries, and conservation. *Ciencia y Conservación* 2008(1): 1–23.

Academia Sinica Institute of Zoology. 1962. *The fishes of South China Sea*. Beijing: Science Press. 1183 pp.

Acero PA, Garzon FJ. 1982. Meros, chernas y cabrillas del caribe Colombiano (Pisces: Serranidae: Epinephelinae: Epinephelini). *Caldasia* 16(78): 355–376.

Acero PA, Garzon FJ. 1982. Un revision de las fichas de identificacion de la FAO para peces marinos del Atlantico centro-occidental, incluyendo nuevos registros para el cabribe Colombiano. *Anales del Instituto de Investigaciones Marinas* 12: 33–39.

Adams S. 2002. The reproductive biology of three species of *Plectropomus* (Serranidae) and responses to fishing. PhD thesis, James Cook University, Queensland, Australia.

Adams S. 2003. Morphological ontogeny of the gonad of three plectropomid species through sex differentiation and transition. *Journal of Fish Biology* 63: 22–36.

Adams S, Mapstone BD, Russ GR, Davies CR. 2000. Geographic variation in the sex ratio, sex specific size, and age structure of *Plectropomus leopardus* (Serranidae) between reefs open and closed to fishing on the Great Barrier Reef. *Canadian Journal of Fisheries and Aquatic Sciences* 57: 1448–1458.

Adams S, Williams AJ. 2001. A preliminary test of the transitional growth spurt hypothesis using the protogynous coral trout *Plectropomus maculatus*. *Journal of Fish Biology* 59: 183–185.

AFCD (Agriculture, Fisheries and Conservation Department). 2004. Port Survey 2001/02. Unpublished report, Agriculture, Fisheries and Conservation Department, Hong Kong.

AFD (Agriculture and Fisheries Department). 1998. Port survey 96/97. Prepared by the Capture Fisheries Division, Agriculture and Fisheries Department, Hong Kong.

Afonso P, Porteiro FM, Santos RS, Barreiros JP, Worms J, Wirtz P. 1999. Coastal marine fishes of São Tomé Island (Gulf of Guinea). *Arquipélago* 17(A): 65–92.

Agembe S, Mlewa CM, Kaunda-Arara B. 2010. Catch composition, abundance and length-weight relationships of groupers (Pisces: Serranidae) from inshore waters of Kenya. *Western Indian Ocean Journal of Marine Science* 9(1): 91–102.

Aguilar-Perera A, Aguilar-Davila W. 1996. A spawning aggregation of Nassau grouper *Epinephelus striatus* (Pisces: Serranidae) in the Mexican Caribbean. *Environmental Biology of Fishes* 45: 351–361.

Aguilar-Perera A, Gónzalez-Salas C, Villegas-Hernández H. 2008. Fishing, management, and conservation of the Nassau grouper, *Epinephelus striatus*, in the Mexican Caribbean. *Proceedings of the Gulf and Caribbean Fisheries Institute* 61: 313–319.

Aguilera O, Rodrigues de Aguilera D. 2004. Goliath grouper (Pisces: Serranidae) from the Upper Miocene Urumaco Formation, Venezuela. *Journal of Paleontology* 78(6): 1202–1206.

Al-Kershi A, Al-Yemeni M, Zoom A. 2004. Marine biodiversity in Al-Luhayyah-Midi Region. Ministry of Water and Environment, Environment Protection Department, Yemen.

Allen GR. 2003. Chapter 3: Coral reef fishes of the Raja Ampat Islands. In: Donnelly R, Neville D, Mous PJ (eds), Report on a rapid ecological assessment of the Raja Ampat Islands, Papua, Eastern Indonesia, held October 30–November 22, 2002. Sanur, Bali, Indonesia: The Nature Conservancy, Southeast Asia Center for Marine Protected Areas. pp 42–58.

Allen GR. 2003. Chapter 4: Reef fishes of Milne Bay Province, Papua New Guinea. In: Allen GR, Kinch JP, McKenna SA, Seeto P (eds), *A rapid marine biodiversity assessment of Milne Bay Province, Papua New Guinea—Survey II (2000)*. RAP Bulletin of Biological Assessment 29. Washington, DC, USA: Conservation International. pp 46–55.

Allen GR, Adrim M. 2003. Coral reef fishes of Indonesia. *Zoological Studies* 42(1): 1–72.

Allen GR, McKenna SA (eds). 2001. *A marine rapid assessment of the Togean and Banggai Islands, Sulawesi, Indonesia*. RAP Bulletin of Biological Assessment 20. Washington, DC, USA: Conservation International.

Allen GR, Munday PL. 1994. Kimbe Bay rapid ecological assessment: the coral reefs of Kimbe Bay (West New Britain, Papua New Guinea), Volume 3: fish diversity of Kimbe Bay. Auckland, New Zealand: The Nature Conservancy, South Pacific Program Office. 107 pp.

Allen GR, Robertson DR. 1999. *Epinephelus clippertonensis*, a new species of serranid fish from the tropical eastern Pacific. *Revue Française d'Aquariologie* 19(1–2): 37–40.

Allen GR, Smith-Vaniz WF. 1994. Fishes of the Cocos (Keeling) Islands. *Atoll Research Bulletin* 412: 1–21.

Allen GR, Steene R. 1987. Reef fishes of the Indian Ocean: a pictorial guide to the common reef fishes of the Indian Ocean (Pacific marine fishes). Neptune City, New Jersey, USA: TFH Publications. 240 pp.

Allen GR, Steene RC. 1988. *Fishes of Christmas Island Indian Ocean*. Christmas Island Natural History Association, Christmas Island Indian Ocean 6798, Australia.

Allen GR, Stone GS (eds). 2005. Rapid assessment survey of tsunami-affected reefs of Thailand. In: Allen GR, Stone GS (eds), Rapid assessment survey of tsunami-affected reefs of Thailand. New England Aquarium Final Technical Report 02-05. Boston, USA: New England Aquarium Global Marine Programs Office. pp 7–15.

Allen GR, Swainston R. 1988. *The marine fishes of north-western Australia: a field guide for anglers and divers*. Perth, Australia: Western Australia Museum.

Almany GR. 2004. Differential effects of habitat complexity, predators and competitors on abundance of juvenile and adult coral reef fishes. *Oecologia* 141: 105–113.

Amaoka K, Randall JE. 1990. *Epinephelus suborbitalis*, a new species of grouper (Perciformes: Serranidae) from the Kyushu-Palau Ridge. *JLB Smith Institute of Ichthyology, Special Publication* No. 49: 1–9.

Andrade AB, Machado LF, Hostim-Silva M, Barreiros JP. 2003. Reproductive biology of the dusky grouper *Epinephelus marginatus* (Lowe, 1834). *Brazilian Archives of Biology and Technology* 46(3): 373–382.

Anon. 1993. Fishing Industry Organisation and Marketing Amendment Regulation (No. 3). Subordinate Legislation No. 235. State of Queensland, Australia.

Anon. 1997. Bulletin Statistique des Pêches, Année 1996. Bissau, Guinea-Bissau: Centre National des Sciences Halieutiques de Boussoura.

Anon. 1998. Grouper market big in Philippines. *Fish Farming International* 25(7): 22.

Anon. 2002. Public Consultation Document. Environmental Impact Statement on the Ocean Hauling Fishery – Appendix F. Department of the Environment and Heritage, Canberra, Australia.

Anon. 2003. News from Japan. In: IUCN Specialist Group of Groupers & Wrasses Newsheet No. 6. IUCN Groupers & Wrasses Specialist Group. p 3.

Anon. 2006. *Deepwater fishing guide*. Brisbane, Queensland, Australia: Department of Primary Industries and Fisheries.

Anon. 2007. Threatened species in NSW. Black cod *Epinephelus daemelii*. *Primefact* 189. New South Wales, Australia: NSW Department of Primary Industries. 2 pp. Accessed at http://www.dpi.nsw.gov.au/__data/assets/pdf_file/0019/144154/black-cod.pdf.

Appeldoorn RS, Dennis GD, Lopez OM. 1987. Review of shared demersal resources of Puerto Rico and the Lesser Antilles region. In: Mahon R (ed.), *Report and proceedings of the expert consultation on shared fishery resources of the Lesser Antilles Region. FAO Fisheries Report* No. 338. Rome, Italy: Food and Agriculture Organization. pp 36–104.

Araújo JN, Martins AS. 2006. Age and growth of coney (*Cephalopholis fulva*), from the central coast of Brazil. *Journal of the Marine Biological Association UK* 86: 187–191.

Arena PT, Quinn TP, Jordan LKB, Sherman RL, Harttung FM, Spieler RE. 2004. Presence of juvenile blackfin snapper, *Lutjanus buccanella*, and snowy grouper, *Epinephelus niveatus*, on shallow-water artificial reefs. *Proceedings of the Gulf and Caribbean Fisheries Institute* 55: 701–712.

Arreguin-Sánchez F. 1987. Present status of the red grouper fishery of the Campeche Bank. *Proceedings of the Gulf and Caribbean Fisheries Institute* 38: 498–509.

Arreguin-Sánchez F, Contreras M, Moreno V, Burgos R, Valdes R. 1996. Population dynamics and stock assessment of red grouper (*Epinephelus morio*) fishery on Campeche Bank, Mexico. In: Arreguin-Sanchez F, Munro JL, Balgos MC, Pauly D (eds), *Biology, fisheries and culture of tropical groupers and snappers. ICLARM Conference Proceedings* 48. Manila, Philippines: International Center for Living Aquatic Resources Management. pp 202–217.

Arreguin-Sanchez R, Pitcher TJ. 1999. Catchability estimates and their application to the red grouper (*Epinephelus morio*) fishery of the Campeche Bank, Mexico. *Fishery Bulletin* 97(4): 746–757.

Ashworth JS, Bruce OE, Mohammed EH. 2006. Fish assemblages of Red Sea backreef biotopes. *Aquatic Conservation* 16(6): 593–609.

Audibert K. 1988. Natal offshore Sciaenidae and Serranidae. In: Research report, April 1987–1988.

Unpublished Report No. 54. Durban, South Africa: Oceanographic Research Institute. pp 24–29.

Ault J[S], Bohnsack J, Meester G. 1998. A retrospective (1979–1996) multispecies assessment of coral reef fish stocks in the Florida Keys. *Fishery Bulletin* 96(3): 395–414.

Ault JS, Smith SG, Bohnsack JA. 2005. Evaluation of average length as an estimator of exploitation status for the Florida coral-reef community. *ICES Journal of Marine Science* 62: 417–423.

Aumeeruddy R, Robinson J. 2006. Closure of the live reef food fish fishery in Seychelles. *SPC Live Reef Fish Information Bulletin* No. 16: 3–9.

Australian Bureau of Agricultural and Resource Economics. 2003. *Australian fisheries statistics 2002*. Canberra, Australia: ABARE.

Ayling AM, Ayling AL. 1986. Coral trout survey, Great Barrier Reef. Unpublished report to the Great Barrier Reef Marine Park Authority, Townsville, Queensland, Australia.

Ayling T, Cox GJ. 1982. *Collins guide to the sea fishes of New Zealand*. Auckland, New Zealand: Collins. 343 pp.

Ayling T, Samoilys M, Ryan D. 2000. Trends in common coral trout populations on the Great Barrier Reef. Information Series QI00063. Brisbane, Queensland: Department of Primary Industries.

Azemia R, Assan C. 2006. Seychelles artisanal fisheries statistics for 2005. Victoria, Seychelles: Seychelles Fishing Authority, Government of Seychelles.

Azzurro E, Andaloro F, Marino G. 2000. Presence of Haifa grouper, *Epinephelus haifensis* (Serranidae: Epinephelinae) in the central Mediterranean Sea. *Biologia Marina Mediterranea* 7(1): 786–789.

Bacchet P, Lefevre Y, Zysman T. 2007. *Guide des poissons de Tahiti et ses îles*. Papeete, Tahiti: Éditions Au Vent des Îles.

Bacheler NM, Buckel JA. 2004. Does hook type influence the catch rate, size and injury of grouper in a North Carolina commercial fishery? *Fisheries Research* 69: 303–311.

Bagnis R. 1968. Clinical aspects of ciguatera (fish poisoning) in French Polynesia. *Hawai'i Medical Journal* 28: 25–28.

Bagnis R, Kuberski T, Laugier S. 1979. Clinical observations on 3,009 cases of ciguatera (fish poisoning) in the South Pacific. *American Journal of Tropical Medicine and Hygiene* 28: 1067–1073.

Bagnis R, Mazellier P, Bennett J, Christian E. 1972. *Fishes of Polynesia*. Papeete, Tahiti: Les editions du Pacifique.

Baillie J, Groombridge B (compilers and editors). 1996. *1996 IUCN Red List of threatened animals*. Gland, Switzerland: International Union for Conservation of Nature.

Baldwin CC, Johnson GD. 1993. Phylogeny of the Epinephelinae (Teleostei: Serranidae). *Bulletin of Marine Science* 52: 240–283.

Balguerías E, Quintero ME, Hernández-González CL. 2000. The origin of the Saharan Bank cephalopod fishery. *ICES Journal of Marine Science* 57: 15–23.

Bannerot SP, Fox WW, Powers JE. 1987. Reproductive strategies and the management of snappers and groupers in the Gulf of Mexico and Caribbean. In: Polovina JJ, Ralston S (eds), *Tropical snappers and groupers: biology and fisheries management*. Boulder, USA/London, UK: Westview Press. pp 561–603.

Barreiros JP, Bertoncini AA, Machado LF, Hostim-Silva M, Santos RS. 2004. Diversity and seasonal changes in the ichthyofauna of rocky tidal pools from Praia Vermelha and São Roque, Santa Catarina. *Brazilian Archives of Biology and Technology* 47(2): 291–299.

Bauchot M-L, Pras A. 1980. *Guide de poissons marins d'Europe*. Lausanne-Paris, France: Delachaux & Niestlé.

Béarez P, Jiménez Prado P. 2003. New records of serranids (Perciformes) from the continental shelf of Ecuador with a key to the species, and comments on ENSO-associated fish dispersal. *Cybium* 27(2): 107–115.

Beets J, Friedlander A. 1999. Evaluation of a conservation strategy: a spawning aggregation closure for red hind, *Epinephelus guttatus*, in the US Virgin Islands. *Environmental Biology of Fishes* 55: 91–98.

Beger M. 2002. The diversity and status of coral reef fishes of eastern Kimbe Bay. Brisbane, Australia: The Nature Conservancy-Melanesia Program.

Ben Abdallah A, Ben Souissi J, Méjri H, Capapé C, Golani D. 2007. First record of *Cephalopholis taeniops* (Valenciennes) in the Mediterranean Sea. *Journal of Fish Biology* 71: 610–614.

Beukers-Stewart BD, Jones GP. 2004. The influence of prey abundance on the feeding ecology of two piscivorous species of coral reef fish. *Journal of Experimental Marine Ecology and Biology* 299: 155–184.

Bielsa LM, Labisky RF. 1987. Food habits of blueline tilefish, *Caulolatilus microps*, and snowy grouper, *Epinephelus niveatus*, from the lower Florida Keys. *Northeast Gulf Science* 9: 77–87.

Biusing ER, Phillips M, Cabanban AS. 1999. Aquaculture of coral reef fishes and other nearshore fishes in Sabah, Malaysia. In: Cabanban AS, Phillips M (eds), *Aquaculture of coral reef fishes. Proceedings of Workshop on Aquaculture of Coral Reef Fishes and Sustainable Reef Fisheries*. Sabah, Malaysia: Institute of Development Studies. pp 74–86.

Blaber SJM, Milton DA, Rawlinson NJF, Tiroba G, Nichols PV. 1990. Diets of lagoon fishes of the Solomon Islands: predators of tuna baitfish and trophic effects of baitfishing on the subsistence fishery. *Fisheries Research* 8: 263–286.

Bloch ME. 1793. *Naturgeschichte der ausländischen fische*, Vol. VII. Berlin. 144 pp.

Bodilis P, Ganteaume A, Francour P. 2003. Presence of 1 year-old dusky groupers along the French Mediterranean coast. *Journal of Fish Biology* 62(1): 242–246.

Bohnsack JA. 1996. Marine reserves, zoning, and the future of fishery management. *Fisheries* 21(9): 14–16.

Bohnsack JA. 1999. Incorporating no-take marine reserves into precautionary management and stock assessment. In: Restrepo VR (ed.), Proceedings of the Fifth National NMFS Stock Assessment Workshop: providing scientific advice to implement the precautionary approach under the Magnuson-Stevens Fishery Conservation and Management Act. NOAA Technical Memorandum NMFS-F/SPO-40. US Department of Commerce. pp 8–16.

Bohnsack JA. 2000. A comparison of the short-term impacts of no-take marine reserves and minimum size limits. *Bulletin of Marine Science* 66(3): 635–650.

Bohnsack JA, Harper DE, McClellan DB. 1994. Fisheries trends from Monroe County, Florida. *Bulletin of Marine Science* 54(3): 982–1018.

Bolden SK. 2000. Long-distance movement of a Nassau grouper (*Epinephelus striatus*) to a spawning aggregation in the central Bahamas. *Fisheries Bulletin* 98(3): 642–645.

Bostrom MA, Collete BB, Luckhurst BE, Reece KS, Graves JE. 2002. Hybridization between two serranids, the coney (*Cephalopholis fulva*) and the creole-fish (*Paranthias furcifer*), at Bermuda. *Fishery Bulletin* 100(4): 651–661.

Bouain A. 1980. Sexualité et cycle sexuel des merous (Poisson, Téleostéens, Serranides) des Côtes du Sud Tunisien. *Bulletin Office National Pêches (Tunisia)* 4(2): 215–229.

Bouain A. 1986. Croissance linéaire des mérous du Golfe de Gabes (Tunisie). *Cybium* 10(3): 299–302.

Bouain A, Siau Y. 1983. Observations on the female reproductive cycle and fecundity of three species of groupers (*Epinephelus*) from the southeast Tunisian seashores. *Marine Biology* 73: 211–220.

Bouain A, Siau Y, Quignard JP. 1983. Les mérous des côtes sud-est de la Tunisie. Deuxième partie: pêche. *La Pêche Maritime* 1263: 337–342.

Bouchon-Navaroa Y, Bouchona C, Louisa M, Legendre P. 2005. Biogeographic patterns of coastal fish assemblages in the West Indies. *Journal of Experimental Marine Biology and Ecology* 315: 31–47.

Boulenger GA. 1895. *Catalogue of the perciform fishes in the British Museum*. London, UK: Taylor & Francis.

Bowater RO, Thomas A, Shivas RG, Humphrey JD. 2003. Deuteromycotic fungi infecting barramundi cod, *Cromileptes altivelis* (Valenciennes), from Australia. *Journal of Fish Diseases* 26: 681–686.

Boyan S. 2000. Federal regs won't protect spawning grouper. *Salt Water Sportsman* May 2000: 148.

BRASIL MMA. 2004. Lista de espécies de peixes e invertebrados aquáticos ameaçados de extinção, sobrexplotados ou ameaçados de sobrexplotação no Brasil. Sao Paulo, Brazil: Ministério do Meio Ambiente.

BRASIL MMA. 2005. Estabelece a proibição da pesca do cherne poveiro (*Polyprion americanus*), nas águas jurisdicionais brasileiras, por um período de 10 anos. Instrução Normativa. Sao Paulo, Brazil: Ministério do Meio Ambiente.

Bremer JRA, Vinas J, Mejuto J, Ely B, Pla C. 2005. Comparative phylogeography of Atlantic bluefin tuna and swordfish: the combined effects of vicariance, secondary contact, introgression, and population expansion on the regional phylogenies of two highly migratory pelagic fishes. *Molecular Phylogenetics and Evolution* 36: 169–187.

Brodie J, Al-Sorimi M, Turak E. 1998. Chapter 4: Fish and fisheries of Yemen's Red Sea. In: DouAbal A et al. (eds), *Protection of marine ecosystems of the Red Sea coast of Yemen*. United Nations Development Programme. pp 51–71.

Brown IW, McPherson G, Samoilys MA, Doherty PJ, Russ G. 1991. Growth, reproductive strategies and recruitment of the dominant demersal food-fish species on the Great Barrier Reef. Progress Report on FIRDC Project 90/18. Department of Primary Industries, James Cook University of North Queensland, and Australian Institute of Marine Science.

Brulé T, Colás-Marrufo T. 1997. Explotacion de los serránidos en el sureste del Golfo de México: evolución y situación actual. *Proceedings of the Gulf and Caribbean Fisheries Institute* 49: 173–214.

Brulé T, Colás-Marrufo T, Tuz-Sulub A, Déniel C. 2000. Evidence for protogynous hermaphroditism in the serranid fish *Epinephelus drummondhayi* (Perciformes: Serranidae) from the Campeche Bank in the southern Gulf of Mexico. *Bulletin of Marine Science* 66(2): 513–521.

Brulé T, Déniel C. 1994. Synoptic overview of the biological data on the red grouper, *Epinephelus morio* (Valenciennes, 1828), from the Gulf of Mexico. *FAO Fisheries Synopsis* No. 155. Rome, Italy: Food and Agriculture Organization. 39 pp.

Brulé T, Déniel C. 1996. Biological research on the red grouper (*Epinephelus morio*) from the southern Gulf of Mexico. In: Arreguin-Sanchez F, Munro JL, Balgos MC, Pauly D (eds), *Biology, fisheries and culture of tropical groupers and snappers. ICLARM Conference Proceedings* 48. Manila, Philippines: International Center for Living Aquatic Resources Management. pp 28–42.

Brulé T, Déniel C, Colás-Marrufo T, Renán X. 2003.

Reproductive biology of gag in the southern Gulf of Mexico. *Journal of Fish Biology* 63: 1505–1520.

Brulé T, Déniel C, Colás-Marrufo T, Sánchez-Crespo M. 1999. Red grouper reproduction in the southern Gulf of Mexico. *Transactions of the American Fisheries Society* 128(3): 385–402.

Brulé T, Maldonado-Montiel T del NJ, Rodriguez-Canche LG, Mexicana-Cintora G. 1996. Aspects of the reproductive and trophic biology of the red grouper, *Epinephelus morio* (Valenciennes, 1828), from the Campeche Bank, Yucatan, Mexico. *Proceedings of the Gulf and Caribbean Fisheries Institute* 44: 39–55.

Brulé T, Renán X, Colás-Marrufo T, Hauyon Y, Tuz-Sulub AN, Déniel C. 2003. Reproduction in the protogynous black grouper (*Mycteroperca bonaci* (Poey)) from the southern Gulf of Mexico. *Fisheries Bulletin* 101(3): 463–475.

Bruslé J. 1985. Exposé synoptique des données biologiques sur les mérous *Epinephelus aeneus* (Geoffroy Saint Hilaire, 1809) et *Epinephelus guaza* (Linnaeus, 1758) de l'Océan Atlantique et de la Méditerranée. *FAO Fisheries Synopsis* No. 129. Rome, Italy: Food and Agriculture Organization. 64 pp.

Bruslé J, Bruslé S. 1976. Contribution à l'étude de la reproduction de deux espéces de Mérous (*Epinephelus aeneus* et *Epinephelus guaza*) des côtes de Tunisie. *Rapport de la Commission Internationale pour l'Exploration Scientifique de la Mer Méditerranée* 23(8): 49–50.

Bruslé J, Bruslé-Sicard S, Fourcault B. 1998. Gonads of groupers (Serranidae) from Mururoa and Hao (French Polynesia) used as indicators of the quality of the coral reef environment: histological and ultrastructural aspects. *Cybium* 22(4): 371–382.

Bryant D, Burke L, McManus JW, Spalding M. 1998. *Reefs at risk: a map-based indicator of potential threats to the world's coral reefs*. Washington, DC, USA: World Resources Institute. 56 pp.

Bueno E. 1998. *Capitães do Brasil: a saga dos primeiros colonizadores*. Rio de Janeiro, Brazil: Objetiva. 287 pp.

Bullen EM, Fennessy ST. 2000. Sedgwick's/ORI/WWF tagging programme: catface rockcod (*Epinephelus andersoni*) (code 16634). Data Report. Oceanographic Research Institute (2000.10), Durban, South Africa. 6 pp.

Bullock LH, Godcharles MF, Crabtree RE. 1996. Reproduction of yellowedge grouper, *Epinephelus flavolimbatus*, for the eastern Gulf of Mexico. *Bulletin of Marine Science* 59(1): 216–224.

Bullock LH, Murphy MD. 1994. Aspects of the life history of the yellowmouth grouper, *Mycteroperca interstitialis*, in the eastern Gulf of Mexico. *Bulletin of Marine Science* 55(1): 30–45.

Bullock LH, Murphy MD, Godcharles MF, Mitchell ME. 1992. Age, growth and reproduction of jewfish *Epinephelus itajara* in the eastern Gulf of Mexico. *Fisheries Bulletin* 90: 243–249.

Bullock LH, Smith GB. 1991. *Seabasses (Pisces: Serranidae). Memoirs of the Hourglass Cruises*, Volume VIII, Part II. St Petersburg, Florida, USA: Florida Marine Research Institute, Department of Natural Resources. 243 pp.

Burgess W, Axelrod HR. 1973. *Pacific marine fishes*. Hong Kong: TFH Publications.

Burgos JM, Sedberry GR, Wyanski DM, Harris PJ. 2007. Life history of red grouper (*Epinephelus morio*) off the coasts of North Carolina and South Carolina. *Bulletin of Marine Science* 80: 45–65.

Burgos R. 1999. Modelo bioeconómico captura-mortalidad para la pesquería de mero *Epinephelus morio* del Banco de Campeche: un marco de manejo precautorio. Tesis de Maestría en Ciencias, Cinvestav-IPN, Mexico. 95 pp.

Burgos R, Defeo O. 2000. Un marco precautorio para la pesquería de mero (*Epinephelus morio*) del Banco de Campeche, México. *Oceánides* 15: 129–140.

Burgos R, Defeo O. 2004. Long-term population structure, mortality and modeling of a tropical multi-fleet fishery: the red grouper *Epinephelus morio* of the Campeche Bank, Gulf of México. *Fisheries Research* 66: 325–335.

Burke L, Selig E, Spalding M. 2002. *Reefs at risk in Southeast Asia*. Washington, DC, USA: World Resources Institute.

Burnett-Herkes J. 1975. Contribution to the biology of the red hind, *Epinephelus guttatus*, a commercially important serranid fish from the tropical western Atlantic. PhD dissertation, University of Miami, Coral Gables, Florida, USA.

Bush PG, Lane ED, Ebanks-Petrie GC, Luke K, Johnson B, McCoy C, Bothwell J, Parsons E. 2006. The Nassau grouper spawning aggregation fishery of the Cayman Islands — an historical and management perspective. *Proceedings of the Gulf and Caribbean Fisheries Institute* 57: 515–524.

Bustamante RH. 1998. The artisanal fishing sector of the Galápagos and the 1997 fishing season. In: World Wildlife Fund–Fundación Natura (eds), *Galápagos report 1997–1998*. Quito, Ecuador: Trama Publishers. pp 25–29.

Bustos R, Luque A, Pajuelo JG. 2009. Age estimation and growth pattern of the island grouper, *Mycteroperca fusca* (Serranidae) in an island population on the northwest coast of Africa. *Scientia Marina* 73(2): 319–328.

Cabanban AS. 1999. Coral reef fish resources of Darvel Bay: an analysis of coral reef fish data from 1981–

1990 and reports of diversity from a market survey in Lahad Datu. In: Ranjith MW, DeSilva N, Rahman RA, Mustafa S, Cabanban AS (eds), *Ekspedisi Galaxea*. Kota Kinabalu, Malaysia: University of Malaysia. pp 85–95.

Cabanban AS. 2006. Diversity, abundance, and biomass of groupers (Serranidae) on Lankayan Reefs, Sabah, Malaysia – fisheries benefits of conservation. Report to World Wide Fund for Nature-Malaysia. 20 pp.

Caillart B, Harmelin-Vivien ML, Galzin R, Morize E. 1994. PART III. Reef fish communities and fishery yields of Tikahau Atoll (Tuamotu Archipelago French Polynesia). *Atoll Research Bulletin* 415: 1–38.

Carleton Ray G, McCormick-Ray MG, Layman CA, Silliman BR. 2000. Investigations of Nassau grouper breeding aggregations at High Cay, Andros: implication for a conservation strategy. Final report. The Department of Fisheries, Ministry of Agriculture and Fisheries, Nassau, The Bahamas.

Carlin JL, Robertson DR, Bowen BW. 2003. Ancient divergences and recent connections in two tropical Atlantic reef fishes *Epinephelus adscensionis* and *Rypticus saponaceous* (Percoidei: Serranidae). *Marine Biology* 143: 1057–1069.

Carter J. 1988. Grouper mating ritual on a Caribbean reef. *Underwater Naturalist* 17: 8–11

Carter J, Perrine D. 1994. A spawning aggregation of dog snapper, *Lutjanus jocu* (Pisces: Lutjanidae) in Belize, Central America. *Bulletin of Marine Science* 55: 228–234.

Carvalho-Filho A. 1999. *Peixes da costa Brasileira* (3rd edition). São Paulo, Brazil: Melro.

Cass-Calay SL, Bahnick M. 2002. Status of the yellowedge grouper fishery in the Gulf of Mexico: Assessment 1.0. Sustainable Fisheries Division Contribution SFD-02/03-172. Miami, Florida, USA: Southeast Fisheries Science Center.

Cass-Calay SL, Schmidt TW. 2003. Standardized catch rates of juvenile goliath grouper, *Epinephelus itajara*, from the Everglades National Park Creel Survey, 1973–1999. Sustainable Fisheries Division Contribution SFD-2003-0016. Miami, Florida, USA: Southeast Fisheries Science Center. 17 pp.

Castro-Suaste T, Mexicano-Cintora G, Defeo O. 2000. Las pesquerías del estado de Yucatán (México): evolución y manejo durante el periodo 1976–1997. *Oceánides* 15(1): 47–61.

Cervigón F. 1966. *Los Peces Marinos de Venezuela*. Vols I y II. Caracas, Venezuela: Fundación La Salle de Ciencias Naturales. 951 pp.

Cervigón F. 1983. *La acuicultura en Venezuela: estado actual y perspectives*. Caracas, Venezuela: Editorial Arte. 121 pp.

Cervigón F. 1991. *Los peces marinos de Venezuela*. Caracas, Venezuela: Fundación Científica Los Roques.

Cesar H, Lundin C, Bettencourt S, Dixon J. 1997. Indonesian coral reefs: an economic analysis of a precious but threatened resource. *Ambio* 26(6): 345–350.

Chakraborty SK. 1994. Age, growth, mortality and stock assessment of *Epinephelus diacanthus* (Valenciennes) from Bombay waters. *Bulletin of the Central Marine Fisheries Research Institute* 47: 130–133.

Chakraborty SK, Vidyasagar KD. 1996. Growth, mortality and stock assessment of two perches – moontail bull's eye *Priacanthus hamrur* (Perciformes/Pricanthidae) and thornycheek grouper *Epinephelus diacanthus* (Perciformes/Serranidae) from Bombay waters. *Indian Journal of Marine Sciences* 25: 312–315.

Chan PSW. 2000. The industry perspective: wholesale and retail marketing aspects of the Hong Kong live reef food fish trade. *SPC Live Reef Fish Information Bulletin* No. 7: 3–7.

Chan TC. 2000. Reproductive biology, age and growth in the chocolate hind, *Cephalopholis boneak* (Bloch, 1790), in Hong Kong. MPhil thesis, University of Hong Kong, Hong Kong, China.

Chan TTC, Sadovy Y. 2002. Reproductive biology, age and growth in the chocolate hind, *Cephalopholis boenak* (Bloch, 1790), in Hong Kong. *Marine and Freshwater Research* 53(4): 791–803.

Chan WL. 1968. *Marine fishes of Hong Kong, Part I*. Hong Kong: Hong Kong Government Press.

Chang K-H, Jan R-Q, Shao K-T. 1983. Community ecology of the marine fishes of Lutao Island, Taiwan. *Taiwan Bulletin of the Institute of Academia Sinica* 22(2): 141–155.

Chao TM, Lim LC, Khoo LT. 1993. Studies on the breeding of brown-marbled grouper (*Epinephelus fuscoguttatus* Forsskål) in Singapore. *Tungkang Marine Laboratory Conference Proceedings* 3: 143–156.

Chapman LB, Cusack P. 1990. *South Pacific Commission Deep Sea Fisheries Development Project report on second visit to Tuvalu: 30 August–7 December 1983*. Noumea, New Caledonia: South Pacific Commission. 60 pp.

Chapman RW, Sedberry GR, Koenig CC, Eleby BM. 1999. Stock identification of gag, *Mycteroperca microlepis*, along the southeast coast of the United States. *Marine Biotechnology* 1: 137–146.

Chater SA, Beckley LE, Garratt PA, Ballard JA, van der Elst RP. 1993. Fishes from offshore reefs in the St Lucia and Maputaland Marine Reserves, South Africa. *Lammergeyer* 42: 1–17.

Chater SA, Beckley LE, van der Elst RP, Garratt PA. 1995. Underwater visual census of fishes in the St Lucia marine reserve, South Africa. *Lammergeyer* 43: 15–23.

Chauvet C. 1988. Etude de la croissance du merou *Epinephelus guaza* (Linne, 1758) des cotes tunisiennes. *Aquatic Living Resources* 1: 277–288.

Chauvet CA. 2005. New Caledonia. *Society for the Conservation of Reef Fish Aggregations* [SCRFA] *Newsletter* 8: 9–11. Accessed at http://www.scrfa.org/images/stories/pdf/education/newsletter/newsletter8_bwl.pdf.

Chauvet C, Francour P. 1990. Les mérous du parc national de Port-Cros: aspects socio- démographiques. *Bulletin de la Societe Zoologique de France* 114(4): 5–13.

Chavance P, Damiano A, Diallo A, Doumbouya A, Drapeau L, Koulibaly M, Morize E, Sidibe A, Soumah M, Traoré S. 1996. *Bulletin Statistique des Pêches. Année 1995.* Bissau, Guinea-Bissau: Centre National des Sciences Halieutiques de Boussoura.

Chen CP, Hsieh HL, Chang KH. 1980a. Age and growth of the grouper, *Epinephelus diacanthus* (Cuvier et Valenciensis) in the waters of northern Taiwan. *Bulletin of the Institute of Zoology Academia Sinica* 19: 1–9.

Chen CP, Hsieh HL, Chang KH. 1980b. Some aspects of the sex change and reproductive biology of the grouper, *Epinephelus diacanthus* (Cuvier et Valenciensis). *Bulletin of the Institute of Zoology Academia Sinica* 19: 11–17.

Chen J-P, Shao K-T, Lin C-P. 1995. A checklist of reef fishes from the Tungsha Tao (Pratas Island), South China Sea. *Acta Zoologica Taiwanica* 6(2): 13–40.

Chen Q-C, Cai Y-Z, Ma X-M (eds). 1997. *Fishes from Nansha Islands to South China coastal waters* 1. Beijing, China: Science Press. 222 pp.

Cheung WL. 1998. Fecundity of *Epinephelus awoara* (yellow grouper). Unpublished report, Department of Ecology and Biodiversity, University of Hong Kong, Hong Kong, China.

Cheung WL. 2001. Restocking – an effective measure to restore the depleted fishery stocks in Hong Kong. *Porcupine!* [newsletter of the Department of Ecology & Biodiversity, The University of Hong Kong, Hong Kong, China] 24: 12–14.

Ching K, Sito A, To A. 2006. A survey of reef fish diversity in Port Shelter (Hong Kong). *Porcupine!* [newsletter of the Department of Ecology & Biodiversity, The University of Hong Kong, Hong Kong, China] 34: 14–16.

Chou LM, Tuan VS, Reefs P, Yeemin T, Cabanban A, Suharsono, Kessna I. 2002. Status of southeast Asia coral reefs. In: Wilkinson CR (ed.), *Status of coral reefs of the world*. Townsville, Australia: Australian Institute of Marine Science, GCRMN Report. pp 123–152.

Chu Y-T (ed.). 1984. *Fishes of Fujian Province* (1st edition). Fujian, China: Fujian Science and Technology Press.

Claro R, Baisre J, Garcia-Arteaga JP (eds). 1994. Evolución y manejo de los recursos pesqueros. In: Claro R (ed.), *Ecologia delos peces marinos de Cuba*. Quintana Roo, Mexico: CIQRO (Centro de Investigaciones de Quintana Roo). pp 435–492.

Claro R, Lindeman KC. 2003. Spawning aggregation sites of snapper and grouper species (Lutjanidae and Serranidae) on the insular shelf of Cuba. *Gulf and Caribbean Research* 14(2): 91–106.

Claro R, Lindeman KC, Parenti LR. 2001. *Ecology of the marine fishes of Cuba*. Washington, DC, USA: Smithsonian Institution Press.

Claro R, Sadovy de Mitcheson Y, Lindeman KC, García-Cagidea AR. 2009. Historical analysis of Cuban commercial fishing effort and the effects of management interventions on important reef fishes from 1960–2005. *Fisheries Research* 99: 7–16.

Coelho FN. 2001. Distribuição, abundância e biologia reprodutiva da garoupinha, *Cephalopholis fulva* (Linnaeus, 1758) (Periciformes, Serranidae, Epinephelinae), na costa central do Brasil. PhD thesis, Universidade Federal Fluminense, Rio de Janeiro, Brazil.

Coello S, Grimm AS. 1993. The reproductive biology of *Mycteroperca olfax* (Jenyns) (Pisces Serranidae): protogyny and breeding season. *Revistas Cientificas Marina y Limnologia* 3: 115–128.

Coleman FC, Baker P, Koenig CC. 2004. A review of Gulf of Mexico marine protected areas: successes, failures, and lessons learned. *Fisheries* 29(2): 10–21.

Coleman FC, Koenig CC, Collins LA. 1996. Reproductive styles of shallow-water groupers (Pisces: Serranidae) in the eastern Gulf of Mexico and the consequences of fishing spawning aggregations. *Environmental Biology of Fishes* 47(2): 129–141.

Coleman F[C], Koenig C[C], Eklund A. 2002. Report to the Curtis and Edith Munson Foundation on the preliminary studies using pop-up archival satellite tags to track movements of adult goliath grouper *Epinephelus itajara* populations in the eastern Gulf of Mexico. 16 pp.

Coleman FC, Koenig CC, Huntsman GR, Musick JA, Eklund AM, McGovern JC, Chapman RW, Sedberry GR, Grimes CB. 2000. Long-lived reef fishes: the grouper-snapper complex. *Fisheries* 25(3): 14–20.

Coleman FC, Williams SL. 2002. Overexploiting marine ecosystem engineers: potential consequences for biodiversity. *Trends in Ecology and Evolution* 17: 40–44.

Colin PL. 1992. Reproduction of the Nassau grouper, *Epinephelus striatus* (Pisces: Serranidae) and its relationship to environmental conditions. *Environmental Biology of Fishes* 34: 357–377.

Colin PL. 1994. Preliminary investigations of reproductive activity of the jewfish, *Epinephelus itajara* (Pisces: Serranidae). *Proceedings of the Gulf and Caribbean Fisheries Institute* 43: 138–147.

Colin PL, Shapiro DY, Weiler D. 1987. Aspects of the reproduction of two groupers, *Epinephelus guttatus* and *E. striatus* in the West Indies. *Bulletin of Marine Science* 40(2): 220–230.

Collins MR, Waltz CW, Roumillat WA, Stubbs DL. 1987. Contribution to the life history and reproductive biology of gag, *Mycteroperca microlepis* (Serranidae), in the South Atlantic Bight. *Fishery Bulletin* 85(3): 648–653.

Connell SD. 1998. Patterns of pisciviory by resident predatory reef fish at One Tree Reef, Great Barrier Reef. *Marine and Freshwater Research* 49(1): 25–30.

Contreras M, Arreguin-Sánchez F, Sánchez JA, Moreno V, Cabrera MA. 1994. Mortality and population size of the red grouper (*Epinephelus morio*) fishery from the Campeche Bank, Mexico. *Proceedings of the Gulf and Caribbean Fisheries Institute* 43: 393–401.

Cook DC, Chan ALK, Fok KCC, Wilson KDP. 1997. A fisheries resource profile prior to artificial reef deployment in Hong Kong. PACON 97 Symposium on Resource Development – Environmental Issues and the Sustainable Development of Coastal Waters, August 6–8, 1997. The Chinese University of Hong Kong, Hong Kong, China.

Cook M, Fitzhugh GR, Franks JS. 2009. Validation of yellowedge grouper, *Epinephelus flavolimbatus*, age using nuclear bomb-produced radiocarbon. *Environmental Biology of Fishes* 86: 461–472.

Cornish AS, Peres MB. 2003. *Polyprion americanus* (Brazilian subpopulation). In: IUCN 2003, *IUCN Red List of Threatened Species 2003*. Accessed at www.iucnredlist.org.

Costa PAS, Braga AC, Frota LO. 2003. Reef fisheries in Porto Seguro, eastern Brazilian coast. *Fisheries Research* 60: 577–563.

Crabtree RE, Bullock LH. 1998. Age, growth, and reproduction of black grouper, *Mycteroperca bonaci*, in Florida waters. *Fisheries Bulletin* 96(4): 735–753.

Craig MT. 2007. Preliminary observations on the life history of the white spotted grouper, *Epinephelus ongus*, from Okinawa, Japan. *Ichthyological Research* 54: 81–84.

Craig MT, Bartsch P, Wirtz P, Heemstra PC. 2006a. Redescription and validation of *Alphestes afer* (Bloch 1793) as an amphi-Atlantic grouper species (Perciformes: Serranidae). *Cybium* 30(4): 327–331.

Craig MT, Graham RT, Torres RA, Hyde JR, Freitas MO, Ferreira BP, Hostim-Silva M, Gerhardinger LC, Bertoncini AA, Robertson DR. 2009. How many species of goliath grouper are there? Cryptic genetic divergence in a threatened marine fish and the resurrection of a geopolitical species. *Endangered Species Research* 7: 167–174.

Craig MT, Hastings PA. 2007. A molecular phylogeny of the groupers of the subfamily Epinephelinae (Serranidae) with a revised classification of the Epinephelini. *Ichthyological Research* 54(1): 1–17.

Craig MT, Hastings PA, Pondella DJ II. 2004. Speciation in the Central American Seaway: the importance of taxon sampling in the identification of trans-ismuthian germinate pairs. *Journal of Biogeography* 31: 1085–1091.

Craig MT, Hastings PA, Pondella DJ II, Robertson DR, Rosales-Casian A. 2006b. Phylogeography of the flag cabrilla *Epinephelus labriformis* (Serranidae): implications for the biogeography of the tropical Eastern Pacific and the early stages of speciation in a marine shore fish. *Journal of Biogeography* 33(6): 969–979.

Craig MT, Pondella DJ II, Hafner JC. 1999. Analysis of age and growth in two Eastern Pacific groupers (Serranidae: Epinephelinae). *Bulletin of Marine Science* 65(3): 807–814.

Craig MT, Pondella DJ II, Lea RN. 2006c. New records of the flag cabrilla, *Epinephelus labriformis* (Serranidae: Epinephelinae), from the Pacific coast of Baja California, Mexico, and San Diego, California, USA, with notes on the distribution of other groupers in California. *California Fish and Game* 92(2): 91–97.

Cribb TH, Bray RA, Wright T, Pichelin S. 2002. The trematodes of groupers (Serranidae: Epinephelinae): knowledge, nature and evolution. *Parasitology* 124(Suppl.): S23–S42.

Cunningham S, Bodiguel C. 2006. Subregional review: Southwest Indian Ocean. In: De Young C (ed.), *Review of the state of world marine capture fisheries management: Indian Ocean. FAO Fisheries Technical Paper* No. 488. Rome, Italy: Food and Agriculture Organization. pp 67–84.

Cury P, Roy C. 1988. Migration saisonnière du thiof (*Epinephelus aeneus*) au Sénégal: influence des upwellings sénégalais et mauritanien. *Oceanologica Acta* 11(1): 25–36.

Cury P, Worms J. 1982. Pêche, biologie et dynamique du thiof (*Epinephelus aeneus* E. Geoffroy Saint-Hilaire, 1817) sur les côtes sénégalaises. Documents Scientifiques, Centre de Recherches Oceanographiques de Dakar-Thiaroye, Senegal.

Dahlgren CP. 1998. Population dynamics of early juvenile Nassau grouper: an integrated modelling and field study. PhD thesis, North Carolina State University, USA.

Dahlgren CP, Eggleston DB. 2001. Spatio-temporal variability in abundance, size and microhabitat associations of early juvenile Nassau grouper *Epinephelus striatus* in an off-reef nursery system. *Marine Ecology Progress Series* 217: 145–156.

Dai Q, Zhang Q, Cai Y, Zhang J. 1988. Age and growth of the red grouper *Epinephelus akaara* from the coastal waters of Fujian. *Oceanologia et Limnologia Sinica* 19(3): 215–224.

Dalzell P, Adams PJH, Polunin NVC. 1996. Coastal fishes in the Pacific islands. *Oceanography and Marine Biology: an Annual Review* 34: 395–531.

Dalzell P, Preston G. 1992. *Deep reef slope fishery resources of the South Pacific: a summary and analysis of the dropline fishing survey data generated by the activities of the SPC fisheries programme between 1974 and 1988*. Noumea, New Caledonia: South Pacific Commission.

Danulat E, Edgar GJ (eds). 2002. Indicadores socio-economics de la pesca. In: *Pesca blance y pesquerias que duran todo el ano en Galapagos*. Santa Cruz Island, Galapagos, Ecuador: Fundación Charles Darwin/ Servicio Parque Nacional Galápagos. pp 441–449.

Darar A. 1994. An account of fisheries development in the Republic of Djibouti with notes on the growth and mortality of three species of groupers. *Naga, the WorldFish Center Quarterly Newsletter* 17(2): 30–32.

da Silva Monteiro VM. 1998. *Peixes de Cabo Verde*. Ministério do Mar, Gabinete do Secretário de Estado da Cultura. Lisbon, Portugal: M2-Artes Gráficas. 179 pp.

Davies CR. 2000. Inter-reef movement of the common coral trout, *Plectropomus leopardus*. Research Publication No. 61. Great Barrier Reef Marine Park Authority, Townsville, Australia. 50 pp.

Davies CR, Choat JH, Samoilys M, Mapstone BM, Benzie J, Russ GR. 1999. Stock structure and regional variation in population dynamics of the red throat emperor and other target species of the Queensland tropical reef line fishery. Progress report for FRDC Project 98-13, December 1999. Fisheries Research and Development Corporation, Deakin, Australian Capital Territory, Australia.

Davies CR, Williams AJ, Mapstone BD, Benzie J, van Herwerden L, Choat JH, Adams S, Murchie CD, Bean K, Carlos G, Tobin A, Ackerman J. 2006. *Stock structure and regional variation in population dynamics of the red throat emperor and other target species of the Queensland tropical reef line fishery*. Technical Report No. 61. Townsville, Australia: CRC Reef Research Centre. 119 pp.

Daw T. 2004. Reef fish spawning aggregations in Sabah, Eastern Malaysia. A report on stakeholder interviews conducted for the Society for the Conservation of Reef Fish Aggregations. In: *Western Pacific Survey Series (Volume 5)*. San Diego, USA: SCRFA. pp 1–63.

Debelius H. 1993. *Indian Ocean: tropical fish guide*. Neu Isenburg, Germany: Aquaprint Verlags GmBH. 321 pp.

De Bruin GHP, Russell BC, Bogusch A. 1994. *The marine fishery resources of Sri Lanka. FAO Species Identification Guide for Fishery Purposes*. Rome, Italy: Food and Agriculture Organization. 400 pp, 32 plates.

Delgado de Molina A, Santana JC. 1985. Estudio de los recursos demersales de la plataforma continental de la Republic de Gabon. *Informe Técnico del Instituto Español de Oceanografía* 31: 1–110.

Dengo A, David B. 1993. The linefishery in southern Mozambique: fish, fishers and fisheries. In: Beckley LE, van der Elst RP (eds), *Proceedings from the Second South African Marine Linefish Symposium, Durban, 23–24 October 1992*. Special Publication No. 2. Durban, South Africa: Oceanographic Research Institute. pp 104–106.

Department of Primary Industries. 2003. Fisheries Act 1994 Fisheries (Coral Reef Fin Fish) Management Plan 2003, Regulatory Impact Statement for SL 2003 No. 212. Brisbane, Australia: Queensland Department of Primary Industries.

Dettaï A, Lecointre G. 2004. In search of notothenioid (Teleostei) relatives. *Antarctic Science* 16: 71–85.

Dettaï A, Lecointre G. 2005. Further support for the clades obtained by multiple molecular phylogenies in the acanthomorph bush. *Comptes Rendus Biologies* 328: 674–689.

Devaraj M, Murty VS. No date. The fisheries for and the mariculture potential of groupers, red snappers, breams and other coral reef fishes in India. Accessed at http://www.library.enaca.org/Grouper/Research/Species/2000/0807.htm.

Diario Oficial de la Federación (Mexico). 2003. Tomo DXCIII. 13 de Febrero. pp 48–49.

Diatta Y, Bouaïn A, Clotilde-Ba F, Capapé C. 2003. Diet of four serranid species from the Senegalese coast (eastern tropical Atlantic). *Acta Adriatica* 44(2): 175–182.

Díaz-Uribe JG, Elorduy-Garay JF, González-Valdovinos MT. 2001. Age and growth of the leopard grouper, *Mycteroperca rosacea*, in the southern Gulf of California, México. *Pacific Science* 55: 171–182.

Diggles BK, Ernst I. 1997. Hooking mortality of two species of shallow-water reef fish caught by recreational angling methods. *Marine and Freshwater Research* 48: 479–483.

Direccion del Parque Nacional Galápagos. 2003. Pesquerias artesanales de Galápagos que duran todo el ano. Informe del tercer trimester de 2003. Ecuador: Direccion del Parque Nacional Galápagos.

Doak W. 1978. *Fishes of the New Zealand region*. Auckland, New Zealand: Hodder & Stoughton. 132 pp.

Doi T, Mendizábal D, Contreras M. 1981. Análisis preliminar de la población de mero, *Epinephelus morio*, en el Banco de Campeche. *Ciencia Pesquera* 1(1): 1–15.

Domeier ML, Colin PL. 1997. Tropical reef fish spawning aggregations: defined and reviewed. *Bulletin of Marine Science* 60: 698–726.

Domeier ML, Colin PL, Donaldson TJ, Heyman WD, Pet JS, Russell M, Sadovy Y, Samoilys MA, Smith A, Yeeting BM, Smith S. 2002. Transforming coral reef conservation:

reef fish spawning aggregations component. Spawning Aggregations Working Group Report. Hawaii, USA: The Nature Conservancy. 85 pp.

Donaldson TJ. 1989. Pair spawning of *Cephalopholis boenak* (Serranidae). *Japanese Journal of Ichthyology* 35(4): 497–500.

Donaldson TJ. 1995a. Courtship and spawning behavior of the pygmy grouper, *Cephalopholis spiloparaea* (Serranidae: Epinephelinae), with notes on *C. argus* and *C. urodeta*. *Environmental Biology of Fishes* 43: 363–370.

Donaldson TJ. 1995b. Partitioning behaviour and intra- and interspecific interactions: a comparison between male and female groupers, *Cephalopholis spiloparaea* (Pisces: Serranidae: Epinephelinae). *Marine Biology* 121(4): 581–584.

Donaldson TJ. 2002. Habitat association and depth distribution of two sympatric groupers of the genus *Cephalopholis* (Serranidae: Epinephelinae). *Ichthyological Research* 49: 191–193.

Done TJ, Williams DMcB, Speare PJ, Turak EI, Davidson J, Devantier LM, Newman SJ, Hutchings JB. 1994. Surveys of coral and fish communities at Scott Reefs and Rowley Shoals. Produced in accordance with the Scott Reef Research Services agreement between Woodside Offshore Petroleum Pty Ltd and the Australian Institute of Marine Science. 94/183. Australian Institute of Marine Science. 112 pp. Accessed at http://data.aims.gov.au/extpubs/do/viewPub.do?articleId=3154.

Donnelly R. 2001. Socio-economic environmental and the effect of the live reef food fish trade in Marovo Lagoon, Roviana Lagoon and Ontong Java, Solomon Islands. Discussion Paper No. 5. Report to the Australian Centre for International Agricultural Research (ACIAR), Canberra, Australia. 91 pp.

Donnelly R, Neville D, Mous PJ (eds). 2003. Report on a rapid ecological assessment of the Raja Ampat islands, Papua, Eastern Indonesia, held October 30–November 22, 2002. Sanur, Bali, Indonesia: The Nature Conservancy – Southeast Asia Center for Marine Protected Areas. 249 pp.

Durville P, Chabanet P, Quod JP. 2003. Visual census of the reef fishes in the natural reserve of the Glorieuses Islands (Western Indian Ocean). *Western Indian Ocean Journal of Marine Science* 2(2): 95–104.

Edgar GJ, Banks S, Farina JM, Calvopina M, Martínez C. 2004b. Regional biogeography of shallow reef fish and macro-invertebrate communities in the Galápagos archipelago. *Journal of Biogeography* 31: 1107–1124.

Edgar GJ, Bustamante RH, Farina J-M, Calvopina M, Martinez C, Toral-Granda MV. 2004a. Bias on evaluating the effects of marine protected areas: the importance of baseline data for the Galápagos Marine Reserve. *Environmental Conservation* 31(3): 212–218.

Eggleston DB. 1995. Recruitment in Nassau grouper *Epinephelus striatus*: post-settlement abundance, microhabitat features and ontogenetic habitat shifts. *Marine Ecology Progress Series* 124: 9–22.

Eklund AM, McClellan DB, Harper DE. 2000. Black grouper aggregation in relation to protected areas within the Florida Keys National Marine Sanctuary. *Bulletin of Marine Science* 66: 721–728.

Elmer M. 1998. The live reef fish debate: what are the facts? *Queensland Fisheries News* 2 June.

El-Sayed AFM. 1999. Reproductive biology of grouper fish *Epinephelus tauvina* (Family Serranidae) in the Arabian Gulf waters. *Indian Journal of Marine Sciences* 28: 89–91.

Epperly SP, Dodrill JW. 1995. Catch rates of snowy grouper, *Epinephelus niveatus*, on the deep reefs of Onslow Bay, southeastern USA. *Bulletin of Marine Science* 56(2): 450–461.

Erdman DS. 1976. Spawning patterns of fishes from the northeastern Caribbean. *Agriculture and Fisheries Contributions* VIII (2). San Juan, Puerto Rico: Departamento de Agricultura. 36 pp.

Erdman DS. 1985. *Common names of fishes in Puerto Rico. Informe Técnico* 3(2). Puerto Rico: CODREMAR. 44 pp.

Erisman BE. 2008. Reproductive biology and evolution of epinephelid and serranid fishes (Perciformes, Epinephelidae, Serranidae). PhD dissertation, University of California, San Diego, USA.

Erisman BE, Buckhorn ML, Hastings PA. 2007. Spawning patterns in the leopard grouper, *Mycteroperca rosacea*, in comparison with other aggregating groupers. *Marine Biology* 151(5): 1849–1861.

Erisman BE, Craig MT, Hastings PH. 2010. Reproductive biology of the Panama graysby *Cephalopholis panamensis* (Teleostei: Epinephelidae). *Journal of Fish Biology* 76: 1312–1328.

Erisman BE, Konotchick TH, Blum S. 2009. Observations of spawning in the leather bass, *Dermatolepis dermatolepis* (Teleostei: Epinephelidae), at Cocos Island, Costa Rica. *Environmental Biology of Fishes* 85: 15–20.

Erisman BE, Rosales-Casián JA, Hastings PH. 2008. Evidence of gonochorism in a grouper, *Mycteroperca rosacea*, from the Gulf of California, Mexico. *Environmental Biology of Fishes* 82: 23–33.

Errity CM. 2003. A description of the Northern Territories finfish trawl fishery. Fishery Report. Northern Territory Department of Business, Industry and Resource Development, Darwin, Australia.

Eschmeyer WN. 1990. *Catalog of the genera of recent fishes*. San Francisco, USA: California Academy of Sciences. 697 pp.

Eschmeyer WN, Herald ES, Hammann H. 1983. *A field*

guide to Pacific coast fishes of North America. Boston, USA: Houghton Mifflin.

Espino-Barr E, Ruiz-Luna A, Garcia-Boa A. 2004. Changes in tropical fish assemblages associated with small-scale fisheries: a case study in the Pacific off central Mexico. *Reviews in Fish Biology and Fisheries* 12(4): 393–401.

Everson AR. 1992. Sexual maturity and seasonal spawning of hapu'upu'u, *Epinephelus quernus*, in Hawaii. Southwest Fisheries Science Center Administrative Report H-92-13, National Marine Fisheries Service, NOAA Honolulu, Hawaii, USA.

Ezzat AA, Wadie WF, Mikail MY, Hashem MT. 1981. Age and growth of *Epinephelus aeneus* in the Egyptian Mediterranean waters. *Bulletin of the Institute of Oceanography and Fisheries* 7(3): 395–406.

Fallatah AA. 2005. Trends in catches of *Plectropomus pessuliferus* (Fowler) and *Plectropomus areolatus* (Ruppel) in the Red Sea coast of Saudi Arabia from 1995 to 2002. *Egyptian Journal of Aquatic Research* 31(Special Issue): 256–285.

FAO. 1998. Fishery country profile: Republic of Korea. FAO report FID/CP/ROK. Rome, Italy: Food and Agriculture Organization.

FAO. 2002a. *FAO yearbook. Fishery statistics: catches and landings*. Vol. 94/1. Rome, Italy: Food and Agriculture Organization.

FAO. 2002b. *The state of world fisheries and aquaculture*. Rome, Italy: Food and Agriculture Organization. Accessed at http://library.enaca.org/NACA-Publications/World_Fisheries_Aquaculture_2002.pdf.

FAO. 2003. FAO nominal catches of *Epinephelus guaza* (=*marginatus*). Vol. 96/1. Rome, Italy: Food and Agriculture Organization.

Feitoza BM, Rocha LA, Luiz-Júnior O, Floeter SR, Gasparini JL. 2003. Reef fishes of St Paul's Rocks: new records and notes on biology and zoogeography. *Aqua, Journal of Ichthyology and Aquatic Biology* 7(2): 61–82.

Fennessy ST. 1998. Biology and management of some sex-changing rockcods (Serranidae) from southern Africa. In: *African Fishes and Fisheries Diversity and Utilisation, Grahamstown, South Africa, 1998*. Book of Abstracts compiled by the JLB Institute of Ichthyology, for the International Conference for the Paradi Association and the Fisheries Society of Africa. p 112.

Fennessy ST. 2000. Comparative life histories and stock assessments of rock cods (Family Serranidae) from the east coast of South Africa. PhD thesis, University of Natal, Durban, South Africa. 165 pp.

Fennessy ST. 2006. Reproductive biology and growth of the yellowbelly rockcod *Epinephelus marginatus* (Serranidae) from South-East Africa. *African Journal of Marine Science* 28(1): 1–11.

Fennessy ST, Sadovy Y. 2002. Reproductive biology of a diandric protogynous hermaphrodite, the serranid *Epinephelus andersoni*. *Marine and Freshwater Research* 53(2): 147–158.

Ferreira BP. 1995. Reproduction of the common coral trout *Plectropomus leopardus* from the central and northern Great Barrier Reef. *Bulletin of Marine Science* 56(2): 653–669.

Ferreira BP, Cava FC, Ferraz AN. 1998. Relações morfométricas em peixes recifais na Zona Econômica Exclusiva Brasileira, região nordeste. *Boletim Tecnico e Cientifico do CEPENE* 6(1): 71–76.

Ferreira BP, Maida M. 1995. Projeto meros: apresentação e resultados preliminares. *Boletim Tecnico e Cientifico do CEPENE-Ibama* 3(1): 201–210.

Ferreira BP, Maida M. 2006. *Monitoramento dos recifes de coral do Brasil. Serie Biodiversidade* 18. Brasília, Brazil: Ministério do Meio Ambiente.

Ferreira BP, Maida M, Castro CB, Pires DO, D'Amico T, Prates APL, Marx D. 2004. The status of coral reefs in Brazil. In: McManus JW (ed.), *Proceedings of the 10th International Coral Reef Symposium,* Okinawa, Japan, 28 June–2 July, 2004. pp 1011–1016.

Ferreira BP, Russ GR. 1992. Age, growth and mortality of the inshore coral trout *Plectropomus macualtus* (Pisces: Serranidae) from the central Great Barrier Reef, Australia. *Australian Journal of Marine and Freshwater Research* 43: 1301–1312.

Ferreira BP, Russ GR. 1994. Age validation and estimation of growth rate of the coral trout, *Plectropomus leopardus*, (Lacepede 1802) from Lizard Island, northern Great Barrier Reef. *Fishery Bulletin* 92: 46–57.

Ferreira CEL, Floeter SR, Gasparini J, Ferreira BP, Joyeux JC. 2004. Trophic structure patterns of Brazilian reef fishes: a latitudinal comparison. *Journal of Biogeography* 31: 1093–1106.

Ferreira CEL, Goncalves JEA. 1999. The unique Abrolhos reef formation (Brazil): need for specific management strategies. *Coral Reefs* 18: 352.

Figuerola-Fernández M, Matos-Caraballo D, Torres W. 1997. Maturation and reproductive seasonality of four reef fish species in Puerto Rico. *Proceedings of the Gulf and Caribbean Fisheries Institute* 50: 938–968.

Fine JC. 1990. Groupers in love: spawning aggregations of Nassau grouper in Honduras. *The Explorers Journal* Fall 1990: 131–134.

Fine JC. 1992. Greedy for groupers. *Wildlife Conservation* May/June 1992: 1–5.

Fischer W, Sousa I, Silva C, de Freitas A, Poutiers JM, Schneider W, Borges TC, Feral JP, Massinga A. 1990. *Fichas FAO de identificaçao de espécies para actividades de pesca. Guia de campo das espécies comerciais marinhas e de águas salobras de Moçambique*. Rome, Italy: Food and Agriculture Organization. 424 pp.

Fisheries Scientific Committee. No date. Recommendation *Epinephelus daemelii* – black cod. Ref. No. FR4 File No. FSC 99/12. Accessed at http://www.dpi.nsw.gov.au/__data/assets/pdf_file/0004/208246/FR4-black-cod.pdf.

Fitch JE. 1949. Cabrilla and grouper. In: Staff of California Bureau of Marine Fishes (eds), *The commercial fish catch of California for the year 1947 with an historical review 1916–1947. California Division of Fish and Game Fish Bulletin* No. 74: 80–82.

Fitzhugh GR, Lombardi-Carlson LA, Evou NM. 2003. Age structure of gag (*Mycteroperca microlepis*) in the eastern Gulf of Mexico by year, fishing mode, and region. *Proceedings of the Gulf and Caribbean Fisheries Institute* 54: 538–549.

Fitzpatrick BM, Penrose HM (eds). 2002. *A preliminary marine ecological survey of Bateman Bay, Ningaloo Reef*. Perth, Australia: Oceanwise Environmental Scientists. 58 pp.

Flewwelling P, Hosch G. 2006. Country Review: Malaysia. In: De Young C (ed.), *Review of the state of world marine capture fisheries management: Indian Ocean. FAO Fisheries Technical Paper* No. 488. Rome, Italy: Food and Agriculture Organization. 458 pp.

Floeter SR, Halpern BS, Ferreira CEL. 2006. Effects of fishing and protection on Brazilian reef fishes. *Biological Conservation* 128: 391–402.

Fourmanoir P. 1971. Description de quatre poissons trouves pour le premiere fois dans les Tuamoto et en Nouvelle-Caledonie. *Cahiers du Pacifique* 15: 127–135.

Francesconi JJ, Schwartz FJ. 2000. Jewfish, *Epinephelus itajara*, from North Carolina, with range correction and body comparisons. *Journal of the Elisha Mitchell Scientific Society* 116(2): 167–170.

Francini-Filho RB, Moura RI, Sazima I. 2000. Cleaning by the wrasse *Thalassoma noronhanum*, with two records of predation by its grouper client *Cephalopholis fulva*. *Journal of Fish Biology* 56(4): 802–809.

Francis MP. 1993. Checklist of the coastal fishes of Lord Howe, Norfolk, and Kermadec Islands, southwest Pacific Ocean. *Pacific Science* 47(2): 136–170.

Frédou T, Ferreira BP, Letourneur Y. 2006. A univariate and multivariate study of reef fisheries off northeastern Brazil. *ICES Journal of Marine Science* 63(5): 883–896.

Freegard A (ed.). 1983. *The Times atlas of the oceans*. London: Times Books.

Freitas JC. 1982. Predacao de *Stenorhynchus seticornis* (Crustacea, Majidae) por *Diplasterias brandtii* (Asteroidea) a *Epinephelus morio* (Teleostei). *Ciencia e Cultura* 34(8): 1056–1058.

Freitas JEP, Araújo ME, Solé-Cava AM. 2003. Estruturação genética das populações de duas espécies de peixes recifais do Atol das Rocas e da costa do Ceará. *Tropical Oceanography, Recife* 31(2): 193–201.

Frias-Torres S. 2006. Habitat use of juvenile goliath grouper *Epinephelus itajara* in the Florida Keys, USA. *Endangered Species Research* 2: 1–6.

Frisch A, van Herwerden L. 2006. Field and experimental studies of hybridization between coral trouts, *Plectropomus leopardus* and *Plectropomus maculatus* (Serranidae), on the Great Barrier Reef, Australia. *Journal of Fish Biology* 68(4): 1013–1025.

Froese R. 2004. Keep it simple: three indicators to deal with overfishing. *Fish and Fisheries* 5: 86–91.

Frota LO, Costa PAS, Braga AC. 2004. Length-weight relationships of marine fishes from the central Brazilian coast. *Naga, the WorldFish Center Quarterly Newsletter* 27(1): 20–26.

Fry GC, Brewer DT, Venables WN. 2006. Vulnerability of demersal fishes to commercial fishing: evidence from a study around a tropical seamount in Papua New Guinea. *Fisheries Research* 81: 126–141.

Fundación AZTI. 2003. *Estudio de prospectiva para la cadena productiva pesquera, en la región de la costa del pacífico, en América del Sur*. Document prepared for the United Nations Industrial Development Organization, Vienna.

Furnestin J, Dardignac J, Maurin C, Vincent A, Coupé R, Boutiere H. 1958. Données nouvelles sur les poissons de Maroc atlantique. *Revue des Travaux de l'Institut des Pêches Maritimes Paris* 22(4): 379–493.

Galal N, Ormond RFG, Hassan O. 2002. Effect of a network of no-take reserves in increasing catch per unit effort and stocks of exploited reef fish at Nabq, South Sinai, Egypt. *Marine and Freshwater Research* 53: 199–205.

Galzin R, Lecchini D, Williams JT, Planes S, Menou JL. 2006. Diversite de l'ichtyofaune corallienne a Rapa (Polynesie Fancaise). *Cybium* 3(3): 221–234.

García-Cagide A, Claro R, Koshelev BV. 1994. Reproducción. In: Claro R (ed.), *Ecología de los peces marinos de Cuba*. Instituto de Oceanología Academia de Ciencias de Cuba and Centro de Investigaciones de Quintana Roo (CIQRO), México. pp 187–262.

García-Cagide A, García T. 1996. Reproducción de *Mycteroperca bonaci* y *Mycteroperca venenosa* (Pisces: Serranidae) en la plataforma cubana. *Revista de Biologia Tropical* 44(2): 771–780.

García Charton JA, Williams ID, Pérez Ruzafa A, Milazzo M, Chemello R, Marcos C, Kitsos M-S. 2000. Evaluating the ecological effects of Mediterranean marine protected areas: habitat, scale and the natural variability of ecosystems. *Environmental Conservation* 27: 159–178.

García-Moliner Basora G. 1986. Aspects of the social spacing, reproduction and sex reversal in the red hind *Epinephelus guttatus* (Linnaeus). MSc abstract. Department of Marine Sciences, University of Puerto Rico, Mayagüez, Puerto Rico.

García-Moliner G, Sadovy Y. 2007. The case for regional management of the Nassau grouper, *Epinephelus striatus*. *Proceedings of the Gulf and Caribbean Fisheries Institute* 60: 596–602.

Gardner L, Anderson T, Place AR, Dixon B, Elizur A. 2005. Sex change strategy and the aromatase genes. *The Journal of Steroid Biochemistry and Molecular Biology* 94(5): 395–404.

Gascuel D, Ménar F. 1997. Assessment of a multispecies fishery in Senegal, using production models and diversity indices. *Aquatic Living Resources* 10: 281–288.

Gasparini JL, Floeter SR. 2001. The shore fishes of Trindade Island, western South Atlantic. *Journal of Natural History* 35: 1639–1656.

GEP (Grupo de Estudos Pesqueiros)/UNIVALI (Universidade do Vale do Itajaí). 2006. *Estatística pesqueira industrial de Santa Catarina: 2002–2006*. Itajaí, Brazil.

Gerber LR, Heppell SS, Ballantyne F, Sala E. 2005. The role of dispersal and demography in determining the efficacy of marine reserves. *Canadian Journal of Fisheries and Aquatic Sciences* 62(4): 863–871.

Gerhardinger LC, Marenzi RC, Bertoncini AA, Medeiros RP, Hostim-Silva M. 2006. Local ecological knowledge on the goliath grouper *Epinephelus itajara* (Teleostei: Serranidae) in southern Brazil. *Neotropical Ichthyology* 4(4): 441–450.

Ghorab HM, Bayoumi AR, Bebars MI, Hassan AA. 1986a. Age and growth of the grouper *Epinephelus chlorostigma* (Serranidae) from the Red Sea. *Bulletin of the Institute of Oceanography and Fisheries* (Egypt) 12: 1–12.

Ghorab HM, Bayoumi AR, Bebars MI, Hassan AA. 1986b. The reproductive biology of the grouper *Epinephelus chlorostigma* (Pisces, Serranidae) from the Red Sea. *Bulletin of the Institute of Oceanography and Fisheries* (Egypt) 12: 13–33.

Gibran FZ. 2004. Dying or illness feigning: an unreported feeding tactic of the comb grouper *Mycteroperca acutirostris* (Serranidae) from the Southwest Atlantic. *Copeia* 2004(2): 403–405.

Gibson J. 2007. Lessons learned from Belize Nassau grouper management and possible ways forwards. *Proceedings of the Gulf and Caribbean Fisheries Institute* 60: 603–604.

Gilmore RG, Bullock LH, Berry FH. 1978. Hypothermal mortality in marine fishes of south-central Florida, January 1977. *Northeast Gulf Science* 2(2): 77–97.

Gilmore RG, Jones RS. 1992. Color variation and associated behavior in the epinepheline groupers, *Mycteroperca microlepis* (Goode and Bean) and *M. phenax* (Jordan and Swain). *Bulletin of Marine Science* 51(1): 83–103.

Gladstone, W. 2002. Fisheries of the Farasan Islands (Red Sea). *Naga, the WorldFish Center Quarterly Newsletter* 25(3&4): 30–34.

Glamuzina B, Skaramuca B. 1999. A review: Adriatic groupers – status of natural populations and breeding perspectives. *Acta Adriatica* 40(Suppl.): 79–90.

Glamuzina B, Tutman P, Geffen AJ, Kožul V, Skaramuca B. 2000. First record of white grouper, *Epinephelus aeneus* (Serranidae) in the south eastern Adriatic. *Cybium* 24(3): 306–308.

Gloerfelt-Tarp T, Kailola PJ. 1984. *Trawled fishes of southern Indonesia and northwestern Australia*. Canberra, Australia: Australian Development Assistance Bureau.

GMFMC (Gulf of Mexico Fisheries Management Council). 2001. *Report of the Reef Fish Stock Assessment Panel (final draft – November 26, 2001)*. Miami, Florida, USA: GMFMC.

GMFMC (Gulf of Mexico Fisheries Management Council). 2003. *Commercial fishing regulations for Gulf of Mexico federal waters*. Tampa, Florida, USA: GMFMC.

Gobert B. 1990. Production relative des pêcheries côtières en Martinique. *Aquatic Living Resources* 3(3): 181–191.

Gobert B, Berthou P, Lopez E, Lespagnol P, Turcios MDO, Macabiau C, Portillo P. 2005. Early stages of snapper-grouper exploitation in the Caribbean (Bay Islands, Honduras). *Fisheries Research* 73(1–2): 159–169.

Gomez AE, Zapata LA. 1999. An overview on the fishery and biology of the grouper and of the seabass (Pisces: Serranidae) in the Colombian Pacific. *Boletin Cientifico* 6: 227–246.

Gomon MF, Glover CJM, Kuiter RH. 1994. *The fishes of Australia's south coast*. Adelaide, Australia: State Print.

Gonzalez LW, Celaya J. 1986. Diagnostico socio-economico de la pesqueria de media altura pargo-mero del estado Nueva Esparta. Investigaciones Cientificas, Universidad Oriente Venezuela, Centro de Investigaciones Cientificas, Venezuela.

Gopinadha Pillai CS, Gopakumar G, Mohan M. 1992. Ichthyofauna of the intertidal reef flats of Minicoy Atoll, Lakshadweep: an analysis of its structure, relative abundance and food. *Journal of the Marine Biological Association of India* 34(1&2): 74–83.

Gorshkova GV, Protas Y, Ben-Atia S, Gorshkov S. 2002. Cytogenetic examination of early embryonic development in the white grouper *Epinephelus aeneus* (Pisces, Serranidae). *Journal of Applied Ichthyology* 18: 29–34.

Gosline WA. 1966. The limits of the fish family Serranidae, with notes on other lower percoids. *Proceedings of the California Academy of Sciences* 33: 91–111.

Grana Rafucci F. 1993. *Catalogo de nomenclatura de los peces de Puerto Rico y las Islas Virgenes* (2da edicion). San Juan, Puerto Rico: Departamento de Recursos Naturales. 201 pp.

Grandcourt EM. 2002. Demographic characteristics of a

selection of exploited reef fish from the Seychelles: preliminary study. *Australian Journal of Marine and Freshwater Research* 53: 123–130.

Grandcourt EM. 2005. Demographic characteristics of selected epinepheline groupers (Family: Serranidae; Subfamily: Epinephelinae) from Aldabra Atoll, Seychelles. *Atoll Research Bulletin* 539: 201–216.

Grandcourt EM, Al Abdessalaam TZ, Francis F, Al Shamsi AT, Hartmann SA. 2009. Reproductive biology and implications for management of the orange-spotted grouper *Epinephelus coioides* in the southern Arabian Gulf. *Journal of Fish Biology* 74: 820–841.

Grant E. 1975. *Guide to fishes*. Brisbane, Australia: the Coodinator-General's Department.

Green A, Lokani P, Atu W, Ramohia P, Thomas P, Almany J (eds). 2006. *Solomon Islands marine assessment: technical report of surveys conducted May 13 to June 17, 2004*. TNC Pacific Island Countries Report No. 1/06. Brisbane, Australia: The Nature Conservancy.

Griffiths SP. 2003. Spatial and temporal dynamics of temperate Australian rockpool ichthyofaunas. *Marine and Freshwater Research* 54: 163–176.

Guime MP. 2003. La Reserva Marina de Galápagos. Accessed at http://www.mapa.es/rmarinas/jornada_rrmm/galapagos.pdf.

Gutherz EJ. 1982. Reef fish assessment, snapper/grouper stocks in the western north Atlantic south of Cape Hatteras, NC. In: Huntsman GR, Nicholson WR, Fox WW Jr (eds), Proceedings of a workshop held October 7–10, 1980 at St Thomas, Virgin Islands of the United States. NOAA Technical Memorandum NMFS-SEFC-80. pp 124–141.

GWSG (Groupers & Wrasses Specialist Group). 2007. Workshop for global Red List Assessment of groupers. Final Report, University of Hong Kong, Hong Kong, China. 24 pp. Accessed at http://cmsdata.iucn.org/downloads/final_report_for_gwsg_workshop_feb_2007.pdf.

Haimovici M, Ávila-da-Silva AO, Rossi-Wongtschowski (eds). 2004. *Prospecção pesqueira de espécies demersais com espinhel-de-fundo na zona econômica exclusiva da Região Sudeste-Sul do Brasil*. São Paulo, Brazil: Instituto Oceanográfico da USP (Série Documentos Revizee: Score Sul).

Halley RB, Jaap WC, Mead G, Earle S, Hine AC, Jarret B, Locker SD, Naar DF, Donahue B, Twichell DC. 2003. Pulley Ridge. The US's deepest hermatypic coral reef? In: Best GR (ed.), *Joint Conference on the Science and Restoration of the Greater Everglades and Florida Bay Ecosystem*, April 13–18, 2003, Palm Harbor, Florida, USA. Gainsville, Florida, USA: University of Florida. pp 13–18.

Hamilton F. 1822. *An account of the fishes found in the River Ganges and its branches*. Edinburgh & London.

Hamilton R. 2003. A report on the status of exploited reef fish aggregations in the Solomon Islands and Papua New Guinea – Choiseul, Ysabel, Bougainville and Manus provinces. Prepared for the Society for the Conservation of Reef Fish Aggregations. 52 pp. Accessed at http://www.scrfa.org/images/stories/pdf/scrfa/hamilton_final_report.pdf.

Hamilton RJ, Giningele M, Aswani S, Ecochard J-L. In press. Fishing in the dark – local knowledge, night spearfishing and spawning aggregations in the Western Solomon Islands. *Biological Conservation*.

Hamilton RJ, Kama W. 2004. Spawning aggregations of coral reef fish in Roviana Lagoon, Western Province, Solomon Islands: local knowledge field survey report. TNC Pacific Island Countries Report No. 5/04. Report prepared for the Pacific Island Countries Coastal Marine Program, The Nature Conservancy. 50 pp.

Hamilton RJ, Matawai M. 2006. Live reef food fish trade causes rapid declines in abundance of squaretail coral grouper (*Plectropomus areolatus*) at a spawning aggregation site in Manus, Papua New Guinea. *SPC Live Reef Fish Information Bulletin* No. 16: 13–18.

Hamilton RJ, Matawai M, Potuku T, Kama W, Lahui P, Warku J, Smith AJ. 2005. Applying local knowledge and science to the management of grouper aggregation sites in Melanesia. *SPC Live Reef Fish Information Bulletin* No. 14: 7–19.

Hamilton RJ, Potuku T. 2007. Local knowledge of white-dotted grouper (*Epinephelus polystigma*) aggregations in Melanesia. *SPC Traditional Marine Resource Management and Knowledge Information Bulletin* No. 22: 3–14.

Hamilton RJ, Smith AJ. 2006. Supporting community-based inshore fisheries management in Melanesia to achieve conservation goals. In: Abstracts of the 7th Indo-Pacific Fish Conference, May 16–20, 2005, Howard International House, Taipei, Taiwan. p 53.

Hardy JD Jr (ed.). 1978. *Development of fishes of the mid-Atlantic bight: an atlas of eggs, larval and juvenile stages*. Three volumes. Washington, DC, USA: US Fish and Wildlife Service, Office of Biological Services.

Hariri KI, Nichols P, Krupp F, Mishrigi S, Barrania A, Ali AF, Kedidi SM. 2000. Status of the living marine resources in the Red Sea and Gulf of Aden region and their management. Regional Organisation for the Conservation of the Environment of the Red Sea and Gulf of Aden (PERSGA), Jeddah, Saudi Arabia.

Harmelin JG, Robert P. 2001. Evolution récente de la population du mérou brun (*Epinephelus marginatus*) dans le Parc national de Port-Cros (France, Méditerranée). *Travaux Scientifiques du Parc National de Port-Cros Hyeres*, No. 18: 149–161.

Harmelin-Vivien ML, Bouchon C. 1976. Feeding behaviour of some carnivorous fishes (Serranidae and Scorpaenidae) from Tulear (Madagascar). *Marine Biology* 37: 329–340.

Harper DE, Bohnsack JA, Lockwood BR. 2000. Recreational fisheries in Biscayne National Park, Florida, 1976–1991. *Marine Fisheries Review* 62: 8–26.

Harris PJ, Wyanski DM, White DB, Moore JL. 2002. Age, growth, and reproduction of scamp, *Mycteroperca phenax*, in the southwestern North Atlantic, 1979–1997. *Bulletin of Marine Science* 70(1): 113–132.

Hassin S, De Monbrison D, Hanin Y, Elizur A, Zohar Y, Popper DM. 1997. Domestication of the white grouper, *Epinephelus aeneus*: 1. Growth and reproduction. *Aquaculture* 156: 305–316.

Hawkins JP, Roberts CM, Gell FR, Dytham C. 2007. Effects of trap fishing on reef fish communities. *Aquatic Conservation: Marine and Freshwater Ecosystems* 17: 111–132.

Hearn A. 2008. The rocky path to sustainable fisheries and conservation in the Galápagos Marine Reserve. *Ocean & Coastal Management* 51: 567–574.

Heemstra PC. 1991. A taxonomic revision of the eastern Atlantic groupers (Pisces: Serranidae). *Boletim do Museu Municipal do Funchal* (Funchal, Portugal) 43(226): 5–71.

Heemstra PC, Heemstra E. 2004. *Coastal fishes of southern Africa*. Grahamstown, South Africa: South African Institute for Aquatic Biodiversity (SAIAB) and National Inquiry Services Centre (NISC). 512 pp.

Heemstra PC, Jerson WD, Lobel PS. 2002. Serranidae. In: Carpenter KE (ed.), *The living marine resources of the western central Atlantic. Volume 2. Bony fishes part 1 (Ascipenseridae to Grammatidae). FAO species identification guide for fishery purposes. American Society of Ichthyologists and Herpetologists Special Publication* No. 5. Rome, Italy: Food and Agriculture Organization. pp 1308–1369.

Heemstra PC, Randall JE. 1984. Serranidae. In: Fischer W, Bianchi G (eds), *FAO species identification sheets for fishery purposes: Western Indian Ocean (Fishing Area 51)*. Volume 4. Rome, Italy: Food and Agriculture Organization. 141 unnumbered pages.

Heemstra PC, Randall JE. 1986. Serranidae. In: Smith MM, Heemstra PC (eds), *Smiths' sea fishes*. Berlin, Germany: Springer-Verlag. pp 509–536.

Heemstra PC, Randall JE. 1993. *FAO species catalogue. Vol. 16. Groupers of the world (Family Serranidae, Subfamily Epinephelinae). An annotated and illustrated catalogue of the grouper, rockcod, hind, coral grouper, and lyretail species known to date*. Rome, Italy: Food and Agriculture Organization of the United Nations. 400 pp + colour plates.

Heyman WD, Wade B. 2007. Status of reef fish spawning aggregations in Belize. *Proceedings of the Gulf and Caribbean Fisheries Institute* 58: 301–306.

Hodgson G, Liebeler J. 2002. The global coral reef crisis: trends and solutions. Los Angeles, California, USA: Reef Check Foundation. 77 pp.

Hoese DF, Bray DJ, Paxton JR, Allen GR. 2006. Fishes. In: Beesley PL, Wells A (eds), *Zoological catalogue of Australia*. Vol. 35. Australia: Australian Biological Resources Study and Australia's Commonwealth Scientific and Industrial Research Organisation (CSIRO). 2178 pp.

Hokama Y, Shirai JL, Islam MA. 1994. Evaluation of bioassay for toxicity of ciguateric fish and associated toxins. *Journal of Clinical Laboratory Analysis* 8: 63–69.

Hombron JB, Jacquinot H. 1853. *Reptiles et poissons*. In: Dumont d'Urville J (ed.), Voyage au Pole Sud et dans l'Oceanie sur les corvettes "L'Astrolabe" et "La Zelee"...pendant 1837–40. Paris, France. pp 31–56.

Hood PB, Schlieder RA. 1992. Age, growth, and reproduction of gag, *Mycteroperca microlepis* (Pisces: Serranidae), in the eastern Gulf of Mexico. *Bulletin of Marine Science* 51(3): 337–352.

Huang G. 1999. Shi yong shi ban yu yang zhi/Application of grouper mariculture. Taipei, Taiwan: Shui chan chu ban she. pp 8–9. [In Chinese]

Huang ZG. 1994. *Marine species and their distributions in China's seas*. Beijing, PR China: China Ocean Press.

Huntsman GR, Potts JC, Mays RW, Vaughan DS. 1996. The status of the population of gag, *Mycteroperca microlepis*, in the western Atlantic Ocean from Cape Hatteras, NC to the Dry Tortugas, FL. Report to the South Atlantic Fishery Management Council. Beaufort, North Carolina, USA: National Ocean Service Laboratory.

Huntsman GR, Potts J, Mays RW, Vaughan D. 1999. Groupers (Serranidae, Ephinephelinae): endangered apex predators of reef communities. In: Musick JA (ed.), *Life in the slow lane: ecology and conservation of long-lived marine animals*. Bethesda, Maryland, USA: American Fisheries Society. pp 231–271.

Huntsman GR, Schaaf WE. 1994. Simulation of the impact of fishing on reproduction of a protogynous grouper, the graysby. *North American Journal of Fisheries Management* 14: 41–52.

Hureau JC. 1991. La base de données GICIM: gestion informatisée des collections ichthyologiques du muséum. In: Allardi J, Keith P (eds), *Atlas preliminaire des poissons d'eau douce de France*. Paris, France: Museum D'Histoire Naturelle. pp 225–227.

Hussin MA, Ali A, Nik-Daud NS, Nik-Razali NL. 1996. Natural spawning and larval rearing of tiger grouper, *Epinephelus fuscoguttatus* (Forsskål). A preliminary result. Paper presented during the 5th Fisheries Research Conference at the Fisheries Research Institute, Penang, Malaysia, 8–10 July 1997. 7 pp.

Hutchinson N, Rhodes KL. 2010. Home range estimates for squaretail coralgrouper, *Plectropomus areolatus* (Rüppell 1830). *Coral Reefs* 29(2): 511–519.

IBAMA (Instituto Brasileiro do Meio Ambiente e dos Recursos Naturais Renováveis)–CEPERG (Centro de Pesquisa e Gestão dos Recursos Pesqueiros Lagunares e Estuarinos). 2006. *Estatística de desembarque de pescado no Rio Grande do Sul: 1991–2005*. Rio Grande, Brazil: IBAMA/CEPERG. 45 pp.

Imamura H, Yabe M. 2002. Demise of the Scopaeniformes (Actinopterygii:Percomorpha): an alternative phylogenetic hypothesis. *Bulletin of Fisheries Science Hokkaido University* 53: 107–128.

Institute of Oceanology, Academia Sinica. 1991. *Atlas of the sea fishes of China in live color*. Vols. 1–2. Shanghai, China: Shanghai Scientific & Technical Publishers. [In Chinese]

Instituto de Pesca. 2006. Estatística pesqueira do estado de São Paulo: 1998–2005. Accessed at http://www.pesca.sp.gov.br/estatistica.php.

IUCN. 2003. *2003 IUCN Red List of Threatened Species*. Accessed at http://www.iucnredlist.org.

IUCN. 2004. *2004 IUCN Red List of Threatened Species*. Accessed at http://www.iucnredlist.org.

IUCN. 2006. *2006 IUCN Red List of Threatened Species*. Accessed at http://www.iucnredlist.org.

IUCN. 2008. *2008 IUCN Red List of Threatened Species*. Accessed at http://www.iucnredlist.org.

IUCN. 2009. *2009 IUCN Red List of Threatened Species*. Accessed at http://www.iucnredlist.org.

IUCN. 2011. *2011 IUCN Red List of Threatened Species*. Accessed at http://www.iucnredlist.org.

James CM, Al-Thobaiti S, Rasem BM. 1998. Development of a grouper hybrid for aquaculture. *INFOFISH International* 5: 23–29.

James PSBR, Sriramachandra Murty V, Nammalwar P. 1996. Groupers and snappers of India: biology and exploitation. In: Arreguin-Sanchez F, Munro JL, Balgos MC, Pauly D (eds), *Biology, fisheries and culture of tropical groupers and snappers. ICLARM Conference Proceedings* 48. Manila, Philippines: International Center for Living Aquatic Resources Management. pp 106–136.

Jennings S, Polunin NVC. 1996. Effects of fishing effort and catch rate upon the structure and biomass of Fijian reef fish communities. *Journal of Applied Ecology* 33(2): 400–412.

Jennings S, Polunin NVC. 1997. Impacts of predator depletion by fishing on the biomass and diversity of non-target reef fish communities. *Coral Reefs* 16: 71–82.

Jennings S, Reynolds JD, Polunin NVC. 1999. Predicting the vulnerability of tropical reef fishes to exploitation with phylogenies and life histories. *Conservation Biology* 13(6): 1466–1475.

Johannes RE. 1988. Spawning aggregations of the grouper *Plectropomus areolatus* (Rüppell) in the Solomon Islands. In: Choat JH, Barnes D, Borowitzka MA et al. (eds), *Proceedings of the 6th International Coral Reef Symposium*, Vol. 2, Contributed papers. Townsville, Australia. pp 751–755.

Johannes RE, Kile N. 2001. Protecting grouper spawning aggregations, a potential target of the live reef food fish trade in Ysabel and Wagina, Solomon Islands. *SPC Live Reef Fish Information Bulletin* No. 8: 5–9.

Johannes RE, Lam M. 1999. The live reef food fish trade in the Solomon Islands. *SPC Live Reef Fish Information Bulletin* No. 5: 8–15.

Johannes RE, Riepen M. 1995. Environmental, economic and social implications of the live reef fish trade in Asia and the western Pacific. Report to The Nature Conservancy and the South Pacific Forum Fisheries Agency, Honolulu, Hawaii. 83 pp.

Johannes RE, Squire L, Graham T. 1994. Developing a protocol for monitoring spawning aggregations of Palauan serranids to facilitate the formulation and evaluation of strategies for their management. FFA Report 94/28. South Pacific Forum Fisheries Agency. 25 pp.

Johannes RE, Squire L, Graham T, Sadovy Y, Renguul H. 1999. Spawning aggregations of groupers (Serranidae) in Palau. The Nature Conservancy Marine Conservation Research Series Publication No. 1. The Nature Conservancy and the Forum Fisheries Agency. 144 pp.

Johnson AG, Collins LA. 1994. Age-size structure of Red grouper (*Epinephelus morio*) from the eastern Gulf of Mexico. *Northeast Gulf Science* 13(2): 101–106.

Johnson GD. 1983. *Niphon spinosus*: a primitive epinepheline serranid, with comments on the monophyly and intrarelationships of the Serranidae. *Copeia* 1983: 777–787.

Johnson GD. 1988. *Niphon spinosus*, a primitive epinepheline serranid: corroborative evidence from the larvae. *Japanese Journal of Ichthyology* 35: 7–18.

Jones RS, Gutherz EJ, Nelson WR, Matlock GC. 1989. Burrow utilization by yellowedge grouper, *Epinephelus flavolimbatus*, in the northwestern Gulf of Mexico. *Environmental Biology of Fishes* 26: 277–284.

Jordan DS, Eigenmann CH. 1890. A review of the genera and species of Serranidae found in the waters of America and Europe. *Bulletin of the United States Fish Commission* VIII: 329–441.

Jordan T. 1998. NMFS – a partner for endangered species. *Endangered Species Bulletin* [US Department of the Interior Fish and Wildlife Service] 23(1): 4–9.

Jory DE, Iversen ES. 1989. Species profiles: life histories and environmental requirements of coastal fishes and invertebrates (south Florida). Black, red, and Nassau groupers. Biological Report of the US Fish and Wildlife Service 82(11.110). US Army Corps of Engineers, TR EL-82-4. 30 pp.

Kailola PJ. 1987. *The fishes of Papua New Guinea. A revised and annotated checklist. Volume II: Scorpaenidae to Callionymidae*. Research Bulletin No. 41. Port Moresby, Papua New Guinea: Research Section, Department of Fisheries and Marine Resources.

Kailola PJ, Williams MJ, Stewart PC, Reichelt RE, McNee A, Grieve C. 1993. *Australian fisheries resources*. Canberra, Australia: Bureau of Resource Sciences.

Kalish JM, Newman SJ, Johnston J. 2002. Chapter 18: Use of bomb radiocarbon to validate the age estimation method for *Epinephelus octofasciatus*, *Etelis carbunculus* and *Lethrinus nebulosus* from Western Australia. In: Kalish JM (ed.), *Use of the bomb radiocarbon chronometer to validate fish age*. Final report to the Fisheries Research and Development Corporation (FRDC) on Project No. 93/109. Canberra, Australia: Australian National University. pp 281–298.

Kara MH, Derbal F. 1995. Morphométrie, croissance et mortalité du mérou *Epinephelus marginatus* (Serranidae) des côtes de l'est algérien. *Cahiers de Biologie Marine* 36: 229–237.

Kasteleijin H. 1987. Marine biological research in the Galápagos: past, present, and future. *Oceanus* 30(2): 33–41.

Katayama M. 1959. Studies on the serranid fishes of Japan. *Bulletin of the Faculty of Education Yamaguchi University* 8: 103–180.

Katayama M. 1960. *Fauna Japonica. Serranidae (Pisces)*. Tokyo, Japan: Japanese News Service.

Kato K, Ishimaru K, Sawada Y, Mutsuro J, Miyashita S, Murata O, Kumai H. 2004. Ontogeny of digestive and immune system organs of larval and juvenile kelp grouper *Epinephelus bruneus* reared in the laboratory. *Fisheries Science* 70(6): 1061–1069.

Keener P, Johnson GD, Stender BW, Brothers EB, Beatty HR. 1988. Ingress of postlarval gag, *Myteroperca microlepis* (Pisces: Serranidae) through a South Carolina barrier island inlet. *Bulletin of Marine Science* 42(3): 367–396.

Keener-Chavis P. 1984. Age, growth, and reproductive biology of the yellowedge grouper, *Epinephelus flavolimbatus,* off the coast of South Carolina. MSc thesis, College of Charleston, USA.

Kendall AW Jr. 1976. Predorsal and associated bones in serranid and grammistid fishes. *Bulletin of Marine Science* 26: 585–592.

Kendall AW Jr. 1979. Morphological comparisons of North American seabass larvae (Pisces: Serranidae). *NOAA Technical Report Circular* 428: 1–50.

Kingsley MCS (ed.). 2004. The goliath grouper in southern Florida: assessment review and advisory report. Report prepared for the South Atlantic Fishery Management Council, the Gulf of Mexico Fishery Management Council, and the National Marine Fisheries Service. Data and Assessment Review. vii + 17 pp.

Kirkwood JM, Hooper JNA. 2004. Burnett Mary regional assessment: coastal & marine biodiversity. Burnett Mary Regional Group for Natural Resource Management, Bundaberg, Australia.

Kitajima C, Takaya M, Tsukashima Y, Arakawa T. 1991. Development of eggs, larvae and juveniles of the grouper, *Epinephelus septemfasciatus*, reared in the laboratory. *Japanese Journal of Ichthyology* 38(1): 47–55.

Kitalong A, Dalzell P. 1994. *A preliminary assessment of the status of inshore coral reef fish stocks in Palau*. Inshore Fisheries Research Project (South Pacific Commission), Technical Document No. 6. Noumea, New Caledonia: South Pacific Commission. 37 pp.

Klippel S, Olavo G, Costa PAS, Martins AS, Peres MB. 2005. Avaliação dos estoques de lutjanídeos da costa central do Brasil: análise de coortes e modelo preditivo de Thompson e Bell para comprimentos. In: *Pesca e potenciais de exploração de recursos vivos na região central da Zona Econômica Exclusiva Brasileira*, série livros REVIZEE. Rio de Janeiro, Brazil: Museu Nacional. pp 83–98.

Koenig CC, Coleman FC, Collins LA, Sadovy Y, Colin PL. 1996. Reproduction in gag, *Mycteroperca microlepis* (Pisces: Serranidae) in the eastern Gulf of Mexico and the consequences of fishing spawning aggregations. In: Arreguin-Sanchez F, Munro JL, Balgos MC, Pauly D (eds), *Biology, fisheries and culture of tropical groupers and snappers. ICLARM Conference Proceedings* 48. Manila, Philippines: International Center for Living Aquatic Resources Management. pp 307–323.

Koenig C[C], Coleman F[C], Eklund A. 1997. Report to the Curtis and Edith Munson Foundation on studies of the jewfish (*Epinephelus itajara*) in the eastern Gulf of Mexico. 28 pp.

Koenig CC, Coleman FC, Eklund A-M, Schull J, Ueland J. 2007. Mangroves as essential nursery habitat for goliath grouper (*Epinephelus itajara*). *Bulletin of Marine Science* 80(3): 567–585.

Koenig CC, Coleman FC, Grimes CB, Fitzhugh GR, Scanlon KM, Gledhill CT, Grace M. 2000. Protection of fish spawning habitat for the conservation of warm-temperate reef-fish fisheries of shelf-edge reefs of Florida. *Bulletin of Marine Science* 66(3): 593–616.

Kospartov M, Berger M, Ceccarelli D, Richards Z. 2006. An assessment of the distribution and abundance of seacucumbers, *Trochus*, giant clams, fish and invasive marine species at Ashmore Reef National Nature Reserve and Cartier Island Marine Reserve. Canberra, Australia: Department of Environment and Heritage.

Kramer SH, Kramer SH, Newman SJ. 1994. New deep-water fish records form the Great Barrier Reef, Australia. *Pacific Science* 48(1): 70–79.

Kramer S[H], Kramer S, Newman S. 1993. Potential deep slope fishery resource off the Great Barrier Reef. *Australian Fisheries* 52: 18–21.

Kuiter RH. 1993. *Coastal fishes of south-eastern Australia*. Bathurst, New South Wales, Australia: Crawford House Press.

Kuiter RH, Tonozuka T. 2001. *Pictorial guide to Indonesian reef fishes. Part 1. Eels–snappers, Muraenidae–Lutjanidae*. Australia: Zoonetics. 302 pp.

Kulbicki M. 2006. Ecologie des poissons lagonaires de Nouvelle Calédonie. PhD thesis, Ecole Pratique des Hautes Etudes, University of Perpignan, France. 194 pp.

Kulbicki M, Bozec Y-M, Labrosse P, Letourneur Y, Mou-Tham G, Wantiez L. 2005b. Diet composition of carnivorous fishes from coral reef lagoons of New Caledonia. *Aquatic Living Resources* 18: 231–250.

Kulbicki M, Guillemot N, Amand M. 2005a A general approach to length-weight relationships for New Caledonian lagoon fishes. *Cybium* 29(3): 235–252.

Kulbicki M, Labrosse P, Letourneur Y. 2000. Fish stock assessment of the northern New Caledonian lagoons: 2 – Stocks of lagoon bottom and reef-associated fishes. *Aquatic Living Resources* 13(2): 77–90.

Kunzmann A. 2004. Corals, fishermen and tourists. *Naga, the WorldFish Center Quarterly Newsletter* 27(1/2): 15–19.

Kuo CM, Ting YY, Yeh SL. 1988. Induced sex reversal and spawning of blue-spotted grouper *Epinephelus fario*. *Aquaculture* 74(1/2): 113–126.

Kuo SR, Shao KT. 1999. Species composition of fish in the coastal zones of the Tsengwen Estuary, with descriptions of five new records from Taiwan. *Zoological Studies* 38(4): 391–404.

Kyushin K, Amaoka K, Nakaya K, Ida H. 1977. *Fishes of Indian Ocean*. Tokyo, Japan: Japan Marine Fishery Resource Research Center.

La Mesa G, Vacchi M. 1999. An analysis of the coastal fish assemblage of the Ustica Island Marine Reserve (Mediterranean Sea). *Marine Ecology* 20(2): 147–165.

Lau PPF, Li LWH. 2000. *Identification guide to fishes in the live seafood trade of the Asia-Pacific Region*. Hong Kong: WWF Hong Kong and Agriculture, Fisheries and Conservation Department. 137 pp.

Lau PPF, Parry-Jones R. 1999. *The Hong Kong trade in live reef fish for food*. Hong Kong: TRAFFIC East Asia and World Wide Fund for Nature Hong Kong.

Laurans M, Barry M, Gascuel D. 2003. Revue des connaissances sur la biologie du thiof (Epinephelus aeneus) et diagnostic de l'état du stock au Sénégal. In: Gascuel D, Barry MD, Laurans M, Sidibe A (eds), *Evaluation des stocks demersaux en Afrique du nord-ouest. Travaux du Groupe "Analyses monospecifiques" du projet SIAP. COPACE/PACE Séries* 03/65: 5–70.

Laurans M, Sidibe A, Chassot E, Gascuel D. 2002. Ecosystem effects of a quickly developed fishery: trends in biomass of demersal resources of Senegal and Guinea. Census of marine life: turning concept into reality. In: *ICES CM 2002*. Copenhagen, Denmark: International Council for the Exploration of the Sea. L:21.

Leadbitter D. 1992. Special fauna (B) black cod survey. In: Hutchings P (ed.), *Reef biology: a survey of Elizabeth and Middleton reefs, South Pacific. Kowari* 3. Sydney: Australian Museum, and Canberra: Australian National Parks and Wildlife Service. pp 103–106.

Lee C, Sadovy Y. 1998. A taste for live fish: Hong Kong's live reef fish market. *Naga, the WorldFish Center Quarterly Newsletter* 21(2): 38–42.

Lee YD, Park SH, Takemura A, Takano K. 2002. Histological observations of seasonal reproductive and lunar-related spawning cycles in the female honeycomb grouper *Epinephelus merra* in Okinawan waters. *Fisheries Science* 68: 872–877.

Leis JM (ed.). 1987. Review of the early life history of tropical groupers (Serranidae) and snappers (Lutjanidae). In: Polovina JJ, Ralston S (eds), *Tropical snappers and groupers: biology and fisheries management*. Boulder, USA/London, UK: Westview Press. pp 189–237.

Lessa R, Nóbrega MF de. 2000. Guia de identificação de peixes marinhos da regiao nordeste. Programa REVIZEE/SCORE-NE. Recife, Brazil: UFRPE-DIMAR.

Letourneur Y. 1996. Dynamics of fish communities in Réunion fringing reefs. Indian Ocean 1: patterns of spatial distribution. *Journal of Experimental Marine Biology and Ecology* 195: 1–30.

Letourneur Y, Kulbicki M, Labrosse P. 1998. Length-weight relationships of fish from coral reefs and lagoons of New Caledonia, southwestern Pacific Ocean: an update. *Naga, the WorldFish Center Quarterly Newsletter* 21(4): 39–46.

Leung KMY, Chu JCW, Wu RSS. 1999. Effects of body weight, water temperature and ration size on ammonia excretion by the areolated grouper (*Epinephelus areolatus*) and mangrove snapper (*Lutjanus argentimaculatus*). *Aquaculture* 170(3): 215–227.

Lewis ND. 1986. Epidemiology and impact of ciguatera in the Pacific: a review. *Marine Fisheries Review* 48(4): 6–13.

Lichucha IDLT, Govender A, van der Elst RP, Abdula RG. 2000. *The linefishery of Mozambique: annual report for the years 1998–1999*. Maputo, Mozambique: Instituto de Investigacao Pesquiera.

Lichucha IDLT, van der Elst RP, Govender A, Abdula RG, Cuco CA. 1999. The linefishery of Mozambique: report to the Instituto de Investigaçao Pesqueira for the year 1997. Unpublished Report No. 164, Oceanographic Research Institute, Durban, South Africa.

Licuanan WY, Gomez ED. 2000. *Philippine coral reefs, reef fishes and associated fisheries: status and recommendations to improve their management*. Global Coral Reef Monitoring Network report. Cape Ferguson, Queensland, Australia: Australian Institute of Marine Science.

Lieske E, Myers R. 1994. *Collins pocket guide. Coral reef fishes. Indo-Pacific and Caribbean including the Red Sea*. HarperCollins.

Lieske E, Myers R. 2002. *Collins pocket guide to coral reef fishes*. Princeton, USA: Princeton University Press.

Lieske E, Myers R. 2004. *Coral reef guide: Red Sea*. London, UK: HarperCollins.

Lindberg GU, Orlans D (eds). 1967. *Fishes of the Sea of Japan and the adjacent areas of the Sea of Okhotsk and the Yellow Sea*. Part 3. Washington, DC, USA: Smithsonian Institution.

Lindholm J, Kaufman L, Miller S, Wagschal A, Newville M. 2005. Movement of yellowtail snapper (*Ocyurus chrysurus* Block 1790) and black grouper (*Mycteroperca bonaci* Poey 1860) in the northern Florida Keys National Marine Sanctuary as determined by acoustic telemetry. Marine Sanctuaries Conservation Series MSD-05-4. Department of Commerce, National Oceanic and Atmospheric Administration, Marine Sanctuaries Division, Silver Spring, Maryland, USA. 17 pp.

Liu FH. 1971. Studies on the biology of *Epinephelus awoara* (Temminck and Schlegel) with discovery of the metapterygoid for age-determination. *New Asia College Academic Annual* 13: 49–65.

Liu M, Sadovy Y. 2001. Embryos and early larvae of *Cephalopholis boenak* (Bloch, 1790) (Serranidae). In: Hendry CI, Van Stappen G, Willie M, Sorgeloos P (eds), *Larvi'01 – Fish & Shellfish Larviculture Symposium*. Special Publication No. 30. Oostende, Belgium: European Aquaculture Society. pp 322–325.

Liu M, Sadovy Y. 2004a. Early gonadal development and primary males in the protogynous epinepheline, *Cephalopholis boenak*. *Journal of Fish Biology* 65: 987–1002.

Liu M, Sadovy Y. 2004b. The influence of social factors on adult sex change and juvenile sexual differentiation in a diandric, protogynous epinepheline, *Cephalopholis boenak*. *Journal of Zoology* (London) 264: 239–248.

Liu M, Sadovy Y. 2005. Habitat association and social structure of the chocolate hind, *Cephalopholis boenak* (Pisces: Serranidae: Epinephelinae), at Ping Chau Island, northeastern Hong Kong waters. *Environmental Biology of Fishes* 74: 9–18.

Liu M, Sadovy de Mitcheson Y. 2008. Chapter 7: Grouper aquaculture in mainland China and Hong Kong. In: Liao IC, Leaño EM (eds), *The aquaculture of groupers*. Manila, Philippines: Asian Fisheries Society; Louisiana, USA: World Aquaculture Society; Keelung, Taiwan: The Fisheries Society of Taiwan; Keelung, Taiwan: National Taiwan Ocean University. pp 111–142.

Liu M, Sadovy de Mitcheson Y. 2009a. Gonad development during sexual differentiation in hatchery-produced orange-spotted grouper (*Epinephelus coioides*) and humpback grouper (*Cromileptes altivelis*) (Pisces: Serranidae, Epinephelinae). *Aquaculture* 287: 191–202.

Liu M, Sadovy de Mitcheson Y. 2009b. Exploitation history, mariculture and trade status of the threatened Hong Kong grouper (*Epinephelus akaara*) throughout its geographic range. Final report to Ocean Park Conservation Foundation of Hong Kong. 181 pp.

Loebman D, Vieira JP. 2005. Distribuição espacial das assembléias de peixes na Lagoa do Peixe, RS, Brasil. *Revista Brasileira de Zoologia, Curitiba* 22(3): 667–675.

Loftus WF, Kushlan JA. 1987. Freshwater fishes of southern Florida. *Bulletin of the Florida State Museum, Biological Sciences* 31(4): 147–344.

Lombardi-Carlson LA, Fitzhugh GR, Mikulas JJ. 2002. Red grouper (*Epinephelus morio*) age-length structure and description of growth from the Gulf of Mexico: 1992–2001. Panama City Laboratory Contribution Series 2002-06. National Marine Fisheries Service, Panama City, Florida, USA.

Longhurst AR. 1960. A summary survey of the food of West African demersal fish. *Bulletin de l'Institut Fondamental de d'Afrique Noire (Serie A, Sciences naturelles)* 22(1): 276–282.

Loubens G. 1980a. Biologie de quelques espèces de poissons du lagon Néo-Calédonien. II. Sexualité et reproduction. *Cahiers de l'Indo-Pacifique* 2(1): 41–72.

Loubens G. 1980b. Biologie de quelques espèces de poissons du lagon Néo-Calédonien. III. Croissance. *Cahiers de l'Indo-Pacifique* 2(2): 101–153.

Luckhurst BE. 1996. Trends in commercial fishery landings of groupers and snappers in Bermuda from 1975 to 1992 and associated fishery management issues. [Tendencias en los desembarques de la pesqueria commercial de meros y pargos en Bermuda, de 1975 a 1992 y aspectos asociados al manejo de la pesqueria]. In: Arreguin-Sánchex F, Munro JL, Balgos MC, Pauly D (eds), *Biology, fisheries and culture of tropical groupers and snappers. ICLARM Conference Proceedings* 48. Manila, Philippines: International Center for Living Aquatic Resources Management. pp 277–288.

Luckhurst BE. 2003. Development of a Caribbean regional conservation strategy for reef fish spawning aggregations. *Proceedings of the Gulf and Caribbean Fisheries Institute* 54: 668–679.

Luckhurst BE. 2009. West Africa. *Society for the Conservation of Reef Fish Aggregations Newsletter* No. 12: 7.

Luckhurst BE, Barnes JA, Sadovy Y. 1992. Record of an

unusually large red hind, *Epinephelus guttatus* (Pisces: Serranidae) from Bermuda, with comments on its age. *Bulletin of Marine Science* 51(2): 267–270.

Mace GM, Collar NJ, Gaston KJ, Hilton-Taylor C, Akçakaya HR, Leader-Williams N, Milner-Gulland EJ, Stuart SN. 2008. Quantification of extinction risk: IUCN's system for classifying threatened species. *Conservation Biology* 22(6): 1424–1442.

Machida Y, Sato T, Dasilao JC. 1997. A small collection of groupers from Agdao Fish Market in Davao City, Mindanao, the Philippines (Serranidae, Epinephelinae). *Memoirs of the Faculty of Science Kochi University (Series D Biology)* 18: 47–55.

Mackie MC. 1998. Biology and ecology of the Chinaman cod, *Epinephelus rivulatus*, at Ningaloo Reef, Western Australia. PhD thesis, University of Western Australia, Australia.

Mackie MC. 2000. Reproductive biology of the halfmoon grouper, *Epinephelus rivulatus*, at Ningaloo Reef, Western Australia. *Environmental Biology of Fishes* 57(4): 363–376.

Mackie MC. 2003. Socially controlled sex-change in the half-moon grouper, *Epinephelus rivulatus*, at Ningaloo Reef, Western Australia. *Coral Reefs* 22: 133–142.

Mackie MC. 2006. Anatomical changes to the gonad during protogynous sex-change in the half-moon grouper, *Epinephelus rivulatus* (Valenciennes). *Journal of Fish Biology* 69: 176–186.

Maigret J, Ly B. 1986. *Les poissons de mer de Mauritanie*. Compiègne, France: Centre national de recherches océanographiques et des pêches (Mauritania) and Science Naturales.

Mangi SC, Roberts CM. 2006. Quantifying environmental impacts of artisanal fishing gear on Kenya's coral reef ecosystems. *Marine Pollution Bulletin* 52: 1646–1660.

Manickchand-Heileman SC, Phillip DAT. 2000. Age and growth of the yellowedge grouper, *Epinephelus flavolimbatus*, and the yellowmouth grouper, *Mycteroperca interstitialis*, off Trinidad and Tobago. *Fisheries Bulletin* 98: 290–298.

Mann BQ, Celliers L, Fennessy ST, Bailey S, Wood AD. 2006. Towards the declaration of a large marine protected area: a subtidal ichthyofaunal survey of the Pondoland coast in the Eastern Cape, South Africa. *African Journal of Marine Science* 28(3&4): 535–551.

Mann BQ, Fielding PJ. 2000. Shallow water fishes of the east (Indian Ocean) and north east (Gulf of Aden) coast of Somalia. *East African Natural History Series Bulletin* 30(1): 1–12.

Manooch CS III, Haimovici M. 1978. Age and growth of gag *Mycteroperca microlepis* and size age composition of the recreational catch of the southeastern United States. *Transactions of the American Fisheries Society* 107: 234–240.

Manooch CS III, Mason DL. 1987. Age and growth of the Warsaw grouper and black grouper from the southeast region of the United States. *Northeast Gulf Science* 9(2): 65–75.

Mapstone BD, Choat JH, Davies CR. 1996. A description of the commercial reef line fishery logbook data held by the Queensland Fisheries Management Authority. Report to the QFMA from the CRC Reef Research Centre and the Department of Tropical Environment Studies and Geography, James Cook University, Townsville, Australia. 480 pp.

Marconi MR, Beck S, Majluf P. 2001. *Conservación de ecosistemas transfronterizos y espécies amenazadas*. La Paz, Bolivia: Comunidad Andina, Estrategia Regional de Biodiversidad para los Países del Trópico Andino, GTZ/FUNDECO/IE.

Marino G, Azzuro E, Massari A, Finoia MG, Mandich A. 2001. Reproduction in the dusky grouper from the southern Mediterranean. *Journal of Fish Biology* 58: 909–927.

Martial L, Mariama B, Didier GC. 2002. Diagnostics de cinq stocks senegalais par l'approche globale (*Galeoides decadactylus, Pagellus bellottii, Pseudupeneus prayensis, Sparus caeruleostictus, Epinephelus aeneus*). In: Gascuel D, Barry MD, Laurans M, Sidibe A (eds), *Evaluations des stocks démersaux en Afrique du nord-ouest. Travaux du Groupe "Analyses monospécifiques" du projet SIAP. COPACE/PACE Séries* 03/65: 23–32.

Martins AS, Olavo G, Costa PAS. 2005. Recursos demersais capturados com espinhel de fundo no talude superior da região entre Salvador (BA) e o Cabo de São Tomé (RJ). In: Costa PAS, Martins AS, Olavo G (eds), *Pesca e potenciais de exploração de recursos vivos na região central da Zona Econômica Exclusiva Brasileira*. Rio de Janeiro, Brazil: Museu Nacional. pp 109–128.

Martosubroto P. 2005. Regional review: eastern Indian Ocean. In: *Review of the state of world marine fishery resources. FAO Fisheries Technical Paper* No. 457. Rome, Italy: Food and Agriculture Organization. pp 94–98.

Masuda H, Allen GR. 1993. *Meerfische der welt – groß-Indopazifische*. Herrenteich, Melle, Germany: Tetra Verlag.

Masuda H, Amaoka K, Araga C, Uyeno T, Yoshino T. 1984. *The fishes of the Japanese Archipelago*. Tokyo, Japan: Tokai University Press.

Masuda H, Araga C, Yoshino T (eds). 1980. *Coastal fishes of southern Japan* (2nd edition). Tokyo, Japan: Tokai University Press.

Mateo I, Tobias WJ. 2001. Distribution of shallow water coral reef fishes on the northeast coast of St Croix, USVI. *Caribbean Journal of Science* 37(3–4): 210–226.

Matheson RH III. 1982. Age, growth, and mortality of two groupers, *Epinephelus drummondhayi* Goode and

Bean and *Epinephelus niveatus* (Valenciennes), from North Carolina and South Carolina. MSc dissertation, North Carolina State University, Raleigh, USA. 37 pp.

Matheson RH III, Huntsman GR. 1984. Growth, mortality, and yield-per-recruit models for speckled hind and snowy grouper from the United States South Atlantic Bight. *Transactions of the American Fisheries Society* 113: 607–616.

Matheson RH, Huntsman GR, Manooch CS. 1986. Age, growth, mortality, food and reproduction of the scamp, *Mycteroperca phenax*, collected of North Carolina and South Carolina. *Bulletin of Marine Science* 38(2): 300–312.

Mathews CP, Samuel M. 1991. Growth, mortality and length-weight parameters for some Kuwaiti fish and shrimp. *Fishbyte* 9(2): 30–33.

Mathews CP, Samuel M, Baddar MK. 1986. Sexual maturation, length and age in some species of Kuwait fish related to their suitability for aquaculture. *Kuwait Bulletin of Marine Science* 8: 243–256.

Matos D. 1997. Status of the groupers in Puerto Rico, 1970–1995. *Proceedings of the Gulf and Caribbean Fisheries Institute* 49: 340–353.

Matos D, Padilla W. 1995. Description of the tiger grouper (*Mycteroperca tigris*) fishery of Vieques Puerto Rico, during 1995. *Proceedings of the Gulf and Caribbean Fisheries Institute* 48: 143–153.

Matos D, Posada J. 1998. Current status of the tiger grouper (*Mycteroperca tigris*) fishery at Vieques Island, Puerto Rico. *Proceedings of the Gulf and Caribbean Fisheries Institute* 51: 182–194.

Matos-Caraballo D. 2002. Portrait of the commercial fishery of the red hind *Epinephelus guttatus* in Puerto Rico during 1992–1999. *Proceedings of the Gulf and Caribbean Fisheries Institute* 53: 446–459.

Matthieu LF, Allison EH, Tinch R. 2002. Seeking patterns of population variability from fish catch and stock biomass time series. University of East Anglia, Overseas Development Group, Norwich, UK.

Mazurek R. 2004. Commercially important groupers of the Gulf of Mexico & south Atlantic regions. Seafood Watch, Seafood Report, Monterey Bay Aquarium, California, USA.

McClenachan L. 2009. Documenting loss of large trophy fish from the Florida Keys with historical photographs. *Conservation Biology* 23(3): 636–643.

McCosker JE, Rosenblatt RH. 1975. Fishes collected at Malpelo Islands. In: Graham JB (ed.), *The biological investigations of Malpelo Island, Colombia. Smithsonian Contributions to Zoology* 176(25): 91–93.

McCulloch AR. 1922. *Checklist of the fishes and fish-like animals of New South Wales*. Sydney, Australia: Royal Zoological Society of New South Wales.

McErlean AJ, Smith CL. 1964. The age of sexual succession in the protogynous hermaphrodite, *Mycteroperca microlepis. Transactions of the American Fisheries Society* 93: 301–302.

McGovern JC, Wyanski DM, Pashuk O, Manooch CS II, Sedberry GR. 1998. Changes in the sex ratio and size at maturity of gag, *Mycteroperca microlepis*, from the Atlantic coast of the southeastern United States during 1976–1995. *Fishery Bulletin* 96(4): 797–807.

McIlwain J, Hermosa GV, Claereboudt M, Al-Oufi HS, Al-Awi M. 2003. Spawning and reproductive patterns of six exploited finfish species from the Arabian Sea, Sultanate of Oman. *Journal of Applied Ichthyology* 22: 167–176.

McKenna SA, Allen GR, Suryadi S (eds). 2002. *A marine rapid assessment of the Raja Ampat Islands, Papua Province, Indonesia.* RAP Bulletin of Biological Assessment 22. Washington, DC, USA: Conservation International. 191 pp.

Mees CC. 1992. *Seychelles demersal fishery: an analysis of data relating to four key demersal species.* SFA Technical Report. Victoria, Seychelles: Seychelles Fishing Authority. 143 pp.

Mees CC. 1996. Demersal fish stock assessment in Seychelles: an analysis of a mothership/catcher boat fishery. In: Arreguin-Sanchez F, Munro JL, Balgos MC, Pauly D (eds), *Biology, fisheries and culture of tropical groupers and snappers. ICLARM Conference Proceedings* 48. Manila, Philippines: International Center for Living Aquatic Resources Management. pp 254–265.

Mellin C, Ferraris J, Galzin R, Kulbicki M, Ponton D. 2006. Diversity of reef fish assemblages: modelling of the species richness spectra from multi-scale environmental variables in the Tuamotou Archipelago (French Polynesia). *Ecological Modelling* 198(3–4): 409–425.

Mendoza JJ, Larez A. 1996. Abundance and distribution of snappers and groupers targeted by the artisanal, medium-range fishery off northeastern Venezuela. In: Arreguin-Sanchez F, Munro JL, Malgos MC, Pauly D (eds), *Biology, fisheries and culture of tropical snappers and groupers. ICLARM Conference Proceedings* 48. Manila, Philippines: International Center for Living Aquatic Resources Management. pp 266–275.

Mendoza JJ, Larez A. 2004. A biomass dynamics assessment of the southeastern Caribbean snapper-grouper fishery. *Fisheries Research* 66: 129–144.

Menzel DW. 1959. Utilization of food by a Bermuda reef fish, *Epinephelus guttatus. Journal du Conseil International pour l'Exploration de la Mer* 24(2): 308–313.

Moe MA. 1969. Biology of the red grouper, *Epinephelus morio* (Valenciennes 1828), from the eastern Gulf of Mexico. Florida Department of Natural Resources Marine Research Laboratory, Professional Papers Series No. 10. 95 pp.

Mohsin AKM, Ambak MA. 1996. *Marine fishes and fisheries of Malaysia and neighbouring countries*. Serdang, Malaysia: University of Pertanian Malaysia Press.

Mohsin AKM, Ambak MA, Salam MNA. 1993. Malay, English and scientific names of the fishes of Malaysia. Occasional Publication No. 11. Faculty of Fisheries and Marine Science, University of Pertanian Malaysia, Selangor Darul Ehsan, Malaysia.

Monroy C. 1998. Análisis bioeconómico de la pesquería de mero (*Epinephelus morio*) en el Banco de Campeche bajo condiciones de riesgo e incertidumbre. Tesis de Maestría en Ciencias, Cinvestav-IPN, Mexico. 118 pp.

Montiero-Neto C, Cunha FEdeA, Nottingham MC, Araújo ME, Rosa IL, Barros GML. 2003. Analysis of the marine ornamental fish trade at Ceara State, northern Brazil. *Biodiversity and Conservation* 12: 1287–1295.

Moon TS, Park SR, Choi HS. 1999. Study on the development of culture technology for Serranidae. (2) Rearing density of cultured sevenband grouper, *Epinephelus septemfasciatus* in marine floating net cage. *Bulletin of the National Fisheries Research and Development Institute* 56: 33–141.

Moore CM, Labisky RF. 1984. Population parameters of a relatively unexploited stock of snowy grouper in the lower Florida Keys. *Transactions of the American Fisheries Society* 113(3): 322–329.

Morales-Nin B, Bauza D, Grau AM. 2005. Otolith characteristics and age estimation of an exceptionally old dogtooth grouper (*Epinephelus caninus*) captured off Majorca Island (western Mediterranean). *Cybium* 29(1): 100–102.

Moran MJ, Jenke J, Burton C, Clarke D. 1988. The Western Australian trap and line fishery on the north-west shelf. Final Report. FIRTA Project 86/28. Western Australian Marine Research Laboratories. 79 pp.

Morandini AC, Martorelli SR, Marques AC, Silveira FL. 2005. Digenean metacercaria (Trematoda, Digenea, Lepocreadiidae) parasitizing "coelenterates" (Cnidaria, Scyphozoa and Ctenophora) from southeastern Brazil. *Brazilian Journal of Oceanography* 53(1–2): 39–45.

Moreno V, Burgos R, Contreras M, Mena J. 1996. Composición por espécies de la captura comercial de mero en Yucatán durante 1995. Resúmenes del II Simposio en Biología y Oceanografía Pesquera en México. La Paz, BCS, México: CICIMAR-IPN. 8 pp.

Morgan GR. 1983. A preliminary multi-species assessment of Kuwait's gargoor (fish trap) fishery. *Annual Research Report, Kuwait Institute for Scientific Research* 8: 65–66.

Morgan G[R]. 2006. Subregional review: Northwest Indian Ocean. In: De Young C (ed.), *Review of the state of world marine capture fisheries management: Indian Ocean. FAO Fisheries Technical Paper* No. 488. Rome, Italy: Food and Agriculture Organization. pp 51–66.

Morgans JFC. 1964. *A preliminary survey of bottom fishing on the North Kenya Banks*. London, UK: HM Stationery Office.

Morgans JFC. 1982. Serranid fishes of Kenya and Tanzania. *Ichthyological Bulletin of the JLB Smith Institute of Ichthyology* 46(1): 1–44.

Morris AV, Roberts CM, Hawkins JP. 2000. The threatened status of groupers (Epinephelinae). *Biodiversity and Conservation* 9(7): 919–942.

Mosse JW, Adams S, Welch DJ. 2002. Bommie cod, (*Cephalopholis cyanostigma*): a big surprise from a little fish. In: Williams AJ, Welch DJ, Muldoon G. Marriott, R, Kritzer JP, Adams SA (eds), *Bridging the gap: a workshop linking student research with fisheries stakeholders*. Technical Report No 48. Townsville, Australia: CRC Reef Research Centre. pp 94–104.

Moura RL, Menezes NA. 2003. Serranidae. In: Buckup PA, Menezes NA (eds), *Catálogo dos peixes marinhos e de água doce do Brasil* (2nd edition). São Paulo, Brazil: Museu de Zoologia.

Moussac G de. 1986. Evidence of protogynous hermaphroditism of *Epinephelus chlorostigma* (Valenciennes, 1828) in Seychelles (Pisces, Serranidae). *Cybium* 10(3): 249–262.

MRAG (Marine Resources Assessment Group). 1996. The status of Seychelles demersal fishery. Seychelles Fishing Authority Technical Report. Government of Seychelles, Victoria/MRAG, London.

Mullaney MD Jr. 1994. Ontogenetic shifts in the diet of gag, *Mycteroperca microlepis* (Goode and Bean) (Pisces: Serranidae). *Proceedings of the Gulf and Caribbean Fisheries Institute* 43: 432–445.

Munasinghe NLR. 1972. The catch rate in relation to depth and its seasonal variation in two groups of commercially important fish from the Wadge Bank trawl fishery. *Indo-Pacific Fisheries Council* 13(3): 340–353.

Munro JL, Gaut VC, Thompson R, Reeson PH. 1973. The spawning seasons of Caribbean reef fishes. *Journal of Fish Biology* 5: 69–84.

Munro JL, Williams DMcB. 1985. Assessment and management of coral reef fisheries: biological, environmental and socio-economic aspects. In: Gabrie C, Salvat B (eds), *Proceedings of the 5th International Coral Reef Congress, Tahiti, 27 May–1 June 1985*. Vol. 4: Symposia and Seminars (B). pp 545–578.

Murillo JC, Espinoza E, Edgar GJ, Nicolaides F and 9 others. 2002. *La pesca artesanal en Galápagos: comparación de indicadores entre 1997–2001. Informe Galápagos 2001/2002*. Quito, Ecuador: Fundación Natura and WWF.

Musick JA, Harbin MM, Berkeley SA, Burgess GH, Eklund AM, Findley L, Gilmore RG, Golden JT, Ha DS, Huntsman GR, McGovern JC, Parker SJ, Poss SG, Sala E, Schmidt TW, Sedberry GR, Weeks H, Wright SG.

2000. Marine, estuarine, and diadromous fish stocks at risk of extinction in North America (exclusive of Pacific salmonids). *Fisheries* 25(11): 6–30.

Myers RF. 1989. *Micronesian reef fishes: a practical guide to the identification of the inshore marine fishes of the tropical central and western Pacific.* Barrigada, Guam: Coral Graphics.

Myers RF. 1999. *Micronesian reef fishes: a comprehensive guide to the coral reef fishes of Micronesia* (3rd edition). Barrigada, Guam: Coral Graphics.

Nader IA, Jawdat S. 1977. First records of fishes from Iraqi waters. *Bulletin of the Faculty of Science Riyadh University* 8: 429–443.

Nagelkerken I, Vermonden K, Moraes OC, Debrot AO, Nagelkerken WP. 2005. Changes in coral reef communities and an associated reef fish species, *Cephalopholis cruentata* (Lacépède), after 30 years on Curaçao (Netherlands Antilles). *Hydrobiologia* 549: 145–154.

Nagelkerken WP. 1979. Biology of the graysby, *Epinephelus cruentatus*, of the coral reef of Curaçao. *Studies on the Fauna of Curaçao and other Caribbean Islands* 60: 1–118.

Nagelkerken WP. 1981. *Distribution and ecology of the groupers (Serranidae) and snappers (Lutjanidae) of the Netherland Antilles.* Contributions LVV, No. 1, Sep. 1981. Curacao, Netherlands Antilles: Department of Agriculture, Fisheries and Husbandry. 71 pp.

Nakabo T (ed.). 2000. *Fishes of Japan, with pictorial keys to the species.* Tokyo, Japan: Tokai University Press.

Nakai T, Sano M. 2002. Evidence of protogynous hermaphroditism in the darkfin hind *Cephalopholis urodeta* (Serranidae) at Iriomote Island, southern Japan. *Fisheries Science* 68: 697–699.

Nakai T, Sano M, Kurokura H. 2001. Feeding habits of the darkfin hind *Cephalopholis urodeta* (Serranidae) at Iriomote Island, southern Japan. *Fisheries Science* 67: 640–643.

Nelson JS, Crossman EJ, Espinosa-Perez H, Findley LT, Gilbert CR, Lea RN, Williams JD. 2001. Recommended change in the common name for a marine fish: goliath grouper to replace jewfish (*Epinephelus itajara*). *Fisheries* 26(5): 31–34.

Nemeth RS. 2005. Population characteristics of a recovering US Virgin Islands red hind spawning aggregation following protection. *Marine Ecology Progress Series* 286: 81–97.

Newman SJ, Williams DMcB, Russ GR. 1997. Patterns of zonation of assemblages of the Lutjanidae, Lethrinidae and Serranidae (Epinephelinae) within and among mid-shelf and outer-shelf reefs in the central Great Barrier Reef. *Marine and Freshwater Research* 48: 119–128.

Nichols RS, DeMartini E. 2008. Preliminary estimates of age and growth for the endemic Hawai'ian grouper (Hapu'upu'u, *Epinephelus quernus*, F. Serranidae). Pacific Islands Fisheries Science Center Administrative Report H-08-06. Honolulu, Hawaii, USA: National Marine Fisheries Service, National Oceanic and Atmospheric Administration. 19 pp.

Nicolaides F, Murillo JC. 2001. Efectos de El Ninõ 1997–1998 y post Ninõ 1990 en la pesca del bacalao *Mycteroperca olfax* y la lisa rabo negro (*Mugil cephalus*) de las Islas Galápagos. In: Falconí C, Ruiz RE, Valle C (eds), *Informe Galápagos 2000–2001.* Quito, Ecuador: Fundación Natura and WWF. pp 104–109.

Nicolaides F, Murillo JC, Toral MV, Reck G. 2002. Bacalao. In: Danulat E, Edgar GJ (eds), *Reserva marina de Galápagos, línea base de la biodiversidad.* Galápagos, Ecuador: Charles Darwin Foundation and Galápagos National Park Service. pp 146–165.

NMFS (National Marine Fisheries Service). 2003. Final Amendment 13A to the Fishery Management Plan for the Snapper Grouper Fishery of the South Atlantic Region. Includes an Environmental Assessment, Initial Regulatory Flexibility Analysis, Regulatory Impact Review and Social Impact Assessment/Fishery Impact Statement. Charleston, South Carolina: NMFS. 268 pp.

NMFS (National Marine Fisheries Service) and Gulf of Mexico Fishery Management Council. 2003. Revised secretarial amendment 1 to the reef fish fishery management plan to set a 10-year rebuilding plan for red grouper, with associated impacts on gag and other groupers and supplemental environmental impact statement.

NMFS (National Marine Fisheries Service)/NOAA (National Oceanographic and Atmospheric Administration). 1997–2006. Status of fisheries (1997–2006). Electronic report. Accessed at http://www.nmfs.noaa.gov/sfa/statusoffisheries/SOSmain.htm.

NOAA (National Oceanic and Atmospheric Administration). 2003. Annual commercial landing statistics. Online report. Accessed at http://www.st.nmfs.noaa.gov/st1/commercial/index.html.

Nowara GB, Newman SJ. 2001. A history of foreign fishing activities and fishery-independent surveys of the demersal finfish resources in the Kimberley region of Western Australia. *Fisheries Research Report, Fisheries Western Australia* 125: 1–84. Accessed at http://www.fish.wa.gov.au/docs/frr/frr125/frr125.pdf.

Nzioka RM. 1979. Observations on the spawning seasons of East African reef fishes. *Journal of Fish Biology* 14(4): 329–342.

Ohman MC, Rajasuriya A, Olafsson E. 1997. Reef fish assemblages in north-western Sri Lanka: distribution patterns and influences of fishing practices. *Environmental Biology of Fishes* 49: 45–61.

Olsen DA, LaPlace JA. 1979. A study of a Virgin Islands grouper fishery based on a breeding aggregation.

Proceedings of the Gulf and Caribbean Fisheries Institute 31: 130–144.

Olsen DA, Nellis DW, Wood RS. 1984. Ciguatera in the eastern Caribbean. *Marine Fisheries Reviews* 46(1): 13–18.

Opic P, Conand F, Bourret P. 1994. *Poissons commerciaux du sud-ouest de l'océan Indien*. Paris, France: ORSTOM Éditions.

Opitz S. 1996. *Trophic interactions in Caribbean coral reefs*. ICLARM Technical Report No. 43. Manila, Philippines: International Centre for Living Aquatic Resources Management. 341 pp.

Orts S. 1993. *Field guide to Seychelles commercial fishes: Lutjanidae, Lethrinidae, Serranidae: their range in the southwestern Indian Ocean*. Victoria, Seychelles: Seychelles Fishing Authority (SFA). 171 pp.

Oxley WG, Ayling AM, Cheal AJ, Osborne K. 2004. Marine surveys undertaken in the Elizabeth and Middleton Reefs Marine National Nature Reserve, December 2003. Report produced for Department of the Environment and Heritage. Australian Institute of Marine Science, Townsville, Australia. Accessed at http://www.aims.gov.au/pages/research/marine-surveys/elizabeth-middleton-reefs-surveys-december2003/pdf/elizabeth-middleton-reefs-surveys-december2003.pdf.

Padilla JE, Mamauag S, Braganza G, Brucal N, Yu D, Morales A. 2003. *Sustainability assessment of the live reef-fish for food industry in Palawan Philippines*. Quezon City, Philippines: WWF-Philippines. 109 pp.

Paiva MP, Andrade-Tubino MF. 1998. Distribuicao e abundancia de peixes bentonicos explotados pelos linheiros largo do sudeste do Brasil (1986–1995). *Revista Brasileira de Biologia* 58(4): 619–632.

Paiva MP, Garcez DS. 1998. Distribuicao batimetrica de recursos pesqueiros no talude continental do sudeste do Brasil. *Arquivos de Ciências do Mar* 31(1–2): 107–110.

Parenti P. 2003. On the status of species classified in the genus *Perca* by Johann Julius Walbaum (1792). *Zoological Studies* 42(4): 491–505.

Parker RO, Mays RW. 1998. Southeastern US deepwater reef fish assemblages, habitat characteristics, catches, and life history summaries. *NOAA Technical Report NMFS* 138. Seattle, Washington, USA: US Department of Commerce. 41 pp.

Parrish JD. 1987. The trophic biology of snappers and groupers. In: Polovina JJ, Ralston S (eds), *Tropical snappers and groupers: biology and fisheries management*. Boulder, USA/London, UK: Westview Press. pp 405–463.

Passfield K. 1996. Notes on grouper spawning aggregations in Tongareva, Cook Islands, June 1995. *SPC Traditional Marine Resource Management and Knowledge Information Bulletin* 7: 20.

Pastor OT. 2002. Life history and stock assessment of the African hind *Cephalopholis taeniops* (Valenciennes, 1828), in Sao Vicente-Sao Nicolau insular shelf of the Cape Verde. Final Project 2002. United Nations University, Fisheries Training Programme, Reykjavik, Iceland. 45 pp.

Paulin C, Roberts C. 1992. *The rockpool fishes of New Zealand*. Wellington, New Zealand: Museum of New Zealand. 177 pp.

Paxton JR, Hoese DF, Allen GR, Hanley JE. 1989. *Zoological Catalogue of Australia,* Vol. 7. *Pisces. Petromyzontidae to Carangidae*. Canberra, Australia: Australian Government Publishing Service/Australian Biological Resources Survey. 665 pp.

Paz G, Grimshaw T. 2001. Status report on Nassau groupers for Belize, Central America. Scientific report of the Green Reef Environmental Institute, San Pedro Town, Ambergris Caye, Belize.

Pears RJ. 2005. Comparative demography and assemblage structure of serranid fishes: implications for conservation and fisheries management. PhD thesis, James Cook University, Townsville, Australia.

Pears RJ, Choat JH, Mapstone BD, Begg GA. 2006. Demography of a large grouper, *Epinephelus fuscoguttatus*, from Australia's Great Barrier Reef: implications for fishery management. *Marine Ecology Progress Series* 307: 259–272.

Pears RJ, Choat JH, Mapstone BD, Begg A. 2007. Reproductive biology of a large, aggregation-spawning serranid, *Epinephelus fuscoguttatus* (Forsskål): management implications. *Journal of Fish Biology* 71(3): 795–817.

Pereira MAM. 2003. Recreational scuba diving and reef conservation in southern Mozambique. MSc thesis, University of Natal, Durban, South Africa.

Pet JS. 1999. Marine resource utilization in Komodo National Park. Monitoring Report 1997–1998. The Nature Conservancy, Jakarta Selatan, Indonesia. 38 pp.

Pet JS, Mous PJ, Muljadi AH, Sadovy YJ, Squire L. 2005. Aggregations of *Plectropomus areolatus* and *Epinephelus fuscoguttatus* (groupers, Serranidae) in the Komodo National Park, Indonesia: monitoring and implications for management. *Environmental Biology of Fishes* 74(2): 209–218.

Pina-Amargós F, González-Sansón G. 2008. Movement patterns of goliath grouper *Epinephelus itajara* around southeast Cuba: implications for conservation. *Endangered Species Research* 7: 243–247.

Pittman SJ. 1996. Coral reef fish assemblages of Baie Ternay National Marine Park and Baie Beau Vallon, Mahe, Seychelles. Seychelles Marine Parks Expedition, Mahe, Seychelles.

Planes S, Lecaillon G. 1998. Consequences of the founder effect in the genetic structure of introduced island coral reef fish population. *Biological Journal of the Linnean Society* 63: 537–552.

Pogonoski JJ, Pollard DA, Paxton JR. 2002. *Conservation overview and action plan for Australian threatened and potentially threatened marine and estuarine fishes*. Canberra, Australia: Environment Australia. 375 pp.

Poll M. 1954. Poissons IV. Téléostéens Acanthoptérygiens (Première Partie). Résultats scientifiques de l'expédition océanographique Belge dans les eaux côtières Africaines de l'Atlantique sud (1948–49). *Mémoires de l'Institut Royal des Sciences Naturelles de Belgique* 4(3A): 1–390.

Polovina JJ. 1987. Assessment and management of deepwater bottom fishes in Hawaii and the Marianas. In: Polovina JJ, Ralston S (eds), *Tropical snappers and groupers: biology and fisheries management*. Boulder, USA/London, UK: Westview Press. pp 505–532.

Polovina JJ, Ralston S (eds). 1987. *Tropical snappers and groupers: biology and fisheries management*. Boulder, USA/London, UK: Westview Press.

Pomeroy RS, Sugama K, Slamet B, Tridjoko. 2006. The financial feasibility of small-scale grouper *Cromileptes altivelis* aquaculture in Indonesia. *Asian Fisheries Science* 19: 27–42.

Popple ID, Hunte W. 2005. Movement patterns of *Cephalopholis cruentata* in a marine reserve St Lucia, WI, obtained from ultrasonic telemetry. *Journal of Fish Biology* 67(4): 981–992.

Porch CE, Eklund AM. 2004. Standardized visual counts of goliath grouper off South Florida and their possible use as indices of abundance. *Gulf of Mexico Science* 22: 155–163.

Porch CE, Eklund AM, Scott GP. 2003. An assessment of rebuilding times for goliath grouper. SEDAR Goliath Grouper Data Workshop Report, March 2003. 11 pp.

Posada JM. 1996. Life-history, reproductive biology and sexual pattern of the Creole fish, *Paranthias furcifer* (Valenciennes, 1828) (Pisces: Serranidae). PhD thesis, University of Puerto Rico, Mayaguez campus, Puerto Rico.

Postel E, Fourmanoir P, Guézé P. 1963. Serranides de la Réunion. *Mémoire Institut Fondamental d'Afrique Noire* 68: 339–384.

Pothin, K, Letourneur, Y, Lecompte-Finiger, R. 2004. Age, growth and mortality of the tropical grouper *Epinephelus merra* (Pisces: Serranidae) on Réunion Island, SW Indian Ocean. *Vie et Milieu* 54(4): 193–202.

Potts JC, Brennan K. 2001. Trends in catch data and estimated static SPR values for fifteen species of reef fish landed along the southeastern United States. Report submitted to the South Atlantic Fishery Management Council, Charleston, South Carolina, USA. 41 pp.

Potts JC, Manooch CS III. 1995. Age and growth of red hind and rock hind collected from North Carolina through the Dry Tortugas, Florida. *Bulletin of Marine Science* 56(3): 784–794.

Potts JC, Manooch CS III. 1999. Observations on the age and growth of graysby and coney from the southeastern United States. *Transactions of the American Fisheries Society* 128: 751–757.

Potts JC, Manooch CS III. 2001. Differences in the age and growth of white grunt (*Haemulon plumieri*) from North Carolina and South Carolina compared with southeast Florida. *Bulletin of Marine Science* 68: 1–12.

Pratt V, Mamauag S, Alban J, Parfan E, Donaldson T. 2000. Status report on the Philippine live reef fish trade and strategies to combat its destructive fishing practices. Report on the status of coral reefs in the Philippines. Marine Science Institute, University of the Philippines, Diliman, Quezon City, Philippines.

Prokop F. 2002. *Australian fish guide* (2nd reprint). Victoria, Australia: Australian Fishing Network. 224 pp.

Queensland Fisheries Management Authority. 1999. Draft management plan and regulatory statement. Queensland coral reef fin fish fishery. Prepared by the Queensland Fisheries Management Authority with the Reef Fishery Management Advisory Committee.

Queensland Fisheries Service. 2003. Fisheries (Coral Reef Fin Fish) Management Plan 2003. Queensland Fisheries Service, Department of Primary Industries, Brisbane, Australia.

Quick P. 2006. A lexicographical introduction and inventory of Pendau fish names. Tenth International Conference on Austronesian Linguistics, 17–20 January 2006, Palawan, Philippines. Puerto Princesa City, Philippines: Linguistics Society of the Philippines and SIL International. 62 pp. Accessed at http://www.sil.org/asia/philippines/ical/papers.html.

Rajasuriya A, Perera N, Fernando M. 2005. Status of coral reefs in Trincomalee, Sri Lanka. In: Souter D, Lindén O (eds), *Coral reef degradation in the Indian Ocean*. Status Report 2005. Kalmar, Sweden: CORDIO, University of Kalmar.

Ramesh DA. 1996. Association of the coral colony (*Montipora digitata*) and the fish fry (*Epinephelus* sp.) on the western side of Puli Island (lagoon and reef crest) of Gulf of Mannar. *Indian Hydrobiology* 2(1): 20.

Rámirez-Rodriguez M. 1996. Pesquería de escama. In: Casas Valdez M, Ponce Díaz G (eds), *Estudio del potencial pesquero y acuícola de Baja California Sur*. Vol. 1. La Paz, Baja California, Mexico: SEMARNAP, Gobierno del Estado de Baja California Sur, FAO, Instituto Nacional de la Pesca, UABCS, CIBNOR, CICIMAR, CETMAR. pp 287–304.

Ranatunga RRMKP. 2007. Demographic patterns in coral reef fishes: Biogeographic trends and fishing effects.

PhD thesis, James Cook University, School of Marine and Tropical Biology, Australia.

Randall JE. 1961. Tagging reef fishes in the Virgin Islands. *Proceedings of the Gulf and Caribbean Fisheries Institute* 14: 201–241.

Randall JE. 1963. An analysis of the fish populations of artificial and natural reefs in the Virgin Islands. *Caribbean Journal of Science* 3(1): 31–47.

Randall JE. 1967. Food habits of reef fishes of the West Indies. *Studies in Tropical Oceanography* 5: 665–847.

Randall JE. 1973. Tahitian fish names and a preliminary checklist of the fishes of the Society Islands. *Occasional Papers of the Bernice Pauahi Bishop Museum* 24(11): 167–214.

Randall JE. 1980. A survey of ciguatera at Enewetak and Bikini, Marshall Islands, with notes on the systematics and food habits of ciguatoxic fishes. *Fishery Bulletin* 78(2): 201–249.

Randall JE. 1987. Introductions of marine fishes to the Hawaiian Islands. *Bulletin of Marine Science* 41(2): 490–502.

Randall JE. 1995a. *Coastal fishes of Oman*. Honolulu, Hawaii, USA: University of Hawaii Press. 439 pp.

Randall JE. 1995b. *Reef and shore fishes of the South Pacific: New Caledonia to Tahiti and the Pitcairn Islands*. Honolulu, Hawaii, USA: University of Hawaii Press. 720 pp.

Randall JE. 1999. Report on fish collections from the Pitcairn Islands. *Atoll Research Bulletin* 461: 1–54.

Randall JE, Allen GR. 1987. Four new serranid fishes of the genus *Epinephelus* (Perciformes: Epinephelinae) from Western Australia. *Records of the Western Australian Museum* 13(3–4): 387–412.

Randall JE, Allen GR, Steene RC. 1997. *Fishes of the Great Barrier Reef and Coral Sea*. Revised and expanded edition. Bathurst, New South Wales, Australia: Crawford House Publishing. 557 pp.

Randall JE, Anderson C. 1993. Annotated checklist of the epipelagic and shore fishes of the Maldives Islands. *Ichthyological Bulletin of the JLB Smith Institute of Ichthyology* 59: 1–47.

Randall JE, Ben-Tuvia A. 1983. A review of the groupers (Pisces: Serranidae: Epinephelinae) of the Red Sea, with description of a new species of *Cephalopholis*. *Bulletin of Marine Science* 33(2): 373–426.

Randall JE, Bishop BP. 2004. Food habits of reef fishes of the West Indies. *Studies in Tropical Oceanography (Miami)* 5: 665–847.

Randall JE, Brock VE. 1960. Observations on the ecology of epinepheline and lutjanid fishes of the Society Islands, with emphasis on food habits. *Transactions of the American Fisheries Society* 89: 9–16.

Randall JE, Heemstra PC. 1991. Revision of the Indo-Pacific groupers: (Perciformes: Serranidae: Epinephelinae): with descriptions of five new species. *Indo-Pacific Fishes* 20: 1–332.

Randall JE, Hoese DF. 1986. Revision of the Indo-Pacific genus *Plectropomus* (Perciformes: Serranidae). *Indo-Pacific Fishes* 13: 1–31.

Randall JE, Hoese DF, Last P. 1993. On the status of the Australian serranid fishes *Epinephelus ergastularius* (Whitley) and *Epinephelus thompsoni* (Whitley). *Records of the Australian Museum* 45(1): 25–31.

Randall JE, Ida H, Kato K, Pyle RL, Earle JL. 1997. *Annotated checklist of inshore fishes of the Ogasawara Islands*. Tokyo, Japan: National Science Museum Monographs.

Randall JE, Satapoomin U. 2000. *Cephalopholis polyspila*, a new species of grouper (Perciformes: Serranidae: Epinephelinae) from southwestern Thailand and Sumatra. *Phuket Marine Biological Center Research Bulletin* 63: 1–8.

Randall JE, Wheeler A. 1991. Reidentification of seven tropical Pacific fishes collected and observed by the Forsters during the voyage of *HMS Resolution*, 1772–75. *Copeia* 3: 760–766.

Randall JE, Williams JT, Smith DG, Kulbicki M, Tham GM, Labrosse P, Kronen M, Clua E, Mann BS. 2003. Checklist of the shore and epipelagic fishes of Tonga. *Atoll Research Bulletin* 502: 1–37.

Rao AC, Drishnan L. 2007. Histological notes on protogynous hermaphroditism in the spinycheek grouper, *Epinephelus diacanthus* (Valenciennes, 1828). *Journal of the Marine Biological Association of India* 49: 243–245.

Rasem BM, James CM, Al-Thobaiti SA, Carlos MH. 1997. Spawning of the camouflage grouper *Epinephelus polyphekadion* (Bleeker) in the hypersaline waters of Saudi Arabia. *Asian Fisheries Science* 9: 251–259.

Reck GK. 1983. The coastal fisheries in the Galápagos Islands, Ecuador: description and consequences for management in the context of marine environmental protection and regional development. Unpublished PhD thesis, Mathematisch-Naturwissenchaftlichen Fakultat der Christian-Albrechts, Universitat zu Kiel, Bremerhaven, Germany.

REEF (Reef Environmental Education Foundation). 2003. Tropical eastern Pacific distribution reports. Online reports. Accessed at http://www.reef.org/db/zonecodes/tep (permission needed).

Reiner F. 1996. *Catálogo dos Peixes do Arquipélago de Cabo Verde.* Publicações avulsas do Lisbon, Portugal: IPIMAR (Instituto de Investigacao das Pescas e do Mar).

Relini G, Relini M, Torchia G. 1995. La barriera artificiale di Loano. *Biologia Marina Mediterranea* 2(1): 21–64.

Renan X, Lecomte-Finiger R, Brule T. 2004. Addressing recruitment in *Mycteroperca microlepis* populations of the north coast of Yucatan Peninsula: an otolith aging approach. *Proceedings of the Gulf and Caribbean Fisheries Institute* 55: 881–889.

Reñones O, Piñeiro C, Mas X, Goñi R. 2007. Age and growth of the dusky grouper *Epinephelus marginatus* (Lowe 1834) in an exploited population of the western Mediterranean Sea. *Journal of Fish Biology* 71(2): 346–362.

REVIZEE-SCORE SUL. 1999. Prospecção pesqueira de recursos demersais com armadilhas e pargueiras na zona econômica exclusiva. Itajaí, Brazil: CEPSUL/IBAMA. 113 pp.

Rhode FC, Ross SW, Epperly SP, Burgess GH. 1995. Fishes new or rare on the Atlantic seaboard of the United States. *Brimleyana* 23: 53–64.

Rhodes KL. 2004. FSM Spawning Aggregation Monitoring Training Workshop Report, Black Coral Island, Pohnpei, Federated States of Micronesia, 31 January–6 February, 2004. Report prepared for the Pacific Island Countries Coastal Marine Program, The Nature Conservancy. TNC Pacific Island Countries Report No. 1/04. Accessed at http://www.reefresilience.org/pdf/Rhodes_2004.pdf.

Rhodes KL, Joseph E, Mathias D, Malakai S, Kostka W, David D. 2005. Reef fish spawning aggregation monitoring in Pohnpei, Federated States of Micronesia, in response to local management needs. *SPC Live Reef Fish Information Bulletin* No. 14: 20–24.

Rhodes KL, Lewis RI, Chapman RW, Sadovy Y. 2003. Genetic structure of camouflage grouper, *Epinephelus polyphekadion* (Pisces: Serranidae), in the western central Pacific. *Marine Biology* 142: 771–776.

Rhodes KL, Sadovy Y. 2002a. Reproduction in the camouflage grouper (Pisces: Serranidae) in Pohnpei, Federated States of Micronesia. *Bulletin of Marine Science* 70(3): 851– 869.

Rhodes KL, Sadovy Y. 2002b. Temporal and spatial trends in spawning aggregations of camouflage grouper, *Epinephelus polyphekadion*, in Pohnpei, Micronesia. *Environmental Biology of Fishes* 63: 27–39.

Rhodes KL, Tupper MH. 2007. A preliminary market-based survey of the Pohnpei, Micronesia, grouper (Serranidae: Epinephelinae) fishery reveals unsustainable fishing practices. *Coral Reefs* 26(2): 335–344.

Rhodes KL, Tupper MH. 2008. The vulnerability of reproductively active squaretail coralgrouper (*Plectropomus areolatus*) to fishing. *Fishery Bulletin* 106(2): 194–203.

Richard R, Daud A, Jamil M, Biusing R. 1998. Distribution, abundance and biological studies of economically important fishes in the South China Sea, Area II: Sarawak, Sabah, and Brunei Darussalam waters. In: *Proceedings of the SEAFDEC Technical Seminar on the Interdepartmental Collaborative Research Program in the South China Sea, Area II: Sarawak, Sabah, and Brunei Darussalam*. Kuala Lumpur, Malaysia, 14–15 December 1998. Samut Prakan, Thailand: Southeast Asian Fisheries Development Center. pp 353–361.

Richards WJ. 1999. Preliminary guide to the identification of the early life history stages of serranid fishes of the western central Atlantic. *NOAA Technical Memorandum* NMFS-SEFSC-419. 109 pp.

Rilov G, Benayahu Y. 1998. Vertical artificial structures as an alternative habitat for coral reef fishes in disturbed environments. *Marine Environmental Resources* 45(4/5): 431–451.

Rimmer MA, Phillips MJ, Sim SY. 2005. Aquaculture of groupers in Asia and the Pacific. In: Johnston B, Yeeting B (eds), *Economics and marketing of the live reef fish trade in Asia-Pacific*. ACIAR Working Paper No. 60. Canberra, Australia: Australian Centre for International Agricultural Research. pp 116–134.

Rivas LR. 1970. The red grouper of the Gulf of Mexico. *Commercial Fisheries Review* 32(10): 24–30.

Rivera MAJ. 2002. Population structure of the commercially important Hawaiian grouper, *Epinephelus quernus* (Serranidae) using mitochondrial and nuclear microsatellite DNA. PhD thesis, Environmental Science and Policy Management, University of California, Berkeley, USA.

Rivera MAJ, Kelley CC, Roderick GK. 2004. Subtle population genetic structure in the Hawaiian grouper, *Epinephelus quernus* (Serranidae) as revealed by analyses. *Biological Journal of the Linnean Society* 81: 449–468.

Roberts CM, Hawkins JP, Chapman N, Clarke V, Morris AV, Miller R, Richards A. 1998. *The threatened status of marine species*. A report to the World Conservation Union (IUCN) Species Survival Commission, and Center for Marine Conservation, Washington, DC. Gland, Switzerland: IUCN.

Robertson WD, Schleyer MH, Fielding PJ, Tomalin BJ, Beckley LE, Fennessy ST, van der Elst RP, Bandiera S, Macia A, Gove D. 1996. Inshore marine resources aand associated opportunities for development of the coast of southern Mozambique: Ponta do Oro to Cabo de Santa Maria. Unpublished report, Oceanographic Research Institute, Durban, South Africa. 51 pp.

Robins CR, Ray GC. 1986. *A field guide to Atlantic coast fishes of North America*. Boston, USA: Houghton Mifflin.

Robinson J. 2004. News from Seychelles. In: IUCN Specialist Group of Groupers & Wrasses Newsheet No. 8. IUCN Groupers & Wrasses Specialist Group. pp 4–5.

Robinson J, Isidore M, Marguerite MA, Öhman MC, Payet RJ. 2004. Spatial and temporal distribution of reef fish spawning aggregations in the Seychelles – an interview-based survey of artisanal fishers. *Western Indian Ocean Journal of Marine Science* 3: 63–69.

Robinson J, Marguerite M, Payet R, Isidore M. 2007. Investigation of the importance of reef fish spawning aggregations for the sustainable management of artisanal fisheries resources in Seychelles. Final project report to the Marine Science for Management (MASMA)Programme/Western Indian Ocean Marine Science Association (WIOMSA). 62 pp.

Rocha LOF, Costa PAS. 1999. *Manual de identificação de peixes marinhos papa a costa central.* 2a edicao. Programa REVIZEE/SCORE-Central. 66 pp.

Rodriguez WT. 1984. Estudio preliminar para evaluar las caracteristicas biologicas pesqueras de *Mycteroperca olfax* en las islas Galapagos (Ecuador). *Boletin Cientifico y Técnico, Instituto Nacional de Pesca, Guayaquil-Ecuador,* 3: 3–66.

Roff DA. 1992. *The evolution of life histories.* New York, USA: Chapman & Hall.

Rosenblatt RH, Zahuranec BJ. 1967. The eastern Pacific groupers of the genus *Mycteroperca,* including a new species. *California Fish and Game* 53(4): 228–245.

Ross SW. 1988. Xanthic coloration as the normal color pattern of juvenile speckled hind, *Epinephelus drummondhayi* (Pisces: Serranidae). *Copeia* 1988(3): 780–784.

Ross SW, Moser ML. 1995. Life history of juveniles gag, *Mycteroperca microlepis,* in North Carolina estuaries. *Bulletin of Marine Science* 56: 222–237.

Roughley TC. 1916. *Fishes of Australia and their technology.* Technical Education Series No. 21. Sydney, Australia: Government of New South Wales. 296 pp.

Rouphael T, Tourak E, Brodie J. 1998. Chapter 3: Seagrasses and mangroves of Yemen's Red Sea. In: DouAbal A et al. (eds), *Protection of marine ecosystems of the Red Sea coast of Yemen.* United Nations Development Programme. pp 41–49.

Rowling KR. 1996. Assessment of the NSW dropline fishery. Progress Report – 1993 to 1995. NSW Fisheries Research Institute, Cronulla, Austalia. 37 pp.

Ruiz Carus R. 1983. A karyotypic study of *Epinephelus guttatus* (Linnaeus) and *Thalassoma bifasciatum* (Bloch) (Serranidae and Labridae – Pisces). MSc thesis, Department of Marine Sciences, University of Puerto Rico, Puerto Rico.

Russ GR, Alcala AC. 1996. Marine reserves: rates and patterns of recovery and decline of large predatory fish. *Ecological Applications* 6(3): 947–961.

Russell BC, Houston S. 1989. Offshore fishes of the Arafura Sea. *Beagle* 6(1): 69–84.

Russell BC, Vail CC. 1988. Report on traditional Indonesian fishing activities at Ashmore Reef Nature Reserve. Project No. 8687/19 of the Australian National Parks and Wildlife Service Research and Surveys Programme. Australian National Parks and Wildlife Service, Canberra and NT Museum of Arts and Sciences, Darwin. 179 pp.

Russell M. 2003. Reducing the impacts of fishing and tourism on fish spawning aggregations in the Great Barrier Reef Marine Park. *Proceedings of the Gulf and Caribbean Fisheries Institute* 54: 681–688.

Ruttenburg BI. 2001. Effects of artisanal fishing on marine communities in the Galápagos Islands. *Conservation Biology* 15(6): 1691–1699.

Sabetian A. 2003. The association of physical and environmental factors with abundance and distribution patterns of groupers around Kolombangara Island, Solomon Islands. *Environmental Biology of Fishes* 68(1): 93–99.

Sadovy Y. 1992. Preliminary assessment of the marine aquarium export trade in Puerto Rico. In: Richmond RH (ed.), *Proceedings of 7th International Coral Reef Symposium* Vol. 2. Mangilao, Guam, USA. pp 1014–1022.

Sadovy Y. 1997. The case of the disappearing grouper: *Epinephelus striatus,* the Nassau grouper in the Caribbean and western Atlantic. *Proceedings of the Gulf and Caribbean Fisheries Institute* 45: 5–22

Sadovy Y. 1998. A taste for live fish: Hong Kong's live reef fish market. *Naga, the WorldFish Center Quarterly Newsletter* 21(2): 38–42.

Sadovy Y. 2000. Regional survey for fry/fingerling supply and current practices for grouper mariculture: evaluating current status and long-term prospects for grouper mariculture in South East Asia. Final report to the Collaborative APEC Grouper Research and Development Network (FWG 01/99 revised). December, 2000. Singapore: APEC Secretariat.

Sadovy Y. 2003. *Polyprion americanus.* In: IUCN 2011, *2011 IUCN Red List of Threatened Species.* Version 2011.1. Accessed at www.iucnredlist.org.

Sadovy Y, Colin PL. 1995. Sexual development and sexuality in the Nassau grouper, *Epinephelus striatus* (Bloch) (Pisces: Serranidae). *Journal of Fish Biology* 46: 961–976.

Sadovy Y, Colin PL, Domeier ML. 1994. Aggregation and spawning in the tiger grouper, *Mycteroperca tigris. Copeia* 1994: 511–516.

Sadovy Y, Cornish AS. 2000. *Reef fishes of Hong Kong.* Hong Kong: Hong Kong University Press. 320 pp.

Sadovy Y, Cornish A. 2006. The Nassau grouper 16 years on: endangered, not just unlucky. *Proceedings of the Gulf and Caribbean Fisheries Institute* 59: 245–246.

Sadovy YJ, Donaldson TJ, Graham TR, McGilvray F, Muldoon GJ, Phillips MJ, Rimmer MA, Smith A, Yeeting B. 2003. The live reef food fish trade while stocks last. Manila, Philippines: Asian Development Bank. 147 pp.

Sadovy Y[J], Eklund A-M. 1999. Synopsis of biological data on the Nassau grouper, *Epinephelus striatus* (Bloch, 1792) and the jewfish, *E. itajara* (Lichtenstein, 1822). *NOAA Technical Report NMFS* 146. Seattle, Washington,

USA: US Department of Commerce. 68 pp.

Sadovy Y[J], Figuerola M, Roman A. 1992. Age and growth of red hind *Epinephelus guttatus* in Puerto Rico and St Thomas. *Fisheries Bulletin* 90: 516–528.

Sadovy YJ, Lau PPF. 2002. Prospects and problems for mariculture in Hong Kong associated with wild-caught seed and feed. *Aquaculture Economics and Management* 6: 177–190.

Sadovy Y[J], Liu M. 2004. Report on the current status and exploitation history of reef fish spawning aggregations in east Indonesia, West Pacific. Fisher Survey Series: Society for the Conservation of Reef Aggregations Volume 6. Accessed at www.scrfa.org.

Sadovy YJ, Vincent ACJ. 2002. Ecological issues and the trades in live reef fishes. In: Sale PF (ed.), *Coral reef fishes: dynamics and diversity in a complex ecosystem.* San Diego, USA: Academic Press. pp 391–420.

Sadovy de Mitcheson Y. 2009. Chapter 2: Biology and ecology considerations for the fishery manager. In: Cochrane KL, Garcia SM (eds), *A fishery manager's guidebook* (2nd edition). Rome, Italy: Food and Agriculture Organization; UK: Blackwell Publishing. pp 21–51.

Sadovy de Mitcheson Y, Colin PL (eds). In press. *Reef fish spawning aggregations: biology, research and management.* Fish and Fisheries Series 35. Springer Science+Business Media. DOI 10.1007/978-94-007-1980-4_12.

Sadovy de Mitcheson Y, Craig, MT, Bertoncini AA, Carpenter K, Cheung W, Choat JH, Cornish A, Fennessy S, Ferreira B, Heemstra P, Liu M, Myers R, Pollard D, Rhodes K, Rocha L, Russell B, Samoilys M, Sanciangco J. In press. Fishing groupers towards extinction: a global assessment of threats and extinction risk in a billion dollar fishery. *Fish and Fisheries.*

Sadovy de Mitcheson Y, Liu M. 2008. Environmental and biodiversity impacts of capture-based aquaculture. In: Lovatelli A, Holthus PF (eds), *Capture-based aquaculture: global overview. FAO Fisheries Technical Paper* No. 508. Rome, Italy: Food and Agriculture Organization. pp 1–66.

Sadovy de Mitcheson Y, Cornish A, Domeier M, Colin P, Russell M, Lindeman K. 2008. A global baseline for spawning aggregations of reef fishes. *Conservation Biology* 22(5): 1233–1244.

Sáenz-Arroyo A, Roberts CM, Torre J, Cariño-Olvera M. 2005. Using fishers' anecdotes, naturalists' observations and grey literature to reassess marine species at risk: the case of the Gulf grouper in the Gulf of California, México. *Fish and Fisheries* 6(2): 121–133.

SAFMC (South Atlantic Fishery Management Council). 1993. Amendment 6, regulatory impact review, initial regulatory flexibility analysis and environmental assessment for the snapper grouper fishery of the South Atlantic region. Charleston, South Carolina, USA: SAFMC. 155 pp.

SAFMC (South Atlantic Fishery Management Council). 2003. Commercial fishing regulations for South Atlantic federal waters. Miami, Florida, USA: SAFMC.

SAFMC (South Atlantic Fishery Management Council). 2006. Amendment 13C to the Fishery Management Plan for the Snapper Grouper Fishery of the South Atlantic. Charleston, South Carolina, USA: SAFMC. 631 pp.

Sala E, Aburto-Oropeza O, Paredes G, Thompson G. 2003. Spawning aggregations and reproductive behavior of reef fishes in the Gulf of California. *Bulletin of Marine Science* 72(1): 103–121.

Sala E, Ballesteros E. 2000. Conservation status and dynamics of the Glover's Reef, Belize, spawning aggregation. December 1999–January 2000. Scientific report to the Wildlife Conservation Society's Glover's Reef Marine Research Station, Belize, Central America.

Sala E, Ballesteros E, Starr RM. 2001. Rapid decline of Nassau grouper spawning aggregations in Belize: fishery management and conservation needs. *Fisheries* 26(10): 23–30.

Salini JP, Blaber SJ, Brewer DT. 1994. Diets of trawled predatory fish of the Gulf of Carpentaria, Australia, with particular reference to predation on prawns. *Australian Journal of Freshwater Research* 45(3): 397–411.

Samoilys MA. 1997. Periodicity of spawning aggregations of coral trout, *Plectropomus leopardus* (Pisces: Serranidae) on the northern Great Barrier Reef. *Marine Ecology Progress Series* 160: 149–159.

Samoilys MA. 2000. Reproductive dynamics of an exploited serranid on the Great Barrier Reef. PhD thesis, James Cook University, Townsville, Australia. 106 pp + Appendix.

Samoilys MA. 2002. Spawning aggregation closures for the live reef fish fishery in Solomon Islands. *SPC Live Reef Fish Information Bulletin* No. 10 (June 2002): 30.

Samoilys MA. 2004. Managing fishery resources in Tanzania's Exclusive Economic Zone. A report to the World Bank. IUCN Eastern Africa Regional Programme, Nairobi.

Samoilys M[A], Carlos G. 1990. A survey of reef fish stocks in western Samoa: Application of underwater visual census methods. FAO/UNDP SAM/89/002. Field Report No. 6. A report prepared for the Forum Fisheries Agency, Honiara, Solomon Islands and the Government of Samoa.

Samoilys MA, Carlos G. 2000. Determining methods of underwater visual census for estimating the abundance of coral reef fishes. *Environmental Biology of Fishes* 57: 289–304.

Samoilys MA, Donnelly RJ. 1998. Effects of fishing spawning aggregations of tropical marine fishes: implications for management of the live reef food fish trade in Solomon Islands. Discussion Paper No. 5. Report to Australian Centre for International Agricultural Research. Project no. ANRE1/1998/094. Canberra, Australia: ACIAR.

Samoilys MA, Fuentes H, Tuwai I, Tikomainiusiladi B, Leqata J, Mobiha A, Potuku T, Die D, Wilson M, Connell S, Lincoln Smith M, Watson R. 1995. Application of underwater visual census to assessing coral reef fish stocks in the tropical Pacific. Final Project PN9304 Report to the Australian Centre for International Agricultural Research, Canberra.

Samoilys MA, Squire L, Roelofs A. 2001. Long term monitoring of coral trout spawning aggregations on the Great Barrier Reef: implications for fisheries management. Abstract. 6th Indo-Pacific Fish Conference, Durban, South Africa, May 2001.

Sanches JG. 1991. *Catálogo dos principais peixes marinhos da República de Guiné-Bissau. Bissau.* Lisbon, Portigal: Instituto Nacional de Investigação das Pescas. 468 pp.

Sanders MJ, Carrara G, Lablache G. 1987. Preliminary assessment for the brown-spotted grouper *Epinephelus chlorostigma* occurring on the Mahé Plateau (Seychelles). In: Sanders MJ, Sparre P, Venema S (eds), Proceedings of the Workshop on the Assessment of the Fishery Resources in the Southwest Indian Ocean. pp 268–277.

Sattar SA, Adam MS. 2005. Review of the grouper fishery of the Maldives with additional notes on the Faafu Atoll fishery. Marine Research Centre, Malé, Maldives. 54 pp.

Sazaki J, Teruya K, Ueno S. 1999. Embryonic, larval and juvenile deveopment of the grouper *Epinephelus merra* (Teleostei: Serranidae). *Tokai Daigaku Kaiyo Jo Kenkyu Ho Koku* 20: 147–155.

Sazima C, Grossman A. 2005. A non-digging zoobenthivorous fish attracts two opportunistic predatory fish associates. *Neotropical Ichthyology* 3(3): 445–448.

Sazima I. 1986. Similarities in feeding behaviour between some marine and freshwater fishes in two tropical communities. *Journal of Fish Biology* 29: 53–65.

Sazima I. 2002. Juvenile snooks (Centropomidae) as mimics of mojarras (Gerreidae), with a review of aggressive mimicry in fishes. *Environmental Biology of Fishes* 65: 37–45.

Sazima I, Krajewski JP, Bonaldo RM, Sazima C. 2005. Wolf in a sheep's clothes: juvenile coney (*Cephalopholis fulva*) as an aggressive mimic of the brown chromis (*Chromis multilineata*). *Neotropical Ichthyology* 3(2): 315–318.

Scales H, Balmford A, Manica A. 2007. Impacts of the live reef fish trade on populations of coral reef fish off northern Borneo. *Proceedings of the Royal Society B* 274: 989–994.

Schirripa MJ. 2000. Status of the vermillion snapper (*Rhomboplites aurorubens*), red snapper (*Lutjanus campechanus*, and gag (*Mycteroperca microlepis*) in the US Gulf of Mexico. *Proceedings of the Gulf and Caribbean Fisheries Institute* 51: 139–151.

Schirripa MJ, Goodyear CP. 1994. Status of the gag stocks of the Gulf of Mexico: Assessment 1.0. Miami, Florida, USA: Southeast Fisheries Science Center, Miami Laboratory, Coastal Resources Division.

Schirripa MJ, Legault CM. 1997. Status of gag stocks of the Gulf of Mexico. Assessment 2.0. 1 October 1997. Report to the Gulf of Mexico Fishery Management Council. Miami, Florida, USA: National Marine Fisheries Service, Sustainable Fisheries Division.

Schirripa MJ, Legault CM, Ortiz M. 1999. The red grouper fishery of the Gulf of Mexico: Assessment 3.0. Sustainable Fisheries Division Contribution No. SFD-98/99-56. Miami, USA: Southeast Fisheries Science Center, Sustainable Fisheries Division. 58 pp.

Schobernd Z. 2004. Groupers and seabasses: Where do they live? Accessed at http://oceanexplorer.noaa.gov/explorations/04etta/logs/aug26/aug26.html.

Schroeder RE. 1980. *Philippine shore fishes of the western Sulu Sea.* Manila, Philippines: National Media Production Center. 266 pp.

Schwamborn SHL. 2004. Dinâmica e organização trófica de assembléias de peixes associadas aos prados de capim marinho (*Halodule wrightii*) de Itamaracá, Pernambuco. PhD thesis, Universidade Federal de Penambuco, Recife, Brazil. 315 pp.

Schwartz FJ. 1978. Xanthochromism in *Epinephelus drummondhayi* (Pisces: Serranidae) caught off North Carolina. *Northeast Gulf Science* 2(1): 62–64.

Seale A. 1901. New Hawaiian fishes. *Occasional Papers of the Bernice P Bishop Museum* 1901: 1–15.

Sedberry GR, McGovern JC, Pashuk O. 2001. The Charleston Bump: an island of essential fish habitat in the Gulf Stream. In: Sedberry GR (ed.), *Island in the Stream: oceanography and fisheries of the Charleston Bump. American Fisheries Society Symposium* 25: 3–24.

Sedberry GR, Stevenson DE, Chapman RW. 1996. Stock identification in potentially threatened species of grouper (Teleostei: Serranidae: Epinephelinae) in Atlantic and Caribbean waters. Final report to the National Marine Fisheries Service. South Carolina Department of Natural Resources, Marine Resources Research Institute. 51 pp.

SEMARNAP (Secretaría de Medio Ambiente, Recursos Naturales y Pesca). 2000a. *Anuario estadístico de pesca 1999*. Tlalpan, Mexico City, Mexico: SEMARNAP. 271 pp.

SEMARNAP (Secretaría de Medio Ambiente, Recursos Naturales y Pesca). 2000b. *Sustentabilidad y pesca responsible en México: evaluación y manejo 1997–1998*. SEMARNAP-IPN, México. 691 pp.

Senou H, Kodato H, Nomura T, Yunokawa K. 2006. Coastal fishes of Ie-jima Island, the Ryukyu Islands, Okinawa, Japan. *Bulletin of Kanagawa Prefectural Museum (Natural Sciences)* 35: 67–92.

Séret B. 1981. *Poissons de mer de l'oest Africain tropical*. Paris, France: ORSTOM. 416 pp.

Servicio Parque Nacional Galápagos & Fundación Charles Darwin. 2003. Monitoreo de las pesquerias que duran todo el año en Galapágos (Informe del segundo trimester in 2003). Servicio Parque Nacional Galápagos & Fundación Charles Darwin, Santa Cruz, Ecuador.

Shakeel H, Ahmed H. 1997. Exploitation of reef resources: grouper and other food fishes. Paper 3. In: Nickerson DJ, Maniku MH (eds), Report and Proceedings of the Maldives/FAO National Workshop on Integrated Reef Resources Management in the Maldives. Malé, March 1996. Madras, India: Bay of Bengal Program Report No. 76. pp 137–167.

Shao KT. 1996. *Fish and shellfish in Taiwan (II)*. Taipei, Taiwan: Taiwan Fisheries Bureau. (In Chinese)

Shao KT. 2005. Taiwan Fish Database. World Wide Web electronic publication. Version 2005/5. See Taiwan Fish Database. Accessed at http://fishdb.sinica.edu.tw/eng/home.php.

Shao KT, Chen JP, Shen SC. 1992. Marine fishes of the Ken-Ting National Park. Construction and Planning Administration, Ministry of Interior, (Hengchun) Taiwan.

Shapiro DY. 1987. Reproduction in groupers. In: Polovina JJ, Ralston S (eds), *Tropical snappers and groupers: biology and fisheries management*. Boulder, USA/London, UK: Westview Press. pp 295–327.

Shapiro DY, García-Moliner G, Sadovy Y. 1994. Social system of an inshore population of the red hind grouper, *Epinephelus guttatus* (Pisces: Serranidae). *Environmental Biology of Fishes* 41: 415–422.

Shapiro DY, Sadovy Y, McGehee MA. 1993a. Size, composition, and spatial structure of the annual spawning aggregation of the red hind, *Epinephelus guttatus* (Pisces: Serranidae). *Copeia* 2: 399–406.

Shapiro DY, Sadovy Y, McGehee MA. 1993b. Periodicity of sex change and reproduction in the red hind, *Epinephelus guttatus*, a protogynous grouper. *Bulletin of Marine Science* 53(3): 1151–1162.

Sheaves MJ. 1995. Large lutjanid and serranid fishes in tropical estuaries: Are they adults or juveniles? *Marine Ecology Progress Series* 129(1–3): 31–40.

Shen SC. 1984. *Coastal fishes of Taiwan*. Taipei, Taiwan: Taiwan Museum. 380 pp.

Sheppard CRC. 2003. Predicted recurrences of mass coral mortality in the Indian Ocean. *Nature* 435: 294–297.

Shpigel M, Fishelson L. 1991. Territoriality and associated behaviour in three species of the genus *Cephalopholis* (Pisces: Serranidae) in the Gulf of Aqaba, Red Sea. *Journal of Fish Biology* 38: 887–896.

Siau Y. 1994. Population structure, reproduction and sex-change in a tropical East Atlantic grouper. *Journal of Fish Biology* 44(2): 205–211.

Siau Y, Bouain A. 1994. Preliminary indications on growth and reproduction in the protogynous grouper *Mycteroperca rubra* (Pisces, Serranidae). *Journal of African Zoology* 108(4): 354–359.

Silvano RAM, MacCord PFL, Lima RV, Begossi A. 2006. When does this fish spawn? Fishermen's local knowledge of migration and reproduction of Brazilian coastal fishes. *Environmental Biology of Fishes* 76(2–4): 371.

Situ YY, Sadovy YJ. 2004. A preliminary study on local species diversity and seasonal composition in a Hong Kong wet market. *Asian Fisheries Science* 17: 235–248.

Sluka RD. 2000. Grouper and Napoleon wrasse ecology in Laamu atoll, Republic of Maldives: part 2. Timing, location, and characteristics of spawning aggregations. *Atoll Research Bulletin* 492: 1–15.

Sluka RD. 2002. Grouper and Napoleon wrasse ecology in Laamu Atoll, Republic of Maldives: part 3. Fishing effects and management of the live fish-food trade. *Atoll Research Bulletin* 493: 1–18.

Sluka RD, Chiappone M, Sullivan Seally KM. 1994. Comparison of juvenile grouper populations in southern Florida and the central Bahamas. *Bulletin of Marine Science* 54: 871–880.

Sluka R[D], Chiappone M, Sullivan KM, Wright R. 1996. *Habitat and life in the Exuma Cays, the Bahamas: The status of groupers and coral reefs in the northern cays*. Nassau, Bahamas: Media Publishing. 83 pp.

Sluka RD, Lazarus S. 2010. Grouper (Pisces: Serranidae) relative abundance and diversity on the west coast of India. *Marine Biodiversity Records* 3: e71, Cambridge University Press doi:10.1017/S1755267210000606.

Sluka RD, Reichenbach N. 1996. Grouper density and diversity at two sites in the Republic of Maldives. *Atoll Research Bulletin* 438: 1–18.

Sluka RD, Sullivan KM. 1998. The influence of spear fishing on species composition and size of groupers on patch reefs in the upper Florida Keys. *Fishery Bulletin* 96(2): 388–392.

Sly S (compiler). 2002. Appendix 7. Monitoring data catch composition 1990–2001. In: Assessing the sustainability of the Northern Territory Timor reef fishery. A report prepared for Environment Australia (EA) to assess the Timor Reef fishery against the Commonwealth guidelines for ecologically sustainable management of fisheries. Darwin, Australia: Northern Territory Department of Business, Industry and

Resource Development. pp 50–52.

Smith A, Dalzell P. 1991. Fisheries resources and management investigations in Woleai Atoll, Yap State, Federated States of Micronesia. Technical document. Inshore Fisheries Research Project (South Pacific Commission) no. 4. South Pacific Commission, Noumea, New Caledonia.

Smith CL. 1958. The groupers of Bermuda. In: Bardach JE, Smith CL, Menzel DW (eds), Final report of the Bermuda fisheries research program. Hamilton, Bermuda: Bermuda Trade Development Board. pp 37–59.

Smith CL. 1965. The patterns of sexuality and the classification of serranid fishes. *American Museum Novitates* 2207: 1–20.

Smith CL. 1971. A revision of the American groupers: *Epinephelus* and allied genera. *Bulletin of the American Museum of Natural History* 146(2): 69–241.

Smith CL. 1972. A spawning aggregation of Nassau grouper, *Epinephelus striatus* (Bloch). *Transactions of the American Fisheries Society* 101: 257–261.

Smith CL. 1978. Serranidae. In: Fischer W (ed.), *FAO species identification sheets for fishery purposes: Western Central Atlantic (Fishing Area 31)*. Volume 4. Rome, Italy: Food and Agriculture Organization.

Smith, CL. 1997. *National Audubon Society field guide to tropical marine fishes of the Caribbean, the Gulf of Mexico, Florida, the Bahamas, and Bermuda*. New York, USA: AA Knopf.

Smith CL, Tyler JC, Davis WP, Jones RS, Smith DG, Baldwin CC. 2003. Fishes of the Pelican Cays, Belize. *Atoll Research Bulletin* 497: 1–90.

Smith GB. 1976. Ecology and distribution of eastern Gulf of Mexico reef fishes. *Florida Marine Research Publications* No. 19: 1–78.

Smith GB, Austin HM, Bortone SA, Hastings RW, Ogren LH. 1975. Fishes of the Florida Middle Ground with comments on ecology and zoogeography. *Florida Marine Research Publications* 9: 1–14.

Smith GB, Bullock LH. 1979. Xanthic specimens of speckled hind, *Epinephelus drummondhayi* (Serranidae), from the eastern Gulf of Mexico. *Florida Scientist* 42(4): 252–253.

Smith MM, Heemstra PC. 1986. *Smiths' sea fishes*. Berlin, Germany: Spinger-Verlag. 1047 pp.

Smith WL, Craig MT. 2007. Casting the percomorph net widely: the importance of broad taxonomic sampling in the search for the placement of serranid and percid fishes. *Copeia* 2007: 35–55.

Smith WL, Wheeler WC. 2004. Polyphyly of the mail-cheeked fishes (Teleostei: Scorpaeniformes): evidence from mitochondrial and nuclear sequence data. *Molecular Phylogenetics and Evolution* 32: 627–646.

Smith-Vaniz WF, Collette BB, Luckhurst BE. 1999. *Fishes of Bermuda: history, zoogeography, annotated checklist, and identification keys*. American Society of Ichthyologists and Herpetologist Special Publication No. 4. 424 pp.

Solis-Ramírez M. 1970. The red grouper fishery of Yucatan Peninsula, Mexico. *Proceedings of the Gulf and Caribbean Fisheries Institute* 22: 122–129.

Sommer C, Schneider W, Poutiers JM. 1996. *The living marine resources of Somalia. FAO species identification field guide for fishery purposes*. Rome, Italy: Food and Agriculture Organization.

Song YB, Oh S, Seo J, Ji BG, Lim BS, Lee YD, Kim HB. 2005. Larval development and rearing of longtooth grouper *Epinephelus bruneus* in Jeju Island, Korea. *Journal of World Aquaculture Society* 36: 209–216.

Song YB, Park YJ, Takemura K, Kim HJ, Choi MS, Choi YC, Lee YD. 2003. Histological observations of the female cycle of honeycomb grouper, *Epinephelus merra* in Chuuk. *Developmental Reproduction* 7(1): 23–28.

Southeast Fisheries Science Center. 2002. Status of red grouper in United States waters of the Gulf of Mexico during 1986–2001. Sustainable Fisheries Division Contribution No. SFD-01/02 175rev. Miami, Florida, USA: NOAA, NMFS, SFSC, SFD. 65 pp.

Soyano K. Matsumoto T, Tanaka H, Nakamura M. 2003. Lunar-related spawning in honeycomb grouper, *Epinephelus merra. Fish Physiology and Biochemistry* 28: 447–448.

Spach HL, Godenfroid RS, Santos C, Schwarz Jr R, Queiroz GML. 2004. Temporal variation in fish assemblage composition on a tidal reef flat. *Brazilian Journal of Oceanography* 52(1): 47–58.

Spalding M, Blasco B, Field C. 1997. *World mangrove atlas*. Okinawa, Japan: The International Society for Mangrove Ecosystems. 178 pp.

Stephenson PC, Edmunds JS, Moran MJ, Caputi N. 2001. Analysis of stable isotope ratios to investigate stock structure of red emperor and Rankin cod in northern Western Australia. *Journal of Fish Biology* 58: 126–144.

Stevenson DK. 1981. *A review of the marine resources of the Western Central Atlantic Fisheries Commission (WECAFC) region. FAO Fisheries Technical Paper* No. 211. Rome, Italy: Food and Agriculture Organization. 132 pp.

Stewart AL. 1999. Summer visitors '99. *Seafood New Zealand* 7(3): 78–79.

Stiles TC, Burton ML. 1994. Age, growth, and mortality of red grouper, *Epinephelus morio*, from the Southeastern US. *Proceedings of the Gulf and Caribbean Fisheries Institute* 43: 123–137.

Stobutzki I, Miller M, Brewer D. 2001. Sustainability of fishery bycatch: a process for assessing highly diverse and numerous bycatch. *Environmental Conservation* 28: 167–181.

Sudaryanto, Meyer T, Mous PI. 2004. Natural spawning of three species of grouper in floating cages at a pilot broodstock facility at Komodo, Flores, Indonesia. *SPC Live Reef Fish Information Bulletin* No. 12: 21–26.

Sugama K, Ismi S, Kawahara S, Rimmer M. 2003. Improvement of larval rearing technique for humpback grouper, *Cromileptes altivelis*. *Aquaculture Asia* 3: 34–39.

Sumner NR, Williamson PC, Malseed BE. 2002. A 12-month survey of recreational fishing in the Gascoyne bioregion of Western Australia during 1998–99. Perth, Western Australia: WA Marine Research Laboratories, Department of Fisheries.

Sumpton W, Ryan S. 2004. Ecological assessment of the Queensland deepwater finfish fishery. Report to the Australian Government Department of Environment and Heritage by the Queensland Government Department of Primary Industries and Fisheries, Brisbane. 83 pp.

Talwar PK, Kacker RK. 1984. *Commercial sea fishes of India*. Calcutta, India: Zoological Survey of India. 997 pp.

Tamaru CS, Carlstrom-Trick C, Fitzgerald Jr WJ, Ako H. 1996. Induced final maturation and spawning of the marbled grouper *Epinephelus microdon* captured from spawning aggregations in the Republic of Palau, Micronesia. *Journal of the World Aquaculture Society* 27: 363–372.

Tan SM, Lim PY, Tetsushi S, Hooi KK. 1982. *A colour guide to the fishes of the South China Sea and the Andaman Sea*. Singapore: Primary Production Department/ Marine Fisheries Research Department.

Teixeira JB. 2005. Aplicação de métodos de visualização gráfica para divulgação de estudos oceanográficos – a pesca demersal do programa Revizee como estudo de caso. Monografia, Curso de Oceanografia, Universidade Federal do Espírito Santo. São Paulo, Brazil: REVISEE.

Teixeira SF, Ferreira BP, Padovan IP. 2004. Aspects of fishing and reproduction of the black grouper *Mycteroperca bonaci* (Poey, 1860) (Serranidae: Epinephelinae) in the Northeastern Brazil. *Neotropical Ichthyology* 2(1): 19–30.

Tester PA, Wolfe CA, Dixon RL, Huntsman GR. 1983. Reef fish distributions off North Carolina and South Carolina as revealed by headboat catches. NOAA Technical Memorandum NMFS-SEFC-115. Beaufort, North Carolina, USA: National Oceanic and Atmospheric Administration, US Department of Commerce. 12 pp.

Thollot P. 1996. *Les poissons de mangrove du lagon sud-ouest de Nouvelle-Calédonie*. Paris, France: ORSTOM Éditions.

Thompson R, Munro JL. 1978. Aspects of the biology and ecology of Caribbean reef fishes: Serranidae (hinds and groupers). *Journal of Fish Biology* 12: 115–146.

Thompson R, Munro J[L]. 1983. Chapter 7: The biology, ecology and bionomics of the hinds and groupers, Serranidae. In: Munro JL (ed.), *Caribbean coral reef fisher resources*. Manila, Philippines: International Center for Living Aquatic Resources Management. pp 59–81.

Thomson DA, Findley LT, Kerstitch AN. 2000. *Reef fishes of the Sea of Cortez. The rocky-shore fishes of the Gulf of California*. Revised edition. Austin, Texas, USA: The University of Texas Press. 353 pp.

Thomson JM. 1977. *A field guide to the common sea and estuary fishes of non-tropical Australia*. Sydney, Australia: Collins.

To W-L. 2009. The biology, fishery of groupers (Family: Serranidae) in Hong Kong and adjacent waters, and implications for management. PhD thesis, University of Hong Kong, Hong Kong.

To W-L, Sadovy de Mitcheson Y. 2009. Shrinking baseline: the growth in juvenile fisheries, with the Hong Kong grouper fishery as a case study. *Fish and Fisheries* 10(4): 396–407.

Tookwinas S. 1989. Larviculture of seabass (*Lates calcarifer*) and grouper (*Epinephelus malabaricus*) in Thailand. *Aquacop, Ifremer. Actes de Colloque* 9: 645–659.

Tseng WY, Chan KF. 1985. On the larval rearing of the white-spotted grouper, *Epinephelus amblycephalus* (Bleeker), with a description of larval development. *Journal of World Mariculture* 16: 114–126.

Tseng WY, Ho SK. 1988. *The biology and culture of red grouper*. Kaohsiung, Republic of China, Taiwan: Chien Cheng Publisher.

Tuan LA. 2003. Status of aquaculture and associated environmental management issues in Vietnam. *Tropeca Working Paper* No. 4.

Tuan LA, Hambrey J. 2000. Seed supply for grouper cage culture in Khanh Hoa, Vietnam. *Aquaculture Asia* 5(2): 39–41.

Tuan LA, Nho NT, Hambrey J. 2000. Status of cage mariculture in Vietnam. In: Liao IC, Lin CK (eds), *Proceedings of the 1st International Symposium on Cage Aquaculture in Asia*. Manila, Philippines: Asian Fisheries Society; Bangkok, Thailand: World Aquaculture Society – Southeast Asian Chapter. pp 111–123.

Tucker JW, Fitzgerald WJ. 2004. Induced spawning of two western tropical Pacific groupers, *Plectropomus areolatus* and *Epinephelus fuscouttatus*, in Palau. *Asian Fisheries Science* 7: 57–62.

Turnbull C. 1996. Domestic reef fishing industry assessment. Report from Northern Fisheries Centre, Cairns. QDPI Information Series QI96106. 14 pp.

Turnbull CT, Samoilys MA. 1997. Effectiveness of spawning closures in managing the line fishery on the Great Barrier Reef. Report to the Reef Fish Management Advisory Committee of the Queensland Fisheries

Management Authority. 24 pp.

Turner SC, Porch CE, Heinemann D, Scott GP, Ortiz M. 2001. Status of gag in the Gulf of Mexico. Assessment 3.0. Miami, USA: Southeast Fisheries Science Center, Sustainable Fisheries Division.

Turpin RK, Bortine SA. 2002. Pre- and post-hurricane assessment of artificial reefs: evidence for potential use as refugia in a fishery management strategy. *ICES Journal of Marine Science* 59(Suppl.): S74–S82.

Tuya F, Sanchez-Jerez P, Haroun RJ. 2006. Populations of inshore serranids across the Canarian Archipelago; relationships with human pressure and implications for conservation. *Biological Conservation* 128(1): 13–24.

Tuz-Sulub A, Cervera-Cervera K, Colás-Marrufo T, Brulé, T. 2003. First evidences on the formation of spawning aggregations of groupers (Epinephelinae; Epinephelini) from the Campeche Bank, Mexico. *Proceedings of the Gulf and Caribbean Fisheries Institute* 54: 652–667.

Ukawa M, Higuchi M, Mito M. 1966. Spawning habits and early life history of a serranid fish, *Epinephelus akaara* (Teminck & Schlegel). *Japanese Journal of Ichthyology* 13: 156–161.

Uwate KR, Shams AJ. 1997. Bahrain fish stock enhancement: lessons learned and prospects for the future. *SPC Live Reef Fish Information Bulletin* No. 3: 9–13.

Vacchi M, Bussotti S, Guidetti P, Mesa GL. 1998. Study of the coastal fish assemblage in the marine reserve of the Ustica Island (southern Tyrrhenian Sea). *Italian Journal of Zoology* 65: 281–286.

Vadiya V. 1984. Reproductive systems of *Epinephelus aeneus* and *Epinephelus alexandrinus* (Serranidae) from the southeastern Mediterranean. *Journal of Ichthyology* 24(3): 77–81.

van der Elst R. 1988. *A guide to the common sea fishes of Southern Africa* (2nd edition). Cape Town, South Africa: Struik. 398 pp.

van der Elst RP, David B, Govender A. 1994. The marine linefish resources of Mozambique – status, developments and future research. Unpublished Report No. 194. Oceanographic Research Institute, Durban, South Africa.

van Herwerden L, Benzie J, Peplow L, Davies C. 2000. Microsatellite markers for coral trout (*Plectropomus laevis*) and red throat emperor (*Lethrinus miniatus*) and their utility in other species of reef fish. *Molecular Ecology* 9(11): 1929–1931.

van Herwerden L, Choat JH, Dudgeon C, Carlos G, Newman S. 2006. Contrasting patterns of genetic structure in two species of the coral trout *Plectropomus* (Serranidae) from east and west Australia: introgressive hybridisation or ancestral polymorphisms. *Molecular Phylogenetics and Evolution* 41: 420–435.

van Herwerden L, Davies CR, Choat JH. 2002. Phylogenetic and evolutionary perspectives of the Indo-Pacific grouper *Plectropomus* species on the Great Barrier Reef, Australia. *Journal of Fish Biology* 60(6): 1591–1596.

Vidthatanon C, Premcharoen S. 2002. The status of estuarine fish diversity in Thailand. *Marine and Freshwater Research* 53(2): 471–478.

Wantiez L, Thollot P. 2000. Colonization of the F/V Caledonie Toho 2 wreck by a reef-fish assemblage near Noumea (New Caledonia). *Atoll Research Bulletin* 485: 1–21.

Wantiez L, Thollot P, Kulbicki M. 1997. Effects of marine reserves on coral reef fish communities from five islands in New Caledonia. *Coral Reefs* 16: 215–224.

Wass RC. 1984. An annotated checklist of the fishes of Samoa. Special Scientific Report No. 781. Washington, USA: National Oceanic Atmospheric Administration and National Marine Fisheries Service.

Weaver DC, Sedberry GR. 2001. Trophic subsidies at the Charleston Bump: food web structure of reef fishes on the continental slope of the southeastern United States. In: Sedberry GR (ed.), *Island in the Stream: oceanography and fisheries of the Charleston Bump. American Fisheries Society Symposium* 25: 137–152.

Weeber B. 2004. Best fish guide. Part two. Criteria for ecological rankings of New Zealand commercial fisheries. Royal Forest and Bird Protection Society of New Zealand. [See latest version at http://www.forestandbird.org.nz/what-we-do/publications/-best-fish-guide]

Wells S, Burgess N, Ngusaru A. 2007. Towards the 2012 marine protected area targets in eastern Africa. *Ocean and Coastal Management* 50: 67–83.

Werner TB, Allen GR (eds). 2000. *A rapid marine biodiversity assessment of the Calamianes Islands, Palawan province, Philippines.* RAP Bulletin of Biological Assessment 17. Washington, DC, USA: Conservation International. 127 pp.

Westera M, Lavery P, Hyndes G. 2003. Differences in recreationally targeted fishes between protected and fished areas of a coral reef marine park. *Journal of Experimental Marine Biology and Ecology* 294: 145–168.

Western Pacific Regional Fishery Management Council. 2003. Measure to establish eligibility criteria for new entry into the northwestern Hawaiian Islands Mau Zone limited access system. Accessed at http://www.wpcouncil.org/bottomfish/Documents/FEIS2005/AppendixA.pdf.

Wheeler JFG, Ommanney FD. 1953. Report on the Mauritius-Seychelles fisheries survey 1948–1949. *Fisheries Publication, Colonial Office* 1(3): 1–145.

White DB, Wyanski DM, Eleby BM, Lilyestrom CG. 2002.

Tiger grouper (*Mycteroperca tigris*): profile of a spawning aggregation. *Bulletin of Marine Science* 70(1): 233–240.

Whitley GP. 1930. Additions to the check-list of the fishes of New South Wales. *Australian Zoology* 6(2): 117-123.

Whitley GP. 1948. Studies in ichthyology. No. 13. *Records of the Australian Museum Supplement* 22(1): 70–94.

Williams AJ, Currey LM, Begg GA, Murchie CD, Ballagh AC. 2008. Population biology of coral trout species in the eastern Torres Strait: implications for fishery management. *Continental Shelf Research* 28: 2129–2142.

Williams EH, Dixon B. 2003. Preliminary analysis of some deepwater species in the South Atlantic headboat survey data. Working paper SEDAR4-DW-16, South Atlantic and Caribbean deepwater snapper-grouper workshop. Charleston, South Carolina, USA: Southeast Data, Assessment and Review. 23 pp. Accessed at http://www.sefsc.noaa.gov/sedar/download/SEDAR4_DW_16.pdf?id=DOCUMENT.

Williams JT, Wantiez L, Chauvet C, Galzin R, Harmelin-Vivien M, Jobet E, Juncker M, Mou-Tham G, Planes S, Sasa P. 2006. Checklist of the shorefishes of Wallis Islands (Wallis and Futuna French Territories, South-Central Pacific). *Cybium* 30(3): 247–260.

Williams KC, Irvin S, Barclay M. 2004. Polka dot grouper *Cromileptes altivelis* fingerlings require high protein and moderate lipid diets for optimal growth and nutrient retention. *Aquaculture Nutrition* 10(2): 125–134.

Willing RS, Pender PJ. 1989. Length-weight relationships for 45 species of fish and three invertebrates from Australia's northern prawn fishery. Technical Bulletin No. 142. Darwin, Australia: Northern Territory Department of Primary Industry and Fisheries. 57 pp.

Wilson KDP. 2003. *Artificial reefs and reef fish in Hong Kong*. Hong Kong: Friends of the Country Parks and Cosmo Books Ltd. 176 pp.

Wright D. 2005. Report of HURL Cruise KOK0510: submersible dives and multibeam mapping to investigate benthic habitats of Tutuila, American Samoa. NOAA Office of Undersea Research Submersible Science Program, Hawai'i Undersea Research Laboratory.

Wu HL, Shao KT, Lai CF. 1999. *Latin-Chinese dictionary of fishes names*. Keelung, Taiwan: The Sueichan Press.

Wyanski DM, White DB, Barans CA. 1999. Growth, population age structure, and aspects of the reproductive biology of snowy grouper, *Epinephelus niveatus*, off North Carolina and South Carolina. *Fisheries Bulletin* 98(1): 199–218.

Ximenes-Carvalho M, Fonteles-Filho AA, Tubino RA, Andrade-Tubino MF, Paiva MP. 1999. Parametros de crescimento e mortalidade do cherne *Epinephelus niveatus* (Valenciennes) (Osteichthyes: Serranidae), no sudeste de Brasil. [Growth and mortality parameters of the snowy grouper, *Epinephelus niveatus* (Valenciennes) (Osteichthyes: Serranidae), off southeastern Brazil]. *Arquivos de Ciências do Mar* 32: 111–117.

Yashiro R. 1999. Status of grouper breeding and culture in Thailand. In: Proceedings of a Workshop on Aquaculture of Coral Fishes and Sustainable Reef Fisheries, 6–10 December, 1996, Tuaran, Sabah, Malaysia. pp 123–141.

Yeh S, Dai Q, Chu Y, Kuo C, Ting Y, Chang C. 2003. Induced sex change, spawning and larviculture of potato grouper, *Epinephelus tukula*. *Aquaculture* 228: 371–381.

Yeh SP et al. Undated. The aquaculture status and its sustainability in Taiwan. Accessed at http://aquafind.com/articles/seed.php.

Zabala M, Garcia-Rubies A, Louisy P, Sala E. 1997a. Spawning behaviour of the Mediterranean dusky grouper *Epinephelus marginatus* (Lowe, 1834) (Pisces, Serranidae) in the Medes Islands Marine Reserve (NW Mediterranean, Spain). *Scientia Marina* 61(1): 65–77.

Zabala M, Louisy P, Garcia-Rubies A, Gracia V. 1997b. Socio-behavioural context of reproduction in the Mediterranean dusky grouper *Epinephelus marginatus* (Lowe, 1834) (Pisces, Serranidae) in the Medes Islands Marine Reserve (NM Mediterranean, Spain). *Scientia Marina* 61(1): 79–89.

Zacharia PU, Gupta AC, Mahadevaswamy HS. 1995. Exploitation of juveniles of the spinycheek grouper, *Epinephelus diacanthus* by the multi-day trawlers along Dakshina Kannada coast. In: *Marine Fisheries Information Service, Technical and Extension Series* No. 139. Kochi, India: Central Marine Fisheries Research Institute. pp 5–8.

Zacharia PU, Mahadevaswamy HS, Kemparaju S, Sampathkumar G. 1997. High abundance of large sized rock cods (*Epinephelus* spp.) off Karnataka coast during the postmonsoon month of September. In: *Marine Fisheries Information Service, Technical and Extension Series* No. 147. Kochi, India: Central Marine Fisheries Research Institute. pp 10–12.

Zannou LH. 1985. Liste de quelques espèces animales des eaux marines de la République Populaire du Benin. Projet des Pêche Maritime Artisanale. Direction des Pêches, Cotonou, Benin. 6 pp.

Zatcoff MS, Ball AO, Chapman RW. 2002. Characterization of polymorphic microsatellite loci from black grouper, *Mycteroperca bonaci* (Teleostei: Serranidae). *Molecular Ecology Notes* 2: 217–219.

Zeller DC. 1998. Spawning aggregations: patterns of movement of the coral trout *Plectropomus leopardus* (Serranidae) as determined by ultrasonic telemetry. *Marine Ecology Progress Series* 162: 253–263.

Zeller D[C]. 2006. Unpublished database: creel & offshore surveys American Samoa, Guam and the Northern Mariana Islands. Available at http://www.seaaroundus.org (with permission).

Zeller D[C], Booth S, Craig P, Pauly D. 2006. Reconstruction of coral reef fisheries catches in American Samoa, 1950–2002. *Coral Reefs* 25: 144–152.

Zeller DC, Russ GR. 1998. Marine reserves: patterns of adult movement of the coral trout *Plectropomus leopardus* (Serranidae). *Canadian Journal of Fisheries and Aquatic Sciences* 55: 917–924.

Zeller DC, Russ GR. 2000. Population estimates and size structure of *Plectropomus leopardus* (Pisces: Serranidae) in relation to no-fishing zones: mark-release-resighting and underwater visual census. *Marine and Freshwater Research* 51(3): 221–228.

Zeller D[C], Stoute SL, Russ GR. 2003. Movements of reef fishes across marine reserve boundaries: effects of manipulating a density gradient. *Marine Ecology Progress Series* 254: 269–280.

Zhu Q. 1998. *A list of marine fishes in the China seas.* Beijing, China: Xueyuan Press. 293 pp.

Databases

American Museum of Natural History, New York. Fish collection database. Available at http://entheros.amnh.org/db/emuwebamnh/index.php.

Australian Museum Fish Site. Find a fish. Available at http://australianmuseum.net.au/Find-a-fish.

Bernice P Bishop Museum (BPBM), Honolulu, USA. Ichthyology databases. Available at http://www.bishopmuseum.org/research/natsci/fish/fishdb.html.

Dalzell P. 1991. Ciguatera database. Secretariat of the Pacific Community Noumea, New Caledonia. Available at http://www.spc.int/.

Eschmeyer WN (ed.). Catalog of fishes. Electronic version. California Academy of Sciences, San Francisco, USA. Available at http://research.calacademy.org/research/ichthyology/catalog/fishcatmain.asp.

FAO (Food and Agricultural Organization of the United Nations). FishStat. Available at http://faostat.fao.org or at http://www.fao.org/fishery/statistics/software/fishstat/en.

Froese R, Pauly D (eds). FishBase. World Wide Web electronic publication. Available at www.fishbase.org.

IUCN Groupers & Wrasses Specialist Group. See http://www.iucn.org/about/work/programmes/species/about_ssc/specialist_groups/directory_specialist_groups/directory_sg_fishes/groupers_wrasses_sg/about/.

IUCN species assessments. Available at www.iucnredlist.org.

Myers R[F]. Indo-Pacific groupers. Unpublished distributional database.

National Museum of Natural History, Smithsonian Institute, Washington, DC. Fish collection database. Available at http://acsmith.si.edu/vzfishes/pages/nmnh/vz/DtlQueryFishes.php.

Natural History Museum, London. Fish collection database. Available at http://www.nhm.ac.uk/research-curation/collections/departmental-collections/zoology-collections/search/index.php.

NMFS (National Marine Fisheries Service). NOAA landings data. Available at http://www.st.nmfs.noaa.gov/st1/commercial/index.html.

NOAA (National Oceanic and Atmospheric Administration), US Department of Commerce. Coral reef information system (CoRIS). Available at http://coris.noaa.gov/.

NOAA (National Oceanic and Atmospheric Administration), Fisheries Service, Office of Sustainable Fisheries. Status of US fisheries. Available at http://www.nmfs.noaa.gov/sfa/statusoffisheries/SOSmain.htm.

PROCFISH (Pacific Regional Oceanic and Coastal Fisheries). Project Database. Available at http://www.spc.int/coastfish/en/projects/procfish.html.

SAIAB (South African Institute of Aquatic Biodiversity), Grahamstown, South Africa. Fish collection database. Available at http://www.saiab.ac.za.

SCRFA (Society for the Conservation of Reef Fish Aggregations). Global spawning aggregations database. Available at http://www.scrfa.org/database/.

Sea Around Us Project database. Available at http://www.seaaroundus.org.

Seafood Services Australia. Australian fish names list. Available at http://www.seafoodservices.com.au/.

Shao KT. 2005. Taiwan Fish Database. World Wide Web electronic publication. Available at http://fishdb.sinica.edu.tw/eng/home.php.

Zeller D[C]. 2006. Unpublished database: creel & offshore surveys American Samoa, Guam and the Northern Mariana Islands. Available at http://www.seaaroundus.org.

Zoological Museum of Hamburg (ZMH), Germany. The ichthyological collection: a global biodiversity archive of fishes. Available at http://www.uni-hamburg.de/biologie/BioZ/zmh/ich/sam_e.html.

Zoological Museum, University of Copenhagen, Denmark. Fish collection database. Available at http://zoologi.snm.ku.dk/english/collections/Vertebrates/dokument2/.

Note: *Some databases are not in the public domain, and may require special permission for access.*

Index

F

G

H

I

J

L

M

P

S

T

V

Contributors to species accounts

The IUCN Red List Assessment for each species was prepared by one or more 'authors' listed at the end of each species account. Their affiliations are given below. Once completed, the listings were checked by independent assessors organised by the Red List Authority, in this case the IUCN Groupers & Wrasses Specialist Group. The original versions of these contributions have been modified in the preparation of this book.

Bertoncini, AA – Federal University of São Carlos, Brazil
Brulé, T – formerly with University of Hong Kong, China
Cabanban, AS – University of Malaysia, Sabah
Carpenter, KE – Old Dominion University, USA
Chan, C-M – Agriculture, Fisheries and Conservation Department, Hong Kong, China
Chan, T-C – Consultant, Hong Kong, China
Choat, JH – James Cook University, Australia
Cornish, AS – WWF-HK, Hong Kong, China
Craig, MT – University of Puerto Rico, Mayagüez, Puerto Rico
Eklund, AM – formerly with Southeast Fisheries Center, National Marine Fisheries Service, USA
Fennessy, ST – Oceanographic Research Institute, South Africa
Ferreira, BP – Universidade Federal de Pernambuco, Brazil
García-Moliner, G – Caribbean Fishery Management Council, Puerto Rico
Gaspar, ALB – Instituto Recifes Costeiros, Brazil
Gerhardinger, LC – Associação de Estudos Costeiros e Marinhos dos Abrolhos, Brazil
Harmelin-Vivien, ML – Université de la Méditerranée, Station Marine d'Endoume, France
Heemstra, PC – South African Institute for Aquatic Biodiversity, South Africa
Huntsman, GR – formerly with Beaufort Laboratory, Southeast Fisheries Science Center, National Marine Fisheries Service, USA
Kulbicki, M – University of Perpignan, France
Liu, M – formerly with the Swire Institute of Marine Science, University of Hong Kong, China
McIlwain, JL – University of Guam, Guam
Marques, S – Instituto Mar Adentro, Brazil
McGovern, JC – Marine Resources Research Institute, South Carolina Department of Natural Resources, USA
Myers, RF – Coral Graphics, USA
Ng, W-C – Swire Institute of Marine Science, University of Hong Kong, China
Peres, MB – Ministry of the Environment, Brazil
Pollard, DA – Division of Fisheries, New South Wales Department of Agriculture, Australia
Rhodes, KL – Consultant, USA
Robinson, J – Seychelles Fishing Authority, Seychelles
Rocha, LA – University of Hawaii, USA
Russell, BC – Museum & Art Gallery of the Northern Territory, Australia
Sadovy de Mitcheson, YJ – University of Hong Kong, China
Samoilys, MA – IUCN, Kenya
Sampaio, CLS – Universidade Federal de Alagoas, Brazil
Situ, Y-Y – formerly with the Agriculture, Fisheries and Conservation Department, Hong Kong, China
To, W-L – WWF, Hong Kong, China
Yeeting, BM – Secretariat for the Pacific Community, New Caledonia

Acknowledgements

This volume would not have been possible without the help of persons, too numerous to list, who provided support in various ways and over many years. While we cannot thank everyone individually, please know your efforts did not go unnoticed.

Special thanks are due to Rachel Wong for logistic and other support for the IUCN Groupers & Wrasses Specialist Group (GWSG), including her efforts at the 2007 workshop, and to Lily Ng for maintaining the GWSG's website for many years through the Department of Ecology & Biodiversity, University of Hong Kong. The IUCN and University of Hong Kong provided funding for the 2007 workshop, in preparation of which much of the information presented here was gathered. We would like to thank all past and present GWSG members for keeping the group alive and active. The IUCN Global Species Programme has provided support and assistance to the GWSG and its various activities. Kevin Rhodes provided critical help in workshop development and Red List Assessments. We also thank Georgina Mace and those who participated in the first-ever listing of marine fishes on the IUCN Red List in 1996, which led to the formation of the GWSG in 1998.

We are grateful to Janice Chanson and Kent Carpenter who facilitated the 2007 workshop. Elaine Heemstra supplied numerous drawings and related data, and Duane Sanabria and Hector Martínez provided critical support to MTC throughout the process of writing this manuscript. Lastly, we thank all those who generously provided photographs for inclusion in this book.

Sole distribution rights for North America granted to CRC Press/Taylor & Francis Group
6000 Broken Sound Pkwy, NW, Suite 300,
Boca Raton, FL 33487, USA
ISBN 978-1-4665-0602-2
Visit the CRC Press website at: http://www.crcpress.com

First edition 2011

Published by NISC (Pty) Ltd, PO Box 377, Grahamstown, 6140, South Africa

ISBN: 978-1-920033-11-8

The authors and publisher gratefully acknowledge the funding from IUCN, International Union for Conservation of Nature, towards the publication of this work.

Editing, design and production: NISC (Pty) Ltd, with additional page layout by InkDesign
Proofreading: C Kulongowski
Cover design: Advanced Design Group
Photographs and illustrations: cover (front), Keoki Stender; cover (back inside, top to bottom) Luiz A Rocha, George de Mitcheson and Matthew T Craig; p i, Robert F Myers; p ii, Octavio Aburto; p iv, Robert F Myers; pp xvi–xviii, FAO; p xx, Octavio Aburto; all others as credited in the text. Illustrations reproduced from the *FAO Species Catalogue. Vol. 16. Groupers of the World (Family Serranidae, Subfamily Epinephelinae)* are used with permission of the Food and Agricultural Organization of the United Nations (FAO) (permission licence A45/2010).

Printed in South Africa by CADAR Printers (Pty) Ltd, Port Elizabeth